Network Materials

Ideal for entry-level and experienced researchers working in materials science and engineering, this unique book introduces a new subfield of materials science and mechanics of materials: network materials. A comprehensive review of their mechanical behaviors allows readers to understand, design, and enhance the performance of these material systems, across a range of applications including connective tissue, nonwovens, and gels. By introducing simple models, supported by experimental data, the book provides the necessary fundamental knowledge to assist readers to design and develop their own material systems.

By presenting each of these previously disparate material systems within a unified theoretical framework, this book provides a consolidated presentation of the mechanics of networks, introducing parameters that define the stochastic structure of the network, and discussing their mechanical behavior. It is an ideal text for those new to this fast-growing field and for experienced researchers looking to consolidate their knowledge.

Dr. Catalin R. Picu is a professor in the Department of Mechanical Aerospace and Nuclear Engineering at Rensselaer Polytechnic Institute. He is an ASME fellow, recognized for significant contributions to the field of mechanics of materials. He has contributed 20 chapters to various edited works, including *Trends in Nanoscale Mechanics* (Springer, 2014), *Materials with Internal Structure* (Springer, 2016), and *Mechanics of Fibrous Materials and Applications: Physical and Modeling Aspects* (Springer, 2019), and has authored more than 200 journal articles.

Network Materials

Structure and Properties

CATALIN R. PICU
Rensselaer Polytechnic Institute

CAMBRIDGE
UNIVERSITY PRESS

CAMBRIDGE
UNIVERSITY PRESS

University Printing House, Cambridge CB2 8BS, United Kingdom

One Liberty Plaza, 20th Floor, New York, NY 10006, USA

477 Williamstown Road, Port Melbourne, VIC 3207, Australia

314–321, 3rd Floor, Plot 3, Splendor Forum, Jasola District Centre, New Delhi – 110025, India

103 Penang Road, #05–06/07, Visioncrest Commercial, Singapore 238467

Cambridge University Press is part of the University of Cambridge.

It furthers the University's mission by disseminating knowledge in the pursuit of
education, learning, and research at the highest international levels of excellence.

www.cambridge.org
Information on this title: www.cambridge.org/9781108490030
DOI: 10.1017/9781108779920

First published 2022

A catalogue record for this publication is available from the British Library.

Library of Congress Cataloging-in-Publication Data
Names: Picu, Catalin R., author.
Title: Network materials : structure and properties / Catalin R. Picu, Rensselaer Polytechnic Institute,
    New York.
Description: Cambridge, United Kingdom; New York, NY, USA: Cambridge University Press, [2022] |
    Includes bibliographical references and index.
Identifiers: LCCN 2022001138 (print) | LCCN 2022001139 (ebook) | ISBN 9781108490030 (hardback) |
    ISBN 9781108779920 (epub)
Subjects: LCSH: Fibrous materials. | BISAC: TECHNOLOGY & ENGINEERING /
    Materials Science / General
Classification: LCC TA418.9.F53 P53 2022 (print) | LCC TA418.9.F53 (ebook) |
    DDC 620.1/18–dc23/eng/20220407
LC record available at https://lccn.loc.gov/2022001138
LC ebook record available at https://lccn.loc.gov/2022001139

ISBN 978-1-108-49003-0 Hardback

Contents

Preface

The knowledge base of network materials has been developed over more than a century by multiple scientific communities. Rubber elasticity is one of the oldest subjects in mechanics of materials, and investigations in this area have led to key concepts that were later extended to other material systems and given power of generality. Rubber has fascinating properties that few other materials may rival. Biological soft materials have been studied by an entirely different scientific community. Their mechanical properties are largely defined by the mechanics of a network of filaments, as is the case at the tissue level with the extracellular matrix, arteries, and membranes, and at the cellular level, with the cytoskeleton.

These efforts have no common denominator with those aimed to develop engineering materials, like nonwovens, paper, and fiber composites. Paper and textiles are versatile materials developed by humans over millennia. These days, paper is used not only as support for information storage, but also for packaging, filtration, and even structural purposes, and its range of applications is continuously extended by a dedicated community. Nonwoven textiles are the focus of yet another community and find applications in clothing, medicine, filtration, and insulation, to name just a few.

Are these efforts, indeed, as different from each other in scope, methods, and findings as they appear to be? This book provides a negative answer to this question. It attempts to outline a unitary view on the structure and behavior of a broad range of materials that have a common denominator: Their main structural component is a network of filaments. Through this, the book defines the class of network materials.

Network materials are widespread for a good reason: it is cheap to span space with filaments, while it is much more expensive to fill space with a "'continuum" material. Biological systems are particularly inclined to save high cost protein-based construction material. Engineers are constrained less by this requirement, but still comply with it when dealing with large-scale problems, such as when constructing bridges and frames for buildings.

The view that an ensemble of objects in direct interaction with each other may be considered a "material" defies the rooted intuition that materials should be treated as continua. The mechanical behavior of an ensemble of objects is controlled primarily by their interactions and may be entirely different from that of a continuum. In an age in which metamaterials are in fashion, this perspective is adequate and timely. This is the prevailing view taken in this book.

The majority of network materials encountered in the physical world have stochastic structure. Stochasticity ensures insensitivity to structural imperfections and resilience in operation. Stochastic structures of any scale are easier and cheaper to construct compared with periodic or structured configurations.

Every book has a history. The history of this book starts about two decades ago and encompasses the tortuous paths I walked while attempting to understand the complexities outlined above and in the remainder of these pages. Many colleagues, too many to list here, contributed through shared knowledge and wisdom to my own growth, and I thank each of them. The hard work, perseverance, and intelligence of my students who worked on topics related to network materials – H. Hatami-Marbini, A. Shahsavari, L. Zhang, A. Osta, E. Ban, S. Deogekar, M. Islam, V. Negi, A. Sengab, and J. Merson – have made this book possible. I thank all and each of them for the exciting path we walked together. J. Merson and V. Negi also contributed to editorial aspects of the manuscript preparation. The financial support received over the years from the National Science Foundation, the National Institute of Health, Procter&Gamble Inc., and Ecovative Inc. made the work on these topics possible and is gratefully acknowledged.

I wrote the book with the hope that it will serve as a source of information and inspiration for researchers working in the diverse fields for which network materials are relevant. Many open questions are formulated throughout these chapters and many others remain unspecified. Science is a collective and collaborative effort. I can't wait to learn the answers.

1 Introduction
Definitions and Classification

1.1 What Are Network Materials?

The word "network" is important today in many areas of science and engineering. Communication networks, including social networks, transportation systems such as the system of roads and railroads, energy transportation networks, water and sewer systems, and the circulatory system of animals, are examples of networks whose essential role is to transport matter or information. On the other hand, physical networks made from connected filaments are broadly encountered in biology and engineering and perform mechanical functions.

Many biological materials are constructed from fibers. Spanning a volume with fibers is much cheaper than filling the same volume with a continuum material. Since the living organisms of our world build with proteins, and proteinogenic amino acids are expensive to produce, building with less material is a condition for survival of the species. The solution adopted by nature to address this challenge is to build with fibers.

Materials engineering is less constrained by energetic requirements than biology. We build with space-filling materials with little or no free volume, which one may call three-dimensional materials. Structures are also built by assembling plates and membranes, which one may call two-dimensional building blocks, and/or filaments, that is, one-dimensional, fiber-like building blocks. Some of the most versatile and cheapest materials ever developed by humans are made from fibers. And, in most of these materials, the fibers are connected into a network that spans the occupied volume.

The class of Network Materials includes materials that contain a fiber network as the main structural component and in which the network controls the overall mechanical behavior.

The fact that the material contains volume not occupied by fibers is essential in defining its mechanical properties. In the presence of free volume, fibers have sufficient kinematic freedom to change orientations, group and form bundles, or reorganize in other ways, which conveys interesting properties to the ensemble. The discrete nature of the material remains important, as the building blocks and their interaction define the overall response. Collective deformation modes, including in some cases modes of zero stiffness similar in nature with the situation of mechanical mechanisms, become not only possible, but also a common occurrence in network materials. The microstructural multistability of these mechanism-like structures leads to specific mechanical behavior under large deformations.

Interactions are de-localized; they take place at distances comparable with the length of network segments and involve not only forces, but also moments. This is in sharp contrast with monatomic lattices in which interactions are short ranged and generally result from the interplay of strong attractive and repulsive forces. While crystals have high stiffness and strength and may exhibit only those deformation modes allowed by the crystal symmetry, network materials are typically soft and their anisotropy may be tuned by controlling the distribution of fiber orientations.

The open internal structure of network materials allows for the presence of a matrix. If the network is embedded in a solid continuum, the exceptional properties emerging from the collective deformation of network fibers are largely lost, while the mechanical behavior of the composite is defined by the superposition of contributions due to the matrix and individual fibers. However, if the matrix is a fluid, not only is the network behavior preserved, but the interaction between fibers and the fluidic matrix may lead to interesting effects not present in dry networks without a matrix. Also, the network may embed inclusions, entities which modify the local network behavior and lead to significant changes of the global performance. An amazing network material of this type is cartilage, which inherits its properties from the presence of embedded macromolecular entities – the glycosaminoglycans – which are not part of the underlying network.

In all biological network materials, the network structure is stochastic. Likewise, the vast majority of man-made network materials have stochastic network structures. Woven textiles are a notable, but singular exception from this rule. Stochasticity enables a diversity of local deformation modes, including collective modes. Structural stochasticity conveys material resilience and defect tolerance, leading to increased strength and toughness.

Some materials made from fibers do not exhibit the network signature and cannot be classified as network materials. Densely packed fibers held together by strong surface interactions form a continuum. An example of this type is provided by insect cuticles made from chitin fibers arranged approximately parallel to each other, strongly bonded, and tightly stacked, forming anisotropic continuum layers. For the same reason, usual fiber composites cannot be classified as network materials. In short fiber composites, fibers may not interact strongly with each other and may not form a network. In continuous fiber composites, the woven reinforcement is embedded in a stiff matrix aimed to strongly restrict the kinematic freedom of the fibers and to eliminate the signature of the network. In such composites, the goal is for the material to inherit the properties of individual fibers – primarily their high stiffness and strength. Polymeric materials are composed from chains which, if long enough, organize in a network. The signature of the network appears at temperatures above the glass transition temperature, T_g, when filaments are able to move relative to each other. Below this temperature, chain kinetics as well as network collective deformation modes are frozen. Polymers below T_g cannot be classified as network materials due to their dense packing and strong surface interactions, which limit chain mobility.

A large number of examples of network materials can be given. The following list is provided for exemplification, but it is not exhaustive. From the range of man-made engineering network materials, one may mention:

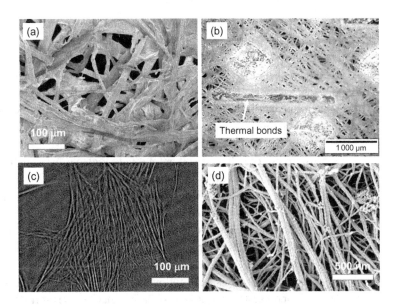

Figure 1.1 Examples of network materials: (a) absorbent, high porosity paper, (b) thermally bonded polypropylene nonwoven with rectangular and square thermal bonds, (c) stress fibers in the cytoskeleton (reproduced with permission from Alioscha-Perez et al., 2016), and (d) collagen structure of arthritic human knee cartilage. Reproduced with permission from Gottardi et al., 2016

▶ Paper, Parchment, and Papyrus. All three are network materials used as supports for writing. The oldest of these is papyrus, which was used in Egypt as far back as 2 500 BC. Papyrus is made from the stem of the *Cyperus Papyrus* plant cut longitudinally and arranged in layers with the direction of the natural fibers orthogonal to each other in subsequent layers. The stack is pressed and dried under pressure to produce a relatively rigid layered fibrous structure. Papyrus was used in Europe until about year 1 000 CE, when it was replaced with parchment. Parchment is made from animal skin processed in a strongly alkaline solution and dried under tension. The process leaves primarily the underlying collagen network of the skin known as the dermis, which is a relatively high strength and high toughness network material in hydrated conditions. Drying makes the material more brittle. Paper dates from about 2 000 years ago and was imported in the Western world from China. It is made from cellulose fibers which are pressed together in the wet state and dried under pressure. Papers with a broad range of mechanical properties are made today, ranging from very soft, low density tissue, to the higher density paper on which this book is printed, and to the resilient multilayered paper of the absorbent kitchen towels. Figure 1.1(a) shows a scanning electron micrograph of an absorbent kitchen paper towel which evidences the porosity of this material. Printing paper is made from similar ribbon-like cellulose fibers packed densely.

▶ Textiles, woven. Weaving is one of the oldest occupations of humans. Woven fabrics are made by interlacing two or multiple threads. The threads are made from natural fibers, such as cotton and hemp, or from polymeric fibers. The threads are continuous, while fibers are short and are twisted to form threads. The structure of fabrics is periodic, which is rather unusual for a network material. The deformation behavior of the fabric is controlled by the mechanics of the periodic unit cell and the essential inter-fiber interactions are topological. Weaving entangles the threads, which ensures that the fabric preserves integrity and sustains loads even though the threads are not crosslinked and are not held together by attractive surface interactions. Today, textile engineering is an important field that assists a vast industry. All of us use these versatile, cheap, and mechanically performant network materials throughout our entire life.

▶ Textiles, nonwoven. Nonwovens are a newcomer to the world of textiles. While older versions such as felt have been produced from natural fibers for a while, the modern nonwovens are made primarily from polymeric fibers, which are spun and deposited on a conveyor belt, followed by calendaring and, in most cases, by thermal bonding. The result is a stochastic mat-like network of "infinite" fibers of 10–50 μm diameter. An as-deposited nonwoven mat has low stiffness and high porosity. To increase the stiffness and strength, fibers may be entangled by exposure to water jets (hydroentangled), may be pressed under moderate heat, which causes weak bonds to form at all fiber contacts (heat sealed), or may be thermally bonded by bringing them in contact with heated punches which melt the fibers they touch. The solid polymeric blocks formed subsequently at the location of thermal bonds create effective crosslinks between fibers (Figure 1.1(b)). Nonwovens are used in many applications, such as geotextiles, packaging materials, air and water filters, structural materials for consumer products, such as wipes, tissue, and disposable diapers, scaffolds for tissue engineering, etc. Nonwovens made from nonpolymeric materials, such as fiberglass, are used for thermal and sound insulation in buildings and automobiles. And how could one not mention the most visible application of nonwovens during the time when this book was written: the surgical and N95 masks worn by everyone during the COVID-19 pandemic of 2019–2022.

▶ Gels. Gels are molecular networks embedded in a liquid. The network is swollen by the liquid, the molecular strands making sparse or no contacts with each other. The network is held together by crosslinks which may be strong chemical bonds (in chemical gels), helical domains (in physical gels), small crystallites, or regions of ordering in which the molecular strands are held together by weak hydrogen bonds. Transient gels form in dense suspensions of high aspect ratio flocculants, which may be macromolecular or fiber-like. Cohesive and dispersive forces acting between suspension components lead to their aggregation in a percolated network structure. Such networks are weakly bonded and may re-organize dynamically during flow. They exhibit the rheological property of thixotropy, that is, behave as solids with a yield point at small loads, but flow like fluids at loads larger than the yield stress. Gel properties cover a broad range from a very soft and brittle, to a relatively stiff and tough, function of the nature and density of the crosslinks, the type of filaments forming the network, and the interaction between filaments and the embedding liquid. Gels are used in many applications in the food and cosmetics industries, in paints and adhesives, and in biomedical applications. In a broader sense, most network materials encountered in biology are hydrogels, in which the network is made from protein fibers and the embedding fluid is water.

▶ Elastomers. Elastomers are networks of long and flexible molecules that interact through strong covalent bonds at a small number of sites, and through weak interactions elsewhere. The strong bonds represent crosslinks which, in the majority of the elastomers, are permanent. Weak interactions hold the molecular strands together at melt densities. The resulting network has low free volume, but inter-molecular interactions at sites other than those of the crosslinks are weak and molecular strands are free to move relative to each other, allowing for network reorganization during deformation. This is a situation somewhat different from that of other network materials in which the large free volume available allows the type of fiber kinematics which ultimately defines the mechanical behavior of the network. While most elastomers are thermosets, some are thermoplastic and may be reshaped at elevated temperatures. This is possible due to the dissociation of the crosslinks at higher temperatures, which transforms the solid network into a viscous melt. The network reforms upon cooling and the elastomeric properties are recovered. Elastomers may sustain large, reversible deformations with limited internal dissipation. They are used in many applications, either as pure network materials, or filled with nanoscale inclusions such as carbon black. The filled elastomers have one of the most impressive sets of mechanical properties of all current materials; consider the extreme loading conditions to which automobile tires are subjected and the excellent lifetime of filled rubber.

▶ Other polymeric networks. Polymeric materials are made from chains packed at high densities. The chains form a load-transmitting and system-spanning network which defines the mechanical properties of the material. The condition for these systems to be classified as network materials is that the chains have sufficient relative mobility to allow network structural evolution during loading. For example, reducing the temperature below the glass transition temperature reduces the mobility of the chains and freezes topological rearrangements. Under such conditions, the macroscopic response is defined primarily by the dense nonbonded interactions, and less by the underlying network. Above the glass transition temperature, dense polymeric solutions and melts may be entangled and exhibit solid-like behavior at probing frequencies above the inverse of the disentanglement time. These materials exhibit the behavior of a Newtonian liquid at probing frequencies below this threshold.

The first three examples in this list – paper, and woven and nonwoven textiles – and the following three examples – gels, elastomers, and polymeric networks – are classified in separate categories based on the sensitivity of their response to tempera-ture variations. The networks in the first group are denoted as *athermal*, while those in the second group are *thermal*. As the name indicates, the behavior of the first group is largely independent of temperature, while the mechanics of networks in the second group are strongly influenced by thermal fluctuations. This distinction can be interpreted in terms of the energy required to deform individual filaments. If filaments are thin and their bending stiffness is low, thermal fluctuations have sufficient energy to perturb filament conformations. This has a strong influence on the mechanical behavior of networks composed from such filaments, which therefore can be classified as thermal.

Network materials are the structural materials of living organisms. A large number (if not the majority) of biological materials have a fiber network as their main structural component:

▶ The cellular cytoskeleton. All eukaryotic cells contain a filamentary network structure in which the cellular organelles are embedded. This network is composed from three types of protein filaments: actin filaments (F-actin), intermediate filaments, and microtubules. The cytoskeleton performs multiple roles: it provides the mechanical structure of the cell, it mediates connections across the cell membrane between neighboring cells, it contributes to mechano-transduction, which converts force into biochemical signals, and it supports transport to and from the nucleus. F-actin filaments are connected by actin-binding proteins and by myosin motors able to apply forces between actin strands, which renders the network active and makes cell migration possible. Under normal conditions, a cell attaches to the neighboring cells and/or to the substrate and pulls; hence the cytoskeleton is under pre-stress due to the action of myosin motors. The F-actin network is also highly transient, as filaments grow at the leading edge of cytoskeleton protrusions and degrade to the constituent monomers at retracting sites. This behavior renders the mechanics of the cytoskeleton highly complex. Due to its key role in sustaining life, the physics of the cytoskeleton is intensely studied at this time. Figure 1.1(c) shows an image of stress fibers formed by the reorganization of actin filaments under stress produced by myosin motors (Alioscha-Perez et al., 2016). The sites where the cell is attached to the substrate are visible along the periphery of the fibrous structure.

▶ Fibrin. Fibrinogen is a protein which exists in blood. Under the action of the enzyme thrombin, it polymerizes, forming fibrin filaments. This process is triggered by a biochemical cascade when the lining of the blood vessels is broken. Platelets in the blood agglomerate at the wound site and trigger the polymerization of fibrinogen. Fibrin filaments grow to connect and embed the platelets, forming a blood clot. This mechanism is the hemostatic first step of wound healing, but it is also central in cardiovascular disease. The fibrin network contracts as it forms due to the dynamic formation of crosslinks and cohesive interactions between filaments. This mechanism densifies the clot and ensures its mechanical integrity and stability.

▶ Collagen networks. Collagen is the most abundant protein in the human and animal bodies. It forms fibers composed from tropocollagen triple molecular strands. Networks of collagen fibers are present everywhere in the body, providing structural and mechanical functions. The collagen network denoted generically as an extracellular matrix provides mechanical and biochemical support to the cells. The mechanical function of membranes and connective tissues is also performed by collagen networks. Cartilage, tendons, and ligaments are collagen networks with specific architecture adapted to the function of the respective tissue. The dermis, the basement membranes, the amnion, and the liver capsule have collagen networks as their main structural component. The elasticity of blood vessels is provided by networks of collagen and elastin. Collagen exists in bones and regulates the bone toughness. It is difficult to overestimate the importance of this type of fiber network in biology. Figure 1.1(d) shows a scanning electron micrograph of cartilage from an arthritic human knee, after enzymatic depletion of proteoglycans and chondrocytes (Gottardi et al., 2016). The image shows thicker bundles, of approximately 200 nm diameter, along with thinner fibers. The presence of the thinner collagen fibers is a result of the degenerative disease.

The first two examples in this list – actin and fibrin – are thermal networks, while the third example – collagen – is of athermal nature. Collagen is a protein fiber composed from molecular strands of tropocollagen which, just like actin, fibrin, and other bio-filaments, are small enough for their mechanics to be affected by thermal fluctuations. However, tropocollagen assembles in larger fibrils, which then bundle to create the collagen fiber bundles that become the building blocks of the extracellular matrix. These collagen bundles are sufficiently large to be athermal (Figure 1.1(d)). Further, it should be noted that most bio-filaments, including actin, are relatively stiff in bending and have large persistence length. In order to differentiate them from flexible polymeric chains with small persistence length, these filaments are called semiflexible. The distinction between flexible and semiflexible filaments is made only in the context of thermal networks and implies a comparison of two length scales: the filament persistence length and the mesh size of the network. If the persistence length is larger or comparable with the mesh size, the semiflexible nature of the filaments has a signature in the overall mechanical behavior of the network. If the mesh size is much larger than the filament persistence length, the mechanical behavior is similar to that of a network of flexible filaments.

1.2   Classification

Network materials may be classified based on various criteria. Three criteria are used here to divide networks into several broad categories that emphasize important commonalities between some apparently very different material systems.

The three criteria used are: (i) the presence in the material of components other than fibers, (ii) the nature of the interactions that stabilize the network, and (iii) the type of fibers forming the network.

Table 1.1 shows the classification in terms of criterion (i). Network materials are divided into three categories: networks composed exclusively from fibers, which are

Table 1.1 Classification of network materials based on the presence of components other than fibers

Dry	Embedded	Embedding
Paper	Gels	Cytoskeleton
Nanopapers	Entangled polymer solutions	Connective tissue
Textiles–woven	Connective tissue	Extracellular matrix
Nonwovens	Dense fiber suspensions	Fibrin
Elastomers		Composites in which a dry networks
Thermoplastics		performs the function of matrix
Thermosets		
Fiberglass/Felt		
Open cell foams		
Mycelium		

Table 1.2 Classification of network materials based on the interactions that stabilize the network

Crosslinked	Entangled	With Surface Interactions
Paper	Textiles–woven	Buckypaper
Nonwovens	Nonwovens	Nanopapers
Elastomers	Fiberglass/felt	Fibrin
Thermosets	Thermoplastics above T_g	Thermoplastics above T_g
Gels	Dense fiber suspensions	Dense fiber suspensions
Connective tissue	Elastomers	Elastomers
Extracellular matrix		
Fibrin		
Open cell foams		
Mycelium		
Cytoskeleton		

denoted as "dry" networks, networks embedded in a continuous matrix, which may be solid or liquid, and networks that embed inclusions, indicated in Table 1.1 as "embedding." Many of the man-made network materials are used in the dry state. All biological network materials are of embedded and/or embedding type. From the mechanical behavior perspective, each of these classes poses specific problems. We take the view that the fundamental network behavior is that of dry networks, while the presence of a matrix and/or inclusions leads to modifications of this basic response in ways dependent on the nature of the added phases. The mechanical response of each of these three types of networks materials is discussed in separate sections of the book.

Table 1.2 shows the classification in terms of criterion (ii): the nature of interactions that stabilize the network. One may distinguish three categories: networks stabilized by crosslinking, networks stabilized by topological interactions between tortuous fibers, and networks stabilized by surface interactions such as cohesion. In most cases, multiple types of interactions operate concurrently and stabilize the network. For example, the mechanical behavior of nonwovens may be controlled by both crosslinks

Table 1.3 Classification of network materials in terms of the types of fibers forming the network

Athermal	Thermal	
	Flexible	Semiflexible
Paper	Elastomers	Cytoskeleton
Textiles–woven	Thermoplastics above T_g	Fibrin
Nonwovens	Gels	Other protein networks
Fiberglass/Felt		
Open cell foams		
Mycelium		
Connective tissue		
Extracellular matrix		
Buckypaper		

and inter-fiber contacts/entanglements (excluded volume interactions), while one of these may dominate in a specific range of network and crosslink densities. Likewise, in the extracellular collagen matrix, bundle crosslinking is essential, but bundles are stabilized by strong surface interactions between fibrils.

Table 1.3 presents the classification based on the third criterion, which takes into account the nature of the fibers composing the network. We differentiate between athermal and thermal networks. Biological and artificial network materials can be classified either as thermal or athermal, depending on the dimensions and bending rigidity of their filaments. Thermal networks are classified further as being composed from flexible and semiflexible filaments. Molecular networks in which the molecules do not bundle are generally thermal. Networks made from fibers of filament bundles of diameter larger than a few hundred nanometers are generally athermal.

A note on the terminology used is necessary at this point. Here and throughout this book the terms "fiber" and "filament" are used somewhat interchangeably. "Filament" is used to denote fibers of small, submicron cross-sectional dimensions, both thermal and athermal. The term "fibers" is used generically but implies athermal behavior. Molecules are referred to either as "strands," "chains," or "filaments."

1.3 Outline of the Book

The objective of this book is to define and establish the class of Network Materials and to review the common features of their mechanical behavior. The focus is on stochastic structures, because these are widespread in nature and in engineering applications, and on mechanical behavior, because the main function of the underlying network is structural. We exclude textiles and continuous fiber composites which have periodic structure and discuss only certain aspects of entangled polymeric solutions and melts; the topics not discussed in detail here are addressed in dedicated texts.

The book is organized into 11 chapters which develop the subject starting from the building blocks of network materials to the complexities of the mechanical behavior of time-dependent and composite networks.

Chapter 2 provides an overview of the mechanical behavior of the three main types of fibers: athermal, thermal flexible, and thermal semiflexible. The linear, nonlinear, and rupture aspects of the behavior are discussed. Since in many practical situations networks are composed from bundles, elements of the mechanics of fiber bundles are presented in the second part of Chapter 2.

Chapter 3 discusses the interactions taking place between fibers. Fibers come into contact either in a crossed geometry or all along their length. Crosslinks or simple contacts with purely repulsive or with cohesive interactions are established in the crossed case. Contacts characterized by a specific work of separation are established when fiber axes are parallel. The mechanics of crosslinks and the separation and relative sliding of various types of contacts are discussed in detail.

Network materials have stochastic structure. Since the number of configurations of fibrous assemblies is infinite, it is necessary to inquire about the minimum set of parameters needed to describe the aspects of the structure most relevant for the mechanical behavior of the network. This is the subject of Chapter 4. Thermal and athermal networks, cellular and fibrous networks with and without fiber tortuosity, defined in two and three dimensions are considered. The porosity of the network – a parameter important both in fluid transport across networks and in mechanics – is analyzed and is placed in relation with other structural parameters.

The deformation of stochastic networks is nonaffine, meaning that local strains are different from the global applied strains and vary spatially. However, assuming that the deformation is affine allows us to derive analytic constitutive descriptions which cannot be obtained in the general, nonaffine case. The affine description may be taken as a reference for the more accurate characterization that accounts for nonaffinity. The affine description is also of historical importance. Due to these reasons, a separate chapter is dedicated to this model (Chapter 5).

Chapter 6 presents the key aspects of the mechanical behavior of thermal and athermal network materials. The chapter is divided into two parts, referring to cross-linked and non-crosslinked networks. The section addressing crosslinked networks discusses uniaxial tension, uniaxial compression, and multiaxial loading situations separately. For each type of loading the generic aspects are presented first, as they emerge from computer models and experiments. The relation between various aspects of the behavior and network parameters is analyzed in detail. The effects of connectivity, fiber alignment, fiber tortuosity, crosslink compliance, and elastic–plastic behavior of the fiber material are discussed. This establishes the structure–properties relations needed in material design. The behavior described is demonstrated using representative experimental data for a variety of network materials. Size effects, that is, the dependence of the mechanical response on the sample size, are also discussed. The section on non-crosslinked networks has two parts outlining the response of athermal non-crosslinked fiber masses to compression, and the role of topological interactions (entanglements) in defining the response in tension.

Chapter 7 presents a review of constitutive descriptions developed for network materials, with emphasis on micromechanics. Constitutive models are required by continuum representations of fibrous materials, but the complex behavior emerging from the collective kinematics of the fibers is generally difficult to capture by mean field representations. The chapter presents a review of the state of the art in this area.

Chapter 8 focuses on the strength of network materials. It has two parts, referring to athermal and thermal networks. Each of these presents two topics: Strength in the absence of pre-existing cracks or damage; and the growth of a pre-existing flaw. The failure modes and mechanisms, the relation between strength and network parameters, the effect of fiber alignment, fiber tortuosity, variability of fiber properties, and the size effect on strength are analyzed. In the context of thermal networks, a summary of key results from the vast literature on rubber failure under monotonic and fatigue loading conditions is presented. Since network materials are occasionally used as adhesives or as a matrix for composites, it is important to explore the design paths that may lead to increased toughness without loss of strength. A section of Chapter 8 is dedicated to the development of tough network materials.

The mechanical behavior of some network materials is time dependent. Network-scale time dependence emerges from a variety of sources, including the fact that the fiber material behavior may be rate-sensitive, the presence of dissipative interactions between fibers, viscous friction at contacts between athermal fibers, the flow of solvent in and out of the network, and the presence of transient crosslinks. These mechanisms, their interaction, and influence on the mechanical behavior of the network are discussed in Chapter 9.

Chapter 10 presents the effect of surface interactions on the mechanical behavior of networks. Surface interactions are important in non-crosslinked networks of flexible filaments. Two types of systems are considered: dense suspensions of filaments that self-organize, forming percolated and transient network structures, and dry networks self-organized by cohesive forces. In suspensions, the fluid develops solid-like behavior manifest at relatively small strains due to the presence of the self-organized network structure. Surface interactions produce fiber bundling and networks formed exclusively from fiber bundles emerge. Networks of fiber bundles are fundamentally different from both the crosslinked and non-crosslinked networks discussed in Chapter 6.

Chapter 11 presents an analysis of composite networks. This class includes network materials reinforced with spheroidal inclusions or with staple fibers much stiffer than the base network fibers. Interpenetrating networks formed by two or multiple cross-linked networks which span the same problem domain are also composite networks and are discussed in this chapter. A crosslinked stochastic network embedded in a solid continuum matrix is a special type of composite, which is marginally related to the usual fiber composites. A section of Chapter 11 is dedicated to this topic.

A number of emerging topics are left out from this book. These include sub-isostatic networks, networks with pre-stress, active networks, tensegrity structures, and the developing sub-field of metamaterials. In sub-isostatic networks the condition of stability of the structure is not fulfilled in the unloaded state. These networks are mechanisms and have zero stiffness when probed at small strains but acquire stiffness upon further straining. Their mechanical behavior is more sensitive to the details of the network architecture than that of a stable isostatic network because such structures function in the proximity of a critical point, that is, the state in which they shift from being sub-isostatic to isostatic.

Pre-stress modifies the mechanical behavior significantly, as expected in a non-linear system. Specifically, pre-stress stiffens the network, renders stable isostatic networks below the stability limit, and decreases the strength of networks. The use of pre-stress to systematically modify network behavior was not explored extensively to date in the context of applications.

The cytoskeleton is an active network material of obvious importance. Many aspects associated with activity in this complex network have been studied, while others await systematic examination. The design of active engineered networks has been discussed in the literature, particularly in the context of periodic network structures. However, this sub-field is emerging and its impact is uncertain at this time.

Metamaterials are generally periodic fiber-based structures whose architecture is designed such as to provide a pre-defined material-scale behavior. The network may undergo internal instabilities both on the repeat unit scale, as well as on the global scale. Since this book focuses on stochastic networks prevalent in the biological world and in contemporary engineering applications, metamaterials are not discussed. However, many results presented in Chapter 6 also apply to this class of structures. Further details are available in recently published texts on metamaterials (e.g., Lakes, 2020).

References

Alioscha-Perez, M., Benadiba, C., Goossens, K., et al. (2016). A robust actin filaments image analysis framework. *PLoS Comput. Biol.* **12**, e1005063.

Gottardi, R., Hansen, U., Raiteri, R., et al. (2016). Supramolecular organization of collagen fibrils in healthy and osteoarthritic human knee and hip join cartilage. *PLoS ONE* **11**, e0163552.

Lakes, R. (2020). *Composites and metamaterials*. World Scientific, Singapore.

2 Fibers and Fiber Bundles

A broad range of fibers are found in biological materials, including collagen, elastin, fibrin, and smaller filaments such as F-actin, microtubules, and amyloids. Cellulose fibers, either extracted from plants or produced by bacteria, are processed and used in artificial network materials such as paper products and textiles. Engineering nonwovens are generally made from polymeric fibers. Glass fibers are used for sound and thermal insulation. The mechanical behavior of several classes of fibers is presented in this chapter.

A distinction is made between fibers with thermal and athermal behavior. Fibers of a diameter sufficiently large such that thermal fluctuations play no role in their mechanics are considered athermal. Flexible filaments with small or vanishing bending stiffness are subjected to thermal fluctuations, and this has a significant impact on their mechanical behavior. These are considered thermal filaments. Semiflexible filaments which exhibit features specific to both of these categories are also important, since many protein networks are of this type. This chapter includes separate discussions of these three fiber categories.

We recall here that the term "fiber" is used for athermal fibers, while "filament" is used for sub-micron fibers, whether of molecular or supra-molecular type.

In some materials, such as geotextiles and other nonwovens, the network is constructed from individual fibers. In some other cases, particularly in biological networks such as the extracellular matrix, the network is composed of filament bundles, where a bundle contains multiple filaments. The mechanical behavior of a bundle depends not only on the properties of the constituent fibers, but also on the way these are bonded together. A bundle in which fibers are weakly connected behaves differently from one in which fiber-to-fiber interfaces are strong. Fiber bundles have more complex mechanical behavior than individual fibers and this is reflected in the network-scale behavior. The mechanical behavior of bundles is discussed in Section 2.5.

2.1 Examples of Fibers and Fiber Properties

The mechanical behavior of fibers depends on their chemical composition and structure and is probed experimentally either in bending or tension. Variability from fiber to fiber is generally large and experiments can only provide ranges for the constitutive parameters. This section reviews fibers encountered in some of the most important current network materials: cellulose fibers, polymeric fibers, and collagen.

2.1.1   Cellulose Fibers

Cellulose is a polysaccharide which consists of D-glucose units linked in linear chains (Figure 2.1(a)). The chains are arranged in crystals stabilized by strong H-bonds established between the hydroxyl groups of neighboring molecules. This parallel molecular arrangement and the high density of intermolecular bonds convey high strength and stiffness to the crystal. The crystal strength increases as the chain length increases. Cellulose is semicrystalline, that is, it contains both crystalline and amorphous sub-domains. The fraction of the crystalline component varies greatly between plant species and is controlled during the subsequent chemical processing of the plant-derived material.

Cellulose is abundant in nature as it is the building block of the load-carrying structure of most plants. It represents 90% of the cotton and about 50% of the wood structure. Natural fibers contain lignin and hemicellulose in addition to cellulose. Hemicellulose is mostly amorphous and contains a variety of sugar monomers and imperfections which prevent the growth of crystals. This makes it less stiff than cellulose and more prone to hydrolysis by dilute acids. Lignin has a complex, disordered molecular structure which prevents crystallization.

Fibrils containing many molecular strands assemble to form fibers of larger than micrometer cross-sectional dimensions and complex structure. Fibers have successive layers, somewhat similar to the growth rings in wood. Cotton and wood-derived cellulosic fibers used to make paper are tubular, in the sense that they present a lumen, which typically collapses during drying such that the resulting fiber cross-section appears ribbon-like (Figure 2.1(b)). The arrangement of fibrils in the various layers of the fiber influences the fiber mechanical properties. The layers providing the mechanical function have fibrils that spiral around the axis of the fiber (Lichtenegger et al., 1999). Layer S2 in Figure 2.1(b) accounts for about 86% of the thickness of the tubular fiber wall and has cellulose fibrils oriented at an angle relative to the fiber axis. The outer layer denoted by P has a random fibril structure, while the thin layer S1 has fibrils oriented at a large angle (about 80°) to the fiber axis (Neagu et al., 2006). The helix angle defines the stiffness and the deformability of the fiber. This aspect of the

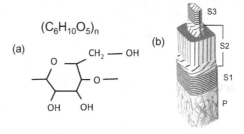

Figure 2.1 (a) Repeat unit of cellulose and (b) schematic representation of the cell wall of a wood fiber showing preferential orientation of fibrils in various layers. Reprinted with permission of the publisher from Neagu et al. (2006)

Table 2.1 Typical properties of common fibers

Fiber Type		Characteristic Cross-section Dimensions (μm)	Young's Modulus, E_f (GPa)	Strength (MPa)
Natural fibers	Cotton	10–20	1.5–3	150–400
	Wool	15–40	2–3.9	120–180
	Jute	~100	20–30	200–350
	Rayon	12–40	0.5–1	100–250
	Softwood	20–70	20–40	100–300
Synthetic polymeric fibers	Nylon	~20	5	~900
	Aramid	~10	60–100	2 000–3 000
	Polyester	10–50	15–20	300–1 000
	Polypropylene	20–50	3–5	200–700
	Polyethylene (high molecular weight)	20–50	5	300–500
Other fibers	Glass	10–20	70–80	1 400–4 000
	Carbon	~10	230–380	1 800–2 600

Note: See a collection of properties of various types of fibers in Mohanty et al. (2000).

mechanics is discussed in Section 2.5.2.2. Here it suffices to indicate the rather obvious fact that a smaller angle (indicating strong alignment in the fiber direction) corresponds to stiffer fibers that deform less before breaking. Controlling the helix angle is an important mechanism by which the plant controls its stiffness. For example, fibril orientation varies greatly in a tree branch leading to a functionally graded material across the cross-section and along the branch, from the tree trunk to the outer extremity of the branch (Farber et al., 2001). The gradient changes spatially, depending on the local mechanical loading, so as to maximize flexibility while retaining strength.

The length of cellulose fibrils and fibers varies greatly as a function of their biological origin and the chemo-mechanical treatment to which they were subjected during processing. Cotton fibers have a length of the order of 3–4 cm, wool fibers range from a few centimeters to tens of centimeters and fibers in pulp and paper have lengths of 0.5–3 mm. Processes such as acid treatment and pulp beating reduce the fiber length and may modify the structure of the fiber wall. The cross-section characteristic dimension of these fibers is of tens of micrometers, such that typical aspect ratios for cotton and flax fibers are ~1 500, while the aspect ratio of wool fibers is ~3 000.

The mechanical properties of natural fibers also vary broadly as a function of the same factors that control the fiber length and as a function of the hydration state. Broad ranges of values for the modulus and strength of cellulose fibers are listed in Table 2.1. The properties of individual wood and pulp fibers used in paper depend on the state of hydration; a fiber wall thickness of 4 μm, fiber width (largest dimension of the cross-section) of 15–30 μm, and axial Young's modulus of 35 GPa are typical values reported in the literature.

2.1.2   Polymeric Fibers

Most artificial fibers made today are polymeric. Polymer viscosity in the melt state is a function of temperature, molecular weight (chain length), and the presence of solvents. This allows us to produce fibers of various diameters, from ~100 nm to hundreds of microns.

The most common fiber manufacturing process is spinning from the melt. This entails forcing the melt through a spinneret by applying pressure and/or suction on the opposite side of the melt bath. Fibers solidify during the flight time between the spinneret and the conveyor belt on which the fibrous mat is deposited. This procedure is used industrially to produce nonwovens and leads to fibers with typical diameter of tens of microns (Figure 2.2). Note that textile fibers also have diameters of tens of microns, while the diameter of a human hair is of the same magnitude (20–50 μm).

Electrospinning is a newer process which was developed to produce polymeric fibers of sub-micron diameter. This method is similar to spinning, but allows drawing the fiber during its flight time, after the fiber forms at the spinneret, by the application of an electric field. The flow undergoes an instability that leads to a substantial reduction of the diameter. The mechanics of spinning and electrospinning are described in a large number of articles (e.g., McHugh and Doufas, 2001) and summarized in recent texts (e.g., Andrady, 2008).

The most common polymers used in fiber production are polypropylene and polyethylene, which are broadly used in nonwovens of various types, and nylon, polyester, and acrylic, which are used in textiles. Fibers are made from other synthetic polymers as well, such as polyurethane and aromatic polyamides (aramids). Polymeric fibers can be spun as single components or as core-shell and side-by-side composite fibers made from two constituents. The side-by-side fibers may be spun with a helical geometry.

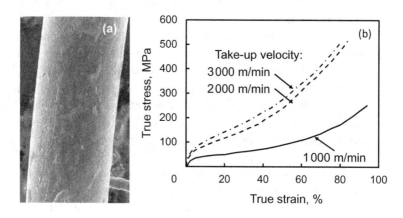

Figure 2.2 (a) Scanning electron micrograph of an iPP fiber of diameter 22 μm. (b) True stress–true strain curves for iPP fibers processed at different take-up velocities. The stiffer response of fibers processed at higher take-up velocities is due to increased chain alignment; the crystallinity is identical in all these fibers. Adapted from Osta et al. (2014)

Fibers made from natural polymers which are chemically decomposed and reconstituted may be produced. The earliest example of this type is rayon (viscose and modal are specific types of rayon), which is obtained from purified cellulose extracted chemically from wood pulp. Cellulose diacetate and cellulose triacetate are obtained by chemically treating cellulose with acetic acid or acetate esters, which preserves the glucose backbone of cellulose (Figure 2.1(a)), but add two and three, respectively, acetate groups to the ring. Fibers are currently being produced from reconstituted collagen, which can be spun in mats supposed to mimic the extracellular matrix and used as scaffolds for tissue engineering. The properties of reconstituted collagen depend on the temperature at which synthesis takes place.

Polymer spinning causes strong uniaxial stretching of the melt, which leads to preferential chain orientation in the flow direction. The alignment favors polymer crystallization. In strong flows, new polymer structures form; an example is the shish-kebab structure composed of a strongly aligned bundle of polymeric chains (the shish) on which disc-like crystalline lamellae form in the direction perpendicular to the fiber direction (the kebab). Such strong alignment and, in particular, the partial crystallization, impart high stiffness and high strength to the fibers, which are highly desirable properties in most applications. In addition, the surfaces of spun and electrospun fibers are generally quite smooth (Figure 2.2(a)), with roughness well below the micron range, which increases the strength of fibers.

Spun polymeric fibers are of "infinite length," which is a result of the continuous nature of the spinning process. When deposited in mat form, fibers are tortuous and have random orientations. Preferential fiber orientations may be obtained by depositing on a spinning mandrel.

The properties of synthetic polymeric fibers depend on the type of polymer and processing method used and have less variability than those of natural fibers. An example of the mechanical response in uniaxial tension of an iPP (isotactic polypropylene) fiber similar to that in Figure 2.2(a) is shown in Figure 2.2(b). The three curves correspond to fibers processed at take-up velocities of 1 000, 2 000, and 3 000 m/min (Osta et al., 2014). Increasing the take-up velocity (i.e., the velocity at which the fiber exits the spinneret) increases chain alignment, the yield stress, and the strength. Richaud et al. (2009) report that increasing the draw ratio of iPP fibers from 4 to 10 increases the modulus from 3 to 15 GPa and the strength from 200 to 600 MPa. Typical values of the stiffness and strength of some common polymeric fibers are reported in Table 2.1.

2.1.3   Collagen

Collagen fibers play an essential mechanical role in tissues of invertebrate and vertebrate organisms. In the human body collagen is present in the extracellular matrix, in connective tissue such as cartilage and tendons, in various membranes and blood vessels and, as mineralized collagen, in bones. Multiple types of collagen exist, of which types I to V are the most prevalent, with some being present only in specific tissues. Type I collagen, which is one of the fibrillar collagen types, is the

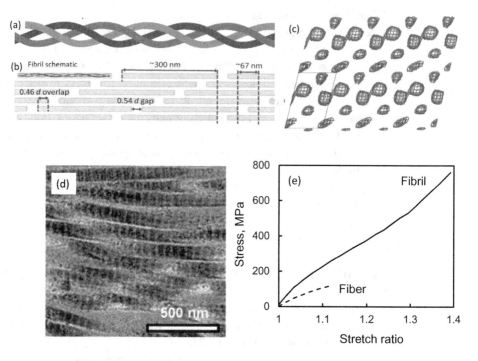

Figure 2.3 Schematic representations of (a) tropocollagen and (b) fibrils showing the characteristic staggering of tropocollagen molecules (reprinted from Sherman et al. (2015) with permission from Elsevier). (c) Arrangement of molecules in the cross-section view of the fibril in (b) (reprinted from Orgel et al. (2001), with permission from Elsevier). (d) Image of fibril bundle forming a fiber (reprinted from Sherman et al. (2015), with permission from Elsevier). The D-band staggering indicated in (b) is visible in this image. (e) Stress–stretch curves of a reconstituted collagen fibril of ∼300 nm diameter (data from Liu et al., 2018), and of a reconstituted collagen fiber of ∼50 μm diameter (data from Sopakayang et al., 2012). As the diameter increases, fibers with hierarchical structure have lower load carrying capacity.

most abundant in the body, being present in tendons, skin/dermis, and arterial walls. Type II, which is also of fibrillar type, is the main component of cartilage.

Collagen has a hierarchical, multiscale structure (Figure 2.3). The present discussion is focused on the structure of fibrillar collagen. The building block is tropocollagen (Figure 2.3(a)), which is a right handed helix made from three polypeptide molecules. This structure was recently imaged in three-dimensions (Orgel et al., 2001). On the next hierarchical level, tropocollagen molecules of ∼300 nm length assemble to form collagen fibrils. The axial arrangement of tropocollagen in the fibril is staggered, that is, tropocollagen strands are arranged in a sequence along the axial direction of the fibril, with gaps between each other, while the gaps of neighboring rows of strands are not aligned (Figure 2.3(b)). The three-dimensional arrangement is complex and it is currently considered that, in the cross-section, the strands have a semi-ordered quasi-hexagonal arrangement, as indicated in Figure 2.3(c) (Orgel et al., 2001). In the simplified view of the axial arrangement shown in Figure 2.3(b) it is suggested that staggering involves a shift of ~67 nm, the so-called D-period, which represents approximately a quarter of the tropocollagen length. The axial gap size is 0.54 of the D-period. When imaged in the direction perpendicular to the fibril axis, a clear periodicity associated with this staggering is observed (Figure 2.3(d)). The diameter of a typical fibril is of about 100–200 nm.

Neighboring tropocollagen strands are connected by enzymatic and nonenzymatic crosslinks which ensure the load transfer between fibrils. Enzymatic crosslinking occurs predominantly at the ends of the tropocollagen strands. Nonenzymatic cross-links due to glycation occur more randomly along the strands and lead to substantial stiffening and associated pathologies.

On the next level of the hierarchy, fibrils form bundles and fibers of diameter on the order of microns and larger. The extracellular matrix and connective tissue are composed from such fibers whose fibril bundles split and merge with neighboring fibers to create a stochastic network (Figure 1.1(d)).

The mechanical properties of collagen are of much interest in biomechanics and have been investigated by many groups. A review is available in Sherman et al. (2015). The effective stiffness of tropocollagen measured on the molecular scale by monitoring thermal fluctuations leading to the evaluation of the persistence length is reported to range from 3 to 4 GPa. Direct stretching of fibrils provides values between 400 and 800 MPa. Testing fibers of above micrometer diameters leads to a broad range of values which depend on the hydration state, degree of tortuosity, and the origin of the respective fiber. The stiffness of the building blocks is larger than that of the larger scale collagen structures, as predicted by the mechanics of discontinuous fiber bundles discussed in Section 2.5.3. Figure 2.3(e) shows stress–strain curves obtained by uniaxial stretching individual collagen fibrils (Liu et al., 2018) and fibers (Sopakayang et al., 2012).

2.2 Energetic and Entropic Mechanical Behavior

Before discussing the mechanical behavior of fibers and networks, it is necessary to clarify the origin of stress in these systems.

Consider a classical system described in terms of a set of kinematic parameters, $\mathfrak{A}$, and the absolute temperature, T. In the canonical ensemble, the representative thermodynamic potential is the Helmholtz free energy:

$$\Psi(\mathfrak{A}, T) = U(\mathfrak{A}, T) - T\Sigma(\mathfrak{A}, T), \tag{2.1}$$

where U and Σ are the internal energy and the entropy of the system. The free energy is related to the partition function, $Z(\mathfrak{A}, T)$, through the relation $\Psi(\mathfrak{A}, T) = -k_B T \log Z(\mathfrak{A}, T)$.

If the system is perturbed by an externally-imposed strain, ε, at least one member of the kinematic parameter set $\mathfrak{A}$, $\mathfrak{A}_i$, varies with the applied strain, that is, $\partial \mathfrak{A}_i / \partial \varepsilon \neq \mathbf{0}$. Then, the quantity

$$\sigma = \frac{\partial \Psi}{\partial \varepsilon} = \frac{\partial \Psi}{\partial \mathfrak{A}_i} \frac{\partial \mathfrak{A}_i}{\partial \varepsilon} \qquad (2.2)$$

is the stress, work conjugate with the prescribed strain.

With Eq. (2.1), one may rewrite the stress as:

$$\sigma = \frac{\partial \Psi}{\partial \varepsilon} = \frac{\partial U}{\partial \varepsilon} - T \frac{\partial \Sigma}{\partial \varepsilon} = \sigma_U + \sigma_\Sigma, \qquad (2.3)$$

where the derivatives are evaluated at constant temperature and number of atoms. Equation (2.3) indicates that stress has two components in this ensemble: one related to the variation of the system energy, σ_U, and the other related to the variation of the entropy, σ_Σ, during deformation.

Since the entropy is given by $\Sigma = -\partial \Psi / \partial T$, it is possible to write $\partial \Sigma / \partial \varepsilon = -\partial^2 \Psi / \partial \varepsilon \partial T = -\partial \sigma / \partial T$. Hence,

$$\sigma_\Sigma = T \frac{\partial \sigma}{\partial T}. \qquad (2.4)$$

Consider a monatomic crystal lattice with no defects at zero Kelvin (static case; Figure 2.4(a)). The lattice exists in only one configuration that, in the absence of defect nucleation or amorphization, remains well-defined during the elastic deformation. Therefore, the entropy vanishes. The internal energy is defined by interactions occurring between atoms. Deformation leads to the variation of the internal energy, since the interatomic distances change with strain. The free energy variation is therefore equal to the variation of the internal energy and stress is purely energetic in this case.

Consider further a string composed of a large number of rigid links (Figure 2.4(b)). The links are pin-jointed and rotate freely relative to each other. Since the links are not deformable, the system stores no internal energy. On the other hand, for any end-to-end distance, the string may take a large number of configurations. This number is maximum when the two ends are co-located. As the string is stretched and the distance between its ends increases, the number of available configurations decreases. In the

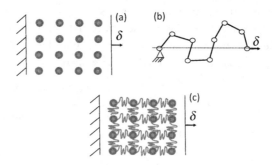

Figure 2.4 Schematic representations of (a) a monatomic lattice, (b) a freely rotating string, and (c) a periodic lattice of "nodes" connected by strings.

limit of a straight string, only one configuration is available. Therefore, the entropy of the string decreases continuously during stretching. The free energy of this system is equal to its entropic component ($\Psi = -T\Sigma$), and stress is purely entropic.

These two examples are idealizations. In real systems both entropic and energetic components are non-zero and in some cases one of the two components dominates. A monatomic crystal develops defects during deformation and the defect density increases with strain. The entropy of the system is not zero and increases as the crystal deforms. At finite temperatures, lattice vibrations lead to another entropic component of the free energy. The variation of the phonon spectrum with the strain is the physical basis of the entropic component of the stress in perfect lattices. For a mathematical description of this term, see Weiner (1983). However, the energetic component of stress remains dominant in most monatomic crystalline solids.

In order to emphasize that the nature of stress depends on the state of the system, consider an imaginary material which is a cross between the two examples discussed in the previous three paragraphs: a periodic lattice of nodes which are connected by strings (Figure 2.4(c)). The links of these strings are allowed to interact energetically with other links which are not immediately bonded to them along the same string. Deformation changes the end-to-end distance of most strings and the entropy of the system. Therefore, the system has non-zero energy and entropy, and both components of the free energy vary during deformation. Both entropic and energetic components of the stress are non-zero. This idealized example is a proxy for rubber, which is a crosslinked molecular network. However, another component of the physics is active in rubber and this substantially modifies the physical picture. The links are quite mobile since the material is above the glass transition temperature in ambient conditions. This allows them to relax in configurations of minimum local energy on time scales shorter than those of network deformation. As the system deforms, the local environment of the links evolves, but these states are energetically equivalent since packing does not change significantly. This implies that the energy of the system is approximately constant during deformation and hence the energetic component of stress is negligible. Although, in principle, the stress has both energetic and entropic components, fast relaxation of the energetic component renders the system approximately entropic.

This example helps in understanding another important characteristic of these physical systems, which points to limitations of the applicability of the entropic stress concept. A string must explore all configurations available to it at all times during deformation in order to probe the variation of its entropy. This implies that (i) the string should be sufficiently mobile to explore its phase space and (ii) the rate of the imposed deformation should be such that the string is given sufficient time to perform the sampling. If these conditions are not fulfilled, the entropy is constant during deformation and the entropic component of stress vanishes.

Polymeric networks are typically considered thermal networks and stress is purely entropic. However, if the deformation rate is high, the energetic stress component is not relaxed effectively by local atomic scale processes and becomes non-negligible relative to the entropic component. Furthermore, reducing the mobility by, for example, reducing the temperature toward the glass transition temperature, reduces the ability of molecular strands to sample their phase space and reduces the importance of the entropic component relative to the energetic counterpart. Stress in polymeric networks below the glass transition temperature is energetic.

This discussion applies equally to individual fibers and to networks and provides the physical background for the distinction made here between thermal and athermal systems. The entropic component of stress is important in thermal networks, while athermal networks are purely energetic. Semiflexible filaments, in which both components are non-negligible, exhibit a richer behavior, which is discussed in Section 2.4.4.

2.3   Athermal Fibers

Athermal fibers have diameters sufficiently large for thermal fluctuations to play no role in their mechanics. Therefore, they can be treated as classical beams. Their elastic and inelastic behavior, stability, and strength are discussed in this section, with an eye on the aspects of the fiber behavior which are important for the mechanical response of the network.

2.3.1   Small Deformations

Beams store energy in the axial, bending, and torsion deformation modes. Under infinitesimal deformations these modes are independent, in the sense that the applied deformation corresponding to one mode does not do work against loads associated with other deformation modes. Energy storage caused by shear associated with transverse forces becomes important for beams of low aspect ratio (i.e., small cross-sectional dimensions relative to the axial length).

The geometric parameters characterizing fibers of circular cross-section are the fiber diameter, d, and the fiber length, L_0. If the deformation is elastic, the only mechanical parameter of importance is the effective stiffness, which is defined as the ratio of the applied load to the conjugated measure of deformation. However, the effective stiffness depends on the boundary conditions. The loading configurations most relevant for networks of fibers are similar to those shown in Figure 2.5, with axial, bending and torsional loads acting at the ends of the fiber. Solutions for these

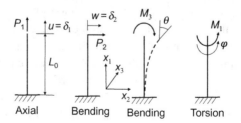

Figure 2.5 Axial, bending, and torsional loading modes of straight beams.

Table 2.2 Effective stiffness of slender beams loaded as shown in Figure 2.5

Loading Mode	Axial	Bending, Transverse Force	Bending, Bending Moment	Torsion
Effective stiffness	$\frac{P_1}{\delta_1} = \frac{E_f A_f}{L_0}$	$\frac{P_2}{\delta_2} = \frac{3 E_f I_f}{L_0^3}$	$\frac{M_3}{\theta} = \frac{E_f I_f}{L_0}$	$\frac{M_3}{\varphi} = \frac{G_f J_f}{L_0}$

boundary value problems are obtained within the linearized beam theory which can be found in many texts (Gere and Timoshenko, 2002). The effective stiffness corresponding to each of these loading modes is listed in Table 2.2. E_f and G_f represent the Young's and shear moduli of the fiber material, and A_f, I_f, and J_f represent the area, axial moment of inertia, and polar moment of inertia of the fiber cross-section relative to its centroidal principal axes, respectively. For the circular cross-section, $A_f - \pi d^2/4$, $I_f = \pi d^4/64$, and $J_f = 2 I_f = \pi d^4/32$.

The Timoshenko model leads to the total strain energy of a beam:

$$U = \frac{1}{2} \int E_f I_f \left(\frac{d\theta(s)}{ds} \right)^2 + E_f A_f \left(\frac{du(s)}{ds} \right)^2 + G_f J_f \left(\frac{d\varphi(s)}{ds} \right)^2$$
$$+ \gamma G_f A_f \left(\frac{dw(s)}{ds} - \theta(s) \right)^2 ds. \tag{2.5}$$

The integral is written in terms of s, a curvilinear coordinate along the fiber, and is performed along the entire fiber length. The four terms represent the bending, axial, torsion, and shear energies, respectively. In this expression, $\theta(s)$, $\varphi(s)$, $u(s)$, and $w(s)$ represent the s-dependent rotation of a plane which remains perpendicular to the neutral axis of the fiber, the torsional rotation of a section about the neutral axis, axial and transverse displacements, respectively. $\frac{dw(s)}{ds} - \theta(s)$ represents the additional rotation of the fiber cross-section due to shear. In the Timoshenko theory the cross-section of the beam is not perpendicular to the beam axis due to the shear loading. The Euler–Bernoulli model is a particular case of the Timoshenko model which neglects the contribution of shear. In this approximation, the last term in Eq. (2.5) is absent. Accounting for the contribution of the shear term is important for beams of small aspect ratio. As discussed in Chapter 4, the distribution of fiber segment lengths in random networks is of Poisson type (exponential) and hence many fibers segments are short. This suggests that the Timoshenko model is preferable to the Euler–Bernoulli model as a description of the mechanical behavior of athermal network fibers (Shahsavari and Picu, 2012).

In complex structures of beams, loading is such that multiple fiber deformation modes are enabled at the same time. The way the strain energy is distributed across these modes is defined by the effective stiffness of the structure. Figure 2.6(a) shows a beam loaded by a force P making an angle of $\pi/4$ with the axis of the beam, such that the axial and bending modes are enabled by forces of equal magnitude. The energy stored in the axial mode is given by $U_{ax} = P^2 L_0/4 E_f A_f$, while the energy stored in the bending mode is $U_{bd} = P^2 L_0^3/12 E_f I_f$. The ratio of the two energies is $U_{ax}/U_{bd} = (3/16)(d/L_0)^2 \ll 1$. Hence, under load control, the fiber stores energy predominantly in the bending mode. Note that the ratio of the effective stiffnesses of

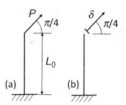

Figure 2.6 Cantilever beam with imposed (a) force and (b) displacement at the free end.

the two modes (Table 2.2) is $k_{bd}/k_{ax} = (3/16)(d/L_0)^2$ in this particular case, that is, $U_{ax}/U_{bd} = k_{bd}/k_{ax}$, which implies that fibers store energy primarily in the softer deformation mode.

Consider now the equivalent displacement boundary value problem (Figure 2.6(b)). A displacement is applied at an angle $\pi/4$ relative to the fiber axis, such to equally activate the two deformation modes. In this case, the energy ratio is $U_{ax}/U_{bd} = (16/3)(L_0/d)^2 \gg 1$. Therefore, under displacement control, the fiber stores energy predominantly in the stiffer, axial mode.

In the load control case, the fiber behaves as if the two modes are connected in series. In displacement control, the two modes appear to be acting in parallel. Since one mode (axial) is much stiffer than the other, the apparent stiffness of the fiber is smaller under load control than when probed in displacement control.

A similar situation is encountered in composites made from stiff and soft phases. In this context, imposing an affine deformation, that is, forcing the two phases to deform by the same strain, leads to the upper bound of the composite stiffness. On the other hand, requiring that each phase carries the same load leads to the lower bound estimate for the effective composite stiffness. These are known as the Voigt and Reuss bounds, respectively. In Chapter 6 it is shown that the same applies on the scale of the network that, if forced to deform affinely, exhibits the largest attainable stiffness, while, if allowed to reduce its energy by nonaffine deformations, has a much lower stiffness.

2.3.2   Large Deformations

The linear beam theory cannot represent the geometric nonlinearity associated with large deformations. Beams undergo large deformations primarily in the bending mode. The geometric nonlinearity associated with large deformations was accounted for in the initial work on the mechanics of slender beams by Jacob Bernoulli (1654–1705), his nephew Daniel Bernoulli (1700–1782), and Leonard Euler (1707–1783). Jacob Bernoulli investigated the deformed shape of a beam loaded in bending and related the axial deformation of planes parallel to the neutral axial to the applied bending moment. Daniel Bernoulli correctly identified that the strain energy of a small segment of the beam is proportional to the square of the beam curvature at the respective site and suggested to Euler that his then new variational calculus can be used to find a solution for the beam deformation problem. Euler's follow-up analysis led to the governing equation which reads:

$$\kappa(x) = \frac{M}{E_f I_f}(x), \tag{2.6}$$

where $\kappa(x)$ is the curvature of the beam at position x of the undeformed configuration, M is the applied bending moment in the same cross-section, and $E_f I_f$ characterizes the beam material and the cross-section geometry. The left hand side of Eq. (2.6) may be written in terms of the deflection of the beam, measured in the direction perpendicular to the initial beam axis, $w(x)$, as

$$\frac{d^2w/dx^2}{\left[1 + (dw/dx)^2\right]^{3/2}} = \frac{M}{E_f I_f}(x). \tag{2.7}$$

Equation (2.7) is a nonlinear differential equation for w and the nonlinearity is of geometric type. This expression can be linearized by neglecting dw/dx relative to 1 in the denominator. Since $dw/dx = \tan\theta$, with θ being the angle between the tangent unit vector and the undeformed beam axis, the respective approximation requires that the tangent slope of the beam is small. Within this approximation, Eq. (2.7) becomes:

$$\frac{d^2w}{dx^2} = \frac{M}{E_f I_f}(x), \tag{2.8}$$

which can be readily integrated to derive the shape of the deformed beam, $w(x)$, for given distribution of moments, M(x). The linearized formulation (2.8) is used to derive the results presented in Section 2.3.1 and Table 2.2.

Since the small deformation formulation (Eq. (2.8)) is entirely linearized, both geometrically and in terms of the beam material behavior, superposition applies. Solutions for complex loading situations can be obtained by superimposing solutions for simpler problems with same boundary conditions. This does not apply in nonlinear systems and, in such cases, Eq. (2.7) has to be solved for each loading and boundary condition separately.

Several features of the nonlinear beam bending problem are of importance for the broad understanding of network mechanics. We recall one of the best known and most relevant (in the context of networks) problems of this type: the bending of a cantilever beam by a concentrated force acting at its end (Mattiasson, 1981; Fertis, 1999). Two possibilities are considered, with the force preserving its initial direction (Figure 2.7(a))

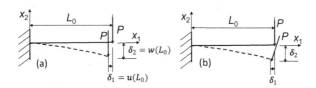

Figure 2.7 Cantilever beam loaded by (a) a force preserving its initial direction and (b) a follower force.

and with the force remaining perpendicular to the beam axis, which is also known as the "follower force" problem (Figure 2.7(b)). Figure 2.8 shows the displacement of the point of application of the force in the two cases of Figure 2.7, and in the two directions, x_1 and x_2. The figure also shows the deflection (displacement in the direction perpendicular to the undeformed beam axis) computed with the small deformation formula (i.e., $\delta_2 = w(L_0) = PL_0^3/3E_fI_f$), and the axes are normalized such to render the small deformation solution a straight line. This emphasizes the difference between the linear and nonlinear solutions and shows the rapid change of the effective, instantaneous stiffness in the nonlinear case.

Figure 2.9 shows the tangent stiffness (derivative of the applied force with respect to the displacement δ_2), versus the applied force computed from the data in Figure 2.8. The force is normalized with $3E_fI_f/L_0^2$ and δ_2 is normalized with L_0, as in Figure 2.8. With this normalization, the small deformation formula gives a horizontal line at 1. The geometric nonlinear effect becomes pronounced for normalized forces

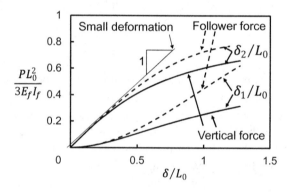

Figure 2.8 Force-displacement relations for a cantilever beam loaded with a vertical force (Figure 2.7(a)) and a follower force which remains normal to the beam at all times (Figure 2.7(b)).

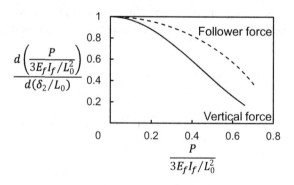

Figure 2.9 Normalized tangent stiffness versus normalized force for the curves in Figure 2.8.

$PL_0^2/3E_fI_f$ larger than 0.1, which corresponds to $\delta > 0.2L_0$. Beyond this threshold, the response becomes softer. The behavior shown in Figure 2.9 is also observed if the tangent stiffness is plotted versus δ/L_0, that is, as if the beam would be loaded in displacement control.

A significant difference between displacement and load control is observed when investigating the increment of work performed (or stored strain energy) associated with an increment of the controlled variable. Under displacement control, the increments of work increase continuously as the beam response becomes stiffer. However, in load control, the increments of work decrease as the load increases.

Before closing this section, the geometrically exact beam formulations should be recalled. The first complete theory that eliminates the approximations of the Euler–Bernoulli theory and which accounts for large deformations and bending, axial, torsion, and shear deformation modes was developed by Reissner for two- and three-dimensional problems in 1972 (Reissner, 1972) and 1981 (Reissner, 1981), respectively. A large number of variants of this theory and the details of their numerical implementation, including the discussion of their objectivity and path dependence, have been published (e.g., Crisfield and Jelenic, 1999; Jelenic and Crisfield, 1999). A comparison of such models is presented in Romero (2008) and a historical perspective on their evolution is provided in Armero and Valverde (2012).

2.3.3 Relevance of Fiber Buckling for the Deformation of Fiber Networks

It is often claimed that fiber buckling takes place during the deformation of random networks and that this has important consequences on the overall network behavior. We indicate here that buckling instabilities of individual fibers are unlikely to occur. However, instabilities involving groups of fibers occur frequently.

Consider the loading shown in Figure 2.10(a) in which a compressive force, P_{ax}, loads the beam axially, along with a transverse force, $P = \alpha P_{ax}$. If only the axial force is applied ($\alpha = 0$), the Euler buckling problem is recovered. The critical buckling

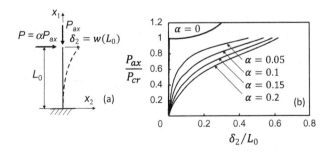

Figure 2.10 (a) Axial loading of a beam in the presence of a lateral force and (b) force versus lateral displacement of the point of application of forces for the pure axial loading (showing buckling instability) and combined loading with several values of α.

force for these boundary conditions is $P_{cr} = \pi^2 E_f I_f / 4L_0^2$. The deflection $w(L_0)/L_0$ for $P_{ax} < P_{cr}$ is zero, but increases rapidly once the critical load is reached, as shown in Figure 2.10(b). The exact solution shown here is given in Bazant and Cedolin (2010). The increment of the axial force versus the increment of $w(L_0)/L_0$ immediately after the critical point is vanishingly small, that is, $(P_{ax} - P_{cr})/P_{cr} \approx (\pi^2/8)(\delta_2/L_0)^2$. The incremental work performed by the axial force changes from being proportional to the square of the incremental axial strain $u(L_0)/L_0$ for $P_{ax} < P_{cr}$ ($u(L_0)$ measured in the x_1 direction), when the beam deforms in the axial mode, to being approximately linear in $u(L_0)/L_0$ for $P_{ax} > P_{cr}$, since the deformation after the instability takes place at essentially constant axial force.

The behavior of the beam changes drastically in the presence of a force acting perpendicular to the beam axis at the free end, $P = \alpha P_{ax}$. The deflection $w(L_0)/L_0$ is shown in Figure 2.10(b) for several non-zero values of α. The instability is removed in the presence of the transverse force. A similar situation results when bending moments are applied in place of the transverse force and acting along with the axial force. This solution is presented in most books on structural stability (Timoshenko and Gere, 1961; Bazant and Cedolin, 2010) and resembles the curves shown in Figure 2.10(b) for $\alpha > 0$.

If a torsion moment, M_1, is applied along with the axial compressive force, P_{ax}, the situation is slightly more complicated but remains of buckling type. Solutions for various boundary conditions are presented in Bazant and Cedolin (2010). Here we confine attention to the boundary conditions that render the problem conservative: the beam is fixed at $x_1 = 0$, while the end at $x_1 = L_0$ is free to shift in the direction perpendicular to the undeformed beam axis, but it is prevented from rotating about axis x_2, $w'(L_0) = 0$. The critical conditions are described by the equation $M_1^2/4E_f I_f + P_{ax} = \pi^2 E_f I_f / L_0^2$ (Timoshenko and Gere, 1961), where P_{ax} is positive in compression.[1] Note that the presence of a torsion moment reduces the critical axial stress for the buckling instability.

In a random network, fibers are loaded in complex ways and it is extremely improbable that any fiber would be loaded exclusively with an axial force or axial moment. Therefore, buckling of individual fibers is largely irrelevant in networks, since the presence of loads other than axial removes the usual beam buckling instabilities.

However, instabilities that involve multiple fibers and are associated with local soft modes are frequent. The critical states for these instabilities cannot be estimated based on simple theoretical considerations due to the stochastic nature of the fiber arrangement and boundary conditions of the group of fibers that undergo the respective instability.

---

[1] The Euler buckling critical load for the respective boundary conditions is recovered for $M_1 = 0$. The torsional moment acting alone (when $P_{ax} = 0$) may produce buckling at a critical value given by this expression. The presence of a tensile axial force ($P_{ax} < 0$, in the notation of Figure 2.10(a)) increases the critical torsional buckling moment, and therefore stabilizes the beam.

2.3.4   Straight Fibers with Noncircular Cross-section

In many networks, fibers have a noncircular cross-section, and this introduces additional complexities. In some real networks, the cross-section is not even constant along any given fiber, with variability in the shape and size of the cross-section being sometimes pronounced, particularly in the case of natural fibers such as cellulose fibers in paper and cotton, flax, and other natural fibers used in textiles. Therefore, it becomes of interest to discuss the effect of the fiber cross-section shape on the mechanics of the network.

Any fiber cross-section has two principal axes of inertia. The principal axial moments of inertia, I_{min} and I_{max}, are computed in terms of the section geometry and dimensions (Gere and Timoshenko, 2002). If the difference between these values is large, $I_{max} \gg I_{min}$, and the fiber is loaded in bending in the stiffer mode, it becomes energetically favorable for the fiber to rotate and bend in the softer mode.

To demonstrate this effect, consider the cantilever beam shown in Figure 2.11(a), loaded by a force P acting in the x_2 direction. The beam has rectangular cross-section and is oriented with its longer edge along x_2, such that x_2 and x_3 are the principal axes of inertia and $I_3 \gg I_2$. As shown, the force bends the beam in the $x_1 - x_2$ plane (i.e., in the stiffer bending mode). However, it is more likely that the beam twists and bends in the softer mode, to reduce the stored strain energy, as shown schematically in Figure 2.11(b). This process is known as "lateral buckling."

The critical load P_{cr} leading to lateral buckling is given by $P_{cr} = \gamma^{**}\sqrt{E_f I_{min} K_t}/L_0^2$, where, for the configuration in Figure 2.11(a), $I_{min} = hb^3/12$, and K_t is the torsional rigidity of the beam given by $K_t = \gamma^* G_f hb^3$. G_f is the shear modulus and parameter γ^* is a function of the ratio $\xi = h/b$ which, for $\xi \geq 1.5$, can be approximated with the function $\gamma^* \approx \xi^2 / \left[3(\xi + 0.4)^2\right]$. Parameter γ^{**} is a slowly varying function of ξ, taking values of 4.013 for $\xi = 10$ and 5.03 for $\xi = 3$ (Timoshenko and Gere, 1961).

If the beam in Figure 2.11(a) is loaded in pure bending with a moment M aligned with axis x3, lateral buckling sets in at a critical moment given by $M_{cr} = \pi\sqrt{E_f I_{min} K_t}/L_0$, and, after this instability, the beam continues to bend in the softer mode characterized by the moment of inertia I_{min} ($I_{min} = I_2$ in Figure 2.11(a)).

The expression of the critical transverse force for lateral buckling may be rearranged as $P_{cr} = \left(\sqrt{2}\gamma^{**}/\sqrt{1+\nu}\right)\left(E_f I_{min}/L_0^2\right)$, where ν is the Poisson ratio of the beam material. The critical force for Euler buckling for the same boundary conditions has a similar expression which differs by a constant:

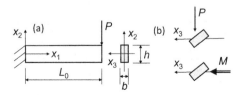

Figure 2.11 (a) Loading of a beam of noncircular cross-section with a transverse force, and (b) lateral buckling during bending with a transverse force or bending moment.

$P_{cr} = \pi^2 E_f I_{min}/4L_0^2$. The critical moment for lateral buckling in pure bending may be arranged as $M_{cr} = \left(\sqrt{2}\pi/\sqrt{1+\nu}\right)\left(E_f I_{min}/L_0\right)$. Therefore, the only moment of inertia relevant for the mechanics of such beams is that of the softer mode, I_{min}. The deformation of the beam after instability takes place in the softer mode and hence I_{min} is also relevant for this situation.

In view of this observation, the mechanics of networks of slender fibers of noncircular cross-section is expected to be approximately similar to that of networks of fibers with circular section and with moment of inertia equal to the I_{min} value of the actual, noncircular fibers. This issue is discussed further in Section 6.1.1.3.7.

2.3.5   Crimped Fibers

Fibers are rarely straight in real networks. Therefore, accounting for the effect of fiber crimp (or tortuosity) in the analysis of fiber mechanical behavior is important.

Crimp is usually quantified using the parameter:

$$c = \frac{L}{L_0}, \tag{2.9}$$

where L is the length of the end-to-end vector of the fiber, while L_0 is the contour length. With a curvilinear coordinate, s, taken along the contour of the fiber, $L_0 = \int ds$. In all cases, $c \le 1$.

To demonstrate the occurrence of crimp in network materials, Figure 2.12 shows the probability density function of collagen fiber crimp in the rabbit adventitia, the outermost layer of the carotid artery (Rezakhaniha et al., 2012). The experimental distribution is well approximated by a beta distribution of coefficients $\alpha = 4.47$ and $\beta = 1.76$. The mean value of the tortuosity parameter in this case is $c = 0.72$.

It is currently thought that the presence of crimp in collagen fibers in arteries is physiologically necessary. Collagen and elastin form co-networks and provide the mechanical function of the tissue but play different roles. Elastin controls the linear elastic response of the tissue at small strains, while the collagen sub-network, which

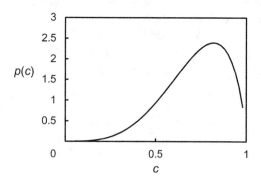

Figure 2.12 Probability density function of collagen fiber crimp in the rabbit adventitia. Adapted from Rezakhaniha et al. (2012)

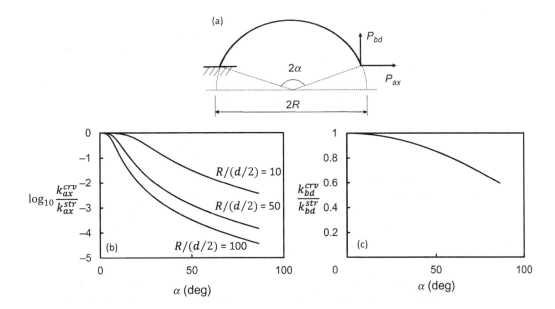

Figure 2.13 (a) Configuration of a beam with crimp (curved beam) probed by forces acting along (axial) and perpendicular (bending) to the end-to-end vector. Variation of the ratio of the effective (b) axial and (c) bending stiffness of the curved and straight beams with angle α.

has tortuous fibers, has limited contribution at small strains while the crimp is stretched out. At larger strains, the collagen network takes over and provides rapid stiffening, which limits tissue deformation and damage. Crimp values are different in different tissues, being smaller in connective tissue such as tendons.

Two cases are considered here in order to demonstrate the effect of crimp on the stiffness of nonstraight fibers, within the small deformation theory. The effect of crimp at constant fiber curvature is discussed first. To this end, a circular beam of radius R is considered (Figure 2.13(a)) and the angle α describing the span of the beam is varied in the range $(0, \pi/2]$. In this problem, the crimp parameter results $c = \sin\alpha/\alpha$ and varies in the range $[2/\pi, 1)$. Second, we consider beams of the same crimp, but different curvature. This is discussed using the family of configurations shown in Figure 2.14(a). These fibers have the same crimp, $c = 2/\pi$, and increasing curvature.

The axial stiffness of the fiber in Figure 2.13(a) is probed by applying a force P_{ax} along the line connecting the two fiber ends and is computed using the Mohr–Maxwell method (Gere and Timoshenko, 2002). The effective stiffness, k_{ax}^{crv}, is compared with that of the straight beam of same end-to-end length, which is given by $k_{ax}^{str} = E_f A_f / 2R \sin\alpha$. Their ratio, $k_{ax}^{crv}/k_{ax}^{str}$, is shown in Figure 2.13(b) as a function of α. k_{ax}^{crv} is also a function of the ratio between the beam radius, R, and the cross-section diameter, d. Results for increasing $R/(d/2)$ are shown in Figure 2.13(b). The stiffness decreases rapidly with increasing α (decreasing c), and the effect is more pronounced as $R/(d/2)$ increases. Slender beams of small cross-section radius are very soft even at modest crimp.

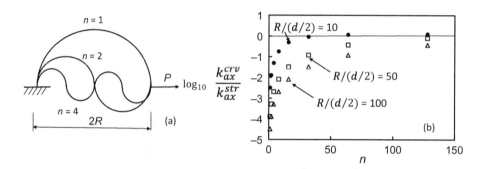

Figure 2.14 (a) Family of tortuous (curved) beams of the same crimp and increasing curvature. (b) Variation of the ratio of the effective axial stiffness (probed by loading along the end-to-end axis) of the crimped and straight beams with increasing n (increasing curvature), as given by Eq. (2.10).

The bending stiffness of the beam in Figure 2.13(a) is probed by applying at the free end a force perpendicular to the end-to-end axis, P_{bd}, and can be evaluated analytically using the same method. The stiffness is compared with that of the straight beam of same end-to-end length, which is given by $k_{bd}^{str} = 3E_f I_f / 8R^3 \sin^3 \alpha$. The ratio $k_{bd}^{crv}/k_{bd}^{str}$ is shown in Figure 2.13(c) and is insensitive to $R/(d/2)$. The bending stiffness reduction with increasing α (decreasing c) is much less pronounced than in the case of the axial stiffness. The bending stiffness of this beam has the same value if the direction of the probing force, P_{bd}, is reversed, despite the geometric asymmetry of the structure. Interestingly, it results that $k_{bd}^{crv}/k_{bd}^{str}$ is approximately proportional to $c = \sin \alpha / \alpha$ in this case.

The next example underlines the fact that crimp is not the only parameter of importance in this discussion; fiber stiffness changes rapidly when the curvature is varied. To demonstrate this effect, consider the family of beams shown in Figure 2.14(a). These beams are constructed by concatenating half-circles, and the number of half-circular units of each beam, n, is varied. The total contour length is πR, independent of n. Since the end-to-end length is $2R$ in all cases, $c = 2/\pi$ is n-independent. The beam curvature increases with n as n/R. The ratio of the effective axial stiffness of the curved and straight beams becomes:

$$\frac{k_{ax}^{crv}}{k_{ax}^{str}} = \frac{1}{\frac{\pi}{4} + \pi \left(\frac{2R}{nd}\right)^2},$$ (2.10)

and is shown in Figure 2.14(b) as a function of n. With n varying from 1 to $4R/d$, the stiffness varies from that of the beam with the smallest curvature, $1/R$ ($\alpha = \pi/2$ in Figure 2.13(a)), to that of the straight beam. It results that the stiffness depends on both crimp and the actual shape of the beam.

The large deformation response of tortuous athermal fibers and the influence of their shape are demonstrated using data from Kabla and Mahadevan (2007). These authors consider fibers of shape described by a cosine function of given frequency and different amplitudes in the unloaded configuration and deform them in tension until

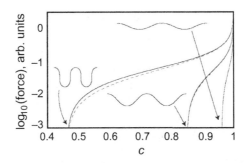

Figure 2.15 Force–crimp (equivalent to force–displacement) curves for fibers of various initial geometry. The crimp parameter, c, represents the current end-to-end length divided by the contour length. Reproduced from Kabla and Mahadevan (2007) with permission from The Royal Society of Chemistry

fully stretched. Figure 2.15 shows the force–crimp (equivalent to force–displacement) curves corresponding to fibers of initial tortuosity parameter $c = 0.46, 0.85$, and 0.96. The effective initial stiffness probed along the end-to-end axis increases with increasing c of the undeformed configuration. The tangent stiffness decreases as the fiber is being pulled and then increases again as the fiber becomes almost straight and $c \to 1$.

If fibers are considered inextensible, the force–crimp curve has a vertical asymptote at a deformation corresponding to the fully stretched configuration. As this limit is approached, the force diverges as $P \sim 1/\sqrt{1-c}$. It is instructive to compare this with similar results for the thermal filaments of freely rotating Langevin and semiflexible types discussed in Sections 2.4.1 and 2.4.4, respectively. In the present notation, the Langevin chain model predicts that the force diverges in the full extension limit as $P \sim 1/(1-c)$ (Eq. (2.33)), while the semiflexible chain model predicts a faster divergence of the form $P \sim 1/(1-c)^2$ (Eq. (2.51)). This comparison is aimed to just indicate the range of behaviors reproduced by these models that, in fact, represent different physics. The divergence of the force at $c = 1$ is eliminated if fibers are extensible.

These examples may appear simplistic in view of the fact that the shape of real fibers is always more complex than considered here. However, realistic fiber shapes may be described using a Fourier series as a superposition of trigonometric functions of increasing frequencies. At small deformations, their response is similar to that of a sinusoidal fiber of wavelength equal to the longest wavelength of the real fiber spectrum. This is due to the fact that fibers subjected to specified loads deform in the softest mode available – see the discussion in relation to Figure 2.6. The high frequency components of the fiber shape spectrum provide larger stiffness, as can be seen in the example of Figure 2.14. Therefore, the deformation begins by pulling out the low frequency waviness, while stretching the higher frequencies takes place at later stages and leads to gradual stiffening.

Closing the discussion of crimp, it is necessary to indicate the limitations of the measure of fiber tortuosity of Eq. (2.9). As discussed in more detail in Section 4.2.1, the crimp parameter is not an intensive quantity. It depends on the length scale of observation, L_0, if the largest wavelength of undulations is larger than or comparable with the contour length of the segment considered when evaluating the crimp parameter. If this wavelength is smaller than the probing length scale, crimp becomes a useful intensive geometric parameter, independent of the probing length. An alternative measure of tortuosity is the persistence length, L_p, which may be defined in both thermal and athermal cases, but is used extensively in thermal models (Section 2.4.2).

2.3.6   Inelastic Fiber Behavior

The mechanical behavior of athermal fibers probed in tension is generally elastic–plastic and may exhibit strain rate dependence. An important characteristic of the inelastic response is the yield stress. Most fibers exhibit yielding and the stress level at which plastic deformation begins depends on hydration, imposed deformation rate, thermal and deformation history of the fiber, etc. The strain rate sensitivity also depends on the state of the fiber and environmental conditions. Given this diversity of behaviors, providing a unified perspective similar to the description of the elastic response is not possible. Two examples are given next which provide an idea about the features of interest in the context of networks.

Figure 2.16(a) shows the stress–strain curve for a cellulose summerwood fiber, adapted from Seth and Page (1983). An initial elastic regime is followed by yielding at a stress of about 100 MPa and by plastic deformation with strain hardening. Residual strains result upon unloading and unloading–reloading cycles lead to limited hysteresis. The response changes greatly with the humidity level.

Figure 2.16(b) shows results from tensile uniaxial tests performed with a reconstituted collagen fiber of diameter $\sim$200 nm (Liu et al., 2018). The fiber exhibits a slightly nonlinear response with no obvious yield point. A small residual strain results upon unloading and part of it recovers after the load is removed. Pronounced hysteresis is observed while performing loading–unloading cycles at constant maximum stretch. The fiber is viscoelastic, but a small plastic component cannot be excluded. The response is sensitive to hydration and temperature.

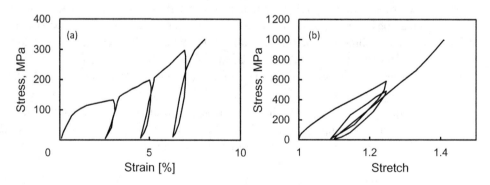

Figure 2.16 (a) Stress–strain curve for a cellulose fiber (adapted from Seth and Page, 1983). (b) Stress–stretch curve for a reconstituted collagen fiber (adapted from Liu et al., 2018).

It is important to asses to what extent the inelastic fiber behavior is relevant for the mechanical behavior of networks. This issue is discussed quantitatively in Section 6.1.4. To address this question, it is necessary to consider the following points:

▶ Fibers of stochastic networks are loaded axially only at large strains and in networks of large fiber density which deform approximately affinely. In relatively sparse networks of thin filaments, such as most biological networks, nonwovens, felt, and other low-density network materials, fibers deform predominantly in their bending mode.

While plastic deformation may take place in bending, the probability of entering the plastic regime depends on the fiber diameter. A cantilever beam subjected to bending enters the plastic regime at a deflection which depends inversely on the fiber diameter. As the diameter is reduced below the micron range, very large deflections can be achieved while the fiber remains elastic. An impressive example of this type is provided by the glass fibers used in fiber optics which sustain very large bending deformations without failure or plastic deformation.

▶ In many network materials, the strain experienced by individual fibers is smaller than the strain imposed on the scale of the network. This is due to the nonaffine nature of deformation observed in most networks, and to the large geometric nonlinearity of the network behavior. The macroscale deformation is accommodated primarily by the reorganization of the network structure. This geometric effect, which allows the fibers to deform little even though the network-scale strains are large, is prevalent in low density networks.

Puxkandl et al. (2002) measured the strain experienced by fibers in a rat tendon sample by evaluating the D-spacing of the collagen fibrils. They compare the applied global strain with the fiber strain and conclude that the fiber strain is much smaller (below 30%) than the strain applied on the scale of the network. Considering that tendon is a strongly aligned collagen structure, the large difference between the local and global strains is particularly telling. In sparse networks with less pronounced preferential fiber alignment, this difference is expected to be even larger.

A similar conclusion was reached by Kabla and Mahadevan (2007) who performed experiments with felt made from randomly oriented polyester fibers of 30–50 μm diameter. Despite the global strains being close to 100%, it is reported that the fiber-scale strains remain limited, such that fibers, which have a yield strain in uniaxial tension of ∼5%, remain elastic.

In the case of tortuous fibers, the gradual reduction of crimp during deformation allows for the accommodation of large network-scale strains with limited fiber-level deformation.

These considerations suggest that fiber inelasticity is expected to reflect in the network behavior only in dense and densely crosslinked networks and in networks of fibers with low critical stress for the onset of inelastic behavior. However, networks of thin fibers, of low density and/or with a low degree of cross-linking – which represent the vast majority of network materials of practical importance – are expected to be less sensitive to the inelasticity of the fiber constitutive behavior.[2]

2.3.7   Fiber Strength and the Weibull Distribution

While the stiffness of a solid is an average property obtained by homogenization over the entire volume, the strength is governed by extreme value statistics. Strength is generally associated with the unstable propagation of a critical flaw, and hence depends on the probability to find such a flaw in a given body. This implies that the variability of the modulus values obtained in experiments with different samples is lower than the variability of the strength of the same sample set.

When testing fibers in uniaxial tension, material sub-domains (fiber segments) are connected in series and are loaded by the same stress. The failure of the weakest of these sub-domains entails the failure of the fiber.

Generally, failure of materials is represented with the Weibull distribution. Consider a fiber segment of length L and of strength described by the cumulative distribution:

$$cp(\sigma) = 1 - \exp\left(-\left(\frac{\sigma}{\sigma_0}\right)^\beta\right), \tag{2.11}$$

where $cp(\sigma)$ represents the probability that failure takes place at a stress smaller or equal to σ. Parameters β and σ_0 are known as the Weibull modulus and the scale factor, respectively. Function $cp(\sigma)$ increases from 0 to 1 as σ increases, with the fastest increase taking place in the vicinity of σ_0 (if $\beta > 1$). The modulus, β, controls how fast the probability increases and hence controls the variability of the strength values.

The probability density function results as the derivative $d(cp)/d\sigma$:

$$p(\sigma) = \frac{\beta}{\sigma_0}\left(\frac{\sigma}{\sigma_0}\right)^{\beta-1}\exp\left(-\left(\frac{\sigma}{\sigma_0}\right)^\beta\right), \tag{2.12}$$

while the mean of the distribution is given by

$$\bar{\sigma} = \frac{\sigma_0}{\beta}\Gamma\left(\frac{1}{\beta}\right), \tag{2.13}$$

and the variance is given by

$$\mathrm{var}(\sigma) = \sigma_0^2\left[\Gamma\left(1 + \frac{2}{\beta}\right) - \Gamma^2\left(1 + \frac{1}{\beta}\right)\right]. \tag{2.14}$$

---

[2] The fact that fibers remain in the elastic range does not mean that the behavior of the network is elastic. Inter-fiber friction, rupture of crosslinks, and the presence of an inelastic matrix may result in inelasticity on the network scale.

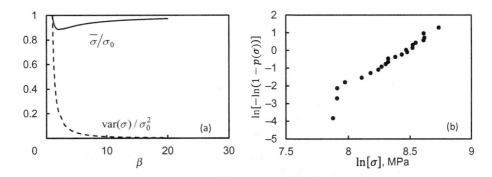

Figure 2.17 (a) Normalized mean and variance of Weibull distribution function of the Weibull modulus, Eqs. (2.13) and (2.14). (b) Weibull plot of the strength of carbon fibers. Data from Wu et al. (1991)

Figure 2.17(a) shows the variation of the normalized mean, $\bar{\sigma}/\sigma_0$, and variance, $\mathrm{var}(\sigma)/\sigma_0^2$, with the Weibull modulus, for $\beta > 1$. The figure shows that the variance decreases fast as β increases above ≈ 5, and the mean approaches σ_0. Hence, the larger the Weibull modulus, the smaller the strength variability.

Figure 2.17(b) shows data for the strength of carbon fibers (Wu et al., 1991) with a 10 mm gauge length L, in a typical Weibull plot. The cumulative distribution of Eq. (2.11) is shown after taking the log twice on both sides, that is, one plots $\log\left[-\log\left(1 - p(\sigma)\right)\right]$ versus $\log\left(\sigma\right)$. The slope of this plot is the Weibull modulus, β.

Size effect: Consider that experiments are performed with samples of length L and a Weibull distribution of parameters β and σ_0 represents the resulting data. It is of interest to inquire whether a different mean strength would be obtained if the experiments were performed with fibers of length nL instead.

The probability of no failure at stresses below σ for a fiber of length L is

$$1 - cp(\sigma) = \exp\left(-\left(\frac{\sigma}{\sigma_0}\right)^{\beta}\right). \tag{2.15}$$

The probability of no failure of all n segments is the product of terms similar to Eq. (2.15). Therefore, the cumulative probability of failure below stress σ of a fiber of length nL is:

$$cp_n(\sigma) = 1 - \exp\left(-n\left(\frac{\sigma}{\sigma_0}\right)^{\beta}\right). \tag{2.16}$$

The strength of the fiber of length nL is characterized by a Weibull distribution of the same modulus, β, and of a scale factor equal to $\sigma_0 n^{-1/\beta}$. The subscript n indicates that the respective quantity refers to a fiber of length nL.

This is an important result. It implies that the mean strength of fibers of length nL is:

$$\overline{\sigma}_n = n^{-1/\beta}\overline{\sigma}. \tag{2.17}$$

Therefore, the strength decreases continuously as the fiber length increases – a result also noted centuries ago by Leonardo da Vinci (see the collected works in da Vinci (1972)). Note that, since the modulus, β, is rather large for most fibers of interest, $1/\beta$ is small and the size effect, although not negligible, may be rather weak.

The variance of the distribution of the strength of fibers of length nL may be computed with Eq. (2.14) and reads:

$$\text{var}_n(\sigma) = n^{-2/\beta}\text{var}(\sigma). \tag{2.18}$$

The distribution becomes narrower as n increases. According to this model, the strength vanishes in the theoretical limit of $n \to \infty$.

2.4   Thermal Filaments

Filaments whose mechanics are influenced by thermal fluctuations are called thermal. The average energy per degree of freedom provided by the thermal bath is $k_B T/2$, where k_B is Boltzmann's constant. In ambient conditions, this is a small energy of the order of 0.025 eV, or 4×10^{-21} J. Therefore, only very thin filaments that require small amounts of energy to deform can be classified as thermal. All molecular filaments in gels and polymeric materials, as well as most biomolecules, are thermal filaments. Even rather large molecules, such as the DNA and protein filaments (e.g., microtubules), are influenced by thermal fluctuations and belong to this class.

Filaments of very different bending stiffness may be classified as thermal. A polyethylene macromolecule is thin and flexible, while a polyisoprene molecule is also flexible, although less so than polyethylene. On the other hand, DNA is much stiffer in bending. For the same contour length, a flexible filament may assume many more configurations than a stiff thermal filament.

With this in mind, we discuss in this section the structure and mechanical behavior of three conceptual models of thermal filaments: (i) filaments made from rigid links which rotate freely relative to each other and may overlap as they explore the phase space of available conformations, (ii) similar filaments for which the links cannot overlap, and (iii) filaments composed of flexible links whose relative rotation is constrained to a narrow angular range. The first category is known as "freely rotating" or "ideal," the second is known as "freely rotating with excluded volume," while the third category is that of "semiflexible" filaments. Above the glass transition temperature, the effective behavior of such filaments is considered to be elastic and the elasticity is of entropic type.

A note on the terminology used is necessary. Since thermal filaments are studied by the polymer physics community, they are called "chains" (as in "polymeric chains") in

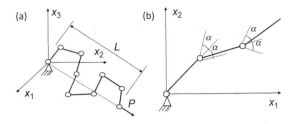

Figure 2.18 (a) Schematic of a freely rotating filament. (b) Two-dimensional version of the filament in (a), in which the angle between subsequent segments is limited to $(-\alpha, \alpha)$. In a polymeric material, these segments represent Kuhn segments, or segments along the backbone equal to the persistence length (see Section 2.4.4 and Eq. (2.49)).

the respective literature. In order to preserve consistency throughout this work, we prefer to refer to them with the more generic term of "filaments," while distinguishing between fibers and filaments only based on the cross-section dimensions, as defined at the beginning of this chapter.

2.4.1 Freely Rotating Thermal Filaments

Consider a filament composed of n rigid links of length a, which are pin-jointed and are free to rotate relative to each other (Figure 2.18(a)). Such filaments take random tortuous configurations under the action of thermal fluctuations. The excluded volume condition is not enforced and links of the filament can overlap other links. We are interested in determining the force required to deform such filaments. In this context, deformation refers to modifying the end-to-end length, L.

Since the links are rigid and no energetic interactions between links (nonbonded interactions) take place, the system has zero internal energy. However, due to the fact that the filament is free to sample all configurations available for given L, the entropy is not zero. The force in this model is purely entropic, that is, it results from the variation of the entropy during stretching. To see this, consider two extreme cases: one with the filament ends being co-located ($L = 0$), and the other with the filament being fully stretched ($L = na$). The number of configurations available to the chain in the first case is largest, while in the second case only one configuration is available. Therefore, the entropy decreases monotonically during stretching and this leads to an entropic force (Section 2.2).

2.4.1.1 Gaussian Model

We evaluate the probability of finding the filament in a configuration with given L. It is easier to begin the discussion by referring to the one-dimensional case and using an analogy with the one-dimensional random walk. Consider a walk of step length a, along axis x. The position of the walker after n steps is given by $x = \sum_{i=1}^{n} q_i a$, where q_i is $+1$ or -1, with equal probability, and the direction of the steps is uncorrelated along the walk. Then, the average position of the walker is $\langle x \rangle = \sum_{i=1}^{n} \langle q_i \rangle a = 0$. The mean square displacement, or the variance of the distribution of trajectory ends, is $\langle x^2 \rangle = \left\langle \sum_{i=1}^{n} q_i a \sum_{j=1}^{n} q_j a \right\rangle = a^2 \sum_{i=1}^{n} \sum_{j=1}^{n} \langle q_i q_j \rangle = na^2$, where only terms with $i = j$ survive in the double sum since the stochastic process q is not correlated.

The distribution of L can be found by simple arguments using the central limit theorem. This theorem indicates that the sum of n independent random variables is normal (Gaussian) distributed in the limit of large n, with mean equal to the sum of the means of each variable in the sum, and variance equal to the sum of the respective variances.

For the one-dimensional random walk, the mean of individual steps is $\langle x \rangle = 0$, and the variance is $\langle x^2 \rangle = a^2$, and, hence, for the entire walk one obtains a mean of zero and variance of na^2. The corresponding distribution function for the one-dimensional walk (the probability to find the walker between x and $x + dx$), p_{1D}, is:

$$p_{1D}(x)dx = \frac{1}{\sqrt{2\pi na^2}} \exp\left(-\frac{x^2}{2na^2}\right) dx. \qquad (2.19)$$

This result may also be obtained by directly counting the number of possible paths leading from the origin to position x, or by Fourier transform techniques, as done in some texts on polymer physics (Rubinstein and Colby, 2003) and in statistics texts (Weiner, 1983).

The result of Eq. (2.19) is restricted by the condition that n should be large. While this is generally true in practice, it is useful to note that, for small n, the sum of independent variables follows an Irwin–Hull distribution. This distribution converges to the Gaussian rather rapidly; the Gaussian approximation is already reasonably accurate for $n \geq 10$.

Moving now to the three-dimensional space, we observe that the probability for a random walk of n steps, with the end-to-end vector being oriented in any direction in space, to end in the spherical layer bounded by L and $L + dL$, is:

$$p_{3D}(L) \cdot 4\pi L^2 dL = \int_0^\pi \int_0^{2\pi} p_{1Dp}(L\sin\theta\sin\varphi)p_{1Dp}(L\sin\theta\cos\varphi)p_{1Dp}(L\cos\theta)\sin\theta L^2 d\theta d\varphi dL,$$

$$(2.20)$$

where θ, φ are the Euler angles of the end-to-end vector. The probability density p_{1Dp} corresponding to the projection of the path along some axis which does not coincide with the end-to-end vector is, by the arguments based on the central limit theorem discussed in relation to Eq. (2.19), a Gaussian of mean zero and variance $na^2/3$. Replacing Eq. (2.19) in Eq. (2.20) leads to the probability density function for the end-to-end length of the walk as:

$$p_{3D}(L) = \left(\frac{3}{2\pi na^2}\right)^{3/2} \exp\left(-\frac{3L^2}{2na^2}\right). \qquad (2.21)$$

Equation (2.21) represents a Gaussian distribution for the end-to-end vector length. The corresponding unit vector is uniformly distributed over the sphere of unit radius. Obviously, the mean of the distribution is zero, while the variance of $L = |\mathbf{L}|$ is na^2. We further use the notation $L^* = a\sqrt{n}$ for the representative measure of the end-to-end distance of the filament.

We are now in the position to evaluate the entropic force required to stretch the filament. We consider first situations in which small variations of L are applied to a filament with $L^* = a\sqrt{n} \ll L_0 = na$. For large n, such configurations represent states in which the end-to-end distance of the filament is much smaller than the fully stretched (contour) length, that is, situations in which the representation of the filament as a random walk is appropriate. Note that the distribution of Eq. (2.21) is only valid for an undirected random walk. If the filament is stretched such that L is smaller than, but comparable with the contour length, L_0, one may use a directed random walk model, which leads to statistics different from Eq. (2.21).

With this limitation in mind, we consider that L is the control parameter and evaluate the variation of the filament entropy associated with a small variation of L. The entropy Σ is computed using:

$$\Sigma = k_B \ln \Omega(L), \tag{2.22}$$

where $\Omega(L)$ is the total number of configurations available to a filament of n links with end-to-end length L. This is related to the probability of Eq. (2.21) as:

$$p_{3D}(L) = \frac{\Omega(L)}{\int \Omega(L)dL}, \tag{2.23}$$

and, therefore,

$$\Sigma = k_B \ln p_{3D}(L) + k_B \ln \int \Omega(L)dL. \tag{2.24}$$

The second term in Eq. (2.24) is only a function of n and remains constant if L changes. With Eq. (2.21), Eq. (2.24) becomes:

$$\Sigma = -\frac{3k_B L^2}{2na^2} + g(n), \tag{2.25}$$

where terms independent of L are represented by $g(n)$.

The ensemble considered here is similar to the canonical ensemble in which the number of particles, the kinematic parameters, and the temperature are controlled. The only kinematic parameter in this problem is L. In these conditions, the appropriate potential function is the Helmholtz free energy, Ψ, seen in Eq. (2.1).

The force required to change the end-to-end length of the filament, $\mathbf{P}$, is obtained using Eq. (2.3). The energy of the filament is zero and hence only the entropic term remains, which leads to:

$$\mathbf{P} = -T\frac{\partial \Sigma}{\partial \mathbf{L}}. \tag{2.26}$$

Equation (2.26) shows that the force is aligned with $\mathbf{L}$. Using Eq. (2.25), the magnitude of the force is

$$P = k_B T \frac{3L}{na^2}. \tag{2.27}$$

This expression exhibits two important features. The force is proportional to the temperature. This is the signature of its entropic nature. To see the difference relative to a fiber with energetic stress, consider a metallic rod held in a stretched configuration and further subjected to an increase of temperature. The temperature has no effect on the stress-production mechanism. However, thermal expansion causes the rod to dilate and hence the axial tensile force decreases with increasing temperature. Equation (2.27) predicts the opposite behavior: the force in the filament increases as the temperature increases. This type of behavior was observed in rubber at the beginning of the twentieth century, which opened the way to the development of the now classical theory of rubber elasticity which assumes that stress in this network material is purely entropic.

The second important feature of Eq. (2.27) is that the force is proportional to the end-to-end distance L. Hence, the filament behaves as if it were a spring of axial stiffness $3k_B T/na^2$ with a force that vanishes only when the two ends of the filament are co-located ($L = 0$). The filament with any nonzero L is subjected to a tensile force. The natural question that arises is why polymeric chains do not always have a zero end-to-end distance. Another force must be present to provide the necessary balance and ensure a distribution of end-to-end vector lengths. It is generally assumed that this reaction force is provided in a dense material by the excluded volume interactions (repulsion between monomers forming the filaments).

The model described in this section is the Gaussian model of the thermal filament. It is limited in scope to small stretches. To develop a more suitable representation, it is necessary to account for large deformations, including situations in which L becomes comparable with the contour length of the filament, L_0.

2.4.1.2 Langevin Model

Consider now the problem of chain extension under force control, with P being the force stretching the filament. The current ensemble is similar to the isothermal–isobaric ensemble in which the number of particles, the stress (or force), and the temperature are the control parameters. The relevant thermodynamic potential in this case is the Gibbs free energy.

In statistical mechanics, the Gibbs free energy, $\overline{G}$, is related to the partition function of the isothermal-isobaric ensemble, Z_f, as:

$$\overline{G} = -k_B T \ln Z_f. \tag{2.28}$$

For a system with zero internal energy, $U = 0$, on which external forces do work, the partition function Z_f is expressed as an integral of the Boltzmann factor computed based on the enthalpy, performed over all possible conformations compatible with the given force:

$$Z_f = \int \exp\left(-\frac{U - \mathbf{P} \cdot \mathbf{L}}{k_B T}\right) d\theta_i = \int \exp\left(\frac{\mathbf{P} \cdot \mathbf{L}(\theta_i, \varphi_i)}{k_B T}\right) \prod_{i=1}^{n} \sin\theta_i \, d\theta_i \, d\varphi_i. \quad (2.29)$$

Here it is made explicit that the variable $\mathbf{L}$ conjugated with the applied force is a function of the spatial orientation of the n links described by the Euler angles, θ_i and φ_i, $i = 1 \ldots n$, measured relative to the direction of the force $\mathbf{P}$. The $2n$ Euler angles define the phase space of the problem. The product $\mathbf{P} \cdot \mathbf{L}$ is the work performed by the force on the system (hence the sign convention). The end-to-end distance can be written in terms of θ_i as $L = a \sum_{i=1}^{n} \cos\theta_i$ and the work can be written $\mathbf{P} \cdot \mathbf{L} = Pa \sum_{i=1}^{n} \cos\theta_i$. Then, the integral in Eq. (2.29) can be evaluated leading to:

$$Z_f = \int \prod_{i=1}^{n} \exp\left(\frac{Pa}{k_B T} \cos\theta_i\right) \sin\theta_i \, d\theta_i \, d\varphi_i = \left[\int \exp\left(\frac{Pa}{k_B T} \cos\theta\right) \sin\theta \, d\theta \, d\varphi\right]^n$$

$$= \left[4\pi \frac{\sinh Pa/k_B T}{Pa/k_B T}\right]^n. \quad (2.30)$$

The elimination of the product in the first equality is possible because all i links sample the same sub-space defined by θ_i and φ_i. Then, the product can be factored out and the integrals remaining under the product are identical.

Using Eqs. (2.28) and (2.30), the Gibbs free energy can be computed as:

$$\widehat{G} = -nk_B T\left[\ln\left(\sinh\frac{Pa}{k_B T}\right) - \ln\left(\frac{Pa}{k_B T}\right) + \ln 4\pi\right]. \quad (2.31)$$

L may be computed from the thermodynamic potential as:

$$L = -\frac{\partial \widehat{G}}{\partial P} = na\left[\coth\left(\frac{Pa}{k_B T}\right) - \frac{k_B T}{Pa}\right] = na\widehat{L}\left(\frac{Pa}{k_B T}\right) = L_0\widehat{L}\left(\frac{Pa}{k_B T}\right), \quad (2.32)$$

where function $\widehat{L}(x) = \coth(x) - 1/x$ is known as the Langevin function. The normalized force, $Pa/k_B T$, is expressed in terms of the inverse Langevin function as $Pa/k_B T = \widehat{L}^{-1}(L/L_0)$. The inverse Langevin function can be approximated with Padé's formula as $\widehat{L}^{-1}(x) \approx x(3 - x^2)/(1 - x^2)$, which provides values within 3% of the actual function for $L/L_0 \leq 0.8$ (Cohen, 1991).

An alternative derivation can be performed in the canonical ensemble, with the control variable being the filament end-to-end distance, as described in, for example, Weiner (1982).

Equation (2.32) describes the constitutive behavior of the thermal filament with no constraints on the magnitude of the deformation. The relation between the force and the mean end-to-end length resulting from the inversion of Eq. (2.32) is shown in Figure 2.19 along with the prediction of the Gaussian filament model, Eq. (2.27). The Langevin model predicts stiffening at large elongations, while the Gaussian model predicts constant linear stiffness of magnitude $3k_B T/na^2$ at all stretches.

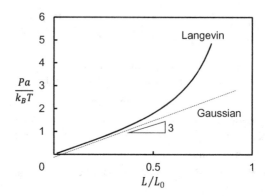

Figure 2.19 Relation between the normalized force Pa/k_BT and the filament end-to-end distance (or crimp parameter L/L_0) as predicted by the Langevin (Eq. (2.32)) and Gaussian (Eq. (2.27)) models.

In both models, the normalized force Pa/k_BT is related to the ratio L/L_0 through a universal relation independent of n, which fully defines the stiffening behavior of the filament in tension.

In the large force limit, the Langevin function can be approximated as $\bar{L}(x) \approx 1 - 1/x$. The normalized force diverges when the end-to-end distance L approaches the filament contour length L_0 as:

$$\frac{Pa}{k_BT} \approx \frac{L_0}{L_0 - L} = \frac{1}{1 - c},$$ (2.33)

an asymptotic behavior which is also independent of n.

The fact that the response stiffens rapidly at large stretches indicates that the variation of the entropy, that is, the variation of the total number of conformations available to the filament corresponding to a given increment of the stretch, is more rapid when the chain is almost fully stretched. Furthermore, in the Gaussian filament case, an increment of stretch reduces the entropy associated with the long wavelength modes faster. This situation is similar to that of athermal fibers in which, likewise, stiffness is primarily controlled by the long wavelength undulations (Section 2.3.5).

It is important to emphasize that the filament only responds to forces aligned with its end-to-end vector. Transverse forces perform no work since the entropy does not vary when the two ends of the filament rotate relative to each other, while maintaining constant their relative distance. It results that thermal filaments can be modeled with linear (Gaussian) or nonlinear (Langevin) axial springs. The choice of the filament model has significant consequences for the mechanics of networks made from thermal filaments.

2.4.2   Persistence Length

In Sections 2.4.1.1 and 2.4.1.2 it was assumed that the filament is composed from links of specified length, a, which are free to rotate relative to each other

(Figure 2.18(a)). It is of interest to analyze cases in which filaments are less flexible. To this end, we consider a slightly modified model in which the angle made by a link relative to the previous link is selected from a uniform distribution in the range $(-\alpha, \alpha)$ (Figure 2.18(b)). If α is relatively small, the filament is less curvy on scales comparable with a, but it resembles the freely rotating model on scales much larger than a. Therefore, a length scale exists which separates these two types of behavior. To define it, we evaluate the rate of decay of the link orientation correlation function along the contour of the filament. In the two-dimensional case, the directional correlation is computed as:

$$C(m) = \langle \mathbf{t}_i \cdot \mathbf{t}_{i+m} \rangle_i = \left\langle \cos \left(\sum_{j=i}^{m} \Delta\theta_{j,j+1} \right) \right\rangle, \tag{2.34}$$

where $\mathbf{t}_i$ is the filament tangent unit vector at the location of link i and $\mathbf{t}_{i+m}$ is the tangent unit vector m links away. The average is performed over all realizations and/or along the filament, for various i values. $\Delta\theta_{j,j+1}$ is the angle between link j and the subsequent link $j + 1$. The random independent variables $\Delta\theta_{j,j+1}$ are distributed with zero mean and variance $\alpha^2/3$.

$\sum_{j=i}^{m} \Delta\theta_{j,j+1}$ follows an Irwin–Hall distribution which converges to the Gaussian distribution for sufficiently large m (approximately $m \geq 10$). The resulting Gaussian has zero mean and variance $m\alpha^2/3$. Hence, the average in Eq. (2.34) can be computed and the correlation function results:

$$C(m) = \exp \left(-\frac{ma}{6a/\alpha^2} \right). \tag{2.35}$$

The correlation function of the tangent vector decays exponentially with the number of steps (each having length a) along the filament. The emerging length scale is $6a/\alpha^2$, which is the correlation length of the filament tangent vector orientation, also known as the persistence length, L_p:

$$\frac{L_p}{a} \approx \frac{6}{\alpha^2}, \tag{2.36}$$

where the approximate sign is introduced to account for the assumption of large m made in the derivation.

For contour lengths smaller than L_p, the filament appears approximately straight, while for much larger contour lengths, it resembles the freely rotating filament discussed in Section 2.4.1. The normalization of all lengths by L_p (more exactly by $2L_p$, which is known in polymer physics as the Kuhn length), renders this type of filament equivalent to the freely rotating one.

The derivation is more complicated in three dimensions due to the additional random rotation of link $j + 1$ relative to link j. It is nevertheless possible to show that the correlation function decays exponentially:

$$C(m) \sim \exp \left(-\frac{ma}{L_p(\alpha)} \right). \tag{2.37}$$

The analysis in this section is purely geometric and is independent of the nature of the filament. It applies equally to thermal and athermal cases. In the thermal case of semiflexible filaments, the persistence length can be computed based on the bending rigidity of the filament (Eq. (2.49)). In the athermal case, in which filaments are continuous (not made from links, as in Figure 2.18), the average of Eq. (2.34) can be computed identically based on the filament tangent unit vectors.

2.4.3   Role of the Excluded Volume

We now take one more step toward eliminating the approximations inherent in the freely rotating model of thermal filaments, which assumes that links are phantom and may freely overlap and cross each other. The self-avoiding random walk model imposes the excluded volume constraint which prevents chain segments from overlapping and crossing. The statistics of the filament end-to-end distance is modified in this representation relative to the freely rotating model, but the entropic nature of stress is preserved.

The self-avoiding random walk was studied by a large number of researchers starting from the middle of the twentieth century. Nevertheless, the number of analytic results available is relatively small and generally applicable only for large n. Numerical evidence and scaling arguments suggest that the representative measure of the end-to-end distance, L^*, results in this case:

$$L^* = \sqrt{\langle L^2 \rangle} \sim an^\gamma, \tag{2.38}$$

where γ is an exponent that depends on the dimensionality of the embedding space. Fisher (1969) used ideas developed by Flory (1949) to provide an approximation for the exponent:

$$\gamma = \begin{cases} 3/4 & 2D \\ 3/5 & 3D \end{cases}. \tag{2.39}$$

Comparing with the freely rotating model for which $\gamma = 1/2$, the larger γ value indicates that the size of the filament increases significantly when the no overlap condition is enforced. The filament remains a fractal object, but the fractal dimension (which is equal to $1/\gamma$) increases from 2 in the freely rotating model, to 5/3 (in 3D) in the self-avoiding representation.

The simplest explanation for the scaling of Eq. (2.38) is given by the model due to Flory (1949). This theory provides a physical justification for the swelling of the self-avoiding filament. Consider a filament made from n links that occupies the sphere of radius equal to the end-to-end length, and of volume proportional to $\sim L^{*3}$. If the volume of one link is v, and all links occupy a volume nv, the probability of overlap is $p_{ov} \sim nv/L^{*3}$. An energy penalty for overlaps is introduced, of magnitude ε per overlap. Therefore, the energy associated with packing is $\sum_{i=1}^{n} \varepsilon p_{ov} \sim \varepsilon n^2 v/L^{*3}$. At the same time, the entropic contribution to the free energy is the entropy of the filament of end-to-end distance L^*. Flory uses the result for the Gaussian filament, Eq. (2.25), to approximate the entropy. The resulting free energy contribution is proportional to k_BTL^{*2}/na^2. The total free energy becomes $q_1 \varepsilon n^2 v/L^{*3} + q_2 k_B TL^{*2}/na^2$, where q_1 and q_2 account for constants not made explicit in the expressions of the two components of the free energy. The equilibrium configuration is obtained by minimizing the free energy with respect to L^*. Taking the derivative of the free energy expression with respect to L^* and setting it to zero leads to the scaling relation $L^* \sim n^{3/5}$; hence the exponent γ of Eq. (2.39) results.

Beginning with Fisher (1966), Chay (1970), des Cloizeaux (1974), and de Gennes (1979), many authors sought to establish the probability density function of the end-to-end distance of the self-avoiding random walk. It is considered that the probability density for large n values is of the form:

$$p_{3D}^{sa}(L) \sim \left(\frac{L}{L^*}\right)^{\beta} \exp\left(-\frac{3}{2}\left(\frac{L}{L^*}\right)^{\delta}\right), \qquad (2.40)$$

Fisher (1966) suggests that the exponent δ is related to exponent γ in Eq. (2.38) as $\delta = 1/(1-\gamma)$.

p_{3D}^{sa} decreases to zero when $L \to 0$, which is expected considering that the energy penalty associated with packing diverges in this limit. p_{3D}^{sa} has a maximum for intermediate values of L and decreases to zero at large L. In this limit, packing is less important and the Gaussian-like statistic prevails, albeit with different exponents reflecting the self-avoiding walk statistics.

With Eq. (2.40), one may follow the procedure described for the Gaussian filament (Eqs. (2.25)–(2.27)) to derive the force required to stretch the filament:

$$\frac{Pa}{k_BT} = \frac{3}{2}\frac{1}{1-\gamma}\left(\frac{L}{L^*}\right)^{\frac{\gamma}{1-\gamma}}\frac{a}{L^*} - \beta\frac{a}{L}. \qquad (2.41)$$

In the freely rotating filament case, $\gamma = 1/2$ and the excluded volume effect disappears, which requires $\beta = 0$, and, with $L^* \sim an^{\gamma}$, Eq. (2.41) reduces to Eq. (2.27). The first term represents the entropic effect and is similar to the Gaussian filament case, while the second term is a consequence of packing and has the effect of swelling the filament. The first term dominates at large L and the second term contribution is important at small L. For $\gamma = 3/5$, Eq. (2.39), Eq. (2.41) predicts a nonlinear increase of the force as $Pa/k_BT \sim (L/L_0)^{3/2}$; note that $L_0 = na$.

It should be observed that the nonlinearity of the present model originates from the statistic of the walk and is different from the nonlinearity of the Langevin filament, which is associated with the nonquadratic variation of the entropy with the filament stretch at large filament extensions.

2.4.4 Semiflexible Filaments

Semiflexible filaments are thermal filaments for which the energetic contribution of the bending modes cannot be neglected. Imagine an initially straight filament which is allowed to fluctuate under the action of thermal excitation. The primary deformation mode is bending, which, for slender filaments of common interest, is much softer than the axial mode. Hence, the filament performs random bending oscillations and these fluctuations influence the response to axial loading. Stretching leads to a reduction of the number of configurations available to the filament. Most biopolymers are semiflexible; examples include DNA, microtubules, actin, and fibrin. Single wall carbon nanotubes may also be regarded as semiflexible filaments. In all these cases, the persistence length is large and often comparable with, or even larger than, the end-to-end length of the filament. The tortuosity (crimp) of semiflexible filaments is reduced relative to the fully flexible case. This strongly contrasts with the case of some synthetic polymers and rubber, in which the persistence length is small and the tortuosity is pronounced. Therefore, it is expected that the mechanics of semiflexible filaments is different from that of flexible filaments.

The classical model for semiflexible filaments is the worm-like chain (WLC) model due to Kratky and Porod (1949). Following the statistical mechanics solutions provided in Marko and Siggia (1995) and Odijk (1995), many researchers contributed to the development of this model (e.g., Moroz and Nelson, 1998; Bouchiat et al., 1999; Dobrynin et al., 2010). Here we present the conceptual outline of the model, without going into the details of the derivation, for which the reader is referred to the cited literature.

Consider a filament of length L_0, which is straight in the absence of thermal fluctuations, fixed at one end and loaded by a force P at the free end (inset to Figure 2.20(b)). The tensile force is aligned with the filament axis and produces no bending if the filament is in the straight, unperturbed configuration. The filament becomes undulated due to thermal excitations and its end-to-end length, L, decreases as the temperature increases. Increasing the axial force leads to the reduction of the fluctuation amplitude and to an increase of L. The objective here is to evaluate the relation between L and the applied force.

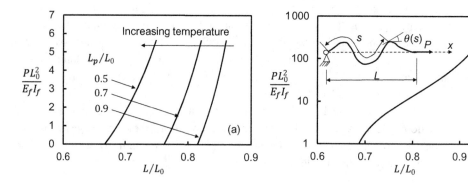

Figure 2.20 (a) Normalized force $PL_0^2/E_f I_f$ versus the chain crimp parameter L/L_0, as predicted by the WLC model of Eq. (2.48). At zero force $L/L_0 < 1$ due to thermal fluctuations and L decreases with increasing temperature. (b) One of the curves in (a) shown in semi-log plot with a larger range of forces to demonstrate the divergence at $L \to L_0$. The inset shows a schematic of the WLC model and the notation used.

The potential energy of the filament includes an energetic component associated with bending, and a work term, and can be obtained from Eq. (2.5):

$$PE = \frac{1}{2} \int\limits_0^{L_0} E_f I_f \left(\frac{d\theta(s)}{ds}\right)^2 ds - \mathbf{P} \cdot \mathbf{u}(L_0). \tag{2.42}$$

The axial deformation mode is neglected here (but considered in the derivation in Odijk (1995)), $d\theta(s)/ds$ is the curvature of the filament, and $\mathbf{u}(L)$ is the displacement at the free end of the filament, at $x = L$ (or $s = L_0$). Note that if large rotations and deformations are allowed, all bending, axial, and torsional modes are engaged and the mechanics of the filament becomes more complex. The statistical mechanics of this case is discussed in Moroz and Nelson (1998). For simplicity, here we consider $\mathbf{P}$ and $\mathbf{u}$ to be aligned in the x-direction and hence $\mathbf{P} \cdot \mathbf{u}(L_0) = P\left(\int_0^{L_0} \cos\theta\, ds - {}^0L\right)$, where $P = |\mathbf{P}|$ and ${}^0L$ is the end-to-end length when $P = 0$ and thermal fluctuations are present (${}^0L < L_0$). Assuming further that the departure from the straight configuration is small, $\cos\theta \approx 1 - \theta^2/2$, the potential energy reads:

$$PE = \frac{1}{2} \int\limits_0^{L_0} E_f I_f \left(\frac{d\theta(s)}{ds}\right)^2 ds + \frac{1}{2} P \int\limits_0^{L_0} \theta^2 ds - P\left(L_0 - {}^0L\right). \tag{2.43}$$

The filament shape is defined by $\theta(s)$. This function is expanded in a Fourier series:

$$\theta(s) = \sum_{n=1}^{\infty} \alpha_n \sin\frac{\pi n s}{L_0} + \beta_n \cos\frac{\pi n s}{L_0}. \tag{2.44}$$

The filament is considered pinned at both ends, which implies that no moments are applied at either end and the curvature, $d\theta/ds$, vanishes at the filament ends. This requires $\alpha_n = 0$, and only the second term in Eq. (2.44) remains. With this, the potential energy can be written as:

$$PE = \sum_{n=1}^{\infty} \beta_n^2 \left[E_f I_f \frac{\pi^2 n^2}{4L_0} + \frac{PL_0}{4}\right] - P\left(L_0 - {}^0L\right). \tag{2.45}$$

Thus far the mechanics of the filament is described in energetic terms. In the next step, an average of Eq. (2.45) is performed and the equipartition theorem is used requiring that each mode of the energetic component carries, on average, an energy equal to $k_B T/2$.[3] Then, $\langle \beta_n^2 \rangle$ can be computed as:

$$\langle \beta_n^2 \rangle = \frac{2k_B T L_0}{E_f I_f \pi^2 n^2 + PL_0^2}. \tag{2.46}$$

---

[3] In 3D, the filament oscillation modes in the two directions orthogonal to the filament axis are statistically independent and hence two modes correspond to each n. Therefore, the energy corresponding to each n is $k_B T$.

The mean end-to-end length of the filament can be computed as:

$$\langle L \rangle = \int_0^{L_0} \langle \cos\theta \rangle ds \approx L_0 - \frac{1}{2}\int_0^{L_0} \langle \theta^2 \rangle ds, \tag{2.47}$$

and with Eqs. (2.44) and (2.46), it results that

$$1 - \frac{\langle L \rangle}{L_0} = \frac{1}{2L_0}\int_0^{L_0} \langle \theta^2 \rangle ds = \frac{1}{4}\sum_{n=1}^{\infty} \langle \beta_n^2 \rangle = \frac{1}{2}\frac{L_0}{L_p}\frac{1}{\sqrt{\widehat{P}}}\left(\coth\sqrt{\widehat{P}} - \frac{1}{\sqrt{\widehat{P}}} \right), \tag{2.48}$$

where $\widehat{P} = PL_0^2/E_f I_f$ is a non-dimensional version of the force P. The last expression in Eq. (2.48) is evaluated with Eq. (2.46).

The quantity

$$L_p = \frac{E_f I_f}{k_B T} \tag{2.49}$$

is the persistence length of a semiflexible filament in 3D. In two dimensions (the derivation presented here is performed in 2D), $L_p = 2E_f I_f/k_B T$.

Expanding in series the right side of Eq. (2.48) in the vicinity of $P = 0$ and retaining only the first term in P leads to $1 - \langle L \rangle/L_0 = L_0/6L_p - PL_0^3/(90E_f I_f L_p)$. Therefore, the effective small strain axial stiffness of the filament is $90E_f I_f L_p/L_0^4$. The mean slack of the filament in the unloaded state and in 2D is $L_0 - \langle {}^0 L \rangle = L_0^2/6L_p = L_0^2 k_B T/12E_f I_f$.

Equation (2.46) indicates that the equipartition theorem establishes a power spectrum of the filament shape which decays as the inverse of the square of the frequency (with and without an applied force). The Wiener–Khinchin theorem relates the auto-correlation function to the power spectrum (the power spectrum is the Fourier transform of the autocorrelation function) and mandates that, for a spectrum of this type, the autocorrelation function is of the exponential type. Hence, the exponentially decaying correlation function of Eq. (2.37) originates from the modal distribution of energies of the flexible filament subjected to thermal fluctuations. The persistence length of Eq. (2.37) results from this analysis and is given by Eq. (2.49). An alternate derivation of the persistence length is provided in Howard (2001).

Eq. (2.48) should be compared with the equivalent equation describing the freely rotating Langevin chain, Eq. (2.32), which can be rearranged in a format similar to that of Eq. (2.48) and reads:

$$1 - \frac{\langle L \rangle}{L_0} = 1 - \coth\widehat{\widehat{P}} + \frac{1}{\widehat{\widehat{P}}}, \tag{2.50}$$

where $\widehat{\widehat{P}} = Pa/k_B T$. The limit of the right side of Eq. (2.50) for $P \to 0$ is 1, which indicates that in the absence of applied force, $\langle L \rangle = 0$, as expected for the freely rotating model.

It is of interest to evaluate the behavior of the semiflexible model in the two limits of small and large P. The assumption of small departures from the straight configuration renders the model inaccurate in the limit of small forces, when slack may be significant. Nevertheless, it can be inferred from Eq. (2.48) that for $\hat{P} \to 0$ the average end-to-end length of the filament, $\langle^0 L \rangle$, is given in the first order by $\langle^0 L \rangle / L_0 \approx 1 - L_0/(6L_p)$.

In the limit of large forces, when $\langle L \rangle$ approaches L_0, the force diverges as:

$$P \sim \frac{1}{(L_0 - \langle L \rangle)^2} \sim \frac{1}{(1 - c)^2}. \tag{2.51}$$

This is in contrast with the behavior of the freely rotating Langevin chain for which the force diverges at large extensions as $P \sim 1/(1 - c)$ (Eq. (2.33)). The force enters the regime described by Eq. (2.51) for $\hat{P} > 5$ or $1 - \langle^0 L \rangle / L_0 < 0.128\, L_0/L_p$, which is quite close to the unloaded state characterized by $1 - \langle^0 L \rangle / L_0 \approx L_0/6L_p = 0.166\, L_0/L_p$. Therefore, the model essentially has no linear elastic regime.

The worm-like chain response of Eq. (2.48) is shown in Figure 2.20(a) for several values of L_p/L_0.

In the derivation of Eq. (2.48), the equipartition theorem is applied to the potential energy given by Eq. (2.45), leading to the set of coefficients $\langle \beta_n^2 \rangle$ of Eq. (2.46). This implies that equipartition applies at all times as the filament deforms under load. It is of some interest to compare this situation with an equivalent athermal case in which a static, nonthermally fluctuating filament of initial, stress-free state defined based on the frequency spectrum of the thermal filament is loaded axially (van Dillen et al., 2008). This athermal filament model leads to a constitutive behavior similar to that of Eq. (2.48), except that the second term in the parentheses, $1/\sqrt{\hat{P}}$, is replaced by $\sqrt{\hat{P}}/\sinh^2\sqrt{\hat{P}}$ and the $1/2$ coefficient in front of the right hand side becomes $1/4$. The athermal filament of shape defined by the same statistics as the WLC filament is softer than its thermal counterpart. It should be noted though that a generic athermal filament is not expected to have a shape defined by a spectrum similar to that of the WLC. Spectra defining the shape of filaments extracted from a felt and their effective force–stretch response are presented in Kabla and Mahadevan (2007).

The behavior of many semiflexible filaments has been measured in recent years by direct probing using optical tweezers. In this type of experiment, two beads are connected to the two ends of the filament and are then trapped in the focal region of two highly focused laser beams. The beads are attracted or repelled by the beams function of the difference of the refractive index of the bead and the surrounding medium. This allows manipulating the beads by moving the laser beams which makes possible applying pN level forces on the probed filament. The experiment allows evaluating of the force–extension curve of the filament.

Typically, the WLC model is fitted to the measured force–extension curve, which provides an estimate of the filament persistence length. An example is shown in Figure 2.21, where experimental data from stretching a DNA molecule of $L_0 = 15.6$ μm in 10 mM PB buffer is presented (Strick et al., 1996). The line represents the fitted WLC model. The persistence length of the filament results as $L_p = 51.3 \pm 2$ nm.

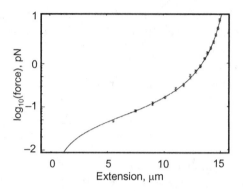

Figure 2.21 Experimental force–extension curve for a DNA molecule and fit with the WLC model. Data from Strick et al. (1996)

A useful approximation to the force–extension relation predicted by the WLC model of Eq. (2.48) was proposed by Bustamante et al. (1994) and has the explicit form:

$$\widehat{P} \approx \left(\frac{L_0}{L_p}\right)^2 \left[\frac{1}{4}\left(\frac{1}{y^2} - 1\right) + 1 - y\right], \tag{2.52}$$

where $y = 1 - x/L_0$ and x is the extension given by $x = \langle L \rangle - \left(L_0 - L_0^2/6L_p\right)$. Another approximation is proposed in Bouchiat et al. (1999). The quality of these approximations depends strongly on L_0/L_p. For $L_0/L_p = 0.5$, Eq. (2.52) is within 10% of Eq. (2.48) only when $\langle L \rangle/L_0 > 0.97$, that is, for almost straight filaments. For a more flexible filament, with $L_0/L_p = 50$, requiring 10% accuracy restricts the range of applicability of Eq. (2.52) to $\langle L \rangle/L_0 > 0.87$.

Comparing the WLC model with other models of thermal filaments, it is useful to recall computer simulation results for filaments with bending stiffness. It has been observed (Livadaru et al., 2003; Rosa et al., 2003) that such chains exhibit two regimes of behavior: the response can be represented by the WLC model at relatively small forces, with the force scaling as $P \sim (L_0 - \langle L \rangle)^{-2} \sim 1/(1 - c)^2$ (Eq. (2.51)), while at forces larger than some threshold, the response is similar to that of the freely rotating chain model, with the force scaling as $P \sim (L_0 - \langle L \rangle)^{-1} \sim 1/(1 - c)$ (Eq. (2.33)). The crossover force increases with increasing the bending rigidity of the filament, $E_f I_f$. An analytic derivation of a model capturing this behavior is presented in Dobrynin et al. (2010) along with results from experiments on isolated synthetic and biological molecules and computer simulation data supporting the proposed concept.

2.5   Mechanics of Fiber Bundles

In most natural and some man-made network materials, the network is composed from fiber bundles, as opposed to individual fibers. Collagen fibers of the extracellular

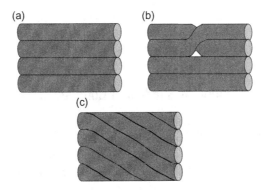

Figure 2.22 Schematic representations of bundles with (a) straight, (b) undulating, and (c) twisted fibers.

matrix and connective tissue are bundles of tropocollagen fibrils. Buckypaper is a stochastic network of bundled carbon nanotubes stabilized by strong cohesive interactions between nanotubes. Virtually all composite materials with continuous fibers are made from fiber bundles (fiber tows), which are woven to produce prepregs. The vast majority of the threads used in the textile industry are bundles (yarns) of discontinuous natural fibers. Therefore, it becomes apparent that in the discussion of the mechanical behavior of network materials, the behavior of fiber bundles is at least as relevant as that of individual fibers.

Figure 2.22 shows schematics of fiber bundles of continuous fibers. An idealized bundle of fibers arranged parallel to each other is shown in Figure 2.22(a). Figure 2.22(b) shows schematically that fibers may exhibit waviness and exchange their relative position within the bundle. Fiber waviness leads to packing defects that increase the total volume of the bundle. Fiber tows used in composites usually exhibit this type of bundle structure. The packing and ordering of fibrils within collagen fibers is poorly described in the current literature. However, it is expected that configurations of the type shown in Figure 2.22(b) are also prevalent in collagen.

Virtually all textile yarns are twisted (Figure 2.22(c)). Twisting increases the total length of fiber per unit length of the bundle, but greatly improves the mechanical properties of the yarn. Fibers in a twisted yarn have approximately helical trajectories. A force acting along the bundle axis tends to reduce the radius of the helix and hence produces a compressive radial stress. This enables frictional interactions between fibers, which assists inter-fiber load transfer. This mechanism provides strength to yarns made from discontinuous fibers which, in the absence of twist, would have no strength at all.

The mechanics of fiber bundles was studied beginning from the first half of the twentieth century in connection with developments in the textile industry and continues today with the development of stochastic models of failure in bundles with and without an embedding matrix, of importance in composites and biological applications. Several central results from this broad literature are summarized in this section. We begin by discussing the axial and bending deformation of bundles of continuous fibers of the type shown in Figure 2.22(a). Further, results on the structure and tensile response of twisted bundles of the type shown in Figure 2.22(c) are presented. The strength of bundles of continuous and discontinuous fibers is also discussed.

2.5.1   The Deformation of Bundles of Continuous Fibers

2.5.1.1   Axial Deformation

When bundles of continuous parallel and straight fibers are loaded axially, all fibers experience the same strain. Hence, the bundle inherits the axial behavior of individual fibers. If these are elastic, the effective bundle modulus if given by:

$$E_{eff} = \langle E_f \rangle \phi_a, \tag{2.53}$$

where $\langle E_f \rangle$ is the mean Young's modulus of individual fibers, while ϕ_a is the area fraction occupied by fibers in the bundle cross-section. The closest packed arrangement of fibers is hexagonal. In this case, and in the limit of large number of fibers in the bundle, $\phi_a = \pi/2\sqrt{3} \approx 0.9$. Equation (2.53) accounts for situations in which fibers have the same cross-sectional area but have different moduli.

In realistic bundles, fibers are not in the idealized close-packed hexagonal configuration. The free volume within bundles depends on the fabrication method. Further, the free volume changes with axial loading. The actual value of ϕ_a of a specific bundle can be evaluated by direct microscopic inspection.

2.5.1.2   Bending Deformation

The behavior of bundles subjected to bending is more complex and depends on the nature of the interfaces between fibers. Figure 2.23 shows several possibilities which are also extreme cases of behavior that may be encountered in practical situations.

Figure 2.23(a) shows a bundle in which inter-fiber interfaces are well bonded and are able to carry shear stress. The strong interfaces also prevent the rearrangement of fibers in the bundle during deformation. The figure shows the deformed configuration, with the bundle cross-section planes at A and B, which are taken perpendicular to the bundle axis before deformation, remaining perpendicular to the bundle axis after deformation.

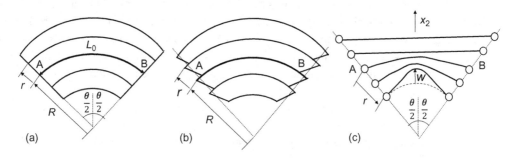

Figure 2.23 Fiber bundles subjected to bending (a) with axial constraint, (b) free to slide axially, and (c) with no constraint in the x_2 direction.

In this case, the bundle bends as a beam of effective moment of inertia, I_{eff}, and the deformation can be described with elementary beam theory. The effective moment of inertia depends on fiber packing. In the limit of large number of fibers in the bundle, n_f, and for close-packed fiber arrangement, the effective moment of inertia is given by $I_{eff} \sim n_f^2 I_f$. If fibers are more sparsely packed, I_{eff} increases with decreasing ϕ_a.

Figure 2.23(b) shows another limit case, in which inter-fiber interfaces are weak and allow relative sliding. However, the packing is dense enough to prevent fiber rearrangement within the bundle during deformation. In this case, each fiber bends about its own centroidal axis. In this configuration, the fibers act as independent springs loaded in parallel and $I_{eff} \sim n I_f$. This bundle has much smaller rigidity compared to the bundle with strong interfaces shown in Figure 2.23(a).

A situation encountered often in practice is that in which the bundle density is low and fiber packing is insufficient to prevent fiber migration and rearrangement within the bundle during bending. In this case, the cross-section of the bundle changes along the bundle length, becoming flatter in the center. This case is shown schematically in Figure 2.23(c).

To reduce complexity, we treat this case under the assumption that fibers are pin-jointed at their ends A and B and hence carry only axial forces. The cross-sections at A and B are given rotations of angles $\theta/2$ and $-\theta/2$, respectively (pure bending of the bundle). Therefore, the fibers above the median plane of the bundle (i.e., the plane defined by the axes of rotation of sections A and B) are subjected to tension and remain straight, while the fibers below the bundle median plane are subjected to compression and buckle, as shown in Figure 2.23(c). This motion leads to the increase of the fiber density in the middle cross-section of the bundle. The maximum deflection of a fiber located at distance r from the median plane of the bundle, w, is given by:

$$\left(\frac{w}{L_0}\right)^2 = \left(\frac{2}{\pi}\right)^2 \frac{r\theta}{L_0}\left(1 - \frac{r\theta}{L_0}\right), \qquad (2.54)$$

where all lengths are normalized by the undeformed bundle length, L_0. The position in the x_2 direction of a fiber located above the bundle median plane shifts down by $r(1 - \cos\theta/2)$. A fiber located below the bundle median plane shifts up by the same amount and also bends up. The maximum displacement in the x_2 direction is $w + r(1 - \cos\theta/2)$ and corresponds to the mid-span point of the respective fiber. Therefore, the density of fibers increases close to the median plane of the bundle and the bundle cross-section flattens out, expanding in the plane perpendicular to the x_2 axis (Figure 2.23(c)).

2.5.2   Structure and Mechanics of Twisted Fiber Bundles

2.5.2.1   Structure of Twisted Bundles

The twisting of fiber bundles has complex mechanics and leads to significant modifications of the bundle structure. This procedure is used with virtually all textiles; twist is applied with concomitant stretching of the yarn. Several methods to produce twisted yarn have been developed, with ring spinning being the oldest method and the procedure that leads to the highest strength and quality of the yarn. The relatively low productivity of ring spinning led to the development of newer techniques such as rotor spinning and air jet spinning. This section presents a simplified description of yarn spinning aimed to outline the complexities that arise in this problem. The structure described here is characteristic primarily for the ring spinning method, while some variations are introduced by the other methods.

Figure 2.24(a) represents an idealized situation in which initially straight fibers parallel to the bundle axis become helices in the twisted configuration. The initial length of the bundle and of all its fibers is L_0, while the final length of the bundle corresponding to one full turn is denoted by L. Figure 2.24(b) shows the unwrapped cylindrical surface containing a helical fiber at distance r from the bundle axis and the outer surface of the bundle. The length of the fiber wrapped around the cylinder of radius r is:

$$l(r) = \sqrt{L^2 + (2\pi r)^2}. \tag{2.55}$$

The angle of the fiber axis with the bundle axis is $\alpha(r)$, which takes the value α^* for $r = R$. Hence, $l \in (L, L/\cos\alpha^*)$ and the mean fiber length can be evaluated as:

$$\langle l \rangle = \frac{L}{2}\left(1 + \frac{1}{\cos\alpha^*}\right). \tag{2.56}$$

The rather restrictive assumptions about the fiber kinematics made here imply that all fibers deform by stretching. Since this is an energetically expensive deformation

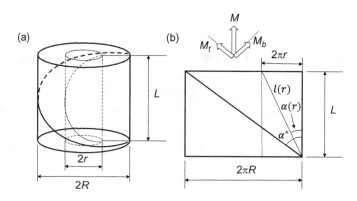

Figure 2.24 (a) Bundle of fibers initially parallel to the bundle axis subjected to twisting, under the assumption that fibers initially at distance r from the bundle axis are constrained to wrap on the surface of the cylinder of radius r. One full turn of two such fibers wrapping on cylinders of radii r and R is shown. (b) Unwrapped cylinders of radii r and R of the deformed configuration showing the two spiral fibers represented in (a), which become straight lines in this representation.

mode, the bundle tends to reduce its length to release some of the axial fiber strain. One may envision that the real deformation is such that

$$\langle l \rangle = L_0. \tag{2.57}$$

This implies that fibers close to the bundle axis are compressed and buckle, while those close to the outer layers of the bundle are stretched. Replacing $\langle l \rangle$ of Eq. (2.57) in Eq. (2.56) and assuming that the mean bundle density remains constant during deformation, such that $\pi R^2 L = \pi R_0^2 L_0$, allows computing of the length of the bundle, L, characterized by the twist of angle α^* on the outer surface:

$$\frac{L}{L_0} = \frac{1}{2} \left(1 + \sqrt{1 - \frac{4\pi^2 R_0^2}{L_0^2}} \right). \tag{2.58}$$

R_0 represents the radius of the bundle of straight fibers before twisting and it is implied that the radius of the bundle increases during twisting since its length decreases.

Equation (2.58) can be used to compute the axial contraction strain of the bundle due to the twist, that is, $(L_0 - L)/L_0$. For small values of the twist (i.e., for small R_0/L_0), this contraction strain can be approximated to the first order as:

$$\frac{L_0 - L}{L_0} \approx \frac{\pi^2 R_0^2}{L_0^2}, \tag{2.59}$$

which indicates that the contraction decreases quadratically as the length of the bundle corresponding to one turn, L_0, increases. The characteristic angle α^* decreases in this process. Figure 2.25 shows experimental data for the contraction strain in twisted cotton yarn (Landstreet et al., 1957). The data indicates that the axial strain scales with L_0 as $(L_0 - L)/L_0 \sim L_0^{-1.66}$. The difference relative to the prediction of Eq. (2.59) is due to the multiple assumptions made in this simplified model, of which the most

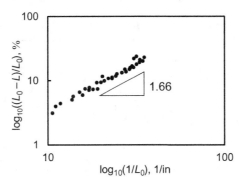

Figure 2.25 Axial strain of a fiber bundle function of the twist (proportional to $1/L_0$) for cotton yarn. The horizontal axis represents the inverse axial length corresponding to one turn. Data from Landstreet et al. (1957)

important is related to the structure of the twisted bundle. The consequences of relaxing these assumptions are discussed next.

The arrangement of fibers in a twisted bundle was studied in detail by Morton (1956), Riding (1964), Treloar (1965), and Hearle et al. (1969). In 1956, Morton proposed that fibers migrate in the radial direction as the bundle is twisted (Morton, 1956). The main argument is geometric: since fibers closer to the axis of the twisted bundle must contract relative to their initial length, L_0, and since axial compressive strains are not supported by thin fibers, these fibers must accumulate slack.

A similar argument can be made based on energetic considerations: the axial deformation mode is energetically expensive and fibers should deform preferentially in the much softer bending mode. This implies that fibers on the outside of the bundle should seek to move closer to the axis of the bundle, while those initially closer to the axis should seek trajectories that move outward in order to keep the bundle density approximately constant and minimize the total strain energy.

This fiber arrangement is, indeed, observed in most twisted bundles and yarns. Figure 2.26 shows experimental data reported in Riding (1964). The curve shows the radial position of a tracer fiber in a twisted bundle along the bundle axis. At the left end of the trace, this fiber is closer to the bundle surface, but it migrates toward the bundle axis. A radial migration pattern, in and out, is observed, with a periodicity of about 4 mm. The periodicity depends on the density of the bundle and on whether twisting is applied with a superimposed stretch or not.

Fiber migration is controlled by packing and energetics. Fibers tend to migrate in order to reduce their axial strain and stored strain energy. However, migration is difficult in a densely packed bundle. The interplay of these mechanisms was studied and an instructive review is presented in Hearle et al. (1969), while a more recent account can be found in Neckar and Das (2018). The situation is qualitatively illustrated in Figure 2.27, which shows the frequency of radial fiber migration in the bundle function of the tension applied during twisting. If the bundle is not stretched, the bundle diameter increases upon twisting, the inner fibers accumulate slack and tend to move outward, while the outer fibers are stretched and tend to move inward.

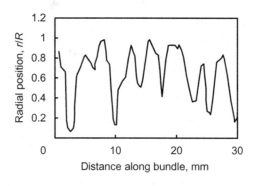

Figure 2.26 Radial position of a tracer fiber in a twisted bundle function of position along the bundle axis. Adapted from Riding (1964)

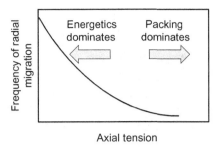

Figure 2.27 Qualitative representation of the variation of the frequency of radial migration of a tracer fiber within the bundle, as a function of the tension applied during twisting.

Radial migration is controlled by this energetic mechanism and fibers develop trajectories within the bundle which allow strain energy minimization. As tension increases, the bundle becomes more compact and the increased packing makes migration difficult. Two mechanisms cause increased packing: the reduction of slack of the central fibers due to the application of external axial loads, and the radial compressive force applied by helical fibers when subjected to a superposition of axial and torsion external loads. The result of the combination of these mechanisms is the continuously decreasing trend of fiber radial migration shown in Figure 2.27. These conclusions are supported by experimental observations with twisted textile yarns.

It is of interest to observe that as an initially straight fiber acquires a helical shape during bundle twisting, it is bent by a moment aligned to the bundle axis (i.e., by the bundle twisting moment). This moment, M, is shown schematically in Figure 2.24(b). Decomposing the moment in the direction perpendicular and parallel to the fiber axis, it results that fibers are both bent (by the component perpendicular to the fiber axis, M_b) and twisted (by the component aligned with the fiber axis, M_t). Fibers closer to the bundle axis are twisted more. This acquired chirality contributes to the inter-fiber load transfer via frictional forces.

2.5.2.2   Mechanical Behavior of Twisted Bundles

Twisted bundles, such as yarns used in the textile industry, twisted cables, and ropes, are mostly loaded in tension along the bundle axis. We focus here on this type of loading.

Twisting reduces the stiffness of the bundle. Consider that the bundle follows the affine kinematics and the strain on the scale of individual fibers is equal to the imposed axial, tensile strain applied to the bundle. Figure 2.28(a) shows the side view of a bundle subjected to an imposed axial strain ε and a resultant force P. Figure 2.28(b) shows one of the fibers of the bundle which forms a helix characterized by the angle $\alpha(r)$. The fiber is strained axially by $\varepsilon_f(r) = \varepsilon \cos^2 \alpha(r)$, which leads to an axial force $P_f(r) = E_f A_f \varepsilon \cos^2 \alpha(r)$. This fiber contributes to the total axial force in the bundle with the projection of P_f in the bundle axis direction, that is, $P_{f\parallel}(r) = E_f A_f \varepsilon \cos^3 \alpha(r)$. The total force in the bundle results by summing up these contributions:

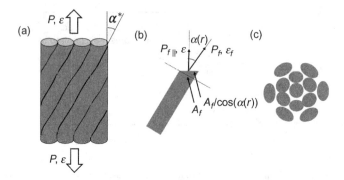

Figure 2.28 Schematic representation of a twisted bundle in (a) longitudinal and (c) cross-section views. (b) Forces acting in the cross-section of a fiber located at distance r from the bundle axis and forming angle $\alpha(r)$ with that axis.

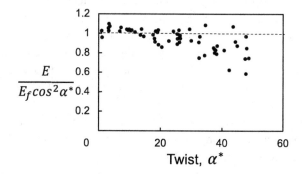

Figure 2.29 Normalized modulus of yarns of various polymeric fibers versus the twist angle (data from Hearle et al., 1959). The prediction of Eq. (2.61) is shown by the horizontal dashed line.

$$P = \int\int P_{f\parallel}(r) \frac{\phi_a}{A_f / \cos \alpha(r)} r \, dr \, d\theta, \qquad (2.60)$$

where ϕ_a represents the area fraction of fibers in the untwisted bundle, while the ratio under the integral represents the area fraction in the bundle cross-section at distance r from the bundle axis. Fibers close to the bundle axis are almost straight (small α) and appear in the bundle cross-section as circles, while those at larger r appear as ellipses, as shown in Figure 2.28(c). The stiffness of the bundle is evaluated using Eq. (2.60) by dividing P by the area of the bundle and the imposed strain, ε. It results that the bundle stiffness is $E = E_f \phi_a \langle \cos^4 \alpha(r) \rangle$, where $\langle \, \rangle$ represents averaging over the cross-section. Performing this average one obtains:

$$E \sim E_f \phi_a \cos^2 \alpha^*. \qquad (2.61)$$

Figure 2.29 shows experimental data from yarns of various polymeric fibers twisted to different degrees (Hearle et al., 1959). The figure supports the scaling of the

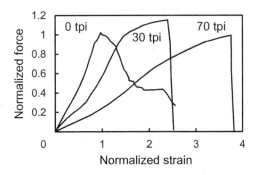

Figure 2.30 Normalized force–normalized strain curves for 100 den nylon yarns with twists of 0, 30, and 70 turns per inch (tpi). The axes are normalized by the strain and force corresponding to the maximum of the 0 tpi curve. Data from Hearle et al. (1959)

modulus with the twist angle indicated by Eq. (2.61), especially at low twists. As the twist increases, fiber migration and the associated disordering of the bundle structure make the physical picture more complex and some departure from the prediction of Eq. (2.61) is observed.

Figure 2.30 shows load–elongation curves for bundles of continuous nylon fibers with 0, 30, and 70 turns per inch twist (Hearle et al., 1959). The two axes are normalized with the peak force and the elongation of the reference untwisted bundle. The decrease of the bundle stiffness with increasing twist is clearly visible. Further, rupture is more gradual in the untwisted bundle compared with the twisted bundles, which exhibit brittle failure. This is due to the inherent presence of some degree of tortuosity in the untwisted bundle. The tortuosity is eliminated as the degree of twist increases. Interestingly, it emerges that the failure stress is largely independent of the twist. The elongation at failure increases as twist increases due to the reduction of the effective bundle stiffness (Eq. (2.61)). These trends have been reported in the literature for different types of bundles.

The situation can be understood qualitatively starting from the following considerations. If the bundle is twisted without being subjected to a tensile force, the fibers close to the bundle axis acquire slack and migrate outward, potentially increasing the free volume of the bundle. The outer fibers are subjected to tension during twisting. If they deform plastically, the elastic rebound during unloading is insignificant, but these fibers have accumulated plastic strain and are closer to their failure strain. In this case, when subjected to subsequent axial testing (without increasing the twist), the outer fibers will break first. Hence, the failure strain should depend on the degree of twist. If fibers remain elastic, the rebound upon unloading is large and puts the fibers close to the bundle axis in compression, therefore increasing the slack. Upon loading, the middle fibers carry little load and, hence, in this case too, the outer fibers break first. This argument implies that the fraction of load bearing fibers decreases with increasing twist and hence the failure strain is expected to depend on the magnitude of twist.

If the bundle is twisted under axial load, slack is reduced. If fibers deform plastically, the elastic rebound is limited and no slack is introduced. Hence, the failure strain is expected to be independent of twist in this case. The situation is opposite if fibers deform elastically at all times. Real situations are in-between these extreme cases.

2.5.3   Bundles of Discontinuous Fibers: The Shear Lag Model

If fibers are discontinuous, the integrity of the bundle hinges on the existence of an appropriate inter-fiber load transfer mechanism. In biological materials and in textile yarns, fibers are of finite length and, in most cases, are much shorter than the length of the bundle. This is in contrast with the situation of structural fiber-reinforced composites and most nonwovens which are made from nominally continuous fibers.

Collagen fibers are bundles of fibrils which are cross-linked within the bundle. In textiles the fibers within yarns are not cross-linked but interact frictionally. Cotton and wool fibers have diameters of tens of micrometers and an aspect ratio larger than 1 000. Carbon nanotubes form bundles with cohesive inter-tube interactions, which is a rather weak load transfer mechanism. The tubes have nanometer cross-sectional dimensions and lengths as large as millimeters, so the aspect ratio can be as high as 10^6.

Fiber–fiber interactions that may lead to effective load transfer include the cross-linking of fibers, inter-fiber friction, and adhesion. It is important in this context to analyze how loads are transmitted between fibers within the bundle. The answer is provided by the shear lag model developed in the literature related to composite materials (Cox, 1952) and subsequently used in other contexts for many material systems.

Consider the configuration in Figure 2.31(a), where two fibers are in contact along a segment of length L and are loaded by a force P which promotes relative sliding in the axial direction. The width of the contact surface is $2a_X$ (see Figure 3.4 for the definition of this parameter). The fiber material is considered linear elastic, of modulus E_f. The interface is subjected to shear (as stated, the problem is one-dimensional) and may be represented either with a continuum or a discrete model. In the continuum representation, an effective thickness, h_X, and a shear modulus, G_X, are assigned to the interface and load transmission is assumed to take place at all points of the contact area. In the discrete interface model, it is assumed that the two fibers are connected at discrete points by springs of stiffness k_X separated in the axial direction by l_X. This representation is adequate for situations in which nanoscale fibers are crosslinked by fibrils of molecular strands (Figure 2.31).

The equilibrium condition for the fiber element of length dx requires:

$$\frac{d\sigma_1}{dx} = -\frac{d\sigma_2}{dx} = \tau \frac{2a_X}{A_f}, \tag{2.62}$$

where σ_1 and σ_2 represent the axial stresses in the two fibers. The constitutive equation of fibers reads:

$$\sigma_i = E_f \frac{du_i}{dx}, \quad i = 1, 2. \tag{2.63}$$

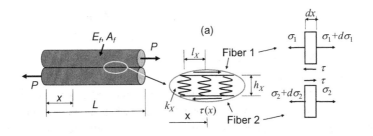

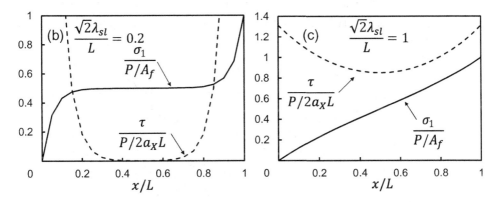

Figure 2.31 (a) Problem set-up and stress distribution predicted by the shear lag model (Eq. (2.68)), for (b) $\sqrt{2}\lambda_{sl}/L = 0.2$ and (c) $\sqrt{2}\lambda_{sl}/L = 1$.

The continuum interface is described by

$$\tau = G_X \frac{u_1(x) - u_2(x)}{h_X}, \tag{2.64a}$$

while the discrete interface is described by

$$\tau = \frac{k_X}{2a_X l_X}(u_1(x) - u_2(x)), \tag{2.64b}$$

with the relation $G_X = k_X h_X / 2a_X l_X$ establishing the equivalence of the two models. Noting that Eq. (2.64b) leads to:

$$\frac{d\tau}{dx} = \frac{k_X}{2a_X l_X E_f}(\sigma_1(x) - \sigma_2(x)), \tag{2.65}$$

the equilibrium equations (Eq. (2.62)) can be rewritten as:

$$\frac{d^2\sigma_1}{dx^2} = \frac{1}{\lambda_{sl}^2}(\sigma_1 - \sigma_2),$$

$$\frac{d^2\sigma_2}{dx^2} = \frac{1}{\lambda_{sl}^2}(\sigma_2 - \sigma_1), \tag{2.66}$$

where

$$\lambda_{sl} = \sqrt{\frac{E_f A_f l_X}{k_X}} = \sqrt{\frac{E_f A_f h_X}{G_X 2 a_X}} \tag{2.67}$$

is known as the shear lag length and is a characteristic length scale emerging from the mechanics of load transmission across the interface. This parameter represents the relative stiffness of the fibers and the interface and has large values when fibers are much stiffer than the interface and small values when fibers are compliant.

Equations (2.66) are solved with boundary conditions representing for each fiber the condition that the stress, P/A_f, is imposed at one end, while the other end is traction free: $\sigma_1(0) = 0$, $\sigma_1(L) = P/A_f$, $\sigma_2(L) = 0$, and $\sigma_2(0) = P/A_f$. The solution reads:

$$\sigma_1(x) = \frac{P}{A_f} \frac{\sinh \dfrac{x/L}{\sqrt{2}\lambda_{sl}/L} \cosh \dfrac{1-x/L}{\sqrt{2}\lambda_{sl}/L}}{\sinh \dfrac{1}{\sqrt{2}\lambda_{sl}/L}}$$

$$\sigma_2(x) = \sigma_1(L-x) \tag{2.68}$$

$$\tau(x) = \frac{P}{2\sqrt{2}a_X \lambda_{sl}} \frac{\cosh \dfrac{2x/L - 1}{\sqrt{2}\lambda_{sl}/L}}{\sinh \dfrac{1}{\sqrt{2}\lambda_{sl}/L}}$$

The solution of Eq. (2.68) is graphically presented in Figure 2.31(b) and (c) for two values of the shear lag parameter, $\sqrt{2}\lambda_{sl}/L$. In the compliant fibers case, when the shear lag length is small, $\lambda_{sl} \ll L$, the interface is loaded only in a region close to the two fiber ends, of length approximately equal to λ_{sl}, as shown in Figure 2.31(b). The interface carries no loads in the central part of the fiber. Consequently, the axial stress is constant in the central section of the fiber and varies close to the fiber ends. The opposite situation is obtained as the fibers become stiffer and λ_{sl} increases: the region of load transmission extends along the interface and, in the limit of rigid fibers and very large λ_{sl}, load is transmitted uniformly along the interface and the axial stress in fibers varies linearly, as shown in Figure 2.31(c). Since fibers slide as rigid bodies in this case, the relative slip is constant along the interface.

2.5.4   Failure of Bundles of Continuous and Discontinuous Fibers

The failure of fiber bundles of continuous and discontinuous fibers subjected to uniaxial tension is discussed in this section. In the case of discontinuous fiber bundles, load transfer between fibers is necessary in order to ensure the integrity and load carrying capacity of the bundle. Load transfer occurs via intra-bundle crosslinks or through the matrix that embeds the fibers, if present.

Bundles of continuous fibers may carry loads in the absence of intra-bundle cross-links. In this case, rupture of a fiber leads to the redistribution of the load to the remaining intact fibers. Load redistribution may take place equally to all remaining fibers of the bundle, or exclusively to the neighbors of the broken fiber. The exact load redistribution modality is generally unknown and many models have been proposed in the literature addressing this problem. Reviews of fiber bundle models are presented in Phoenix (1993) and Pradhan et al. (2010). It should be noted that a continuous fiber bundle with an inter-fiber load transfer mechanism becomes similar to a discontinuous fiber bundle once the fibers start failing. This is because a fiber that breaks can still carry loads at distances along the respective fiber from the rupture site larger than λ_{sl}, provided a mechanism which transfers load from the neighboring fibers operates. Hence, any fiber of the bundle may experience rupture at multiple sites and become fragmented, entailing that the bundle becomes similar to a discontinuous fiber bundle with stochastic fiber lengths.

Due to its prevalence, we focus here on the failure of discontinuous fiber bundles, an example of which is shown schematically in Figure 2.32. It is considered that fibers are connected by inter-fiber crosslinks which behave as linear springs of stiffness k (per unit length of contact) – a situation similar to that discussed in Section 2.5.3 in the context of the shear lag model. Crosslinks are thought to be uniformly distributed in the interface between any two fibers in contact. For simplicity, it is assumed that fibers have identical length, L, and any fiber row in the bundle is shifted relative to its neighbors by a random distance (Figure 2.32).

Loads are transmitted from fiber to fiber across interfaces. If the shear lag length λ_{sl} is much smaller than L, the loads are transmitted between fibers in the vicinity of fiber ends. The bundle may fail by the rupture of fibers only, the rupture of crosslinks only, or by the rupture of both fibers and crosslinks. If fibers do not break, failure takes place along interfaces, followed by fiber pull-out. The bundle ruptures in this case along the shortest path across the bundle running along interfaces, as shown in Figure 2.32. The energy dissipated in this process depends on the length of this path; this failure mode provides the largest toughness of all three failure modes mentioned here. If interface failure does not take place and the bundle ruptures exclusively by the failure of fibers, the only possibility is for all fibers in a given cross-section to fail, which implies that a major crack cuts across the bundle leading to brittle failure. The intermediate case, in which both fibers and interfaces fail, requires fiber pull-out to an extent smaller than that of the situation shown in Figure 2.32. The associated toughness is also intermediate between the other two cases.

This qualitative discussion suggests that the behavior may be described in terms of two non-dimensional parameters: the ratio of the fiber strength to the crosslinks strength, σ_f^c/σ_X^c, and the ratio of the fiber length to the shear lag length, L/λ_{sl}.

Figure 2.32 Schematic showing the localization of relative fiber sliding leading to bundle rupture. The dashed line shows the localization path.

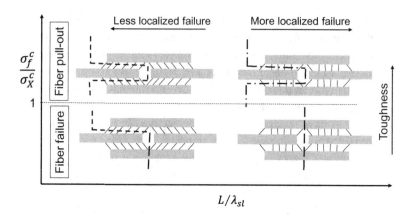

Figure 2.33 Types of interface deformation and bundle failure function of parameters σ^c_f/σ^c_X and L/λ_{sl}. The dashed lines represent expected rupture paths.

Figure 2.33 shows the expected deformation and failure modes in the space of these two parameters. Based on the results of the shear lag analysis of Section 2.5.3, at large values of L/λ_{sl} load transfer is localized at the ends of fibers, while at small values of this parameter load transfer is distributed over larger lengths of the interfaces, starting from fiber ends. If the crosslinks are weaker than the fibers, that is, $\sigma^c_f/\sigma^c_X \gg 1$, bundle failure takes place mainly along interfaces and involves fiber pull-out. If interfaces are strong and fibers have low strength, bundles fail due to fiber rupture. The load transfer mechanism makes a difference in this case: at large values of L/λ_{sl}, when load transfer is localized, fibers carry approximately constant load along segments defined by the position of fiber ends of neighboring fibers. In the low L/λ_{sl} case, the axial force varies continuously along the length of fibers. If fibers have stochastic strength, the probability that a fiber ruptures at multiple sites along its length is lower in the low L/λ_{sl} case than in the high L/λ_{sl} case. Hence, moving from left to right along the horizontal axis at low values of the ordinate, it is expected that bundle failure becomes more localized. The bundle toughness should increase with increasing σ^c_f/σ^c_X due to the change of the failure mode from localized fiber failure to gradual pull-out.

As discussed in Section 2.3.7, in most situations of practical interest the strength of fibers is stochastic and varies along their length. This is particularly important in the case of long, continuous fibers such as those used in structural composites. The fiber strength is generally described by the weakest link theory, which leads to the Weibull statistics. Equation (2.11) provides the cumulative probability that a fiber fails at a stress smaller than σ, $cp_f(\sigma) = 1 - \exp\left(-(\sigma/\sigma_0)^\beta\right)$, with β and σ_0 being the Weibull modulus and the scale factor of the distribution.

In the context of the present discussion, it is of interest to investigate the functional form of the failure probability of the fiber bundle, $cp_b(\sigma)$, given that the individual fiber strength is described by Weibull statistics, $cp_f(\sigma)$. The answer to this question is sought in a vast literature on the "fiber bundle model" which was proposed about

100 years ago in the literature on textile fibers and is still being used in statistical physics for problems such as the occurrence of avalanches and failure of brittle solids. As briefly discussed at the beginning of this section, the basic model assumes that the behavior of all fibers is elastic up to a value of the stress at which failure occurs, after which the load carried by the respective fiber is redistributed to the fibers which are still intact. Models of this type differ with respect to the assumptions regarding load redistribution. Many refinements have been proposed, including accounting for time dependence (Newman and Phoenix, 2001), the hierarchical organization of bundles (Pimenta and Pinho, 2013), fiber and crosslink strength distributions (Swolfs et al., 2015; Kádár et al., 2017), etc. Such models have been applied to a broad range of fibrous systems such as carbon nanotube bundles (Zhang et al., 2014), silk (Bosia et al., 2010), and unidirectional fiber composites (Okabe et al., 2005; Bunsell et al., 2018), and to the failure of nonfibrous materials such as bone and nacre (Zhang et al., 2010). A review of these works with a statistical physics flavor is presented in Pradhan et al. (2010).

To demonstrate the general behavior in the context of the bundle model used in the present section (Figure 2.32), consider first bundles with $\left\langle \sigma_f^c \right\rangle / \sigma_X^c = 0.05$ in which the mean of the strength distribution of fibers, $\left\langle \sigma_f^c \right\rangle$, is 20 times smaller than the (deterministic) strength of inter-fiber crosslinks. In this case, fibers fail first and bundle failure is localized. Figure 2.34(a) shows the cumulative distribution of bundle failure along with the corresponding distribution of fiber failure, which is an input to the model. The mean bundle strength, $\langle \sigma_b^c \rangle$, is somewhat smaller (but comparable to) than the mean fiber strength. However, the more obvious effect is the drastic increase of the Weibull modulus when going from the fiber to the bundle scale. This implies that the variability of the bundle strength is much smaller than that of the individual fiber strength.

Figure 2.34(b) shows a similar result for the bundle with $\left\langle \sigma_f^c \right\rangle / \sigma_X^c = 5$, in which failure is controlled by crosslink rupture and fiber pull-out. The mean bundle strength is significantly smaller than the mean fiber strength, which is a consequence of the fact that bundle failure is related to interface failure. In this case too, the bundle strength variability is much smaller than the variability of the fiber strength.

The functional form of the emerging bundle strength distribution, $cp_b(\sigma)$, cannot be derived, except in a few cases in which the bundle model can be solved analytically. The first such model is due to Daniels (1945) and considers bundles of continuous fibers with no inter-fiber load transmission and equal load sharing among the fibers that remain intact. In this case, the bundle strength follows a Gaussian distribution, with variability decreasing inversely as the square root of the number of fibers in the bundle. Following the work of Daniels, Coleman (1958) derived the mean strength of a bundle of continuous fibers of the same constitutive behavior, but with strengths described by a Weibull distribution of Weibull modulus β (Eq. (2.11)). The analysis is performed under the assumption of equal load sharing, in which all remaining fibers in the bundle share equally the load carried by fibers that break. Coleman's analysis indicates that the mean bundle strength, $\langle \sigma_b^c \rangle$, is smaller than the mean fiber strength, $\left\langle \sigma_f^c \right\rangle$, and the ratio of these two quantities is described by:

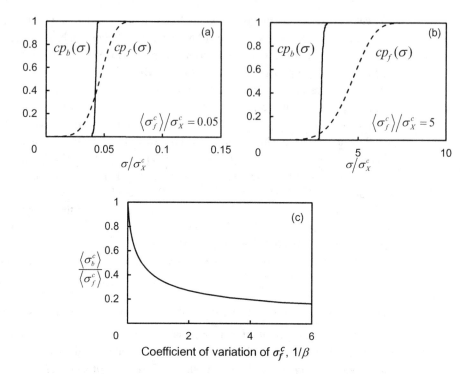

Figure 2.34 Cumulative distribution function of the strength of fiber bundles $cp_b(\sigma)$ (continuous line) for which the equivalent distribution function for individual fibers $cp_f(\sigma)$ is shown with a dashed line. (a) The case in which the mean fiber strength is 1/20 of the crosslink strength. (b) Equivalent results for the case in which the mean fiber strength is five-times larger than the crosslink strength. (c) Variation of the mean bundle strength normalized by the mean fiber strength with the coefficient of variation of the Weibull distribution of fiber strength, $1/\beta$.

$$\frac{\langle \sigma_b^c \rangle}{\langle \sigma_f^c \rangle} = \left[\beta^{1/\beta} \exp\left(\frac{1}{\beta}\right) \Gamma\left(1 + \frac{1}{\beta}\right) \right]^{-1}, \qquad (2.69)$$

where Γ represents the Gamma function. This ratio is a function of the Weibull modulus of the fiber strength distribution only and decreases continuously with decreasing β, that is, with increasing fiber strength variability. Figure 2.34(c) shows the variation of $\langle \sigma_b^c \rangle / \langle \sigma_f^c \rangle$ with the coefficient of variation of the fiber strength distribution $cp_f(\sigma)$ (the coefficient of variation of the Weibull distribution can be approximated as $1/\beta$).

Other analytic and numerical results for the statistics of bundle strength under various types of load redistribution and inter-fiber load transmission are summarized in a review article (Phoenix and Beyerlein, 2000).

The strength of bundles of discontinuous fibers with twist is of great interest in textiles. The mechanism that allows ropes to acquire strengths larger than that of individual fibers is based on the interplay of friction and twist.

A bundle of discontinuous fibers in which the only inter-fiber load transfer mechanism is frictional has no strength if fibers are straight, since no normal forces that may engage friction act between parallel fibers loaded axially. For inter-fiber friction to be engaged, the bundle must be twisted. With an ideal twisting pattern, as shown in Figure 2.24, which requires that fibers become helical and preserve their distance to the bundle axis, it is clear that inter-fiber friction is not engaged. Two features lead to the desired effect: (i) the fact that a helical fiber loaded in tension along the axis of the helix tends to move inward, reducing the radius of the respective helix, and (ii) twist-induced fiber migration. The first mechanism creates radial forces compressing the central region of the bundle, while the second mechanism allows fibers on the outside of the bundle to visit the bundle core, and hence become frictionally engaged with the other fibers. This implies that load transfer takes place when a fiber traverses the core of the bundle, while the other segments of the respective fiber, which are located farther from the bundle axis, carry constant axial force. For the mechanism to be effective with staple fibers, the fiber length must be larger than the distance along given fiber between segments that traverse the bundle core (Figure 2.26).

The bundle strength is limited at low twists by the weak load transfer, since fibers are approximately parallel to the bundle axis and fiber migration is negligible. Under such conditions, the load carried by individual fibers is small and bundle failure takes place primarily by fiber pull-out. At high degrees of twist leading to pronounced fiber migration, load transfer is efficient and strength is controlled by fiber rupture. Overall, the strength versus twist curve is defined by the dependence of the load transfer mechanism and of the stress carried by fibers on the degree of twist. Figure 2.35(a) shows these trends schematically. Load transfer becomes more efficient with increasing twist and this leads to the increase of the bundle strength, as shown by curve AB. The strength of individual fibers limits the bundle strength at large degrees of twist, as shown schematically by curve CD. This decreasing trend can be understood based on the analysis of the simplified bundle model leading to Eq. (2.61). Specifically, in this

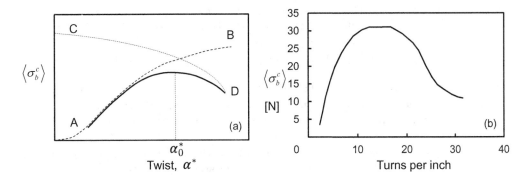

Figure 2.35 Variation of the bundle strength with the angle of twist: (a) schematic representation of trends, and (b) experimental data obtained with cotton yarns. Data from Landstreet et al. (1957)

model, the stress carried by the bundle, σ, is related to the strain on the bundle scale, ε, as $\sigma = E_f \phi_a \varepsilon \cos^2 \alpha^*$ (see Eq. (2.61)), and ε is related to the axial strain in the fiber located at distance r from the bundle axis and forming a helix of angle $\alpha(r)$ by $\varepsilon_f(r) = \varepsilon \cos^2 \alpha(r)$. Since the stress in the fiber is $\sigma_f(r) = E_f \varepsilon_f(r)$, the stress in the fiber and the bundle-scale stress are related as $\sigma = \sigma_f(r) \phi_a \cos^2 \alpha^* / \cos^2 \alpha(r)$. In this simplified model, the fibers close to the bundle axis are the most loaded. Requiring that for these fibers $\sigma_f(r) = \sigma_{fc}$, where σ_{fc} is the fiber strength, it results that the bundle strength, σ_c, scales as $\sigma_c \sim \cos^2 \alpha^*$. This equation quantifies the decreasing trend of curve CD in Figure 2.35(a). The actual dependence of the bundle strength on the twist is the combination of the two aforementioned trends and exhibits a non-monotonic behavior and the emergence of an optimal value of twist corresponding to the position of the maximum strength (Figure 2.35(a)). Figure 2.35(b) shows experimental data obtained with cotton yarns of various degrees of twist (Landstreet et al., 1957) confirming the trends discussed here.

Predicting the optimal twist α_0^* of a fiber bundle is of practical importance. Here it suffices to observe that the position of the peak in Figure 2.35(a) can be adjusted by controlling the rate at which curve AB increases. If twist enables the load transfer mechanism efficiently, AB increases faster with α^*, and consequently α_0^* decreases. This can be achieved by increasing the fiber length and increasing the inter-fiber friction coefficient. Both of these variables have been observed experimentally to reduce the value of the optimal bundle twist.

A number of refinements can be made to this physical picture by taking into account fiber migration, the details of load transfer in view of the shear lag mechanism, the dependence of the fiber strength on fiber length, etc. The reader is referred to the relevant literature for a review of these effects (Hearle et al., 1969; Neckar and Das, 2018).

References

Andrady, A. L. (2008). *Science and technology of polymer nanofibers.* Wiley, Hoboken, NJ.

Armero, F. & Valverde, J. (2012). Invariant Hermitian finite elements for thin Kirchhoff rods. I: The linear plane case. *Comput. Methods Appl. Mech. Engrg.* **213–216**, 427–457.

Bazant, Z. P. & Cedolin, L. (2010). *Stability of structures.* World Scientific Publishing, Singapore.

Bosia, F., Buehler, M. J. & Pugno, N. M. (2010). Hierarchical simulations for the design of supertough nanofibers inspired by spider silk. *Phys. Rev. E* **82**, 056103.

Bouchiat, C., Wang, M. D., Allemand, J. F., et al. (1999). Estimating the persistence length of a worn-like chain molecule from force–extension measurements. *Biophys. J.* **76**, 409–413.

Bunsell, A., Gorbatikh, L., Morton, H., et al. (2018). Benchmarking of strength models for unidirectional composites under longitudinal tension. *Composites A* **111**, 138–150.

Bustamante, C., Marko, J. F., Siggia, E. D. & Smith, S. (1994). Entropic elasticity of λ-Phage DNA. *Science* **265**, 1599–1600.

Chay, T. R. (1970). Distribution function for self-avoiding walks. *J. Chem. Phys.* **52**, 1025–1033.

des Cloizeaux, J. (1974). Lagrangian theory for a self-avoiding random chain. *Phys. Rev. A* **10**, 1665–1669.

Cohen, A. (1991). A Pade approximant to the inverse Langevin function. *Rheol. Acta* **30**, 270–273.

Coleman, B. D. (1958). On the strength of classical fibers and fiber bundles. *J. Mech. Phys. Solids* **7**, 60–70.

Cox, H. L. (1952). The elasticity and strength of paper and other fibrous materials. *British J. Appl Phys.* **3**, 72–81.

Crisfield, M. A. & Jelenic, G. (1999). Objectivity of strain measures in the geometrically exact three-dimensional beam theory and its finite-element implementation, *Proc. R. Soc. London A* **455**, 1125–1147.

Daniels, H. E. (1945). The statistical theory of the strength of bundles of threads. *Proc. R. Soc. London A* **183**, 405.

van Dillen, T., Onck, P. R. & van der Giessen, E. (2008). Models for stiffening in cross-linked biopolymer networks: a comparative study. *J. Mech. Phys. Sol.* **56**, 2240–2264.

Dobrynin, A. V., Carrillo, J. M. Y. & Rubinstein, M. (2010). Chains are more flexible under tension. *Macromolecules* **43**, 9181–9190.

Farber, J., Lichtenegger, H., Reiterer, C., Stanzl-Tschegg, S. & Frazl, P. (2001). Cellulose microfibril angles in a spruce branch and mechanical implications. *J. Mater. Sci.* **36**, 5087–5092.

Fertis, D. G. (1999). *Nonlinear mechanics*, 2nd ed. CRC Press, Boca Raton.

Fisher, M. E. (1966). The shape of a self-avoiding walk or polymer chains. *J. Chem. Phys.* **44**, 616–622.

Fischer, M. E. (1969). Aspects of equilibrium critical phenomena. *J. Phys. Soc. Japan* **26**, 87–93.

Flory, P. J. (1949). The configuration of real polymer chains. *J. Chem. Phys.* **17**, 303–310.

de Gennes, P. G. (1979). *Scaling concepts in polymer physics*. Cornell University Press, Ithaca, NY.

Gere, J. M. & Timoshenko, S. P. (2002). *Mechanics of materials*, 5th ed. Nelson Thornes, Cheltenham.

Hearle, J. W. S., El-Behery, H. M. A. E. & Thakur, V. M. (1959). The mechanics of twisted yarns: Tensile properties of continuous filament yarns. *J. Textile Inst.* **50**, T83–T111.

Hearle, J. W. S., Grosberg, P. & Backer, S. (1969). *Structural mechanics of fibers, yarns and fabrics*. Wiley-Interscience, London.

Howard, J. (2001). *Mechanics of motor proteins and the cytoskeleton*. Sinauer Assoc. Inc., Sunderland, MA.

Jelenic, G. & Crisfield, M. A. (1999). Geometrically exact 3D beam theory: Implementation of a strain-invariant finite element for statics and dynamics. *Comput. Methods Appl. Mech. Eng.* **171**, 141–171.

Kabla, A. & Mahadevan, L. (2007). Nonlinear mechanics of soft fibrous networks. *J. R. Soc. Interface* **4**, 99–106.

Kádár, V., Danku, Z. & Kun, F. (2017). Size scaling of failure strength with fat-tailed disorder in a fiber bundle model. *Phys. Rev. E* **96**, 033001.

Kratky, O. & Porod, G. (1949). Röntgenuntersuchung gelöster Fadenmoleküle *Recueil Traveaux Chim. Pays-Bas* **68**, 1106–1122.

Landstreet, C. B., Ewald, P. R. & Simpson, J. (1957). The relation of twist to the construction and strength of cotton rowings and yarns. *Textile Res. J.*, **27**, 486–492.

Lichtenegger, H., Muller, M., Paris, O., Riekel, C. & Fratzl, P. (1999). Imaging of the helical arrangement of cellulose fibrils in wood by synchrotron X-ray microdiffraction. *J. Appl. Cryst.* **32**, 1127–1133.

Liu, J., Das, D., Yang, F., et al. (2018). Energy dissipation in mammalian collagen fibrils: Cyclic strain-induced damping, toughening and strengthening. *Acta Biomater.* **80**, 217–227.

Livadaru, L., Netz, R. R. & Kreuzer, H. J. (2003). Stretching response of discrete semiflexible polymers. *Macromolecules* **36**, 3732–3744.

Marko, J. F. & Siggia, E. D. (1995). Stretching DNA. *Macromolecules* **28**, 8759–8770.

Mattiasson, K. (1981). Numerical results from large deflection beams and frames problems analyzed by means of elliptic integrals. *Int. J. Numer. Meth. Eng.* **17**, 145–153.

McHugh, A. J. & Doufas, A. K. (2001). Modeling flow-induced crystallization in fiber spinning. *Composites A* **32**, 1059–1066.

Mohanty, A. K., Misra, M. & Hinrichsen, G. (2000). Biofibers, biodegradable polymers and biocomposites: An overview. *Macromolec. Mater. Eng.* 276/277, 1–24.

Morton, W. E. (1956). The arrangement of fibers in single yarns. *Textile Res. J.* **26**, 325–331.

Moroz, J. D. & Nelson, P. C. (1998). Entropic elasticity of twist storing polymers. *Macromolecules* **31**, 6333–6347.

Neagu, R. C., Gamstedt, E. K., Bardage, S. L. & Lindstrom, M. (2006). Ultrastructural features affecting mechanical properties of wood fibers. *Wood Mat. Sci Eng.* **1**, 146–170.

Neckar, B. & Das, D. (2018). *Theory of structure and mechanics of yarns*. Woodhead Publishing India, New Delhi.

Newman, W. I. & Phoenix, S. L. (2001). Time-dependent fiber bundles with local load sharing. *Phys. Rev. E* **63**, 021507.

Odijk, T. (1995). Stiff chains and filaments under tension. *Macromolecules* **28**, 7016–7018.

Okabe, T., Sekine, H., Ishii, K., Nishikawa, M. & Takeda, N. (2005). Numerical method for failure simulation of unidirectional fiber-reinforced composites with spring element model. *Compos. Sci. Technol.* **65**, 921–933

Orgel, J. P. R. O, Miller, A., Irving, T. C., et al. (2001). The in-situ supermolecular structure of type I collagen. *Structure* **9**, 1061–1069.

Osta, A., Picu, R. C., King, A., et al. (2014). Effect of polypropylene fiber processing conditions on fiber mechanical behavior. *Polym. Int.* **63**, 1816–1823.

Phoenix, S. L. (1993). Statistical issues in the fracture of brittle-matrix fibrous composites. *Comp. Sci. Technol.* **48**, 65–80.

Phoenix, S. L. & Beyerlein, I. J. (2000) Statistical strength theory for fibrous composite materials. In *Comprehensive composite materials*, A. Kelly and C. Zweben, eds. Pergamon, New York, Vol. 1, Chap. 1.19.

Pimenta, S. & Pinho, S. T. (2013). Hierarchical scaling law for the strength of composite fibre bundles. *J. Mech. Phys. Solids* **61**, 1337–1356.

Pradhan, S., Hansen, A. & Chkrabarti, B. K. (2010). Failure processes in elastic fiber bundles. *Rev. Mod. Phys.* **82**, 499–555.

Puxkandl, R., Zizak, I., Paris, O., et al. (2002). Viscoelastic properties of collagen: Synchrotron radiation investigations and structural model. *Philos. Trans. R. Soc. London B* **357**, 191–197.

Reissner, E. (1972). On one-dimensional finite-strain beam theory: The plane problem. *Z. Angew. Math. Phys.* **23**, 795–804.

Reissner, E. (1981). On finite deformations of space-curved beams. *Z. Angew. Math. Phys.* **32**, 734–744.

Rezakhaniha, R., Agianniotis, A., Schrauwen, J. T. C., et al. (2012). Experimental investigation of collagen waviness and orientation in the arterial adventitia using confocal laser scanning microscopy. *Biomech. Model. Mechanobiol.* **11**, 461–473.

Richaud, E., Verdu, J. & Fayolle, B. (2009). *Tensile properties of polyproylene fibres.* Handbook of Tensile Properties of Textile and Technical Fibres, Woodhead Publishing Series – Elsevier, Cambridge, pp. 315–331.

Riding, G. (1964). Filament migration in single yarns. *J. Textile Inst.* **55**, T9–T17.

Romero, I. (2008). A comparison of finite elements for nonlinear beams: The absolute nodal coordinate and geometrically exact formulations. *Multibody Syst. Dyn.* **20**, 51–68.

Rosa, A., Hoang, T. X., Marenduzzo, D. & Maritan, A. (2003). Elasticity of semiflexible polymers with and without self-interactions. *Macromolecules* **36**, 10095–10102.

Rubinstein, M. & Colby, R. H. (2003). *Polymer physics.* Oxford University Press, New York.

Seth, R. S. and Page, D. H. (1983). The stress strain curve of paper. In *The role of fundamental research in paper making: Transactions of the symposium held at Cambridge, September 1981*, 2nd ed. (January 1, 1983). Mechanical Engineering Publications Limited, London, pp. 421–452.

Shahsavari, A. & Picu, R. C. (2012). Model selection for athermal cross-linked fiber networks. *Phys. Rev. E* **86**, 011923.

Sherman, V. R., Yang, W. & Meyers, M. A. (2015) The materials science of collagen. *J. Mech. Beh. Biomed. Mater.* **52**, 22–50.

Sopakayang, R., de Vita, R., Kwansa, A. & Freeman J. W. (2012). Elastic and viscoelastic properties of type I collagen fiber. *J. Theor. Biol.* **293**, 197–205.

Strick, T. R., Allemand, J.-F., Bensimon, D., Bensimon, A. & Croquette, V. (1996). The elasticity of a single supercoiled DNA molecule. *Science* **271**, 1835–1837.

Swolfs, Y., Verpoest, I. & Gorbatikh, L. (2015). Issues in strength models for unidirectional fibre-reinforced composites related to Weibull distributions, fibre packings and boundary effects. *Compos. Sci. Technol.* **114**, 42–49.

Timoshenko, S. P. & Gere, J. M. (1961). *Theory of elastic stability.* Dover, Mineola, NY.

Treloar, L. R. G. (1965). A migrating-filament theory of yarn properties. *J. Textile Inst.* **56**, T359–T380.

da Vinci, L. (1972) *I Libri Di Meccanica*, reconstructed from the original notes by Arturo Uccelli. Nendeln, Liechtenstein.

Weiner, J. H. (1982). Use of S= k log(p) for stretched polymers. *Macromolecules* **15**, 542–544.

Weiner, J. H. (1983). *Statistical mechanics of elasticity.* John Wiley, New York.

Wu, H. F., Biresaw, G. & Laemmle. J.T. (1991). Effect of surfactant treatments on interfacial adhesion in single graphite-epoxy composites. *Polymer Compos.* **12**, 281–288.

Zhang, Z., Liu, B., Zhang, Y. W., Hwang, K. C. & Gao, H. (2014). Ultra-strong collagen-mimic carbon nanotube bundles. *Carbon* **77**, 1040–1053.

Zhang, Z. Q., Liu, B., Huang, Y., Hwang, K. C. & Gao, H. (2010). Mechanical properties of unidirectional nanocomposites with non-uniformly or randomly staggered platelet distribution. *J. Mech. Phys. Solids* **58**, 1646–1660.

3   Fiber Interactions

3.1   Types of Fiber Interactions in Networks

This chapter describes the various types of interactions between fibers encountered in network materials. Inter-fiber interactions ensure load transfer, provide structural integrity to the network and, to a large extent, define the network mechanical behavior.

In broad terms, three types of interactions can be identified: crosslinks, contacts, and fiber bundles. In addition, if the network is embedded in a matrix, fibers interact with the embedding medium. Figure 3.1 shows this classification.

Crosslinks (type 1) represent connectors that transmit forces and moments between fibers. A crosslink forms when fibers interpenetrate, therefore sharing a certain volume, or when fibers are in contact over a large enough area to develop tractions that may effectively sustain and transmit moments between fibers. A small contact area requires large tractions in order to transmit large moments, while a large contact area may allow moment transmission at relatively small traction magnitudes. This type of interaction is referred to as "welded joint," using an analogy with macroscopic structural assemblies. Crosslinks of welded type are the most effective load transmission fiber connectors. They are present in many thermal and athermal networks, including molecular networks.

Type 2 shown in Figure 3.1 represents cases in which fibers are connected by links that carry forces, but not moments. These are denoted as "rotating joints," which entails that the respective fibers are free to rotate relative to each other. Since fibers are continuous across the crosslink, moments are transmitted along the fibers, but not across the crosslink from one fiber to the other.

The third type of crosslink shown in Figure 3.1 is a "pin-joint," which transmits only forces (type 3). The essential difference between crosslinks of rotating and pin-joint types is that the pin-joints do not allow moment transmission along fibers. In this representation no moments are transmitted across the crosslink between any of the segments forming the joint. Considering that, in most practical situations, crosslinks do transmit moments, this may be viewed as an idealization of cases in which the transmitted moments are sufficiently small to be inconsequential for the mechanics of the network. While types 1 and 2 represent crosslinks between two continuous fibers, with an effective connectivity (number of segments emerging from the crosslink) $z = 4$, a pin-joint may, in principle, connect any number of segments, and may have any z value.

Crosslinks may be permanent or transient. Permanent crosslinks, such as covalent bonds between molecules in molecular networks and crosslinks resulting from the

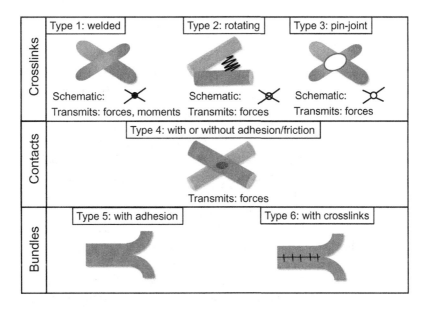

Figure 3.1 Classification of fiber interaction types.

interpenetration of polymeric micro-fibers in nonwovens, may break and do not reform after failure. Transient crosslinks may be broken by thermal fluctuations or by the combined effect of thermal fluctuations and mechanical forces and may reform elsewhere in the network where adequate binding sites exist. The presence of transient crosslinks imparts time-dependent mechanical behavior to the network (Chapter 9).

In most networks fibers come in contact at sites other than those of the crosslinks (type 4 in Figure 3.1). The excluded volume condition requiring that fibers do not interpenetrate leads to compressive stresses in the contact and associated repulsive forces between fibers. The contact does not support tension across the contact plane, unless cohesive interactions are present. Cohesion leads to relatively small forces and is less important in networks of athermal fibers of diameter larger than one micron. In such situations, the tensile, contact-opening force that may be carried is much smaller than the compressive, contact-closing force associated with the excluded volume constraint. Cohesion forces that develop at contacts play an essential role in the mechanics of molecular networks and in athermal networks of nanofibers. It should be noted that relative fiber sliding is not prevented by type 4 contacts, but it is fully inhibited by crosslinks of types 1, 2, and 3. The transmission of moments across a contact requires tensile, contact-opening tractions acting across the contact plane. Therefore, moments cannot be transmitted by contacts without cohesion. Likewise, if cohesion is weak, only small moments can be transmitted between the two fibers, and these may be negligible relative to the moments transmitted along the fibers.

In many network materials fibers form bundles, in which fibers are approximately aligned with the bundle axis and form contacts along their length. Two such cases are shown in Figure 3.1. In type 5, the bundle is held together by cohesion forces acting between fibers, while type 6 represents situations in which crosslinks are established between fibers within the bundle. Whether these crosslinks are of welded or rotating type is not important in this case, since relative rotations of the respective fibers are prevented by the bundle topology. Fibers within the bundle may slide relative to each other if no intra-bundle crosslinks are present. Relative sliding may take place with or without friction.

Crosslinks of types 1 and 2 are prevalent in network materials. In molecular networks, such as rubber, epoxy, and other thermosets, molecular strands are connected by crosslinks. While the chemical nature of the crosslinks differs from material to material, their mechanical function is similar. In all cases the crosslinks transmit forces and, if rigid enough in bending, may also transmit moments. In a densely crosslinked network of strands relatively stiff in bending, the moments are important. However, in a sparsely crosslinked, but densely packed network, such as rubber, whether the crosslinks transmit moments is largely inconsequential for network mechanics. It should be observed in this context that, with the exception of swollen gels, molecular networks are densely packed and the free volume is small. Hence, the excluded volume interactions between strands become dominant and the nature of load transmission at crosslinks is less important.

Strong crosslinks hold together cellulose networks, such as paper. Cellulose fibers are hollow and become flattened, or ribbon-like, upon processing. Therefore, the crosslinks in paper are established between ribbon-like fibers that come in contact over an area approximately equal to the square of the ribbon width. These crosslinks transmit forces and moments and can be classified as type 1.

Nonwovens of polymeric fibers also derive their mechanical properties from crosslinks of type 1. A common method used to crosslink nonwovens is thermal bonding. In the industrial process, a nonwoven mat is created by spinning polymeric fibers onto a conveyor belt. The fibers are generally solid by the time they reach the mat, and hence little to no bonding results at fiber contacts. Commercial nonwovens are made from fibers of diameter larger than 10 μm for which cohesion is insufficiently strong to create mechanically effective crosslinks. The mat is then calendered by passing it between two cylinders whose surfaces are covered by a large number of heated punches of a diameter of hundreds of microns. The regions of the mat that come in contact with punches are melted. Upon solidification, these sub-domains become effective crosslinks that connect many fibers (Figure 1.1(b)). The crosslinks are distributed in the plane of the nonwoven mat in a pattern designed to provide the desired stiffness and strength to the network. Mats of polymeric fibers may also be crosslinked by compression at temperatures between the glass transition and the melting temperature in order to favor polymer interdiffusion at contacts between fibers (heat sealing). The resulting crosslinks connect pairs of fibers and are of type 1. The disadvantage of this method is that it produces mats with large crosslink density and reduced overall deformability. Furthermore, the strength of the resulting crosslinks has large variability. These limitations led to the development of the thermal bonding technology using heated punches, which is more controllable and is preferred in industry.

Crosslinks of type 2 are encountered in molecular networks. This model may also represent crosslinks in athermal networks in which the inter-fiber connectors are flexible in bending and torsion. An example is provided by some actin-binding proteins that hold F-actin filaments together in the cytoskeleton. Filamin has a V-shape and binds pairs of F-actin filaments, effectively forming a crosslink between them. The molecule is flexible and imposes weak kinematic constraints on the actin filaments. A type 2 representation for this type of crosslink is adequate.

Contacts (type 4) form between fibers in most networks. The density of contacts increases as the free volume of the network decreases, which happens either upon network compression or due to the compacting action of inter-fiber cohesion (non-bonded interactions). For example, the mechanics of polymeric chains in thermoplastics is controlled by excluded volume interactions. The kinematics of a given molecule is limited by the presence of neighboring molecules, and this restriction becomes more pronounced as the density increases. Such constraints, which primarily limit the motion of a filament in directions perpendicular to its axis, are known as entanglements. They do not prevent the snaking motion of filaments along their contour, known in the polymer physics literature as reptation, which hence becomes the dominant relaxation mode available to molecular filaments in thermoplastics. In the macroscopic world, contacts control the mechanics of woven networks, such as textiles, and of athermal, non-crosslinked nonwovens.

As already mentioned in this section and further discussed in Chapter 10, nanofibers and molecular fibrils tend to bundle under the action of cohesion. Bundling reduces the total free surface in the system, but generally mandates fiber bending. The internal structure of a bundle is not perfect; fibers are not always close packed and do not have their axes exactly parallel to the bundle axis. Packing defects are inherent in such stochastic assemblies of filaments. However, the stronger the inter-fiber attraction, the more aligned the fibers are within the bundle axis. Once formed, a bundle becomes quite resilient against splitting and separation into sub-bundles. If intra-bundle crosslinks form between fibers (type 6), the energetic cost of bundle splitting increases and bundles behave as effective fibers of diameter larger than that of the individual constituent fibers.

Most systems of nanofibers exhibit bundling. F-actin forms networks of large free volume, in which fibers cross at relatively large angles and are crosslinked by actin binding proteins. The actin fibers merge into bundles under stress; such bundles are known as "stress fibers." This dynamic bundling is an effective way by which the network gains the ability to carry large loads. Bundles of microtubules are encountered in neuronal axons. Microtubules are fibers of large persistence length, composed from proteins (tubulin) arranged in an almost perfect spiral pattern. These bundles are held together by protein crosslinkers. Collagen is also organized in bundles with a hierarchical structure (Figures 1.1(d) and 2.3). Nanofibers of nonbiological origin also bundle, as for example polymeric fibers of sub-micron diameter and carbon nanotubes. Cohesion is particularly strong in the case of carbon nanotubes and helps stabilize large bundles of hundreds or thousands of tubes, which then organize on larger scales into networks of bundles known as buckypaper.

3.2   Mechanics of Crosslinks

3.2.1   Mechanical Behavior

The mechanical behavior of crosslinks depends on their actual geometry. In a type 1 configuration, fibers merge or are intimately connected across an interface of in-plane dimensions comparable with the fiber diameter. The interface being well-bonded, the relative displacements and rotations of the two fibers vanish. The specific geometry of the crosslink may influence the local stress distribution (Berhan and Sastry, 2003), but the overall behavior is simple: Displacements and rotations of the fiber pair at the crosslink site are identical, and load transmission is controlled by this kinematic constraint.

If the crosslink is formed by an additional entity present between the respective fibers, the mechanical behavior of the connector controls the crosslink behavior. In molecular networks crosslinks may be introduced by crosslinker molecules. Figure 3.2(a) shows two molecules that may be used to crosslink diglycidyl ether of bisphenol A strands in commercial epoxies. The difunctional extender bisphenol A provides a link between two strands, $z = 2$, while the trifunctional crosslinker 1,1,1-tris(4-hydroxyphenyl)ethane (THPE) creates crosslinks with $z = 3$. These molecules can be used to control the network architecture (Thompson et al., 2009).

A model used occasionally to represent crosslinks of athermal fibers is shown in Figure 3.2(b). It is envisioned that the two fibers in contact are connected by nanofibrils that can be viewed as a set of parallel springs. This allows the fibers to have different displacements and rotations. The crosslink has its own constitutive behavior resulting from the collective deformation of the springs. Rupture of the crosslink happens gradually, by the sequential failure of connectors. This representation is adequate if the elastic deformation of the fiber material in the vicinity of the contact can be neglected relative to the deformation of the "bed of springs." This model has experimental support from direct observations of crosslinks between cellulose fibers (Schmied et al., 2013) and was used to represent crosslinks in paper models (Heyden and Gustafsson, 1998). However, note that cellulose fibers in paper

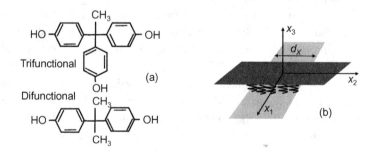

Figure 3.2 (a) Examples of trifunctional and difunctional crosslinkers for diglycidyl ether of bisphenol A strands in epoxies. (b) Schematic of a crosslink of type 1 (Figure 3.1) between two athermal fibers shown by the two rectangles. The crosslink is formed by multiple parallel fibrils connecting the two surfaces in contact.

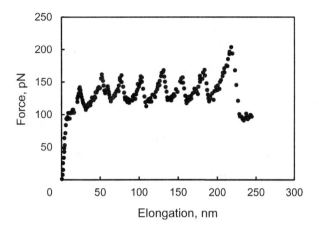

Figure 3.3 Force–stretch curve for filamin, the crosslinker protein of the cytoskeletal F-actin network. Adapted from Schwaiger et al. (2004) with permission from Springer Nature

are held together by van der Waals bonds, H-bonds, and fibrils connecting the fibers in contact, and the relative contribution of these interactions to the mechanics and strength of crosslinks is still a matter of debate.

Type 2 crosslinks provide a less constraining kinematic tie between fibers since the relative rotations are entirely free and no moments are transmitted. If the connector is a single molecule (Figure 3.1), the constitutive behavior of the crosslink is identical to that of the respective molecule. An example is provided in Figure 3.3, which shows the force–stretch response of a filamin molecule measured with an atomic force microscope tip (Schwaiger et al., 2004). This represents the effective constitutive behavior of the crosslinks of F-actin cytoskeletal network. The protein undergoes successive unfolding events during stretching, leading to the serrated force–stretch curve shown in Figure 3.3. The mechanical behavior of a network with this type of compliant crosslinks is different from that of the same network with type 1, welded crosslinks, particularly at larger strains (DiDonna and Levine, 2006; Kasza et al., 2009; Zagar et al., 2015).

In general, if the crosslinks are compliant and undergo large deformations, the behavior of the network is defined by the behavior of the crosslinks. This is due to the fact that random networks tend to deform in the softest mode available to them (Chapter 6). If the crosslinks are rigid, network deformation is supported by the deformation of fibers.

Note that if the crosslinks are not compliant, the mechanical behavior of a fibrous networks with crosslinks of type 1 is not significantly different from that of the same networks with crosslinks of type 2.

3.2.2   Failure

Establishing a criterion for crosslink failure is desirable when studying network rupture and damage accumulation. Crosslink rupture can be described based on the four relative motion modes of the two fibers forming the respective crosslink. Borrowing the nomenclature from Fracture mechanics, relative fiber displacement in the direction normal to the crosslink plane (defined by the tangents to the axes of the two contacting fibers at the crosslink site) leads to mode I opening, while relative sliding in the crosslink plane leads to failure in mode II. The relative rotation of the two fibers about the normal to the crosslink plane at the crosslink site loads the bond in torsion, while rotation of fibers about their own axes leads to peeling and mode I crosslink failure.

Considering these deformation modes, and in the context of the crosslink model shown in Figure 3.2(b), failure can be evaluated based on the effective force which may be approximated by taking into account the relative kinematics of fibers in contact as:

$$F_{Xeq} = \sqrt{F_{X1}^2 + F_{X2}^2 + \left\langle\!\left\langle F_{X3} - \frac{6}{d_X}\sqrt{M_{X1}^2 + M_{X2}^2} \right\rangle\!\right\rangle^2}, \qquad (3.1)$$

where F_{Xi} and M_{Xi} represent forces and moments acting on the crosslink in the coordinate system of Figure 3.2(b), d_X is the characteristic in-plane dimension of the crosslink, and $\langle\!\langle\,\rangle\!\rangle$ denotes here the Macaulay bracket, which vanishes if the quantity in the bracket is negative, and is equal to the respective quantity when it is positive. The term in the bracket accounts for the combined effect of relative fiber rolling and separation in the direction of the crosslink normal, and the negative sign accounts for the fact that separation happens when all fibrils/springs forming the crosslink are ruptured. It is assumed that no crosslink failure is caused by the torsional moment acting in the local x_3 direction. Crosslink failure takes place when

$$F_{Xeq} = f_{Xc}, \qquad (3.2)$$

where f_{Xc} denotes the strength of the crosslink. This model of crosslink failure was used in Deogekar and Picu (2018) and Deogekar et al. (2019). A crosslink failure criterion in terms of the effective applied force is also appropriate for type 2 crosslinks (Figure 3.1). Note that the criterion of Eqs. (3.1) and (3.2) does not predict failure due to the relative fiber rolling and the moment provides only a correction to the total net force applied.

An empirical criterion used to model athermal crosslink rupture in terms of the transmitted moments and forces is given by Heyden and Gustafsson (1998) and Magnusson (2016):

$$\frac{\sqrt{F_{X1}^2 + F_{X2}^2 + F_{X3}^2}}{f_{Xc}} + \frac{\sqrt{M_{X1}^2 + M_{X2}^2 + M_{X3}^2}}{m_{Xc}} = 1. \qquad (3.3)$$

f_{Xc} and m_{Xc} in Eq. (3.3) are the critical failure force and moment of the crosslink. The calibration of this criterion requires the experimental evaluation of both these material parameters, which is not always a straightforward task.

The selection of the crosslink failure criterion depends on the type of crosslink considered. A force-based criterion is adequate for type 2 crosslinks. If fibers are bonded over a larger area, such as in the case of cellulose fibers in paper and of polymeric fibers in nonwovens, crosslink failure is caused by both forces and moments and involves complex processes associated with the nucleation and propagation of a separation front. In these situations, a criterion based on both the maximum force and the total energy required to separate the bond seems more appropriate. A cohesive zone model defined in terms of two parameters – the total energy and the maximum force required to separate the interface – is adequate for this purpose. Cohesive zone representations have been used to represent crosslink failure in networks in Chen and Silberstein (2018) and Borodulina et al. (2016). However, the experimental evaluation of the critical force and crosslink toughness, which allows the calibration of this failure criterion, is difficult.

An athermal crosslink subjected to load control fails once the applied force reaches f_{Xc}, irrespective of how tough it is. In this case, a force-based criterion, such as Eq. (3.2), is adequate. The bond toughness comes into play when loading is performed in displacement control on the crosslink scale. In general, it is difficult to ascertain whether crosslink loading in a stochastic network is of one type or the other.

In molecular networks, crosslinks may exhibit time-dependent behavior and reversibility. In transient polymer networks, bonds such as ionic complexes, hydrogen bonds, and metal ligand coordination complexes have finite lifetimes and may break and reform during loading. Their transient dissociation and reassociation may be triggered by mechanical forces, pH, and temperature. Such transient networks exhibit self-healing and complex rheology, and are currently intensely studied (Kloxin and Bowman, 2013; Balkenende et al., 2014) – see Section 9.5. The frequency at which transient crosslinks separate is typically modeled with an Arrhenius equation:

$$v_X = v_{0X} \exp\left(-\frac{E_X - f_X \delta_X}{k_B T}\right),\tag{3.4}$$

where E_X is the free energy barrier associated with crosslink rupture, f_X is the magnitude of the applied (tensile) force, and δ_X is a characteristic activation length proportional to d_X (Figure 3.2(b)). Inverting Eq. (3.4) leads to a logarithmic dependence of the rupture force on the deformation rate. This provides the thermal counterpart of the athermal failure criterion of Eq. (3.2).

3.2.3 Experimental Evaluation of Crosslink Strength

Direct measurements of the crosslink strength are not straightforward due to the difficulty to isolate crosslinks from networks and to load them in a controlled way. Due to the importance of crosslinks in paper- and cellulose-based products, most studies of this type focus on testing bonds of two cellulose fibers. Magnusson and collaborators (Magnusson and Ostlund, 2013; Magnusson et al., 2013) tested bonds formed by two crossed fibers in which one fiber is directly loaded using a miniature load cell, while the transverse fiber is fixed at both ends. The load is applied in the plane defined by the two fibers, leading to a nominal shear loading of the bond, or perpendicular to the respective plane, causing peeling. This experiment presents two major difficulties. Load transmission from the loading apparatus to the crosslink is mediated by the fibers. Hence, the crosslink is loaded by a compliant mechanism. Furthermore, the crosslink may rotate during loading and its true loading mode is not controlled in this type of test. The second issue was addressed in Magnusson and Ostlund (2013) by using a finite element model of the actual fiber and crosslink geometry to evaluate the local loads applied to the bond at failure. Schmied et al. (2012, 2013) developed a testing method using an atomic force microscope tip which applies force directly to the crosslink. The crosslink strength exhibits high sample to sample variability and, unfortunately, cannot be used to validate any of the crosslink failure criteria discussed in Section 3.2.2. However, it was possible to determine that failure in the peeling mode takes place at smaller forces than in the shear mode.

3.3   Fiber Bundles

Fiber bundles are stabilized by cohesive interactions and/or intra-bundle crosslinks, types 5 and 6 in Figure 3.1. In a network of bundles, the relevant bundle deformation modes amount to two processes: splitting into sub-bundles and sliding of fibers within the bundle. This section presents results needed for the analysis of these deformation modes. The discussion focuses on fibers with circular cross-sections, but the concepts presented apply to bundles of fibers with other cross-section geometries.

3.3.1   Fiber Contacts within Bundles

The geometry of interest is shown in Figure 3.4(a). In the case without cohesion, when the two fibers are pushed against each other by a distributed force P (per unit length of fibers), they establish contact over a rectangular region of width $2a_X$. P is considered positive when pushing the fibers against each other, as shown in Figure 3.4(a). The width of the Hertzian contact depends on the applied force and is given by Johnson (1985):

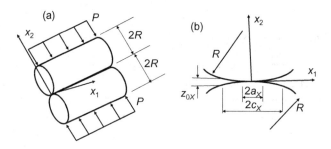

Figure 3.4 (a) Geometry of two fibers with parallel axes in contact. (b) Detail of the contact region.

$$a_X = \sqrt{\frac{4PR_{eq}}{\pi E_{eq}}}, \tag{3.5}$$

where

$$\frac{1}{R_{eq}} = \frac{1}{R_1} + \frac{1}{R_2}, \tag{3.6}$$

and

$$\frac{1}{E_{eq}} = \frac{1 - v_1^2}{E_1} + \frac{1 - v_2^2}{E_2}. \tag{3.7}$$

E_i and v_i, $i = 1, 2$ are the Young's moduli and Poisson ratios of the materials of the two fibers. Equation (3.5) also gives the half width of the contact between a rigid half plane and an elastic cylinder of radius R_{eq} and Young's modulus E_{eq}.

Cohesion is characterized by the work required to separate the unit area of contact between fibers, $\delta\gamma$. This work of cohesion is the difference between the sum of the specific free surface energies of the two fibers and the energy per unit area of the interface. Cohesion holds fibers in contact even when $P = 0$. The contact opens when a negative force P is applied.

Figure 3.4(b) shows a detail of the geometry in Figure 3.4(a). The width of the contact is $2a_X$, but the attractive forces between the two surfaces have a finite range, z_{0X}, which leads to tractions acting on fibers outside the contact area, over a larger width, $2c_X$. The separation at $x_1 = \pm c_X$ is z_{0X}. This contact problem has two limits, one in which $c_X \approx a_X$, and the other in which $c_X \gg a_X$. The first is referred to as the Johnson–Kendall–Roberts (JKR) limit (Johnson et al., 1971), while the second is the Derjaguin–Muller–Toporov (DMT) limit (Derjaguin et al., 1994). They are distinguished by the Tabor parameter (Tabor, 1977; Maugis, 1992), which reads:

$$\lambda_X = \left(\frac{R_{eq}\delta\gamma^2}{E_{eq}^2 z_{0X}^3} \right)^{1/3}. \tag{3.8}$$

The JKR limit is reached for soft bodies with $R_{eq} \gg z_{0X}$, when λ_X is large, while the DMT limit is reached for rigid bodies and small λ_X. Experience indicates that the JKR solution applies with good accuracy for $\lambda_X > 3$. All polymeric fibers used in nonwovens belong to this category (Shi et al., 2012). The transition toward the DMT limit is slow and it is unlikely that network systems of practical interest fall in this category. Consider the contact of fibrils having $R_{eq} = 100$ nm, $E_{eq} = 10$ GPa, interaction range $z_{0X} = 1$ nm and work of cohesion $\delta\gamma = 1$ J/m^2. For these values the Tabor parameter is $\lambda_X = 1$, which is close to the range of validity of the JKR model. Applying these models to contact problems involving molecular fibrils is not appropriate since continuum elasticity does not apply at this scale. Hence, cases in which $R_{eq} \approx z_{0X}$ are outside the range of applicability of this theory.

The relation between the applied force and the contact width in the JKR limit is expressed in terms of the normalized force, $P^* = P/\left(\pi E_{eq}R_{eq}\delta\gamma^2\right)^{1/3}$, and the normalized contact half-width, $a^* = a_X \left(\pi E_{eq} / 8 R_{eq}^2 \delta\gamma \right)^{1/3}$, as (Barquins, 1988; Johnson and Greenwood, 2008):

$$P^* = a^{*2} - 2\sqrt{a^*}. \tag{3.9}$$

Equation (3.5) for the noncohesive Hertzian contact can be written using these normalized quantities as $P^* = a^{*2}$. Equation (3.9) may be used to determine the half width of the JKR contact for the case when no force is applied, which results in $a^* = 2^{2/3}$ or

$$a_X = 2 \left(\frac{4 R_{eq}^2 \delta\gamma}{\pi E_{eq}} \right)^{1/3}. \tag{3.10}$$

In order to break the contact, a pull-out force of magnitude $P = -(3/4)\left(4\pi E_{eq} R_{eq} \delta\gamma^2 \right)^{1/3}$ per unit length of contact must be applied. The contact area at pull-out is finite and is given by $a^* = 2^{-2/3}$.

For values of the Tabor parameter outside the range of validity of the JKR solution, $\lambda_X < 3$, the equivalent of Eq. (3.9) becomes:

$$P^* = a^{*2} - a^* f(\lambda_X), \tag{3.11}$$

where $f(\lambda_X)$ is a function of the Tabor parameter, which is defined in Johnson and Greenwood (2008). In the DMT limit, the pull-out force is $P = -2\pi R_{eq}\delta\gamma$ and the contact width at pull-out vanishes.

Figure 3.5(a) shows the variation of the normalized contact half width with the normalized applied force for the JKR contact ($\lambda_X > 3$) and for a value of the Tabor parameter in the transition to the DMT limit, $\lambda_X = 0.5$. The Hertzian solution, Eq. (3.5), is shown for reference. Figure 3.5(b) shows the characteristic dimensions of the contact, a^* and c^*, function of the Tabor parameter at zero load ($P = 0$) and at pull-out. In the JKR limit, the a^* and c^* curves merge, while as the Tabor parameter decreases, c^* increases rapidly and a^* decreases. At pull-out, a^* becomes zero as the Tabor parameter decreases. Figure 3.5(c) shows the variation of the absolute value of the pull-out force with the Tabor parameter. The critical force gradually decreases as λ_X decreases below the JKR range.

3.3.2   Bundle Splitting in the Presence of Cohesion

The separation of a bundle into two sub-bundles is considered in this section. To begin, consider the situation of Figure 3.6(a) in which two fibers of equal length are pinned at both ends, as shown on the left, and allowed to stick starting from end B. They form a cohesive contact along segment BD. The length y of BD depends on the strength of cohesion and on the fiber bending stiffness. The solution to this problem reads:

$$\frac{y}{L} = \cos\frac{\theta}{2} - \left(\frac{9}{\Psi_{coh}} \sin^2\frac{\theta}{2} \right)^{1/4}. \tag{3.12}$$

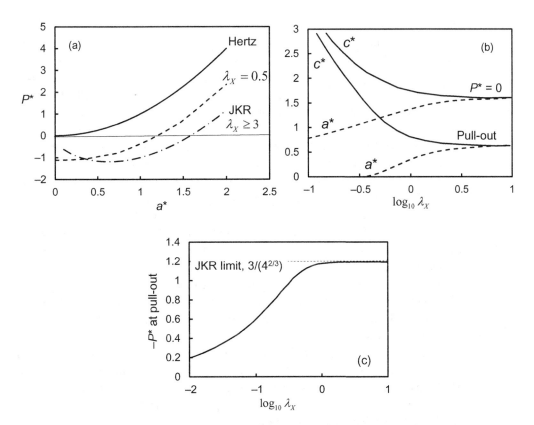

Figure 3.5 (a) Normalized force–contact width for two cylindrical fibers in the configuration of Figure 3.4(a), for the case with no cohesion (Hertz contact) and two cases with cohesion: the JKR limit and a case with Tabor parameter $\lambda_X = 0.5$. The force is positive in compression. (b) Normalized contact widths of the cohesive contact with no applied load and at pull-out, function of the Tabor parameter. (c) Variation of the absolute value of the pull-out force with the Tabor parameter.

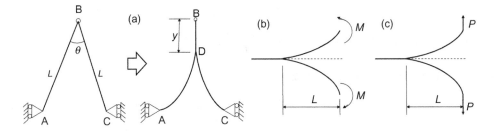

Figure 3.6 Schematics of (a) bundling and (b, c) unbundling problems involving two fibers.

The solution is a function of the geometric parameter θ and a system-specific nondimensional parameter Ψ_{coh} given by:

$$\Psi_{coh} = \frac{\delta \bar{\gamma} L^2}{E_f I_f} = \left(\frac{L}{L_{EC}}\right)^2, \tag{3.13}$$

where $\delta\bar{\gamma} = \delta\gamma 2a_X$ is the work of cohesion per unit length of contact. Ψ_{coh} may be expressed in terms of the elastocapillarity length defined by $L_{EC} = \sqrt{E_f I_f / \delta\bar{\gamma}}$ which expresses the interplay between the bending and surface energies and finds diverse applications in the physics of soft materials. Ψ_{coh} is the ratio of a characteristic geometric length of the problem and the physics-based elastocapillarity length. Cohesion becomes important in the mechanics of the structure if Ψ_{coh} is large. In fact, solutions for the problem shown in Figure 3.6(a) exist ($y \geq 0$) only for large enough Ψ_{coh}.

To see how Ψ_{coh} enters the mechanics of splitting a cohesive contact, consider the configurations shown in Figure 3.6(b) and (c). In both cases, the bundle is composed from two fibers which separate under the action of bending moments (Figure 3.6(b)) and forces (Figure 3.6(c)). The moment and the force leading to the separation of the bundle over a length L in the two configurations are given by $M = \Psi_{coh}^{1/2} E_f I_f / L$ and $P = \sqrt{3}\Psi_{coh}^{1/2} E_f I_f / L^2$, respectively. It results that, in this case too, Ψ_{coh} is the key parameter of the problem. This analysis is similar to the Griffith solution for crack growth in Fracture mechanics.

Since we are primarily interested in the splitting of a bundle of many fibers into two sub-bundles, it is necessary to reconsider the problems shown in Figure 3.6 from this perspective. In a close-packed bundle of n_f fibers with no defects, there are n_c pairwise contacts between fibers, with n_c given by $n_c = 3n_f - \sqrt{12n_f - 3}$ (Harborth, 1974). The first term in this expression represents the number of contacts per fiber in an infinite hexagonal close packed arrangement, while the second term represents the effect of the bundle surface. The total cohesion energy per unit length of the bundle is $n_c \delta\bar{\gamma}$. The work of cohesion for splitting a bundle of $2n_f$ fibers into two sub-bundles each having n_f fibers results:

$$\delta\bar{\gamma}_b = \delta\bar{\gamma}\left(2\sqrt{12n_f - 3} - \sqrt{24n_f - 3}\right) \approx 2\delta\bar{\gamma}\sqrt{n_f}. \qquad (3.14)$$

If fibers are free to slide axially within the bundle, the effective moment of inertia of the bundle of n_f fibers is $I_b = n_f I_f$. If fibers are prevented from sliding axially, $I_b \sim n_f^2 I_f$ for large n_f. Therefore, the effective non-dimensional cohesion parameter for the bundle, Ψ_{coh}^b, becomes:

$$\Psi_{coh}^b = \frac{\delta\bar{\gamma}_b L^2}{E_f I_b} \approx \frac{2}{\sqrt{n_f}} \Psi_{coh}, \qquad (3.15a)$$

for the free axial sliding case, and

$$\Psi_{coh}^b \approx \frac{2}{n_f^{3/2}} \Psi_{coh}, \qquad (3.15b)$$

for the case in which axial sliding is prevented.

3.3.3 Relative Fiber Sliding with Longitudinal Contact

Relative axial displacements of fibers within bundles may take place if axial loads are applied, such as when the bundle is loaded axially in tension or when it is loaded in

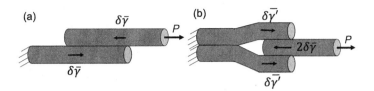

Figure 3.7 Two configurations of fibers in contact undergoing axial relative sliding.

bending. The elementary problem that requires attention in this context is the relative sliding of two parallel fibers in cohesive contact loaded axially (Figure 3.7(a)).

If the interface provides negligible resistance to sliding (this is, for example, the case of nondefective carbon nanotubes in buckypaper), resistance to axial deformation emerges exclusively from the creation of free surface. The energy required to produce a free surface corresponding to the unit length of contact is $\delta\bar{\gamma}$ and hence an effective force of this magnitude acts at sites of free surface creation, as shown in Figure 3.7(a). If a fiber is pulled out from between two stationary fibers, as in Figure 3.7(b), the energy varies since on one side of the moving fiber end there are two interfaces, while only one interface remains on the opposite side. The energy variation in this case is associated with the elimination of interface area as the moving fiber is pulled out, $\delta\bar{\gamma}'$. These forces are important in bundles of nanofibers but are often neglected relative to the other forces involved in bundles of fibers of diameter larger than 1 μm.

In the macroscopic world, the prevalent type of interaction of moving solid objects in contact is Coulomb friction. The friction force is proportional to the load applied normal to the interface and is independent of the nominal contact area and of the relative velocity. This type of friction is caused by the interaction of asperities on the two surfaces. With micron-scale and sub-micron fibers, and in particular when the fiber material is soft, the interaction is better described by a viscous model of friction. In this representation, the interface is viewed as undergoing viscous sliding with a relative velocity proportional to the applied shear stress. In this case, the force required to slide the two objects in contact scales linearly with the contact area. Further, the effective viscosity of the interface may be temperature dependent, which implies that the deformation rate at an imposed shear stress becomes temperature dependent. Yet another model used to represent interfaces is similar to the elastic–perfectly plastic model in plasticity: the opposing surfaces slide once the applied shear stress reaches a characteristic "yield stress," and are elastically coupled at smaller values of the stress. As in the viscous friction case, this model also predicts that the force at which sliding begins scales linearly with the contact area.

Load transmission between fibers sliding relative to each other in the longitudinal direction is described by the shear lag model presented in Section 2.5.3. The constitutive equation used there to describe the interface is linear (Eq. 2.64). This equation may be replaced by the viscous friction representation case in which the distribution of stress and interface tractions remains the same but the magnitude of the stress depends on the deformation rate. The shear lag model may also be used to understand the behavior of the bundle with elastic–perfectly plastic interfaces. In the close-to-rigid

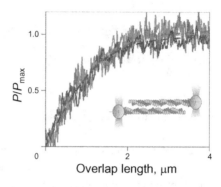

Figure 3.8 Force required for the relative sliding of two F-actin filaments placed in longitudinal contact. Adapted from Ward et al. (2015) with permission from Springer Nature

fiber case, when λ_{sl} is large, the interface is loaded uniformly and sliding takes place all along the interface at any given time. Therefore, if the interface behavior is elastic–perfectly plastic with yield stress τ_0, the pull-out force is proportional to τ_0 multiplied by the total contact area, $2a_X L$. The force scales linearly with the length of the contact between fibers. In the opposite case of relatively compliant fibers and small λ_{sl}, sliding is confined to the regions close to the fiber ends where the shear stress is large. A strong size effect is expected in this situation. If the fiber length L is smaller or comparable to λ_{sl}, the rigid fiber behavior is recovered and the pull-out force scales with L. In the large L limit, sliding takes place by the nucleation of dislocation-like defects representing localized relative sliding, from the loaded shear lag region. After nucleation, the defects propagate along the interface, causing the overall relative slip of the two fibers. In this case, the pull-out force is proportional to the critical stress required for defect nucleation and is independent of the interface length, L. It is important to emphasize that the nucleation and propagation of dislocation-like interfacial defects requires that fibers are axially compliant, so they can support the localization of fiber axial deformation.

This behavior is observed experimentally. Figure 3.8 shows the force required to slide two F-actin filaments relative to each other, as measured using an optical trap-based technique (Ward et al., 2015). The three curves corresponding to different imposed sliding velocities overlap once the force axis is normalized with the maximum force (which depends on the sliding velocity). The curve shows the features predicted by the shear lag model. The pull-out force scales linearly with the overlap length at small overlaps but is independent of this parameter at large overlaps.

3.4   Contact of Nonparallel Fibers

The contact of fibers with nonparallel axes, with and without cohesion, is discussed in this section. The objective is to establish the relation between the work of cohesion,

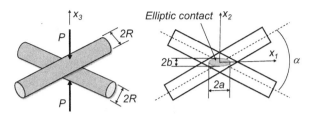

Figure 3.9 Contact between two cylindrical fibers whose axes form an angle α.

the dimensions of the contact area, and the force applied normal to the contact plane. The relative sliding of fibers in the crossed configuration is of interest in some applications and remarks related to this issue are made. Since analytic results for this problem have been derived for cylindrical bodies, only fibers with circular cross-section are considered here.

3.4.1 Contact of Crossed Fibers

Figure 3.9 shows two cylindrical bodies of the same radius in contact, with axes making an angle α. The contact area for this configuration is elliptical, of semiaxes a and b (Johnson, 1985). The size of the contact increases as the force, P, pushing the two fibers against each other increases, but the eccentricity of the contact ellipse, $e = \sqrt{1 - (b/a)^2}$, remains independent of the applied load. The case in which the two fibers forming the contact have identical radii is discussed here, as it is most relevant for networks of fibers. The result that the contact area is elliptical remains valid in the more general case in which the fiber radii are not identical; this configuration is discussed in Johnson (1985).

The ellipticity of the contact area depends on the relative curvature of the two surfaces. The relative surface is defined as the difference between the outer surfaces of the two bodies in contact relative to a frame tied to the contact plane or, in the case of two fibers, tied to the plane $x_1 - x_2$ defined by the axes of the fibers at the contact point (Figure 3.9). The principal curvatures of the relative surface at the contact are the relative curvatures. For the configuration in Figure 3.9, the principal radii of relative curvature can be computed as $R' = R/(1 - \cos\alpha)$ and $R'' = R/(1 + \cos\alpha)$. When the axes of the two fibers are orthogonal, $\alpha = \pi/2$, it results that $R' = R'' = R$ and the contact area becomes circular.

The relation between the applied force and the contact area results (Johnson, 1985):

$$P = \frac{4E_{eq}}{3\sqrt{R'R''}F_1^3(\alpha)}a_X^3,\qquad (3.16)$$

where E_{eq} is the equivalent stiffness given by Eq. (3.7), and $a_X = \sqrt{ab}$. Note that the area of the elliptical contact, πab, can be written as πa_X^2 and, hence, a_X is the effective radius of the contact. The function $F_1(\alpha)$ is shown in Figure 3.10.

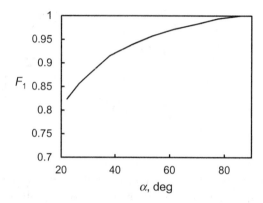

Figure 3.10 Variation of function F_1 of Eq. (3.16) with the angle α formed by the two fibers. Data from Johnson (1985)

If the two fibers are orthogonal, Eq. (3.16) becomes $P = \left(4E_{eq}/3R\right)a_X^3$, which is identical to the relations derived for the contact of two spheres of identical radius, R, and the contact of a sphere of radius R with a half plane. Therefore, the contact problem involving two identical orthogonal cylinders is equivalent to that of two spheres of radius equal to the radius of each of the two cylinders.

In the presence of cohesive interactions, the problem is more complex. Only approximate solutions are available at this time for this problem (Johnson and Greenwood, 2005; Zini et al., 2018). A simple way to address this is to establish a relationship between the cohesive elliptical contact between two cylinders and the cohesive circular contact between two spheres (or a sphere and a flat surface). We review first the exact solution for the circular contact with cohesion. This applies to the contact of two spheres as well as to the contact of two identical cylinders with orthogonal axes.

The relation between the force applied normal to the contact and the contact radius a_X can be written in nondimensional form, in a manner similar to Eq. (3.9). The applied force P is normalized as $P^* = P/(3\pi R \delta \gamma)$, and the contact radius is normalized as $a^* = a_X \left(4E_{eq}/9\pi R^2 \delta \gamma\right)^{1/3}$. The desired relation reads (Johnson and Greenwood, 2005):

$$P^* = a^{*3} - \sqrt{2a^{*3}}. \tag{3.17}$$

This implies that, in the absence of an applied force ($P^* = 0$), the cohesive contact has radius $a^* = 2^{1/3}$. The contact is broken by a pull-out force of magnitude $P^* = -1/2$, which results as the minimum of expression $a^{*3} - \sqrt{2a^{*3}}$. $P^* = -1/2$ is equivalent to $P = -1.5\pi R \delta \gamma$. Note that in the case of the contact between cylinders with parallel axes discussed in Section 3.2.2, P represents the force per unit length of the contact and is measured in N/m, while here P represents the net force applied, measured in N.

It turns out that this solution provides a reasonable approximation for the elliptical contact, provided the ellipticity is moderate, that is, for cases in which the ratio of the

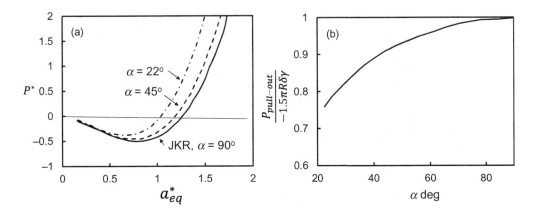

Figure 3.11 (a) Relation between the normalized normal force P^* and the equivalent normalized contact radius a_{eq}^* for several values of the angle between fiber axes. The JKR curve corresponds to Eq. (3.17), with $a^* = a_{eq}^*$ (data from Johnson and Greenwood, 2005). (b) Variation of the normalized pull-out force with angle α (data from Johnson and Greenwood, 2005). The orthogonal cylinders case, $\alpha = 90°$, corresponds to a normalized pull-out force of 1 in this plot.

two principal relative curvatures R'/R'' is smaller than approximately 2. In the case of interest here, in which the contact is established between cylindrical fibers of identical radius, $R'/R'' = (1 + \cos\alpha)/(1 - \cos\alpha)$. It results that Eq. (3.17) applies when $70° < \alpha < 90°$ and provided one replaces R with $\sqrt{R'R''}$ in the normalization factors of the variables appearing in Eq. (3.17), while a_x is interpreted as $\sqrt{ab}$. Further, if no force is applied, the contact area is given by $\pi ab = \pi\left(9\pi R'R''\delta\gamma/2E_{eq}\right)^{2/3}$ and the pull-out force is approximated as $P = -1.5\pi R'R''\delta\gamma$. The pull-out force gradually decreases as the angle α between fibers decreases.

If α is outside the range of validity of this approximation, that is, $\alpha < 70°$, Eq. (3.17) cannot be used to approximate the solution for the cohesive contact of two cylinders. Figure 3.11(a) shows the prediction of the model proposed in Johnson and Greenwood (2005) for the relation between the normalized force P^* and the equivalent normalized contact radius, $a_{eq}^* = \sqrt{ab}\left(4E_{eq}/9\pi R^2\delta\gamma\right)^{1/3}$. The curve labeled $\alpha = 90°$ is the JKR solution for the circular contact of two orthogonal cylinders (Eq. (3.17)). As α decreases, the pull-out force (minimum of P^*) decreases in absolute value, indicating that it is easier to break a cohesive elliptical contact than a circular contact by applying a tensile normal force to it. Furthermore, the contact area in the absence of a normal force, $P^* = 0$, decreases gradually as α decreases.

Figure 3.11(b) shows the normalized pull-out force function of α. The pull-out force is normalized by $-1.5\pi R\delta\gamma$, which is the pull-out force for the orthogonal cylindrical fiber configuration. The figure reinforces the observation that the pull-out force predicted by Eq. (3.17) is an adequate approximation as long as $\alpha > 70°$, and indicates the deviation from this prediction for smaller angles. The pull-out force decreases rapidly as α decreases below $\pi/4$.

A distinct problem is that of inelastic contacts between athermal fibers that undergo plastic deformation. No general solutions similar to the ones discussed in this section for elastic contacts is available for the case in which plastic flow takes place at the contact site. However, a general understanding of the effect of plastic contacts on the mechanics of the network may be obtained by comparison with results presented in Section 6.1.1.3.10 related to compliant crosslinks in networks of elastic fibers. The fact that plastic deformation takes place at contacts is expected to be important only during the compression of fiber masses, at high degrees of compaction, when the density of contacts is large.

3.4.2   Sliding at Contacts of Crossed Fibers

Relative sliding of fibers (as in Figure 3.9) takes place during the deformation of fiber networks and makes an important contribution to the overall stress and the constitutive behavior of the fibrous assembly; see Section 9.4.2. Two types of contact friction can be envisioned. Friction is expected to be Coulombic in the case of fibers of diameter larger than 10 μm. In this case, the friction force depends only on the force applied normal to the contact plane and is independent of the area of the contact. This is expected to be the case in textiles and in nonwovens of large diameter polymeric fibers.

The friction force is expected to scale linearly with the contact area for fibers of sub-micron and micron diameter, cases in which friction is of viscous type. Fibers of such a small diameter also have small bending rigidity, $E_f I_f$, and, for given cohesion strength $(\delta\gamma)$, the value of Ψ_{coh} (Eq. (3.13)) is large. Therefore, cohesion plays a central role in the mechanics of these networks.

Most contacts in random networks are established between fibers that form angles α other than $\pi/2$. As discussed in Section 3.4.1, the contact area decreases monotonically with α, at any given applied force. Therefore, a broad distribution of frictional forces is expected at contacts in a fibrous assembly.

The measurement of friction forces at contacts of isolated pairs of fibers (Das and Chasiotis, 2020) meets with the same challenges encountered when attempting to evaluate the strength of crosslinks (Section 3.2.3). The difficulty associated with handling very thin fibers is obvious. In addition, since contact loading is intermediated in experiments by flexible fibers, unstable behavior is promoted. To measure the real behavior of the interface, rigid loading of the contact is needed, as realized, for example, with the surface force apparatus (Israelachvili, 1987; Parker et al., 1989).

References

Balkenende, D. W. R., Coulibaly, S., Balog, S., et al. (2014). Mechanochemistry with metallo-supramolecular polymers. *J. Am. Chem. Soc.* **136**, 10493–10498.

Barquins, M. (1988). Adherence and rolling kinetics of a rigid cylinder in contact with a natural rubber surface. *J. Adhesion* **26**, 1–12.

Berhan, L. & Sastry, A. M. (2003). On modeling bonds in fused porous networks: 3D simulations of fibrous-particulate joints. *J. Comp. Mater.* **37**, 715–740.

Borodulina, S., Motamedian, H. R. & Kulachenko, A. (2016). Effect of fiber and bond strength variations on the tensile stiffness and strength of fiber networks. *Int. J. Sol. Struct.* **154**, 19–32.

Chen, N. & Silberstein, M. N. (2018). Determination of bond strengths in non-woven fabrics: A combined experimental and computational approach. *Exp. Mech.* **58**, 343–355.

Das, D. & Chasiotis, I. (2020). Sliding of adhesive nanoscale polymer contacts. *J. Mech. Phys. Sol.* **140**, 103931.

Deogekar, S., Islam, M. R. & Picu, R. C. (2019). Parameters controlling the strength of random fiber networks. *Int. J. Sol. Struct.* **168**, 194–202.

Deogekar, S. & Picu, R. C. (2018). On the strength of random fiber networks. *J. Mech. Phys. Sol.* **116**, 1–16.

Derjaguin, B. V., Muller, V. M. & Toporov, Y. P. (1994). Effect of contact deformations on the adhesion of particles. *Prog. Surf. Sci.* **45**, 131–143.

DiDonna, B. A. & Levine, A. J. (2006). Filamin cross-linked semiflexible networks: Fragility under strain. *Phys. Rev. Lett.* **97**, 068104.

Harborth, H. (1974). Lösung zu Problem 664A. *Elem. Math.* **29**, 14–15.

Heyden, S. & Gustafsson P. J. (1998). Simulation of fracture in a cellulose fibre network. *J. Pulp Paper Sci.* **24**, 160–165.

Israelachvili, J. (1987). Direct measurements of forces between surfaces in liquids at the molecular level. *Proc. Natl Acad. Sci. USA* **84**, 4722–4724.

Johnson, K. L. (1985). *Contact mechanics*. Cambridge University Press, Cambridge.

Johnson, K. L. & Greenwood, J. A. (2005). An approximate JKR theory for elliptical contacts. *J. Phys. D* **38**, 1042–1046.

Johnson, K. L. & Greenwood, J. A. (2008). A Maugis analysis of adhesive line contact. *J. Phys. D* **41**, 155315.

Johnson, K. L., Kendall, K. & Roberts, A. D. (1971). Surface energy and the contact of elastic solids. *Proc. R. Soc. London A* **324**, 301–313.

Kasza, K. E., Koenderink, G. E., Lin, Y. C., et al. (2009). Nonlinear elasticity of stiff biopolymers connected by flexible linkers. *Phys. Rev. E* **79**, 041928.

Kloxin, C. J. & Bowman, C. N. (2013). Covalent adaptable networks: Smart, reconfigurable and responsive network systems. *Chem. Soc. Rev.* **42**, 7161–7173.

Magnusson, M. S. (2016). Investigation of interfibre joint failure and how to tailor their properties for paper strength. *Nord. Pulp Pap. Res. J.* **31**, 109–122.

Magnusson, M. S. & Ostlund, S. (2013). Numerical evaluation of interfibre joint strength measurements in terms of three-dimensional resultant forces and moments. *Cellulose* **20**, 1691–1710.

Magnusson, M. S., Zhang, X. & Ostlund, S. (2013). Experimental evaluation of the interfibre joint strength of papermaking fibres in terms of manufacturing parameters and in two different loading directions. *Exper. Mech.* **53**, 1621–1634.

Maugis, D (1992). Adhesion of spheres: The JKR–DMT transition using a Dugdale model. *J. Colloid Interface Sci.* **150**, 243–269.

Parker, J. L., Christenson, H. K. & Ninham, B. W. (1989). Device for measuring the force and separation between two surfaces down to molecular separations. *Rev. Sci. Instrum.* **60**, 3135–3138.

Schmied, F. J., Teichert, C., Kappel, L., et al. (2013). What holds paper together: Nanometre scale exploration of bonding between paper fibres. *Sci. Reports* **3**, 2432.

Schmied, F. J., Teichert, C., Kappel, L., Hirn, U. & Schennach, R. (2012). Joint strength measurements of individual fiber–fiber bonds: An atomic force microscopy based method. *Rev. Sci. Instrum.* **83**, 073902.

Schwaiger, I., Kardinal, A., Schleicher, M., Noegel, A. A. & Rief, M. (2004). A mechanical unfolding intermediate in an actin-crosslinking protein. *Nat. Struct. Mol. Biol.* **11**, 81–85.

Shi, Q., Wong, S. C., Ye, W., et al. (2012). Mechanism of adhesion between polymer fibers at nanoscale contacts. *Langmuir* **28**, 4663–4671.

Tabor, D. (1977). Surface forces and surface interactions. *J. Colloid Interface Sci.* **58**, 2–13.

Thompson, Z. J., Hillmyer, M. A., Liu, J., et al. (2009). Block copolymer toughened epoxy: Role of cross-link density. *Macromolecules* **42**, 2333–2335.

Ward, A., Hilitski, F., Schwenger, W., et al. (2015). Solid friction between soft filaments. *Nature Mater.* **14**, 583–588.

Zagar, G., Onck, P. R. & van der Giessen, E. (2015). Two fundamental mechanisms govern the stiffening of cross-linked networks. *Biophys. J.* **108**, 1470–1479.

Zini, N. H. M., de Rooij, M. B., Fadafan, M. B. A., Ismail, N. & Schipper, D. J. (2018). Extending the double-Hertz model to allow modeling of an adhesive elliptical contact. *Tribol. Lett.* **66**, 30–43.

4 Network Structure and Geometric Parameters

4.1 Classification of Networks Based on Their Structure

The objective of this chapter is to introduce quantities and parameters used to describe the structure of network materials. The need for such a description becomes obvious when working with stochastic networks, as in such cases the meaning of "network structure" is not as obvious as in the case of periodic networks.

The development of new materials is generally driven by the relation between material structure and properties. While the structure can be defined in simple terms when the constituent entities – atoms, molecules, or fibers – are periodically arranged in space, the situation is more complex when these are stochastically distributed. Examples of materials with stochastic structure include metallic and polymeric glasses and melts, classical composites with stochastic microstructure, granular media, and, of course, network materials. In such cases, the relevant structural parameters of importance for the system-scale behavior are generally identified empirically, based on generic arguments. In principle, if a material with stochastic structure is described in terms of a finite set of stochastic variables, the geometry should be defined exclusively in the space of these stochastic variables, which also become the relevant parameters to be placed in relation with the system-scale behavior. Further, it is of practical importance to empirically associate the structural stochastic variables with the parameters of the material production process. If this is possible, material structure may be controlled during fabrication or experiment. While this may be feasible for periodic structures with simple geometry, it is not possible to demonstrate that any given finite (and, desirably, small) set of parameters provides a full description of a stochastic material structure. With this caveat in mind, we discuss in this chapter the most commonly used descriptors of structure for network materials, relevant for the mechanical and transport functions of the network.

In order to place the discussion in perspective, it is necessary to begin with a classification of networks based on structural/geometric considerations. Figure 4.1 shows a classification of networks in broad categories. It differentiates between periodic (P) and stochastic (S) structures. Periodic networks exhibit a motif which is repeated such to fill the embedding space. The network geometry may be defined in this case using geometric descriptors of the unit cell and of the operations used to tile the space, such as translations and rotations. Stochastic networks exhibit random arrangement of fibers, random crosslinking (or no crosslinking at all), and stochastic fiber orientations.

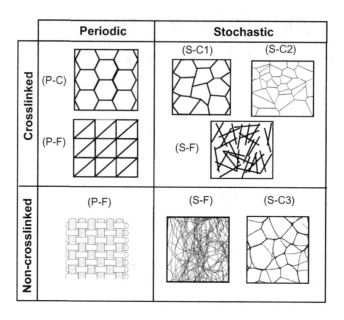

Figure 4.1 Classification of network architectures.

The classification differentiates further between cellular (S-C) and fibrous (S-F) networks. Cellular networks include several categories: (S-C1) networks obtained from a periodic structure whose crosslink positions are perturbed, (S-C2) cellular networks of Voronoi type resembling open cell foams, and (S-C3) cellular networks resulting from the self-organization of filaments of non-crosslinked assemblies in the presence of inter-filament adhesion (Chapter 10). In cellular networks of type S-C1 and S-C2, fibers are crosslinked at the ends, while in the S-C3 case no crosslinks are present and the cellular structure is entirely stabilized by adhesive interactions within bundles and/or intra-bundle crosslinks (e.g., collagen networks, buckypaper). Fibrous networks (S-F) are composed from fibers which are crosslinked or make contacts with other fibers at more than two locations along their length. The majority of network materials are fibrous, of S-F type: the cytoskeleton, nonwovens of various types, paper and cardboard, insulation materials made from entangled fibers, etc. Materials resembling open cell foams represent the most important member of the S-C family. Figure 4.1 indicates that periodic structures may also be classified as cellular (P-C) and fibrous (P-F). In fibrous networks each fiber has multiple crosslinks, while in cellular crosslinked networks each fiber has two crosslinks, at the two ends (dangling ends are excluded from this discussion). If non-crosslinked, the network integrity is preserved by adhesive, frictional, or topological (excluded volume) interactions between fibers. Woven fabrics are examples of periodic fibrous networks with no crosslinks. Stochastic mats of entangled filaments are typical examples of fibrous non-crosslinked network materials. In all these cases, the constituent filaments or fibers may be either straight or tortuous.

In the remainder of this chapter we introduce key parameters describing these structures. Although the focus here is on stochastic networks (due to their far larger practical importance), all parameters discussed apply equally to periodic networks. These parameters may be divided into two categories. Those that may be controlled in the process of network production are placed in the "primary" category, which includes the fiber and crosslink densities, fiber orientation, and network connectivity. Parameters that result from the network generation process are considered "secondary," as these are not directly controlled and depend indirectly on the primary parameters. The mean segment length, fiber tortuosity, flocculation, and network porosity are secondary parameters.

4.2   Fiber-Scale Parameters

4.2.1   Density and Fiber Orientation

4.2.1.1   Density

The most important primary parameter is the network density, ρ, defined as the total length of fiber per unit volume (or area, in 2D) of the network material. The density has units of inverse length square in 3D and of inverse length in 2D. In the quasi-2D case of fibrous mats or membranes, both the 3D density, ρ, and the density in the 2D projection of the 3D structure onto the plane of the mat, ρ_{2D}, are used. In this case, ρ_{2D} is useful since its use in place of ρ eliminates the need to specify the mat thickness. The density is typically controlled during network production and may be evaluated by measuring the network volume fraction or the network mass density.

The network volume fraction represents the fraction of the total material volume occupied by the network, and can be evaluated as:

$$f = \rho A_f, \tag{4.1}$$

where A_f is the cross-sectional area of fibers. It may also be evaluated as $f = \rho_m/\rho_f$, where ρ_f is the density of the fiber material and ρ_m is the network mass density. These relations apply for the case in which all fibers of the network have the same properties (A_f and ρ_f). If this is not the case, Eq. (4.1) becomes $f = \rho\langle A_f\rangle$.

4.2.1.2   Crosslink Density

The crosslink number density, ρ_b, is defined as the number of crosslinks per unit volume, in 3D, or per unit area, in 2D. In the case of non-crosslinked networks, ρ_X is used to represent the number density of inter-fiber contacts. The density of contacts in an unloaded network results from the geometric analysis presented in Section 4.2.3 in terms of the network density, ρ, and the fiber diameter, d. ρ_X changes significantly during deformation, while ρ is constant. In cases in which the network is sparsely crosslinked, contacts form between fibers at sites other than the crosslinks. In such situations, it is necessary to distinguish between the number density of crosslinks, ρ_b, and the number density of contacts, ρ_X, which motivates the use of two distinct parameters.

4.2.1.3   Fiber Geometric Parameters

Fibers are described by their diameter, d, and length, L_0. In 2D cases, or when projecting a 3D network onto a plane, the width of the fiber in projection is also denoted by d. If fibers are not straight, the tortuosity can be defined with the crimp parameter of Eq. (2.9), c, or by using the persistence length, L_p, defined in Section 2.4.2 and Eq. (2.49).

The persistence length is generally a better measure of tortuosity since the crimp parameter depends on the scale of observation. For example, if the filament is a random walk of n steps each of length L_p, the contour length is given by $L_p n$, while the end-to-end length is proportional to $L_p \sqrt{n}$, and hence $c \sim 1/\sqrt{n}$. In this example, the tortuosity is described on scales larger than L_p by the scaling properties of the walk and not by parameters L_p or c. This happens because the random walk has waviness of all length scales (it is a fractal) and no scale separation exists between the "characteristic length scale of tortuosity" and the probing length scale on which crimp is measured. A simpler example in which such scale decoupling exists and the crimp parameter becomes an intensive quantity is that of a thread in a woven fabric. In this case, the longest wavelength describing the fiber shape is equal to the periodic length of the fabric. Crimp becomes independent of the probing length on scales larger than the fabric periodicity. Scale separation generally exists in the case of athermal filaments with stochastic configurations, and L_p may be used to describe tortuosity in such situations, provided $L_0 > L_p$.

4.2.1.4   Orientation

Fibers may be preferentially oriented, and the degree of orientation may be either controlled in the network production process (e.g., in nonwoven production, by adjusting the speed of the substrate on which fibers are deposited relative to the spinneret) or measured for given structure (e.g., by evaluating the network birefringence (Quinn and Winkelstein, 2008; Chen et al., 2016)). The orientation tensor (or structure tensor) is used to describe this aspect of network geometry. Consider a set of straight fibers whose directions in space are defined by the unit vectors $\mathbf{N}^i$, of components $\left\{N^i_j\right\}, j = 1 \ldots 3$. The structure tensor is defined as:

$$\mathbf{A} = \left\langle \mathbf{N}^i \otimes \mathbf{N}^i \right\rangle_i, \tag{4.2}$$

where the average of the tensor product is performed over all i fibers in the set. Tensor $\mathbf{A}$ is symmetric by construction and the sum of the diagonal components $(\mathrm{tr}(\mathbf{A}))$ is 1. The diagonal components of $\mathbf{A}$ represent the system average of the square of the cosines of angles made by fibers with the three coordinate axes, respectively. The structure tensor becomes $\mathbf{A} = 1/3\mathbf{I}$ if fibers are randomly oriented in 3D. The eigenvectors of $\mathbf{A}$ indicate the principal axes of orientation, which are also the axes of the ellipsoid representing the orientation distribution function. The ratio between the largest and smallest of the eigenvalues of $\mathbf{A}$ (all eigenvalues are real since the tensor is symmetric) may be used to characterize the degree of orientation.

If only the preferential orientation relative to a specific axis, $\mathbf{w}$, is needed, the scalar parameter

$$P_2 = \frac{1}{2}\left(3\langle \cos^2\theta^i \rangle_i - 1\right) \tag{4.3}$$

can be used, where θ is the angle between $\mathbf{N}^i$ and $\mathbf{w}$, and since both $\mathbf{w}$ and $\mathbf{N}^i$ have unit length, $\cos\theta^i = \mathbf{N}^i \cdot \mathbf{w}$. P_2 is zero when fibers are randomly oriented, and 1 and –0.5 when they are all parallel or perpendicular to $\mathbf{w}$, respectively. In 2D, the expression equivalent to Eq. (4.3) is $P_2 = 2\langle \cos^2\theta^i \rangle_i - 1$, and P_2 takes values of 0, 1 and, –1 for random, parallel, and perpendicular arrangements relative to axis $\mathbf{w}$, respectively.

If fibers are tortuous, one may evaluate the orientation tensor (or P_2) on multiple scales. For example, one may consider $\mathbf{N}^i$ either as the fiber tangent unit vector or as the unit vector of the end-to-end vector of the fiber. In general, in networks of tortuous fibers the values of P_2 and $\mathbf{A}$ depend on the length scale of observation. The orientation appears more pronounced when probed on larger length scales. Therefore, different fiber ensembles may be compared directly provided the orientation is measured on the same scale.

4.2.2 Network Connectivity

Crosslinked networks may be regarded as graphs with specified position of the nodes (crosslinks, in the present terminology). In graph theory only the connectivity of the network matters. The connectivity is described by the connectivity (or adjacency) matrix, $\mathbf{\Omega}$, which is a square matrix of number of columns and lines equal to the number of nodes in the network, n_b. $\Omega_{ij} = 1$ if node i is connected to node j and $\Omega_{ij} = 0$ otherwise. Hence, $\mathbf{\Omega}$ is a symmetric matrix.

The sum of all entries on line or column i represents the connectivity z_i of node i, that is, $z_i = \sum_{j=1}^{n_b} \Omega_{ij}$. z is the number of fiber segments meeting at a crosslink. The average connectivity becomes:

$$\langle z \rangle = \frac{1}{n_b} \sum_i^{n_b} \sum_j^{n_b} \Omega_{ij}. \tag{4.4}$$

Network materials exhibit relatively simple connectivity compared with the connectivity of random generic graphs and social networks. Cellular networks have $\langle z \rangle \approx 3$ in 2D and $\langle z \rangle \approx 4$ in 3D (see Figure 4.1). The approximate sign accounts for the presence of free boundaries and defects, where the connectivity is lower than the nominal value. Higher connectivity is possible in periodic networks (e.g., the cross-linked P-F example shown in Figure 4.1), but such situations are not common in biology or engineering. In computer simulations, one may generate cellular networks of Voronoi type which have $z = 3$ in 2D and $z = 4$ in 3D, or Delaunay networks – the dual of the Voronoi networks – which have larger z values.

Crosslinks in fibrous networks have $z = 4$, in general. Such crosslinks result by the interpenetration or the contact of two fibers that form a crossed configuration. If fibers are of finite length, crosslinks with $z = 3$ and $z = 2$, corresponding to T-like and L-like junctions, respectively, appear with low, but non-zero probability (Zagar et al., 2011); see also Eq. (6.8). Hence, in fibrous networks $\langle z \rangle \leq 4$. In the cytoskeleton,

$z = 3$ at F-actin branching points, but $z = 4$ at locations where F-actin strands are crosslinked by actin-binding proteins. Collagen fibers of the extracellular matrix organize in bundles and the bundles split and merge, similar to the situation of adhesion-stabilized networks of fiber bundles shown schematically as S-C3 in Figure 4.1, with the bundle connectivity being $z = 3$. Networks of CNT bundles are adhesion-stabilized and exhibit a similar structure with $z = 3$ at bundle splitting/ merging points. In nonwovens, continuous fibers come in contact and may develop inter-fiber bonds, which become effective crosslinks with $z = 4$. This is also the case in woven structures in which bundles/yarns develop pairwise contacts.

In a graph, the connectivity defines the "length of paths." The path length between two nodes is defined as the shortest path in the graph linking the respective nodes. The "diameter" of the graph is the longest path connecting any two nodes of the graph. In other words, it is the longest shortest path connecting any two nodes. This parameter is expressed in terms of the number of segments traversed.

Random graphs of n_b nodes are defined by a probability p to have a connector between a given pair of nodes. It was shown that in these networks the connectivity z is Poisson distributed in the limit of large n_b, and that the graph diameter scales as $\ln(n_b)$ (Newman, 2018). This weak divergence of the graph size with n_b causes the "small world" effect in social, transport, and information networks, which indicates that a small number of steps has to be taken to reach any point in the graph starting from any node. This is a result of the presence of a large number of nodes with high connectivity. In physical networks encountered in network materials, the graph diameter scales as $n_b^{1/D}$, where D is the dimensionality of the embedding space of the network. This puts stochastic physical networks in sharp contrast with random graphs (Erdős and Rényi, 1959).

4.2.3   Mean Segment Length

The distance between successive crosslinks or contacts measured along the contour of a given fiber is referred to as the segment length, l. The mean segment length, l_c, is one of the most important length scales of the network. Since it is not directly controlled in the process of network generation, it is a secondary parameter. The relation between l_c and the primary parameters described in Sections 4.2.1 and 4.2.2 is discussed next.

4.2.3.1   Two-Dimensional Networks
4.2.3.1.1   *Fibrous Networks*

The simplest representation of a 2D network is the Mikado model (shown as an example of stochastic crosslinked fibrous networks, S-F, in Figure 4.1). In this model, straight fibers of length L_0 are deposited in a 2D domain with random orientations and random positions of their centers of mass. The number of fibers per unit area of the model is $\rho_\# = \rho/L_0$. For the purpose of the derivation presented next, we denote the total number of fibers in the network by N_f and the area of the network by A. Fibers cross and are crosslinked at random locations and, if the density is larger than the

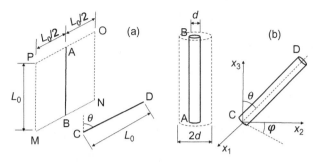

Figure 4.2 Schematics representing the procedure used to evaluate the fiber excluded volume for (a) 2D and (b) 3D networks.

percolation density (see Section 4.5 for specifics), a network forms. We are interested in evaluating the mean segment length resulting from this Poisson process.

To this end, we follow the method used by Kallmes and Corte (1960) and begin by evaluating the probability of crossing of two fibers, AB and CD, that make an angle θ with each other (Figure 4.2(a)). Fiber CD intersects AB if its center of mass is located within the parallelogram MNOP shown by the dashed line in Figure 4.2(a). The probability of crossing is given by the ratio of the area of MNOP to the total area of the network, that is, $(L_0^2/A)\sin\theta$. The probability that a fiber of any orientation intersects AB is given by the integral:

$$\int_0^\pi \frac{L_0^2 \sin\theta}{A} p(\theta)d\theta = \frac{1}{\pi}\int_0^\pi \frac{L_0^2 \sin\theta}{A} d\theta = \frac{2}{\pi}\frac{L_0^2}{A}. \tag{4.5}$$

The number of pairs of fibers in the network is $N_f(N_f - 1)/2$, and the total number of crossings is $N_f(N_f - 1)L_0^2/(\pi A) \approx N_f^2 L_0^2/(\pi A)$. Therefore, the number of crossings per fiber is $2N_f L_0^2/(\pi A)$, and the mean segment length results:

$$l_c = \frac{\pi}{2}\frac{A}{N_f L_0} = \frac{\pi}{2\rho}, \tag{4.6}$$

which is known as the Kallmes–Corte relation (Kallmes and Corte, 1960).

The Kallmes–Corte relation remains valid if the fibers are not straight. Numerical results supporting this statement are presented in van Dillen et al. (2008).

This type of relation between l_c and ρ is actually expected based on the observation that the only length in the problem is $1/\rho$ and, hence, by necessity, $\rho l_c = q$, where q is a numerical constant which cannot be obtained based on a simple dimensional argument and results from the detailed derivation leading to Eq. (4.6). As apparent from the examples discussed next, constant q is related to the specific network geometry considered.

Consider now that fibers have non-zero width, d, and appear as rectangles in the plane of the network. This model was used to represent the microstructure of paper. It is useful in these situations to introduce a new measure similar to the density, known as "coverage," g, which represents the expected number of fibers covering a point of the 2D network domain. The coverage may also be evaluated as the total area associated with fibers divided by the area of the network. The mean coverage is:

$$\langle g \rangle = \rho d, \tag{4.7}$$

and the distribution of coverage is of Poisson type and is given by:

$$p(g) = \frac{\langle g \rangle^g \exp\left(-\langle g \rangle\right)}{g!}. \tag{4.8}$$

Since g is always a positive integer, Eq. (4.8) is a discrete Poisson distribution. However, $\langle g \rangle$ is, in general, a real value. The variance of the coverage is equal to the mean, $\langle g \rangle$.

The mean segment length can be computed following a procedure similar to that used to derive the Kallmes–Corte relation and results in

$$l_c = \frac{\pi}{2} \frac{d}{\langle g \rangle}. \tag{4.9}$$

With Eq. (4.7), it is immediately obvious that Eq. (4.9) reduces to Eq. (4.6), which indicates that the mean segment length is independent of the fiber width (Miles, 1964).

The distribution of segment lengths, $p(l)$, is also of Poisson type, which is to be expected considering that the network is generated using an uncorrelated Poisson process:

$$p(l) = \frac{1}{l_c} \exp\left(-\frac{l}{l_c}\right). \tag{4.10}$$

This distribution is characterized by one parameter – its mean, l_c – and has variance l_c^2. Note the difference between the discrete distribution of Eq. (4.8) and the continuous distribution of Eq. (4.10). The continuous distribution of segment lengths is wider.

In this discussion, fibers are assumed to be randomly oriented. However, preferential orientation is observed in many physical networks and it is necessary to inquire to what extent preferential fiber orientation modifies the distribution of segment lengths and the mean segment length. Furthermore, the above derivation does not take into account the excluded volume of fibers of finite diameter. In this representation, fibers may be placed in any position in the 2D domain and intersect, forming arbitrarily short segments. This implies that the distribution of Eq. (4.10) remains valid even in the limit of very short segments, $l \rightarrow 0$. Clearly, this is not the case in realistic networks in which the distance between contacts along a given fiber cannot be smaller than one fiber diameter. Excluded volume leads to a modification of the segment length distribution for small l values, while the exponential decay of Eq. (4.10) is recovered at larger l. These effects are encountered in both 2D and 3D networks and are discussed further in Sections 4.2.3.3 and 4.2.3.4.

4.2.3.1.2 *Cellular Networks*

In cellular networks each fiber has exactly two crosslinks. The formation of a crosslink does not depend on the probability of intersection of fibers, as is the case with stochastic fibrous networks. Therefore, the mean segment length and the fiber diameter are independent parameters. In fact, one may envision an extreme case in which the fiber diameter becomes arbitrarily small and the statistics of segment lengths is not perturbed. It results that the only relevant length scale of the problem is $1/\rho$, which may also be considered the unit of length. Therefore, the following relation must hold:

$$\rho l_c = const. \tag{4.11}$$

The constant present on the right side of Eq. (4.11) depends on the structure of the network.

In the cellular materials literature, it is common to construct open cell foams using the Voronoi algorithm. In this procedure, a number of seed points, equal to the number of desired cells are distributed randomly in the problem domain and are used to create a Voronoi tessellation. The edges of the tessellation are retained as fibers. In 2D the average number of edges per cell is six. The number density of seed points, ρ_{seed}, is related to the mean segment length as $l_c = 4/(6\sqrt{\rho_{seed}})$ (Meijering, 1953). The combination of this expression with Eq. (4.11) establishes the relation between the seed number density and the mean segment length in 2D. The constant in Eq. (4.11) for Voronoi networks of straight fibers is $4/3$. However, this value depends on the exact geometry of the network and, hence, any significant geometric modification also entails the modification of the constant.

4.2.3.2 Three-Dimensional Networks
4.2.3.2.1 Fibrous Networks

Mikado networks of fibers of zero diameter cannot be created in three dimensions using the procedure described for 2D since the probability of crossing of zero thickness lines of arbitrary orientations in space vanishes. The more physically meaningful problem of fibers of finite diameter, d, must be considered in 3D. The evaluation of the number of contacts per fiber and of the mean segment length follows a similar path with that outlined in Section 4.2.3.1 for the equivalent 2D problem. Specifically, consider straight fibers of length L_0 placed with random orientations and random positions of their centers of mass in a volume V, which becomes the volume occupied by the network. Fibers are considered slender, that is, $L_0 \gg d$. Figure 4.2(b) shows two such fibers, AB and CD, whose relative orientation is defined by the Euler angles θ and φ. They cross if fiber CD penetrates the cylindrical volume shown with a dashed line around fiber AB, also known as the excluded volume of fiber AB. The probability of crossing is given by the ratio of this volume to the total network volume, that is, $(2dL_0^2/V) \sin\theta$. Note that the excluded volume includes the physical volume of fiber AB, since the two fibers are allowed to interpenetrate in this process. The probability that a fiber of any orientation intersects AB is given by the integral:

$$\int_0^{2\pi} \int_0^{\pi} \frac{2dL_0^2 \sin\theta}{V} p(\theta, \varphi) \sin\theta \, d\theta \, d\varphi = \frac{\pi}{2} \frac{dL_0^2}{V}, \tag{4.12}$$

where the orientation probability of fiber CD, $p(\theta, \varphi)$, was considered isotropic (equal probability for all orientations), that is, $p(\theta, \varphi) = 1/4\pi$. The total number of crossings in the network results in $\pi N_f^2 dL_0^2/(4V)$, and the number of crossings per fiber is $\pi N_f dL_0^2/(2V)$. The mean segment length reads

$$l_c = \frac{2}{\pi} \frac{V}{N_f dL_0} = \frac{2}{\pi} \frac{1}{\rho d}. \tag{4.13}$$

While the inverse scaling of l_c with the fiber volume fraction ($l_c \sim 1/\rho d \sim d/f$) is quite general, the $2/\pi$ factor is associated with the specific geometry considered: a network of randomly oriented slender fibers. A different method to generate the network which, for example, would constrain the relative positions of fibers should lead to another constant, but to the same dependence of l_c on ρd. Examples supporting this statement are provided in the remainder of this section, for example, Eq. (4.17).

The arguments presented in this section are purely geometric. Topology controls the formation of contacts in real networks only in the idealized unloaded state. Once load is applied (such load may be as small as the weight of the respective fibrous mass), many of the contacts identified through this geometric analysis disappear, since they become incompatible with the specific boundary conditions applied and the associated network deformation. New contacts that transmit loads between fibers form. It is not immediately obvious that the statistical results discussed here remain valid in these conditions. This issue is further discussed in Section 6.2.1, where it is shown that the scaling of Eq. (4.13) remains valid during the initial stages of loading, but as the network continues to deform, relation $\rho l_c = const$, Eq. (4.13), changes to $\rho l_c^2 = const$, Eq. (6.28). This change of the scaling is not emphasized in the literature on network materials.

4.2.3.2.2 *Cellular Networks*

The considerations presented in Section 4.2.3.1 for 2D cellular networks remain valid for 3D networks of cellular type. The relevant length scale of the 3D problem is $1/\sqrt{\rho}$, and Eq. (4.14) must hold:

$$\rho l_c^2 = const. \tag{4.14}$$

As in the 2D case, the constant in Eq. (4.14) depends on the structure of the network. In the case of Voronoi networks of straight fibers, the constant may be determined numerically and takes the value of approximately 1. The relation between the density of seeds and the network density for 3D Voronoi networks was determined analytically (Meijering, 1953) and reads:

$$\rho = 5.83\rho_{seed}^{2/3}. \tag{4.15}$$

In periodic cellular networks of cubic type, with l_c being the length of the cube edge, Eq. (4.14) applies and the respective constant is 3. In periodic networks of face centered cubic unit cells, the constant takes the value of $3(1 + 2\sqrt{2})$, while in networks with body-centered cubic unit cells, it is $3(1 + 4/\sqrt{3})$.

A special case is that of a Voronoi network perturbed by random displacements applied to the crosslinks after the Voronoi structure was constructed. The configuration labeled S-C1 in Figure 4.1 shows schematically a network of this type. The displacements applied to the crosslinks may be selected from a uniform (or normal) distribution in a specified range (or with specified variance). While in the unperturbed Voronoi structure all cells are convex by construction, such a perturbation may produce non-convex cells, although it does not change the number of faces of a cell. The network density increases rapidly with increasing the magnitude of the perturbation, but the mean segment length increases more slowly. Equation (4.14) holds in this case too, and the right hand side constant increases with increasing perturbation, as reported in Deogekar et al. (2019).

4.2.3.3   Effect of Preferential Alignment

Equation (4.13) was derived under several simplifying assumptions. The most important of these is the assumption of random fiber orientations. Several authors considered the effect of preferential orientation on the distribution of segment lengths (Komori and Makishima, 1977); here we present the results of Toll (1993), which account not only for generic fiber orientations but also for the contribution of crossings at fiber ends.

Toll's result for the density of crossings (number of fiber crossings per unit volume) is given by:

$$\rho_X = \rho^2 d\alpha + \frac{\pi}{2}\rho^2 \frac{d^2}{L_0}(\beta + 1), \tag{4.16}$$

and the number of crossings per fiber results in $2\rho L_0 d\alpha + \pi \rho d^2(\beta + 1)$. In this model, the fibers are allowed to interpenetrate at crossings. The mean segment length results:

$$l_c = \frac{1}{\rho d} \frac{1}{2\alpha + \pi(d/L_0)(\beta + 1)}. \tag{4.17}$$

Equations (4.16) and (4.17) are written in terms of two non-dimensional parameters accounting for fiber orientation: α and β. The orientation is evaluated based on the fiber tangent unit vectors, $\mathbf{t}$, and is defined by a probability density $p(\mathbf{t})$ over the sphere of unit radius.[1] Parameters α and β are given by:

$$\alpha = \int\int |\mathbf{t} \times \mathbf{t}'| p(\mathbf{t})p(\mathbf{t}')dt\,dt',$$

$$\beta = \int\int |\mathbf{t} \cdot \mathbf{t}'| p(\mathbf{t})p(\mathbf{t}')dt\,dt'. \tag{4.18}$$

---

[1] The orientation tensor corresponding to the given configuration can be computed with Eq. (4.2) as $\mathbf{A} = \int (\mathbf{t} \otimes \mathbf{t})p(\mathbf{t})\,dt$, where the integration is performed over the 2D domain of variation of $\mathbf{t}$ expressed, for example, in terms of the two Euler angles.

These expressions may be evaluated for any distribution of fiber orientations. In the random orientation case in which $p(\mathbf{t}) = 1/4\pi$ is uniform over the unit sphere, one obtains $\alpha = \pi/4$ and $\beta = 1/2$.

For preferentially oriented long, slender fibers of large aspect ratio, L_0/d, the second term in the denominator of Eq. (4.17) may be neglected relative to 2α and Eq. (4.17) becomes:

$$l_c = \frac{1}{2\alpha\rho d},\qquad(4.19)$$

which is a generalization of Eq. (4.13). Equation (4.19) reduces to Eq. (4.13) when fibers are randomly oriented ($\alpha = \pi/4$). The correction introduced by the β–dependent second term in the denominator of Eq. (4.17) becomes important when the fiber aspect ratio is small and/or in the presence of strong preferential alignment.

Just as with Eq. (4.13), Eq. (4.19) results from a geometric analysis of the network in the undeformed state. When subjected to loading, some of the initial contacts open and new, load carrying contacts form, which leads to a gradual transition to the $\rho l_c^2 = const$ scaling.

4.2.3.4   Effect of Excluded Volume

The effect of excluded volume is introduced in Section 4.2.3.1 at the end of the discussion related to fibrous networks. The distribution of segment lengths, $p(l)$, in 2D and 3D networks relevant for most network materials is Poisson (Eq. (4.10)). However, the small l end of distributions evaluated experimentally departs from the exponential form due to the fact that topological interactions between fibers prevent the formation of arbitrarily small fiber segments. Hence, the distribution exhibits a downturn in the small segment length limit.

Figure 4.3 shows the distribution of segment lengths, $p(l)$, in a packing of cylindrical fibers whose structure was inspected and reconstructed by X-ray tomography (Ekman et al., 2014). The distribution decays exponentially over a broad range of segment lengths, but the excluded volume effect is observed in the small l limit.

The cut-off introduced at small l leads to a slight increase of the mean segment length relative to the value predicted by Eq. (4.19). This correction was estimated for 3D randomly oriented networks in Neckar and Das (2012), who propose Eq. (4.20) for the corrected mean segment length, l_c^*, in terms of the mean segment length corresponding to the ideal case without excluded volume effects, l_c:

$$\frac{l_c^*}{l_c} = \frac{\xi}{1 - e^{-\xi}}.\qquad(4.20)$$

Here $\xi = \chi d/l_c$, while χd quantifies the excluded volume effect and represents the length along a given fiber which is occupied by a contact with another fiber. χ takes values close to 1. If ξ is small, the right side of Eq. (4.20) can be expanded in series and retaining the first terms leads to $l_c^* = l_c(1 + \chi d/2l_c)$. A similar correction was proposed in Pan (1993), who obtain the same formula, with $\chi = \pi/2$.

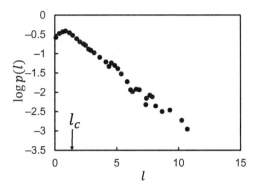

Figure 4.3 Distribution function of segment lengths for a 3D sample of packed fibers whose geometry was determined by tomography. The distribution is Poisson (exponential), except at small segment lengths. Adapted with permission from Ekman et al. (2014), Copyright (2014) by the American Physical Society

The correction is rather small. For example, if the contact length represents 30% of the segment length, that is, $\xi = 0.3$, which corresponds to a very dense distribution of contacts along the representative fiber, the corrected mean segment length l_c^* is larger than the nominal segment length l_c by only 7%.

A much larger screening length appears in fiber mats subjected to compression (Section 6.2.1). This is due to the obvious geometric constraint apparent when fibers are stacked and pressed against each other: in order to form contacts separated by small distances along any given fiber, large deformations of the contacting fibers are needed. If fibers are relatively stiff in bending, the probability to form contacts separated by small distances is small and this leads to the occurrence of a larger cut-off in the distribution function of segment lengths. An example of this type is seen in the segment length distribution shown in Figure 4.4(c).

4.2.3.4.1 *Effect of Fiber Tortuosity*

In most network materials fibers are not straight. Therefore, it is necessary to investigate the effect of fiber tortuosity on the mean segment length.

The arguments presented in this section for both 2D and 3D networks are based on the evaluation of the volume surrounding a representative fiber that needs to be penetrated by another fiber for a contact or crosslink to be established. This volume, and the associated probability of contact formation, depend exclusively on the fiber length and diameter. The result is a function of the volume fraction and it is independent of the fiber shape. Therefore, Eq. (4.17) and its simplified versions (Eqs. (4.6), (4.13), and (4.19)) are expected to hold identically for networks of tortuous fibers. This observation was made by multiple authors and is discussed in Yi et al. (2004).

To demonstrate the independence of l_c on tortuosity, Figure 4.4(b) shows the mean segment contour length (normalized by d) versus $1/\rho d^2$ for multiple simulated 3D networks of nominally infinite fibers and with various persistence lengths and

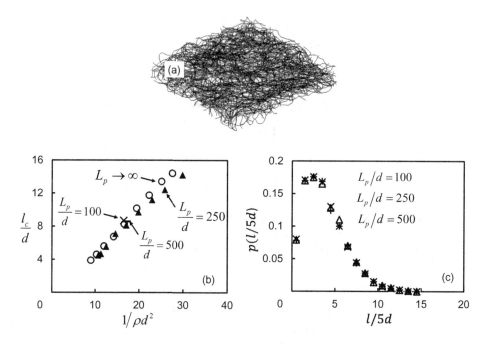

Figure 4.4 (a) Realization of a 3D mat-like non-crosslinked network of tortuous fibers. (b) Variation of the normalized mean segment length with $1/\rho d^2$ for networks of various persistence length, confirming that Eq. (4.19) applies equally in the case of straight and tortuous fibers (data from Negi and Picu, 2019). (c) Probability density distribution function of segment lengths (normalized by $5d$) for networks of fibers with various persistence lengths and same ρd^2. The three distributions shown are identical.

densities. The fibers have complex shapes in 3D and contacts are established as they are compacted to form a mat (Negi and Picu, 2019). Data for networks of straight and tortuous fibers fall on the same curve, indicating that only the excluded volume of fibers defines l_c and Eq. (4.19) holds. Figure 4.4(a) shows a realization of the respective 3D network, while Figure 4.4(c) presents the distribution function of segment lengths for networks of various L_p. The segment length is measured along the fiber contour and is shown normalized by $5d$. The figure shows that the distribution is independent of the degree of fiber tortuosity. The distribution decays exponentially, as expected. However, at segment lengths shorter than $\sim 10d$ the excluded volume effect leads to a downturn of the curve; this is similar to the experimental result shown in Figure 4.3.

4.2.3.4.2 *Thermal Networks*

The relationships derived above for athermal networks (Eqs. (2.13), (2.17), and (2.19)) indicate that, when the fiber diameter becomes negligibly small, the mean segment length diverges, $l_c \to \infty$. This is a direct consequence of the fiber overlap argument and of the static, geometric picture underlying these results.

The situation of thermal networks formed by molecular strands subjected to thermal fluctuations is different. The effective strand diameter is negligibly small compared with the strand length and the molecules are in continuous motion. Contact formation is stochastic both spatially and temporally. Therefore, the overlap argument cannot be used and the molecular diameter becomes irrelevant when discussing the mean segment length. The relevant quantity is the polymer concentration, $\tilde{c}$.

Polymeric chains are generally described as fractal paths for which the end-to-end distance, L, is related to the contour length, L_0, through the relation $L \sim L_0^\varsigma$, where $D = 1/\varsigma$ is the fractal dimension of the chain (see Section 2.4.1 and Figure 2.18). Sub-chain segments are described by the same relation. To determine the mean distance between contacts, we evaluate the largest volume occupied by a sub-chain segment, which is not shared with another segment, leading to the so-called overlap concentration.

A strand of n monomers each of length and diameter a occupies a sphere of diameter L. If no other strand is present within the respective volume, the concentration is $\tilde{c} \sim na^3/L^3 \sim na^3/L_0^{3\varsigma} \sim n^{1-3\varsigma}$, where we use the relation between the contour length and the monomer size, $L_0 = na$. One may now pose the reverse problem, requiring that the concentration is equal to the overall concentration of the polymeric solution and seeking the scale (or the chain segment length) below which volumes are occupied by a single strand. The relation between the concentration and this critical segment length becomes $n_c \sim \tilde{c}^{1/(1-3\varsigma)}$. The end-to-end distance of the critical segment is $\xi \sim n_c^\varsigma \sim \tilde{c}^{-\varsigma/(3\varsigma-1)}$. This is called the "mesh size" and represents the mean distance between contacts in the stochastic and transient network of polymeric chains. To render the notation compatible with that used in the reminder of this section, we observe that $\tilde{c} \sim \rho$ and hence one may infer that for these systems $\rho\xi^{(3\varsigma-1)/\varsigma} = const.$

Replacing the mesh size with the contour length of the corresponding segment, l_c ($l_c \sim n_c \sim \xi^{1/\varsigma}$), one obtains the relation $\rho l_c^{3\varsigma-1} = const.$

For athermal solvents in which the polymer–solvent interaction is purely repulsive (e.g., at high temperatures), the fractal dimension is predicted by Flory's theory as $D = 5/3$ ($\varsigma = 3/5$) and one has $\rho\xi^{4/3} = const$ and $\rho l_c^{4/5} = const$. Replacing the polymeric chains with filaments of smooth contour, that is, nonfractal curves in 3D, the dimension of the respective segments is $D = 1$, and the scaling relation between the density and mean segment length becomes $\rho\xi^2 = \rho l_c^2 = const$. In this case, the end-to-end length of the segment is proportional to the contour length (the constant of proportionality is the tortuosity parameter, Eq. (2.9)) and the distinction between the mesh size and the contour length between contacts disappears.

The relation $\rho l_c^2 = const$ between ρ and l_c also results from dimensional arguments based on the observation that l_c is the only characteristic length of the problem if the filament is not fractal, just as in the case of athermal cellular networks (Eq. (4.14)).

It results that the key relations between density and the mean segment length are $\rho l_c^2 = const$ for the case of filaments with nonfractal geometry, and $\rho\xi^{(3\varsigma-1)/\varsigma} = const$ in the opposite case. The terms "mean segment length" and "mesh size" are used interchangeably in the context of athermal fibers, however, "mesh size" is more appropriate for thermal networks in which the inter-filament contacts fluctuate.

4.2.3.4.3   A Geometric Relation

Before closing the overview of results related to the mean segment length, it is of interest to note a geometric relationship linking some of the parameters discussed in this chapter:

$$\rho = \rho_b \langle z \rangle l_c / 2. \tag{4.21}$$

This relation is exact in the absence of dangling segments and loops (segments that start and end at the same crosslink) and is useful in derivations presented in the subsequent chapters. It results from the observation that the fiber length corresponding to one crosslink is $\langle z \rangle l_c / 2$ and there are crosslinks per unit volume.

4.3   Variation of Network Parameters on Scales Larger Than the Mesh Size: Flocculation

Network parameters are implicitly considered spatially invariant in the discussion of Section 4.2. However, in most network materials one or multiple structural parameters exhibit spatial variability on length scales much larger than the network mesh size, l_c. This introduces new characteristic length scales, which may play a role in the overall mechanical and transport behaviors of the network.

Of the primary parameters of Section 4.2.1, the network density, ρ, is the most commonly observed to exhibit mesoscale variability. While in the network production process the nominal network density is usually defined, the complex dynamics of fiber layout may lead to the emergence of spatial density correlations. This leads to flocculation, that is, the agglomeration of fibers in certain areas and related depletion in some other areas, as commonly observed in nonwovens, paper, and paper products. Flocculation can be easily observed by visual inspection of a sheet of paper in transmitted light.

The mesoscale density variability is generally described using two measures: the pair correlation function, $G_\rho(r)$, and the variance of the density evaluated over subdomains of specified size λ, $\mathrm{var}(\rho_\lambda)$.

Consider a domain of a network material, as shown schematically in 2D in Figure 4.5, and define sub-domains of characteristic size λ, separated by $\mathbf{r}$. The unit vector of $\mathbf{r}$ is denoted by $\mathbf{r_v}$, with $\mathbf{r} = r \mathbf{r_v}$. The nominal, mean density of the network

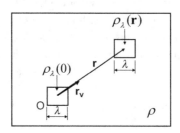

Figure 4.5 Schematic showing the evaluation of the pair correlation function, Eq. (4.22).

is ρ, but the density evaluated over a specific sub-domain of size λ, ρ_λ, may be different from the nominal value. The pair correlation function is defined as:

$$G_\rho(r) = \langle (\rho_\lambda(\mathbf{r}) - \rho)(\rho_\lambda(0) - \rho) \rangle_{\mathbf{r_v},O}, \tag{4.22}$$

where the average is performed over multiple positions of the origin, O, and over all orientations of the unit vector, $\mathbf{r_v}$. The resulting $G_\rho(r)$ is a function of the distance between the two sub-domains. In some cases, the directional pair correlation function is evaluated, $G_\rho(\mathbf{r_v})$, which captures the spatial anisotropy of the correlation. If the network is isotropic, the simpler form of Eq. (4.22) is sufficient. $G_\rho(r)$ reduces to the delta function if no spatial correlations are present. Also, $G_\rho(r)$ computed at $r = 0$, $G_\rho(0) = \mathrm{var}(\rho_\lambda)$, is a measure of the variance of the density field probed on scale λ.

In many situations, the pair correlation function may be approximated (over a range of r) with a power function:

$$G_\rho(r) \sim r^{-\zeta}. \tag{4.23}$$

Consider first the reference situation of a random and isotropic network without measurable flocculation. In such situations the number density of fiber centers of mass does not exhibit spatial variability when measured on length scales larger than $\sim L_0$. It turns out that the pair correlation function is not a delta function in these networks, despite the intuitive expectation. If fibers are straight and of length L_0, the mass distribution on length scales smaller than L_0 is necessarily correlated since fibers may exist that span both sub-domains at O and r. This trivial density correlation has a cut-off at $r \approx L_0/2$, beyond which it decays to zero. For an entirely random network, the exponent in Eq. (4.23) is $\zeta = 1$ in 2D, and $\zeta = 2$ in 3D. More accurate expressions for the 2D correlation function are presented in Sampson (2009) and Picu and Hatami-Marbini (2010). If fibers are not straight, Eq. (4.23) still applies, with the power function cut-off being at $r \approx L_p/2$.

Equation (4.23) provides the baseline correlation function for more complex cases in which fiber centers of mass are distributed in a spatially correlated way. Experiments with laboratory-made paper obtained by sedimentation of cellulose fibers from a suspension evidence strong flocculation (Provatas et al., 2000). The spatial correlation of the density is of the form given by Eq. (4.23), but exponents smaller than 1 result. The data reported in Provatas et al. (2000) can be fitted by choosing $\zeta = 0.37$. The power law fit remains a good approximation for distances r much larger than the fiber length, L_0, which indicates the presence of structuring on large scales due to flocculation (Figure 4.6). Regular office paper also exhibits density fluctuations (which are visible by imaging with light passing through the sheet) with a characteristic length of several millimeters.

A power law correlation indicates that the density function is fractally rough. Specifically, if the correlation function has a power law decay, the fractal dimension of the density function, D_ρ, is related to the exponent ζ as $D + 1 - D_\rho = 1 - \zeta/2$, where D is here the dimensionality of the embedding space (Falconer, 2003; Gneiting and Schlather, 2004). The quantity $D + 1 - D_\rho$ is known as the Hurst coefficient and

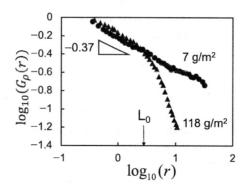

Figure 4.6 Pair correlation function of the density for paper samples of two nominal (average) densities. The distance r is in millimeters. The fiber length is $L_0 = 2.7$ mm ($\log_{10}(L_0) = 0.43$). The correlation function is normalized with the value at the leftmost point. Data from Provatas et al. (2000)

expresses the degree of roughness of the respective manifold. If there are no density fluctuations and G_p is a constant, $\zeta = 0$ and $D_p = D$. In the 2D problem considered in Provatas et al. (2000) and Figure 4.6, $D = 2$, $\zeta = 0.37$, and the fractal dimension results $D_p = 2.18$. The fully random 2D network with no flocculation and $\zeta = 1$ corresponds to a Hurst coefficient of 0.5 and fractal dimension $D_p = 2.5$.

In the world of molecular networks, gels are known to be heterogeneous on scales larger than the mean segment length or mesh size. The degree of heterogeneity, which can be evidenced by small angle neutron scattering experiments, depends on the preparation method and on parameters such as the monomer concentration and the quality of the solvent (Panyukov and Rabin, 1996). Heterogeneity is particularly pronounced close to the gel point. The heterogeneity associated with gelation occurs due to the phase separation of the polymer-rich phase of the growing gel cluster and the surrounding polymer-poor phase. The polymer density in the gel phase is controlled by the crosslink density and must be larger than outside of the growing cluster. This mechanism was described by de Gennes (1979) for low density solutions but applies in denser phases as well (de Molina et al., 2015).

We turn now to the evaluation of the density variance measured over subdomains of size λ (Figure 4.5), and consider first the 2D network problem. Using Eq. (4.6), one may compute the number of contacts per fiber as $N_{Xf} = L_0/l_c = 2\rho L_0/\pi$. Therefore, the probability distribution function of N_{Xf} should be proportional to that of the density. Hence, the variances of N_{Xf} and ρ are related as: $\text{var}(N_{Xf}) = (2L_0/\pi)^2\text{var}(\rho)$. On the other hand, N_{Xf} is a Poisson distributed discrete variable and its variance is equal to the mean of the distribution, that is, $\text{var}(N_{Xf}) = 2\rho L_0/\pi$. Now, moving from the point process to the process defined over sub-domains of size λ, that is, over sub-populations of N_f^λ number of fibers within the sub-domain, the variance decreases. According to the central limit theorem, the variance of the number of contacts per fiber averaged over λ, N_{Xf}^λ, decreases with increasing λ as $\mathrm{var}\left(N_{Xf}^{\lambda}\right) = \mathrm{var}(N_{Xf})/N_f^{\lambda}$. Since $N_f^{\lambda} = \rho\lambda^2/L_0$, the variance of the density can be computed as:

$$\mathrm{var}(\rho_{\lambda}) = \frac{\pi}{2\lambda^2}. \tag{4.24}$$

A similar analysis can be performed in 3D considering that the mean number of contacts per fiber is $N_{Xf} = L_0/l_c = 2\rho A_f L_0/d$. This leads to the relation:

$$\mathrm{var}(\rho_{\lambda}) = \frac{2}{\pi d\lambda^3}. \tag{4.25}$$

Figure 4.7 shows the variance of the density of a paper sample measured by radiography and evaluated over sub-domains of increasing size, λ (top data set) (Sampson, 2009). The lower data set corresponds to a computer generated, fully random distribution of fibers of the same length and width as the cellulose fibers of the respective paper. The scaling predicted by Eq. (4.24) is recovered in the numerical sample once the probing length scale, λ, becomes large enough (here, larger than ~ 1 mm). However, the experimental data does not follow this prediction and exhibits a much slower decaying variance. As in Figure 4.6, this effect may be due to flocculation, which modifies the statistics of the network structure on scales larger than the mesh size, l_c.

It is possible to envision that other network parameters may also exhibit mesoscale fluctuations. The crosslink number density, ρ_b, most likely follows such behavior. This is expected to have a strong effect on the mechanical properties of the network. The connectivity, z, is unlikely to exhibit significant spatial variability in natural network materials, while the fiber length and diameter, although varying from fiber to fiber, are not expected to develop long-range spatial correlations.

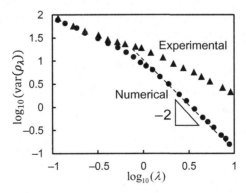

Figure 4.7 Variation of the variance of density with the scale of observation for a sample of paper (experimental) and for a 2D Mikado model of fibers having identical geometry with the cellulose fibers of the experimental sample (numerical). The probing length scale λ is in millimeters and ρ_{λ} is in g/cm². The departure of the experimental data from the numerical model prediction indicates mesoscale flocculation. Adapted from Sampson (2009)

4.4   Porosity of Fibrous Assemblies

The porosity of network materials is important in a number of applications. In many cases, fiber networks are embedded in a fluid medium which fills the volume between fibers and contributes essentially to defining the mechanical behavior of the material. Examples of this type include the cellular cytoskeleton and connective tissue, such as cartilage. The F-actin and, respectively, collagen networks are embedded in an aqueous solution and include other entities such as organelle and proteoglycans. The fluid is free to move within and in and out of the network and interacts mechanically with fibers as the material deforms (Section 9.3). The drag forces associated with internal flow depend on the size and distribution of free volume within the network.

Porosity is important when the network is used for filtration and is supposed to trap and retain particles larger than a certain size. Flow introduces a preferred direction and porosity may be thought of as the distribution of sizes of polygons defined by fibers in the projection of the 3D network onto the plane perpendicular to the flow direction. However, in most applications the network is thick enough for the size of pores measured in projection to vanish. Transport is then associated with flow through tortuous channels which are not necessarily aligned with the nominal flow direction. The three-dimensionality of the network structure becomes important in these applications.

This section presents a summary of results pertaining to the stochastic geometry of the free volume within networks. Motivated by the above considerations, the problem is viewed from two perspectives – that of the distribution of free volume and that of trapping of particles within the network. Only the geometric aspects are discussed here.

The fraction of the material volume not occupied by the network is, generically, $1 - f = 1 - \rho A_f$. $1 - f$ is a measure of porosity evaluated as an average over the entire material. When probed on length scale λ, one may write $1 - f_\lambda = 1 - \rho_\lambda A_f$, where the subscript indicates that the respective quantity is evaluated over sub-domains of the respective characteristic size. This implies that the probability distribution function of the porosity and of ρ_λ are proportional and the variance of porosity probed on scale λ is related to the variance of ρ_λ:

$$\text{var}(1 - f_\lambda) = A_f^2 \text{var}(\rho_\lambda). \tag{4.26}$$

The variance of porosity may be evaluated for 2D and 3D problems using Eqs. (4.24) and (4.25) as:

$$\text{var}(1 - f_\lambda) = \frac{\pi}{2}\left(\frac{d}{\lambda}\right)^2 \quad \text{2D},$$

$$\text{var}(1 - f_\lambda) = \frac{\pi}{8}\left(\frac{d}{\lambda}\right)^3 \quad \text{3D}. \tag{4.27}$$

In networks of thin and slender fibers, porosity is large and the spatial variability of porosity is small. This holds even if the scale of observation, λ, is on the order of the network mean segment length, l_c, since d/l_c is small. Flow through the random network becomes similar to flow through a periodic network of similar density.

In quasi-2D mats, the porosity is related to the coverage $\langle g \rangle$. In projection on the plane of the mat, the porosity is defined as the fraction of free area, which is equal to the probability that the coverage is zero. The coverage is Poisson distributed and is given by Eq. (4.8). Therefore, porosity may be computed as $p(g = 0) = \exp(-\langle g \rangle)$. If the density of fibers is large and $\langle g \rangle \gg 1$, this expression gives zero porosity. While this is exact if the problem is 2D, the porosity in the quasi-2D case of nonwovens used in filtration is not zero in this limit since the layered arrangement of fibers allows paths with 3D structure to percolate across the mat. Transport in these networks has a 3D nature and the 2D projected representation of the mat is not adequate.

We take now the second view on porosity and investigate the distribution of pore sizes and associated geometric characteristics of networks in 2D and 3D. This provides a more detailed, geometric description than the mean field view based on the generic relation between density and porosity described in this subsection.

4.4.1   Fibrous: Two Dimensions

The prototypical example of fibrous 2D networks is the Mikado network constructed by depositing fibers of given length with random orientations and random positions of their centers of mass in a 2D domain. Many exact and numerical results are available for this stochastic geometry. Fibers define polygons of various number of edges with a broad distribution of areas. Considering the polygons to be pores and ignoring the projected fiber width, the mean pore area, $\langle A_p \rangle$, can be expressed in terms of the mean segment length as (Miles, 1964):

$$\frac{\langle A_p \rangle}{l_c^2} = \frac{4}{\pi}. \tag{4.28}$$

The type of polygons present in the 2D Mikado structure is of interest (Crain and Miles, 1976). The polygons that appear with the highest probability are quadrilaterals. These represent 37.59% of the population of polygons. Triangles represent 35.59%, polygons with five edges represent 18.89%, and polygons with six, seven, and eight edges represent 6.08%, 1.3%, and 0.21% of the total number of polygons, respectively. The average number of edges per polygon is four.

An equivalent pore diameter, d_p, may be defined based on the pore area using the relation $A_p = \pi d_p^2 / 4$. This is used as an approximation (although an overestimate) for the diameter of a particle that may pass through the respective pore. Given its importance in various applications, the statistical characteristics of d_p have been studied in detail. The distributions of A_p and d_p are not defined analytically. However, numerical data suggest that both these quantities are approximately gamma distributed (Crain and Miles, 1976; Eichorn and Sampson, 2005) and the variance of the distribution of pore diameters evaluated by Corte and Lloyd (1965) is $\sqrt{16 - \pi^2}/\pi$. With this information and making use of Eq. (4.28), it is possible to evaluate the mean pore diameter as:

$$\langle d_p \rangle = 0.79 l_c = 1.24 \frac{d}{\langle g \rangle}, \tag{4.29}$$

where Eq. (4.9) was used to express the pore diameter in terms of the mean coverage. This relation differs somewhat from that reported in Sampson (2009), which reads $\langle d_p \rangle = 2d/\langle g \rangle$. Equation (4.29) indicates the perhaps surprising result that the pore diameter increases as the fiber diameter increases, at constant coverage. Of course, this is only an apparent contradiction since the condition of constant coverage implies that the number of fibers per unit area should decrease as the fiber diameter increases.

4.4.2   Cellular: Two Dimensions

Similar results have been established for cellular networks of Voronoi type in 2D. The mean pore area for a Voronoi tessellation is given by:

$$\frac{\langle A_p \rangle}{l_c^2} = \frac{9}{4}. \tag{4.30}$$

The Voronoi procedure guarantees that the pores are convex and hence of more roundish shape than those of the Mikado network.

It is also of interest to place this discussion in relation with similar results pertaining to periodic 2D networks. Many periodic 2D network structures can be imagined and their geometry was studied since antiquity. For the purpose of this discussion, we consider only the three regular tessellations of the plane, known also as Euclidean tilings: the triangular, square, and hexagonal tilings. It is a simple matter to show that, not only in these three cases but also for all regular polygons, the ratio of the polygon area to the square of the edge length is a function of the number of edges of the respective polygon, n_e, and is given by:

$$\frac{A_p}{l_c^2} = \frac{1}{4} \frac{n_e}{\tan(\pi/n_e)}. \tag{4.31}$$

More irregular tilings such as the Archimedean and semiregular tilings, which lead to periodic networks with a distribution of segment lengths, have a distribution of polygonal pore areas. In each case, the ratio of the mean pore area and the square of the mean segment length is a constant that depends on the network structure.

2D Voronoi networks have cells with a broad distribution of the number of edges per cell. The cell that appears with the highest probability (29.4%) has $n_e = 6$. Triangular cells appear with a frequency of 1.1%, quadrilateral cells appear with a frequency of 10.7% and pentagonal cells with a frequency of 26%. Cells with $n_e = 7, 8, 9, 10, 11$, and 12 appear with frequencies 19.9%, 9%, 3%, 0.7%, 0.15%, and 0.02%, respectively. The average number of edges per polygon in this type of network is six.

This discussion leads us to the empirical Lewis law, which was proposed in 1928 based on observations of epithelial cells shapes and projected areas (Lewis,

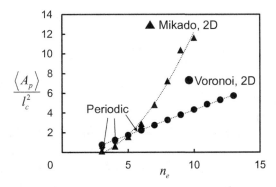

Figure 4.8 Normalized polygon area, $\langle A_p \rangle$, function of the number of edges of the polygon, n_e, for 2D Voronoi (circles), 2D Mikado (triangles), and 2D periodic networks (diamonds) with $n_e = 3, 4, 6$. The two lines represent linear and quadratic relations, and are the best fits to the Voronoi and Mikado data sets, respectively.

1928). It states that the area of a polygon belonging to a space filling tessellation is proportional to the number of its edges. This relation holds for a large number of geometrical patterns, including some of those considered here. Figure 4.8 shows the area of polygons of 2D Mikado and Voronoi networks versus the number of edges of the respective polygons. Data points corresponding to periodic networks with unit cells in the form of the three Euclidean tilings with $n_e = 3, 4$, and 6 (Eq. (4.31)) are included. The figure shows that Lewis law holds for cellular networks but does not apply in the case of the more irregular Mikado networks.

A further interesting observation, which is known as the Aboav–Weaire law (Aboav, 1970), indicates that the cells neighboring a cell with large n_e have, on average, small n_e, and vice-versa. This was initially determined when inspecting the microstructure of polycrystals and was since confirmed in many biological, man-made, and computer-generated tessellations. The mean number of edges, n_e, of a cell neighboring a cell of n_{e0} number of edges is approximately given by $n_e = A + B/n_{e0}$, where A and B depend on the specific geometry considered.

4.4.3 Fibrous: Three Dimensions

The identification of pores in 3D fibrous networks is much less straightforward than in 2D since, in these structures, fiber segments do not define polyhedra of regular geometry. We approach this problem by evaluating the average diameter of a sphere that can be fitted within the network without distorting the fibers.

The mean diameter of the probing sphere, $\langle d_p \rangle$, should scale linearly with the mean segment length, l_c. Therefore,

$$\langle d_p \rangle \sim l_c \sim \frac{1}{\rho d}, \tag{4.32}$$

which indicates that the pore size should scale inversely with the fiber density and that the average normalized pore volume $\langle V_p \rangle / l_c^3$ is a constant which depends on the network geometry. This is similar to the 2D situation quantified by Eqs. (4.28) and (4.30).

A more detailed argument leading to Eq. (4.32) can be developed starting from a mean field description similar to that used to derive the Kallmes–Corte relation of Eq. (4.6). Consider a fibrous network in which a sphere of diameter d_p is inserted. If d_p is much smaller than l_c, the probing sphere makes contact with one fiber at a time or it does not make contact with any fiber, in which case it is clear that d_p is not representative for the inter-fiber spacing. If the probe diameter is much larger than l_c, the sphere overlaps with many fibers, a situation which, again, is not representative for the stated purpose. The intermediate scale situation is that in which the probe makes sufficient contacts with neighboring fibers to just become caged. According to Philipse and Kluijtmans (1999), the minimum number of contacts for proper caging is $N_{Xc} = 7$.

To evaluate the mean number of contacts for a sphere of diameter d_p embedded in a network of straight fibers of diameter d having random distributions of orientations and centers of mass, consider the schematic shown in Figure 4.9. A fiber AB, with the center of mass at O' is in contact with the sphere if it passes through the layer of thickness $d/2$ surrounding the sphere. The locus of fiber positions that fulfill this condition forms the outer surface of the cone O'PP'. The probability that fibers with the center of mass at O' belong to this locus is equal to the ratio between the solid angle shown in gray in Figure 4.9 and 4π, that is, $d_p d / 8\delta^2 \sin \alpha$. The number of fibers with their center of mass in the spherical layer bounded by radii δ and $\delta + d\delta$ from the center of the sphere, O, and which make contact with the sphere is $(\pi \rho d_p d / 2L_0 \sin \alpha) d\delta$. To obtain the total number of contacts, one integrates over δ in the range $D_p/2 < \delta < L_0/2$ to obtain the total number of contacts as:

$$N_X = \frac{\pi}{2} \rho d_p d \sqrt{1 - \left(\frac{d_p}{L_0}\right)^2}. \tag{4.33}$$

Requiring that $N_X = N_{Xc}$ and assuming $d_p \ll L_0$, Eq. (4.33) leads to a relation similar to Eq. (4.32), with the constant of proportionality equal to $N_{Xc}/2 \approx 3.5$.

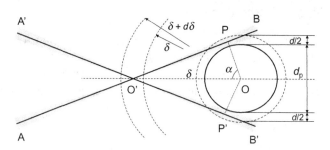

Figure 4.9 Geometric construction used to derive Eq. (4.33).

This corresponds to the following estimate of the pore volume in units of the mean segment length:

$$\frac{\langle V_p \rangle}{l_c^3} \approx 22. \tag{4.34}$$

Once again, the constant in Eq. (4.34) is expected to depend on the specific network geometry. The purpose of the derivation presented here is limited to providing the expected scaling and the order of magnitude of the respective constant.

4.4.4   Cellular: Three Dimensions

Given the analogy between cellular networks and open cell foam structures, the identification of pores in such networks is more straightforward. As in 2D, a well-studied example of cellular network is that of the space filling Voronoi tessellation. Many statistical properties of Voronoi structures are known, some being derived analytically and some resulting from high accuracy numerical simulations (Okabe et al., 1992; Chiu et al., 2013; Lazar et al., 2013). The mean volume of Voronoi cells is given by:

$$\frac{\langle V_p \rangle}{l_c^3} = 14.07. \tag{4.35}$$

This results from the relation between the number density of seeds used to generate the network, which can be written as $1/\langle V_p \rangle$, and the fiber density, Eq. (4.15), which becomes $1/\langle V_p \rangle = (5.38/\rho)^{3/2}$ (Meijering, 1953). Equation (4.35) results by replacing Eq. (4.14) in this expression and taking the constant in Eq. (4.14) to be equal to 0.94.

The distribution of volumes in the Voronoi structure is approximated by a gamma distribution. The cell volume increases with increasing the number of faces of the respective cell, but the relation is nonlinear. Numerical results presented in Sampson (2009) indicate that $V_p \sim n_f^{1.468}$, where n_f is the number of faces of the cell of volume V_p. This suggests that the Lewis law, which represents a linear relation between the area and number of edges of polygons in a 2D tessellation becomes nonlinear when applied to 3D structures.

Diffusion of particles across a cellular 3D network requires penetration across cell faces. This renders the statistics of face dimensions as important as the statistics of pore volumes. This statistic is known for Voronoi networks (Kumar et al., 1992; Lazar et al., 2013). The mean number of edges per face of a Voronoi cell is 5.23, which should be compared with the mean number of edges per cell in the equivalent 2D Voronoi tessellation, which is 6. The face area normalized by the square of the mean edge length, for given number of edges per face, n_{ef}, is $\langle A_f \rangle / l_{cf}^2 (n_{ef}) \approx n_{ef}^2/(4\pi)$. Averaging over the entire distribution of polygons of arbitrary number of face edges, the ratio of the mean face area and the network average mean segment length is:

$$\frac{\langle A_f \rangle}{l_c^2} = 6.32. \tag{4.36}$$

Equations (4.35) and (4.36) shed light on the diameter of a spherical particle that may move freely across the Voronoi network without distorting fibers, d_p. Assuming that d_p is gamma distributed, Eq. (4.36) may be used to restrict the variance of the respective distribution, that is, requiring that the particles may cross cell faces and hence the distribution of particle sizes matches the distribution of cell face sizes. This argument leads to $\langle d_p^2 \rangle = 2.01 l_c^2$. Furthermore, requiring that diffusing particles fit inside the cells of the network, and using Eq. (4.35), leads to a restriction on the third moment of the distribution of d_p: $\langle d_p^3 \rangle = 26.87 l_c^3$. These two conditions can be used to compute the two parameters of the gamma distribution and to evaluate the mean of d_p in terms of the mean segment length of the networks: $\langle d_p \rangle = 0.29 l_c$.

Turning now to periodic networks that result from space filling tilings, we focus on the class of parallelohedra. These are convex geometrical bodies which tile the 3D space by translations only. Retaining only the edges of these constructs leads to periodic networks often used as models for open cells foams.

There are five parallelohedra (Fedorov, 1891): the cube, the hexagonal prism, the rhombic dodecahedron, the elongated dodecahedron, and the truncated octahedron, which have 6, 8, 12, 12, and 14 faces, respectively. With the exception of the hexagonal prism, all other parallelohedra have edges of the same length. Ratio V_p/l_c^3 takes the values of 1, $16\sqrt{3}/9$, 6, and $8\sqrt{2}$ for the cube, rhombic dodecahedron, elongated dodecahedron, and truncated octahedron, respectively. The hexagonal prism has two different types of edges, forming the hexagonal base and the height of the prism. If the two are equal in length, V_p/l_c^3 for this cell is $3\sqrt{3}/2$. It appears that V_p/l_c^3 generally increases with increasing the number of faces, but no universal relation can be defined. This should be compared with the result for the Poisson Voronoi network which has a mean number of faces per cell of 15.54 (Meijering, 1953) and $V_p/l_c^3 = 14.07$ (Eq. (4.35)).

4.5   Percolation

In network materials the problem of percolation appears in the context of geometric percolation, also referred to as transport percolation, and in the context of stiffness percolation. Consider the process of gradually adding fibers to a predefined domain, hence increasing the fiber density. A critical density exists when a system-spanning connected cluster of fibers is created, which contains at least one continuous path across the network. This critical density corresponds to geometric percolation and the first instantiation of the network. If the fibers are conductive, this path ensures that transport (e.g., of electric current or heat) across the system may take place and, hence, the nomenclature of "transport percolation threshold."

The percolated path may have vanishing stiffness. Whether this is the case or not depends on the mechanical behavior of fibers and their crosslinks. For example, if all crosslinks are pin-joints and fibers carry only axial loads, since the connectivity number z is small, the network has vanishing stiffness at small strains. To reach a configuration with non-zero stiffness, the fiber density or the crosslink density must be increased to the threshold of "stiffness percolation." Stiffness percolation occurs at densities equal to or larger than the density associated with transport percolation. For the two percolation thresholds to coincide, the crosslinks must be of weld type (Figure 3.1).

In this chapter we are concerned with the geometry of the network and consider percolation associated with increasing the fiber density and increasing the density of crosslinks. The two problems are relevant for percolation in athermal networks and for stiffness percolation in polymeric solutions and gelation. Since the network has arbitrary geometry, the analysis is performed in free space (as opposed to the study of percolation on lattices). Fibers must have finite length for the problem of be meaningful. If fibers, whether straight or tortuous, have "infinite" length and traverse the network domain, one fiber is sufficient to reach transport percolation. This is the situation in nonwovens made from continuously spun fibers.

4.5.1　Percolation Associated with Increasing Fiber Density

Percolation is a critical phenomenon which occurs at a critical value of the fiber density. Consider that straight fibers of length L_0 and negligible diameter are placed sequentially in a 2D domain of edge length L, and that fibers form contacts at all crossing points. At small fiber number densities fibers remain isolated or form clusters. A percolated path forms at a critical number density, $\rho_{\#c}$, or, equivalently, at a critical fiber density, $\rho_c = \rho_{\#c} L_0$. The objective here is to evaluate ρ_c in terms of the only geometric parameter of the problem – the fiber length, L_0. The percolation threshold is notoriously system size dependent and, in this discussion, we consider L to be sufficiently large for the reported results to be system size independent.

The central observation is that percolation takes place when the product $\rho_\# \langle A_{ex} \rangle$ reaches a critical value (Balberg and Binenbaum, 1984; Balberg et al., 1984). A_{ex} is the excluded area associated with the representative fiber and $\langle A_{ex} \rangle$ is the network average of the excluded area per fiber. The excluded area is defined as the area surrounding the representative fiber within which the presence of the center of another fiber leads necessarily to the formation of a contact with the representative fiber. The critical value of $\rho_\# \langle A_{ex} \rangle$ at percolation, that is, $\rho_{\#c} \langle A_{ex} \rangle$, is, with few exceptions, determined numerically.

In the Mikado network, the percolation condition is given by:

$$\rho_{\#c} \langle A_{ex} \rangle = 3.58. \tag{4.37}$$

The excluded area of a line of length L_0 in 2D is calculated in Section 4.2.3.1 as $\langle A_{ex} \rangle = 2L_0^2/\pi$ (Eq. (4.5)). Therefore, Eq. (4.37) can be rearranged in terms of the fiber density as (Li and Zhang, 2009; Mertens and Moore, 2012; Tarasevich and Eserkepov, 2018):

$$\rho_c L_0 = 5.637. \tag{4.38}$$

It is important to recall that $\rho_{\#c}\langle A_{ex}\rangle$ represents the number of fibers within the excluded volume of a representative fiber and, hence, it is equal to the mean number of contacts per fiber. This allows interpreting the percolation threshold in terms of a critical required number of contacts per fiber in order to produce the percolated path.

In the case of randomly distributed and randomly oriented ribbon-like fibers of length L_0 and diameter d in 2D, the critical condition for percolation is identical to that of Eq. (4.37). The excluded area in this case is given by Balberg et al. (1984): $\langle A_{ex}\rangle = 2dL_0(1 + 4/\pi^2) + 2(L_0^2 + d^2)/\pi$. In the limit of large aspect ratios this expression reduces to $\langle A_{ex}\rangle \approx 2L_0^2/\pi$, that is, to the result for lines of zero width derived in Section 4.2.3.1 and used to derive Eq. (4.38). This observation is expected since the effect of d on the excluded area vanishes in this limit. Then, the percolation condition of Eq. (4.38) may be written in terms of the critical area fraction of (overlapping) rectangles, f_c, as:

$$f_c \frac{L_0}{d} \xrightarrow[L_0/d \to \infty]{} 5.637.$$
(4.39)

The 3D system equivalent to the network of ribbon-like fibers in 2D is composed from straight fibers (cylinders with semi-spherical caps, known as spherocylinders) of length L_0 and diameter d, which are randomly distributed and randomly oriented in space. For the purpose of evaluating the percolation threshold fibers are placed in the problem domain without restricting the overlaps. The percolation condition is still described in terms of the number density of fibers and the mean excluded volume per fiber, in a manner similar to the discussion related to 2D networks. The percolation condition for this type of stochastic network is:

$$\rho_{\#c}\langle V_{ex}\rangle = 1.41.$$
(4.40)

The excluded volume of a spherocylinder is evaluated with the expression (Balberg and Binenbaum, 1984; Balberg et al., 1984): $\langle V_{ex}\rangle = 4\pi d^3/3 + 2\pi d^2 L_0 + \pi dL_0^2/2$. In the limit of large aspect ratios, the excluded volume per fiber can be approximated as $\langle V_{ex}\rangle = \pi dL_0^2/2$. Equation (4.40) can be written in this limit in terms of the critical fiber volume fraction, f_c, and the fiber aspect ratio as:

$$f_c \frac{L_0}{d} \xrightarrow[L_0/d \to \infty]{} 0.705.$$
(4.41)

The average number of contacts per fiber at percolation is 1.41, as indicated by Eq. (4.40). Figure 4.10(a) shows the variation of the volume fraction at percolation with the aspect ratio of fibers for this 3D case. The convergence to the limit of Eq. (4.41) is much faster than in the 2D case.

The results of Eqs. (4.40) and (4.41) can be placed in relation with a slightly different, but highly relevant percolation problem involving straight cylindrical fibers, which refers to the dense packing limit of 3D spherocylinders. Consider that straight, randomly distributed, and randomly oriented fibers of length L_0 and diameter d are packed without overlap until the maximum density is reached. This is similar to the problem of stiffness percolation of non-crosslinked straight fibers probed in

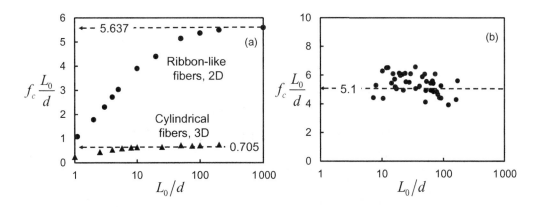

Figure 4.10 (a) Variation of the normalized area and volume fractions at percolation (allowing for overlap), $f_c L_0/d$, with the aspect ratio or fibers, for the 2D problem of ribbon-like fibers and the equivalent 3D problem of spherocylinders with random orientations and random center of mass positions. The limits reached at large L_0/d provided by Eqs. (4.39) and (4.41) are shown. The data used is from Dean and Bird (1967), Hovi and Aharony (1996), Feng et al. (2008), Jacobsen (2014), and Xu et al. (2016). (b) Volume fraction at stiffness percolation of spherocylinders in 3D allowing for inter-fiber contacts, but no overlap and probed in compression. Data from Williams and Philipse (2003)

compression; note that this system has no tensile rigidity at any fiber density. However, since the mechanics of fibers is not considered, the problem remains purely geometric. This is a properly posed percolation problem for fibers if the no overlap condition is imposed.

The solution provided in Williams and Philipse (2003) indicates that the maximum volume fraction that can be reached in such packings is inversely proportional to the fiber aspect ratio, just as predicted by Eq. (4.41), but the related constant is different:

$$f_c \frac{L_0}{d} = 5.1. \tag{4.42}$$

This corresponds to an average number of contacts per fiber at maximum packing of 10.2, which is much larger than the 1.41 value corresponding to transport percolation of overlapping spherocylinders, as expected. It is also comparable with the critical number of contacts estimated in Philipse and Kluijtmans (1999), which would lead to locking of a cylindrical object in a packing of like objects. Numerical and experimental data (Nardin et al., 1985; Evans and Gibson, 1986; Parkhouse and Kelly, 1995; Philipse, 1996; Novellani et al., 2000; Williams and Philipse, 2003) supporting Eq. (4.42) are presented in Figure 4.10(b).

Preferential orientation of fibers modifies the thresholds indicated here, but the condition that the product of fiber volume fraction and fiber aspect ratio reaches a critical value at percolation remains valid. Expressions for the excluded volume and excluded area for networks with preferentially oriented fibers are provided for all cases discussed in this section in Balberg et al. (1984) and Balberg and Binenbaum (1984).

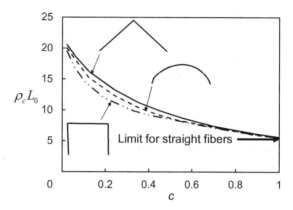

Figure 4.11 Variation of the critical density at percolation with the fiber tortuosity for 2D Mikado networks. The tortuosity is quantified by the crimp parameter of Eq. (2.9). The three curves correspond to the three fiber shapes shown. The percolation density predicted with Eq. (4.38) for 2D Mikado networks of straight fibers is recovered for $c = 1$. Data from Yi et al. (2004)

The critical density $\rho_{\#c}$ at percolation is different when percolation is probed in the direction parallel and perpendicular to the direction of preferential fiber orientation. Specifically, $\rho_{\#c}$ corresponding to probing in the perpendicular direction is larger than that corresponding to probing in the parallel direction. The critical density for the randomly oriented fibers case is in-between these two values. The difference between the two $\rho_{\#c}$ values increases with the degree of orientation.

Fibers are rarely straight in real network materials and tortuosity has a significant effect on the percolation threshold (Yi et al., 2004). Increasing fiber tortuosity, while maintaining random fiber distribution of their centers of mass and random orientations of their end-to-end vectors, leads to an increase of the constant on the right side of Eqs. (4.37)–(4.42). An example provided in Yi et al. (2004) is shown in Figure 4.11. It corresponds to 2D networks of fibers of vanishing thickness and having different degrees of tortuosity. The vertical axis represents $\rho_c L_0$, which is expected to reach 5.637 at percolation (Eq. (4.38)) if fibers are straight, $c = 1$. As the degree of tortuosity increases (the crimp parameter, c, decreases), the critical value $\rho_c L_0$ increases rapidly, indicating that higher densities are needed to reach percolation when fibers are tortuous. This can be understood considering that a tortuous fiber covers a smaller area than a straight fiber of the same contour length and hence the density of such tortuous fibers at percolation should be larger than the density of the equivalent straight fibers. The results in Figure 4.11 are obtained by using triangular, sinusoidal, and U-shaped fibers in different models. The trend is expected to remain unchanged if fibers of other shapes are used, although numerical values may differ.

4.5.2   Percolation Associated with Increasing Crosslink Density: Gelation

In the preceding discussion it is considered that crosslinks are established as the fiber density increases, at all points where fibers touch or overlap. If the fiber density is pre-defined and crosslinks are gradually added to produce a crosslinked network, a critical crosslink density is reached at which a connected system-spanning network first forms. For example, in the gelation process, molecules of connectivity z_M are mixed with crosslinkers of connectivity z_X and the reaction takes place gradually. A system-spanning cluster forms when a fraction p_c of the reaction sites of the molecules are activated. The theory of Flory and Stockmayer (Flory, 1941; Stockmayer, 1943) provides an expression for p_c which is in good agreement with experiments (Macosco and Miller, 1976; Venkataraman et al., 1989):

$$p_c = \frac{1}{\sqrt{(z_M - 1)(z_X - 1)r}}, \qquad (4.43)$$

where $r = z_M m_M / z_X m_X$ is the stoichiometric ratio of M to X, while m_M and m_X are the number of moles of molecules and crosslinkers in the mixture.

If $z_M = z_X = z$ and the stoichiometric ratio is 1,

$$p_c = \frac{1}{z - 1}. \qquad (4.44)$$

This relation applies to the vulcanization of molecular chains and to the end-linking of molecules of low z leading to the formation of tree structures.

Interestingly, Eq. (4.44) describes accurately the bond percolation threshold for a remarkably broad range of 3D lattices of connectivity ranging from $z = 3$ to $z = 24$. Furthermore, $p_c(\langle z \rangle - 1)$ is a constant (close to 1) even for lattices in which not all sites have the same connectivity, and $\langle z \rangle$ represents the mean connectivity of the lattice. This conclusion also applies to 2D lattices. The relation between bond percolation and gelation was pointed out in the work of de Gennes (1979).

References

Aboav, D. A. (1970). The arrangement of grains in a polycrystal. *Metallography* **3**, 383–390.

Balberg, I., Anderson, C. H., Alexander, S. & Wagner, N. (1984). Excluded volume and its relation to the onset of percolation. *Phys. Rev. B* **30**, 3933–3943.

Balberg, I. & Binenbaum, N. (1984). Percolation thresholds in the three-dimensional sticks system. *Phys. Rev. Lett.* **52**, 1465–1468.

Chen, N., Koker, M. K. A., Uzun, S. & Silberstein M. N. (2016). In-situ X-ray study of the deformation mechanisms of non-woven polypropylene. *Int. J. Sol. Struct.* **97–98**, 200–208.

Chiu, S. N., Stoyan, D., Kendall, W. S., & Mecke, J. (2013). *Stochastic geometry and its applications*, 3rd ed. Wiley, Chichester, UK.

Corte, H. & Lloyd, E. H. (1965). Fluid flow through paper and sheet structure. In: Consolidation of the paper web. *Transactions of the 3rd Fundamental Research Symposium.* BPBMA, London, 981–1009.

Crain, I. K. & Miles, R. E. (1976). Monte Carlo estimates of the distributions of the random polygons determined by random lines in the plane. *J. Stat. Comput. Simul.* **4**, 293–325.

Dean, P & Bird, N. F. (1967). Monte Carlo estimates of critical percolation probabilities. *Proc. Camb. Phil. Soc.* **63**, 477–479.

Deogekar, S., Yan, Z. & Picu, R. C. (2019). Random fiber networks with superior properties through network topology control. *J. Appl. Mech.* **86**, 081010.

van Dillen, T., Onck, P. R. & van der Giessen, E. (2008). Models for stiffening in cross-linked biopolymer networks: a comparative study. *J. Mech. Phys. Sol.* **56**, 2240–2264.

Eichorn, S. J. & Sampson, W. W. (2005). Statistical geometry of pores and statistics of porous nanofibrous assemblies. *J. R. Soc. Interface* **2**, 309–318.

Ekman, A., Miettinen, A., Tallinen, T. & Timonen, J. (2014). Contact formation in random networks of elongated objects. *Phys. Rev. Lett.* **113**, 268001.

Erdős, P. & Rényi, A. (1959). On random graphs. *Public. Mathem.* **6**, 290–297.

Evans, K. E. & Gibson, A. G. (1986). Prediction of the maximum packing fraction achievable in randomly oriented short fiber composites. *Compos. Sci. Technol.* **25**, 149–162.

Falconer, K. (2003). *Fractal geometry: Mathematical foundations and applications.* Wiley, Chichester, UK.

Fedorov, E. S. (1891). The symmetry of regular systems of figures, *Proc. Imperial St. Petersburg Mineral. Soc.* **28**, 1–146.

Feng, X., Deng, Y., Henk, W. & Blote, J. (2008). Percolation transitions in two dimensions. *Phys. Rev. E* **78**, 031136.

Flory, P. J. (1941). Molecular size distribution in three dimensional polymers. I. Gelation. *J. Amer. Chem. Soc.* **63**, 3083–3090.

de Gennes, P. G. (1979). *Scaling concepts in polymer physics.* Cornell University Press, Ithaca, NY.

Gneiting, T. & Schlather, M. (2004). Stochastic models that separate fractal dimension and the Hurst effect. *SIAM Review* **46**, 269–282.

Hovi, J. P. & Aharony, A. (1996). Scaling and universality in the spanning probability for percolation. *Phys. Rev. E* **53**, 235–253.

Jacobsen, J. L. (2014). High-precision percolation thresholds and Potts-model critical manifolds from graph polynomials. *J. Phys. A* **47**, 135001.

Kallmes, O. J. & Corte, H. (1960). The structure of paper – The statistical geometry of an ideal two dimensional fiber network. *Tappi J.* **43**, 737–752.

Komori, T. & Makishima, K. (1977). Number of fiber-to-fiber contacts in general fiber assemblies. *Textile Res. J.* **47**, 13–17.

Kumar, S., Kurtz, S. K., Banavar, J. R. & Sharma, M. G. (1992). Properties of a 3D Poisson-Voronoi tesselation: A Monte Carlo study. *J. Stat. Phys.* **67**, 523–551.

Lazar, E. A., Mason, J. K., MacPherson, R. D. & Srolovitz, D. J. (2013). Statistical topology of 3D Poisson-Voronoi cells and cell boundary networks. *Phys. Rev. E* **88**, 063309.

Lewis, F. T. (1928). The correlation between cell division and the shapes and sizes of prismatic cells in the epidermis of cucumis. *Anatomical Record.* **38**, 341–376.

Li, J. & Zhang, S.-L. (2009). Finite-size scaling in stick percolation. *Phys. Rev. E.* **80**, 040104(R).

Macosko, C. W. & Miller, D. R. (1976). A new derivation of average molecular weights of nonlinear polymers. *Macromolecules* **9**, 199–206.

Meijering, J. L. (1953). Interface area, edge length and number of vertices in crystal aggregates with random nucleation. *Philips Res. Rep.* **8**, 270–290.

Mertens, S. & Moore, C. (2012). Continuum percolation thresholds in two dimensions. *Phys. Rev. E* **86**, 061109.

Miles, R. E. (1964). Random polygons defined by random lines in plane. *Proc. Nat. Acad. Sci.* **52**, 901–907.

de Molina, P. M., Lad, S. & Helgeson, M. E. (2015). Heterogeneity and its influence on the properties of difunctional poly(ethylene glycol) hydrogels: Structure and mechanics. *Macromolecules* **48**, 5402–5411.

Nardin, M., Papirer, E. & Schultz, J. (1985). Contribution a l'etude de l'empilements au hazard de fibres et/ou des particules spheriques. *Powder Technol.* **44**, 131–140.

Neckar, B. & Das, D. (2012). *Theory of structure and mechanics of fibrous assemblies.* WPI Press, New Delhi.

Negi, V. & Picu, R. C. (2019). Mechanical behavior of nonwoven non-crosslinked fibrous mats with adhesion and friction. *Soft Matter* **15**, 5951–5964.

Newman M. (2018). *Networks.* Oxford University Press, Oxford.

Novellani, M., Santini, R. & Tadrist, L. (2000). Experimental study of the porosity of loose stacks of stiff cylindrical fibers: Influence of the aspect ratio of fibers. *Eur. Phys. J. B* **13**, 571–578.

Okabe, A., Boots, B. N., Sugihara, K. & Chiu, S. N. (1992). *Spatial tessellations: Concepts and applications of Voronoi Diagrams.* Wiley, Chichester, UK.

Pan, N. (1993). A modified analysis of the microstructural characteristics of general fiber assemblies. *Textile Res. J.* **63**, 336–345.

Panyukov, S. & Rabin, Y. (1996). Polymer gels: Frozen inhomogeneities and density fluctuations. *Macromolecules* **29**, 7960–7975.

Parkhouse, J. G. & Kelly, A. (1995). The random packing of fibers in three dimensions. *Proc. R. Soc. London A* **451**, 737–746.

Philipse, A. P. (1996). The random contact equation and its implications for (colloidal) rods packings, suspension and anisotropic powders. *Langmuir* **12**, 1127–1135.

Philipse, A. P. & Kluijtmans, S. G. J. M. (1999). Sphere caging by a random fiber network. *Physica A* **274**, 516–524.

Picu, R. C. & Hatami-Marbini, H. (2010). Long-range correlations of elastic fields in semi-flexible fiber networks. *Comput. Mech.* **46**, 635–640.

Provatas, N., Haataja, M., Asikainen, J., et al. (2000). Fiber deposition models in two and three spatial dimensions. *Coll. Surf. A* **165**, 209–229.

Quinn, K. P. & Winkelstein, B. A. (2008). Altered collagen fiber kinematics define the onset of localized ligament damage during loading. *J. Appl. Physiol.* **105**, 1881–1888.

Sampson, W. W. (2009). *Modelling stochastic fibrous materials with Mathematica.* Springer, London.

Stockmayer, W. H. (1943). Theory of molecular size distribution and gel formation in branched-chain polymers. *J. Chem. Phys.* **11**, 45–52.

Tarasevich, Y. Y. & Eserkepov, A. V. (2018). Percolation of sticks: Effect of stick alignment and length dispersity. *Phys. Rev. E* **98**, 062142.

Toll, S. (1993). On the tube model for fiber suspensions. *J. Rheol.* **37**, 123–125.

Venkataraman, S. K., Coyne, L., Chambon, F., Gottlieb, M. & Winter, H. H. (1989). Critical extent of reaction of polydimethylsiloxane polymer network. *Polymer* **30**, 2222–2226.

Williams, S. R. & Philipse, A. P. (2003). Random packings of spheres and spherocylinders simulated by mechanical contraction. *Phys. Rev. E* **67**, 051301.

Xu, W., Su, X. & Jiao, Y. (2016). Continuum percolation of congruent overlapping spherocylinders. *Phys. Rev. E.* **93**, 032122.

Yi, Y. B., Berhan, L. & Sastry, A. M. (2004). Statistical geometry of random fibrous networks, revisited: Waviness, dimensionality and percolation. *J. Appl. Phys.* **96**, 1318–1327.

Zagar, G., Onck, P. R. & van der Giessen, E. (2011). Elasticity of rigidly crosslinked networks of athermal filaments. *Macromolecules* **44**, 7026–7033.

5   Affine Deformation

This chapter introduces the concepts of affinity and nonaffinity in quantitative terms and presents the affine models of network mechanics. Although not always accurate, the affine deformation interpretation provides the simplest and most intuitive analytic description of the mechanical behavior of networks. Basic relations from continuum mechanics are reviewed first.

5.1   Continuum Kinematics and Structural Parameters

This section presents a summary of the continuum kinematic descriptors used in this chapter and in Chapter 7 to formulate constitutive laws for network materials. The measures of the fiber orientation used to establish a link between the constitutive description and the network structure are also reviewed.

The deformation is customarily described by the mapping function defining the relation between the current, $\mathbf{x}$, and reference coordinates, $\mathbf{X}$, of a material point. The difference between these two vectors provides the displacement vector, $\mathbf{u}$. Since the displacement incudes the rigid body translation of the respective object which is not related to the stress or the stored strain energy, it is convenient to work instead with the deformation gradient, $\mathbf{F} = \mathrm{Grad}(\mathbf{x})$, of components $F_{ij} = \partial x_i / \partial X_j$, which does not contain information about the rigid body displacement.

The definition of $\mathbf{F}$ indicates that a vector of the reference configuration, $\mathbf{dX}$, is mapped to the current configuration as:

$$\mathbf{dx} = \mathbf{F}\,\mathbf{dX}. \qquad (5.1)$$

Equation (5.1) represents both a rotation and a stretch of vector $\mathbf{dX}$. According to the polar decomposition theorem, $\mathbf{F}$ can be written as a combination of deformation and rotation, $\mathbf{F} = \mathbf{RU}$ (or alternatively $\mathbf{F} = \mathbf{VR}$), with $\mathbf{R}$ being the rigid body rotation and $\mathbf{U}$ (or $\mathbf{V}$) describing the deformation. Then, $\mathbf{dx} = \mathbf{RU}\,\mathbf{dX}$, which indicates that $\mathbf{dX}$ is first stretched and then rotated to obtain $\mathbf{dx}$.

The square of the length of the deformed vector $\mathbf{dx}$ can be computed as:

$$\mathbf{dx} \cdot \mathbf{dx} = F_{ij}\,dX_j\,F_{ik}\,dX_k = \left(\mathbf{F}^{\mathrm{T}}\mathbf{F}\right)_{jk} dX_j\,dX_k = \mathbf{C} : (\mathbf{dX} \otimes \mathbf{dX}), \qquad (5.2)$$

where $\mathbf{C} = \mathbf{F}^T\mathbf{F} = \mathbf{U}^2$ is the right Cauchy–Green deformation tensor which, due to the symmetrization operation by which it is defined, does not contain information about any of the rigid body deformation modes, and hence can be properly related to the stress state. An alternative descriptor of the deformation is the left Cauchy–Green deformation tensor, which is defined as $\mathbf{B} = \mathbf{F}\mathbf{F}^T = \mathbf{V}^2$. $\mathbf{B}$ similarly filters out the rigid body modes. Note that $\mathbf{C} = \mathbf{R}^T \mathbf{B} \mathbf{R}$.

Considering $\mathbf{N}$ to be a unit vector, the tensor $\mathbf{A} = \mathbf{N} \otimes \mathbf{N}$ is known as the structure (or orientation) tensor and defines the orientation of the respective vector in the reference configuration; also defined in Eq. (4.2). From Eq. (5.2) one may infer that $\mathbf{C} : \mathbf{A} = \mathbf{C}\mathbf{N} \cdot \mathbf{N} = \lambda_N^2$ represents the square of the stretch of a segment of unit length oriented in direction $\mathbf{N}$ in the reference configuration, λ_N.

The structure tensor is used to describe the orientation of fibers in network materials and in fibrous composites. For a population of unit vectors $\mathbf{N}$ in 3D, the orientation distribution is represented by a function $p(\Theta, \Phi)$ defined over the reference unit sphere, where Θ and Φ represent the two Euler angles relative to the orthogonal reference frame. Symbols shown in capitals refer to the reference configuration. $p(\Theta, \Phi) \sin \Theta \, d\Theta \, d\Phi$ represents the fraction of these vectors oriented in the angular range $(\Theta, \Theta + d\Theta)$ and $(\Phi, \Phi + d\Phi)$, and fulfills the normalization condition $\int p(\Theta, \Phi) \sin \Theta \, d\Theta \, d\Phi = 1$, where the integration is performed over the reference unit sphere. The structure tensor is computed in the average sense by integrating over all possible orientations, that is,

$$\langle \mathbf{A} \rangle = \int (\mathbf{N} \otimes \mathbf{N}) p(\Theta, \Phi) \sin \Theta \, d\Theta \, d\Phi. \tag{5.3}$$

To simplify notation, in what follows we drop the angular parentheses and we refer to the system average of the orientation tensor by $\mathbf{A}$. In this case, $\mathbf{C} : \mathbf{A}$ represents the average of the square of the stretch ratio for the entire population of unit vectors $\mathbf{N} = (\sin \Theta \cos \Phi, \sin \Theta \sin \Phi, \cos \Theta)$ described by $p(\Theta, \Phi)$.

Preferential fiber orientation can be described using $p(\Theta, \Phi)$. A useful example is provided by the mapping of the normal distribution to the unit sphere. Consider a family of fibers oriented with cylindrical symmetry about a generic direction defined by $\mathbf{N}_0$, such that, when probed in the plane perpendicular to $\mathbf{N}_0$, the distribution appears isotropic. If the axis with respect to which the Euler angle Θ is measured is aligned with $\mathbf{N}_0$, the distribution is independent of Φ and it is only a function of Θ. Then, the normal distribution over the sphere is given by the function:

$$p(\Theta) = \sqrt{\frac{2\zeta}{\pi}} \frac{\exp\left[\zeta \cos(2\Theta) + \zeta\right]}{\mathrm{erfi}\left(\sqrt{2\zeta}\right)}, \tag{5.4}$$

where parameter ζ represents the density of fiber orientations in the vicinity of the direction defined by $\mathbf{N}_0$, and is related to the variance of the equivalent Gaussian distribution. The erfi function in the denominator is the imaginary error function. This distribution is properly normalized as $\int_0^{\pi} p(\Theta) \sin \Theta \, d\Theta = 1$. Eq. (5.4) is the spherical equivalent of the von Mises distribution which is the mapping of the normal distribution onto the unit circle, in 2D. The uniform distribution over the interval $\Theta \in (0, \pi]$ (independent of Θ) is recovered for $\zeta \to 0$.

For the distribution of Eq. (5.4) defined about a generic unit vector $\mathbf{N}_0$, the structure tensor has a particularly simple form (Gasser et al., 2006):

$$\mathbf{A} = \kappa\mathbf{I} + (1 - 3\kappa)(\mathbf{N}_0 \otimes \mathbf{N}_0), \qquad (5.5)$$

where $\kappa = \frac{1}{2}\int_0^\pi p(\Theta)\sin^3\Theta\, d\Theta$ is a decreasing function of ζ. For the uniform distribution (random orientation of segments), with $\zeta \to 0$, $\kappa = 1/3$, and $\mathbf{A} = (1/3)\mathbf{I}$. The perfect orientation in direction $\mathbf{N}_0$ corresponds to $\zeta \to \infty$, $\kappa = 0$, and hence $\mathbf{A} = \mathbf{N}_0 \otimes \mathbf{N}_0$. The functional form of Eq. (5.5) is remarkably simple. It indicates that the structure tensor of this transversely isotropic distribution of fiber orientations is similar to that of a population of fibers perfectly oriented in direction $\mathbf{N}_0$ (represented by the second term on the right side), plus an isotropic component, $\kappa\mathbf{I}$.

Returning to the discussion of the measures of deformation, we recall the Green–Lagrange strain tensor given by $\mathbf{E} = 1/2(\mathbf{C} - \mathbf{I})$, which represents the Lagrangian large strain. Note that the Green–Lagrange strain for an individual fiber (or family of fibers of same alignment) of structure tensor $\mathbf{A}$ is $(\lambda^2 - 1)/2 = (\mathbf{C} : \mathbf{A} - 1)/2$. The stress in the respective fiber can be computed based on this axial strain using the fiber constitutive equation.

The Eulerian version is the Almansi strain computed as $\mathbf{H} = 1/2(\mathbf{I} - \mathbf{B}^{-1})$. Both $\mathbf{E}$ and $\mathbf{H}$ reduce to the small strain tensor, ε, under small deformations. The Jacobian of the transformation which represents the ratio of the current to the reference volumes is given by $J = \det(\mathbf{F})$. An isochoric deformation requires $J = 1$.

If the deformation is viewed in a frame aligned with the eigenvectors of $\mathbf{C}$, tensors $\mathbf{U}$ and $\mathbf{C}$ are diagonal. The diagonal entries of $\mathbf{C}$ are the squares of the principal stretches, $\lambda_i^2, i = 1\ldots 3$. The principal stretch λ_i is the ratio of the lengths in the deformed and reference configurations of a segment aligned with the i-th eigenvector of $\mathbf{C}$. The eigenvalues of $\mathbf{B}$ are also λ_i^2, while the eigenvectors of $\mathbf{B}$ are the eigenvectors of $\mathbf{C}$ rotated with $\mathbf{R}$.

If the deformation is not isochoric, the volumetric component can be separated out by writing:

$$\mathbf{F} = J^{1/3}\overline{\mathbf{F}}. \qquad (5.6)$$

The deformation gradient $\overline{\mathbf{F}}$ corresponds to a constant volume deformation, while the volumetric component is $J^{1/3}\mathbf{I}$. With this, one obtains the corresponding tensors:

$$\mathbf{C} = J^{2/3}\overline{\mathbf{C}}, \text{ and}$$
$$\mathbf{E} = J^{2/3}\overline{\mathbf{E}} + \frac{1}{2}\mathbf{I}\left(J^{2/3} - 1\right). \qquad (5.7)$$

where $\overline{\mathbf{E}} = 1/2(\overline{\mathbf{C}} - \mathbf{I})$.

The relation between kinematics and the stress state is established through the free energy function, $\Psi(\mathfrak{A}, T)$, as discussed in Section 2.2. Here, $\mathfrak{A}$ includes parameters describing the kinematics and parameters describing the structure. If the structure is represented only by the structure tensor, the free energy per unit mass can be written $\Psi(\mathbf{E}, \mathbf{A}, T)$.

The stress is obtained from the free energy density by taking the derivative relative to the conjugated deformation tensor. Specifically, the second Piola–Kirchoff stress (PK2) results as:

$$\mathbf{\Pi} = \rho_0 \frac{\partial \Psi(\mathbf{E}, \mathbf{A}, T)}{\partial \mathbf{E}} = 2\rho_0 \frac{\partial \Psi(\mathbf{E}, \mathbf{A}, T)}{\partial \mathbf{C}}, \tag{5.8}$$

where ρ_0 is the mass density in the reference configuration. Ψ is the free energy density per unit mass and the multiplication by the density provides the free energy per unit volume.

The nominal stress, or the first Piola–Kirchoff stress (PK1), results from $\mathbf{\Pi}$ as:

$$\mathbf{S} = \mathbf{F}\mathbf{\Pi}, \tag{5.9}$$

while the Cauchy stress is related to PK2 as:

$$\boldsymbol{\sigma} = \frac{1}{J} \mathbf{F}\mathbf{\Pi}\mathbf{F}^T. \tag{5.10}$$

5.2   Nonaffinity Measures

The concept of affine deformation is generally an approximation and indicates that the local strain field is identical to the far-field, applied strain. To make this idea more accessible, it is useful to consider two examples: a body made from a homogeneous material, and the same object to which a rigid inclusion is added. In both cases, the boundary conditions are defined in terms of displacements, which are selected such to represent a specific strain, ε^∞. In the case of the homogeneous material, these boundary conditions lead to a strain field of constant magnitude throughout the domain, and equal to ε^∞. This type of deformation is called "affine." By contrast, the body containing the inclusion has a spatially varying strain field which reaches the imposed value at the domain boundaries, as required by the boundary conditions. This deformation is nonaffine and, in this case, nonaffinity is due to heterogeneity.

Figure 5.1 shows schematically a similar comparison involving a random network. A square domain containing a planar network in shown in Figure 5.1(a), on which horizontal and vertical lines that pass through the crosslinks are superimposed. Displacement-imposed boundary conditions leading to uniaxial extension are applied and two configurations of the resulting network are shown in Figure 5.1(b) and (c). The fiduciary lines are mapped affinely. In the case shown in Figure 5.1(b), the crosslinks are still on the respective lines, which indicates that the deformation of the network is affine on scales comparable to and larger than the mean distance between crosslinks, l_c. Since fibers are not constrained at points other than the crosslinks, the deformation is not necessarily affine on finer scales. The case in which fibers deform affinely on all scales, larger and smaller than l_c, is denoted as "strict affine deformation." Figure 5.1(c) shows a network configuration corresponding to a fully nonaffine deformation.

The nonaffinity may be quantified using several measures. If the displacement field is known, one may use NA_u:

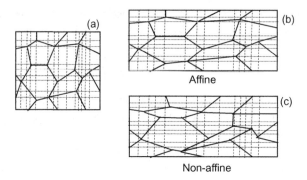

Affine

Non-affine

Figure 5.1 Schematic representation of the affine (b) and nonaffine (c) deformation of the random network in (a) subjected to biaxial deformation. The dashed lines pass through the crosslinks of the configuration in (a) and move affinely. If the network deforms affinely on scales larger than l_c, the crosslinks remain on the fiduciary lines (b), while in the opposite case, the crosslinks move to positions away from the lines (c).

$$NA_u = \left\langle \frac{\left\| \mathbf{u} - \mathbf{u}_{aff} \right\|^2}{\left\| \mathbf{u}_{aff} \right\|^2} \right\rangle, \qquad (5.11a)$$

or

$$NA_u = \frac{\left\langle \left\| \mathbf{u} - \mathbf{u}_{aff} \right\|^2 \right\rangle}{\left\| \boldsymbol{\varepsilon}^\infty \cdot \mathbf{X} \right\|^2}, \qquad (5.11b)$$

where $\mathbf{u}$ represents the actual displacement field and $\mathbf{u}_{aff}$ is the affine displacement predicted based on the far field strain using:

$$\mathbf{u}_{aff}(\mathbf{x}) = \int_0^{\mathbf{x}'} \boldsymbol{\varepsilon}^\infty(\mathbf{x}') \cdot d\mathbf{x}'. \qquad (5.12)$$

In Eq. (5.11), $\| \; \|$ represents the norm of the respective vector and $\langle \; \rangle$ indicates averaging performed over the entire problem domain. If nonaffinity is computed based on crosslink displacements, the average is performed over the set of crosslinks. The difference between the two forms of Eq. (5.11) emerges from the normalization. Equation (5.11a) provides the system average of the relative magnitudes of the nonaffine displacement $\left\| \mathbf{u} - \mathbf{u}_{aff} \right\|^2$ to the local affine displacement. The normalization in Eq. (5.11b) is made with a position-independent displacement computed using the affine strain and a vector $\mathbf{X}$ tied to the structure. For example, in uniaxial deformation, $\mathbf{X}$ may be a system-spanning vector oriented in the loading direction, such that $\left\| \boldsymbol{\varepsilon}^\infty \cdot \mathbf{X} \right\|$ is the largest displacement applied. Equation (5.11a) is more physically relevant, but more susceptible to numerical uncertainties due to the large contribution to the average of domains in which $\left\| \mathbf{u}_{aff} \right\|$ is small. Due to this reason, the form of Eq. (5.11b) is more often encountered in the literature.

An alternative measure of nonaffinity is computed based on the strain field and is similar to that of Eq. (5.11):

$$NA_\varepsilon = \frac{\left\langle \|\boldsymbol{\varepsilon} - \boldsymbol{\varepsilon}^\infty\|^2 \right\rangle}{\|\boldsymbol{\varepsilon}^\infty\|^2}. \tag{5.13}$$

This measure is useful in the case of continua, in which the strain is defined at all points of the problem domain. In the case of networks, the position-dependent strain is only defined in an average sense, for example by fitting a displacement function of position to the actual displacements of the network crosslinks (Hatami-Marbini and Picu, 2008). If the network is first divided in sub-domains and this procedure is applied to each sub-domain separately, it leads to a piecewise constant description of the strain, somewhat similar to the representation of the actual displacement field with constant-strain elements in a finite element model. In this case, it is implicit that the nonaffinity refers to a specific length scale corresponding to the average size of the respective sub-domains.

Both nonaffinity measures of Eqs. (5.11) and (5.13) represent the fluctuations of the respective fields. The mean value of the strain field must fulfill $\langle \boldsymbol{\varepsilon} \rangle = \boldsymbol{\varepsilon}^\infty$.

Since all materials are, at one scale or another, heterogeneous, the affine deformation is necessarily an approximation which is sometimes made in order to facilitate the evaluation of the homogenized, system scale material properties.

A classic example of this type is found in the literature on composite materials. In this context one seeks to evaluate the homogenized properties of the composite (e.g., its stiffness) given the properties of its constituents. Making the affine assumption, that is, implying that each constituent of the composite is subjected to the nominal strain applied macroscopically, leads to an expression for the effective stiffness which is the weighted average of the stiffness values of the constituents. The weights are the volume fractions of the constituents. This is known as the Voigt average and represents the upper bound of the composite stiffness for all possible geometric arrangements of the constituents and for the given volume fractions. The lower bound of the composite stiffness is provided by the Reuss model, which assumes that all constituents are subjected to the same stress. In this case the effective compliance of the composite results as the weighted average of the constituent compliances.

The affine deformation assumption is occasionally used in the context of fibrous materials to facilitate the evaluation of the effective behavior of the network and to derive macroscopic constitutive descriptions. Section 5.3 presents the derivation of the effective constitutive response of affinely-deforming athermal and thermal random networks. The application of these representations to various types of networks is reviewed in Section 7.4.

5.3   Affine Deformation Models

5.3.1   General Formulation

Under the affine assumption, all fibers move as dictated by the applied, network-scale deformation gradient, $\mathbf{F}$. $\mathbf{F}$ is typically applied to the fiber end-to-end vectors, which is equivalent to moving the crosslinks affinely and hence imposing affine deformation on scales larger than l_c. If the fiber response is axial, meaning that the effect of a fiber on the two crosslinks it connects is to apply a force directed along its end-to-end vector, the network is of "central force" type. All models of polymeric chains presented in Section 2.4, including the Gaussian, Langevin, and worm-like chain models, are of this type. Athermal fibers store energy in both axial and bending modes and can be included in this category only in the limit in which the bending contribution is neglected, or when the crosslinks are of pin-joint type (type 3 in Figure 3.1), and hence fibers are not loaded by moments. We consider in this section only interactions of central force type.

Consider an ensemble of N_f fibers of different unit vectors $\mathbf{N}$ in the reference configuration, where the orientation of $\mathbf{N}$ is described by a distribution function $p(\mathbf{N}) = p(\Theta, \Phi)$ over the unit sphere, with $\int p(\mathbf{N})d\mathbf{N} = \int p(\Theta, \Phi) \sin \Theta \, d\Theta \, d\Phi = 1$. The stretch of a fiber of unit vector $\mathbf{N}$ is $\lambda(\mathbf{N})$ and is given by Eq. (5.2). The orientation unit vector in the current configuration corresponding to $\mathbf{N}$ is $\mathbf{n} = \mathbf{FN}/|\mathbf{FN}|$. The resulting force in the fiber is $f_f \mathbf{n}$ and is a function of $\lambda(\mathbf{N})$. The reference orientation distribution is mapped to the current distribution $p^*(\mathbf{n})$, which is normalized such that $\int p^*(\mathbf{n})d\mathbf{n} = \int p^*(\theta, \varphi) \sin \theta \, d\theta \, d\varphi = 1$. The average Cauchy stress in the network can be computed with the virial formula which, for this system of central force interactions (forces act only between crosslinks connected by fibers and along the line connecting the respective crosslinks) has the form:

$$\boldsymbol{\sigma} = \frac{N_f}{V} \int L f_f(\mathbf{n})(\mathbf{n} \otimes \mathbf{n}) p^*(\mathbf{n}) p_L^*(L) d\mathbf{n} \, dL = \rho \int f_f(\mathbf{n})(\mathbf{n} \otimes \mathbf{n}) p^*(\mathbf{n}) d\mathbf{n}, \qquad (5.14)$$

where V is the current volume of the network, $L = \lambda(\mathbf{N})L_0$ is the current length of a fiber of length L_0 and orientation $\mathbf{N}$ in the reference frame, $d\mathbf{n}$ stands for the element of area of the unit sphere in the current configuration, $\sin \theta \, d\theta \, d\varphi$, and p_L^* represents the distribution of fiber lengths.[1] In the second equality it was assumed that the fiber length and orientation are uncorrelated random variables, such that the length and orientation-dependent terms can be separated.[2] The density in the current configuration appears on the right side as $\rho = N_f \langle L \rangle / V$ after the integration over the distribution of fiber lengths in the current frame, $p^*(L)$.

Moving to the reference configuration, the PK2 stress can be computed using Eqs. (5.10) and (5.14):

$$\boldsymbol{\Pi} = \rho_0 \int f_f(\lambda(\mathbf{N})) \frac{\mathbf{N} \otimes \mathbf{N}}{\lambda^2(\mathbf{N})} p(\mathbf{N}) d\mathbf{N}, \qquad (5.15)$$

where $\rho_0 = \rho J$ is the fiber density in the reference configuration (total length of fiber per reference volume). The force in the fiber is kept generic in Eq. (5.15), and may be

---

[1] One may consider instead a joint distribution function of lengths and orientations in the reference configuration (see, e.g., Lee and Carnaby, 1992).

[2] A situation in which the fiber length and orientation are not independent variables was considered in Komori and Itoh (1991a, 1991b).

computed based on the free energy function of individual filaments. This free energy may contain entropic terms as in the Gaussian (Section 2.4.1.1) and Langevin (Section 2.4.1.2) thermal models, or may be purely energetic and, hence, athermal. This formulation was used in a more general setting for other materials as well (Bazant and Oh, 1985), and was generalized in Carol et al. (2004).

The stress state can be evaluated with Eqs. (5.14) and (5.15) provided the distribution functions of fiber orientation are known. In order to avoid the evaluation of the distribution function in the current configuration, one may choose to compute the PK2 stress with Eq. (5.15) based on the (usually known) distribution function in the reference configuration, $p(\mathbf{N})$, and then evaluate the Cauchy stress with Eq. (5.10).

The distribution function $p^*(\mathbf{n})$ is not identical to the reference distribution $p(\mathbf{N})$. To clarify their relation, consider the 2D case in which $p(\mathbf{N})$ can be written as $p(\theta)$, where θ is the angle in the reference configuration between the fiber direction $\mathbf{N}$ and the X_1 axis of the reference frame in which $\mathbf{F}$ is defined, that is, the distribution is axisymmetric with respect to X_1. The orientation distribution in the deformed configuration, $p^*(\mathbf{n})$, can be written in terms of the equivalent angle θ^*. Then, $p^*(\theta^*) = p(\theta)/(d\theta^*/d\theta)$. Since $\mathbf{n} = \mathbf{FN}/|\mathbf{FN}| = (\cos\theta^*, \sin\theta^*)$, the expression $d\theta^*/d\theta$ can be computed in terms of the components of $\mathbf{F}$. The distribution $p^*(\theta^*)$ results:

$$p^*(\theta^*) = p(\theta(\theta^*))J\frac{1+\tan^2\theta^*}{(F_{22}-F_{12}\tan\theta^*)^2+(F_{21}-F_{11}\tan\theta^*)^2},\qquad(5.16)$$

where $J = \det(\mathbf{F})$. If the reference coordinate system is aligned with the principal directions of $\mathbf{C}$, which implies $F_{11} = \lambda_1, F_{22} = \lambda_2, F_{12} = F_{21} = 0$, the distribution becomes:

$$p^*(\theta^*) = p(\theta(\theta^*))\frac{\lambda_1}{\lambda_2}\frac{1}{\cos^2\theta^* + (\lambda_1/\lambda_2)^2\sin^2\theta^*}.\qquad(5.17)$$

To provide an example, consider an isotropic reference orientation distribution with $p(\theta) = 1/2\pi$ subjected to uniaxial stretch in direction X_1 such that the x_1 axis does not rotate during deformation. The preferential alignment in the stretch direction can be evaluated using the 2D scalar orientation parameter $P_2 = 2\langle\cos^2\theta^*\rangle - 1$, introduced in Section 4.2.1, which becomes:

$$P_2 = \int_0^{2\pi}(2\cos^2\theta^* - 1)p^*(\theta^*)d\theta^* = \frac{\lambda_1-\lambda_2}{\lambda_1+\lambda_2}.\qquad(5.18)$$

Consider now the same problem in 3D, with $F_{11} = \lambda_1, F_{22} = \lambda_2, F_{33} = \lambda_3$ and $F_{ij} = 0, i \neq j$. The orientation distribution in the deformed configuration, $p^*(\mathbf{n})$, can be written as $p^*(\theta^*, \varphi^*) = p(\theta,\varphi)/(d\theta^* d\varphi^* \sin\theta^*/d\theta d\varphi \sin\theta)$, or:

$$p^*(\theta^*,\varphi^*)=p(\theta,\varphi)\left(\frac{\lambda_1^2}{\lambda_2\lambda_3}\right)\frac{1}{\left(\cos^2\theta^* + (\lambda_1/\lambda_2)^2\sin^2\theta^*\left(\cos^2\varphi^* + (\lambda_2/\lambda_3)^2\sin^2\varphi^*\right)\right)^{3/2}}.$$

$$(5.19)$$

If the orientation distribution in the reference configuration is uniform, $p(\theta, \varphi) = 1/4\pi$, and the deformation is uniaxial with $\mathbf{F}$ of components $F_{11} = \lambda_1$ and $F_{22} = F_{33} = \lambda_2$, the 3D orientation parameter P_2 given by Eq. (4.3) can be computed using Eq. (5.19) in terms of the ratio $y = \lambda_1/\lambda_2$ as:

$$P_2 = \int_0^{2\pi} \frac{1}{2}(3\cos^2\theta^* - 1)p^*(\theta^*)d\theta^* = \frac{3}{2}\frac{y^2}{y^2-1}\left(1 - \frac{\arctan\sqrt{y^2-1}}{\sqrt{y^2-1}}\right) - \frac{1}{2}.$$

$$(5.20)$$

The degree of alignment predicted by the affine model depends on the lateral contraction. Figure 5.2 shows the variation of $P_2(\lambda)$ during a uniaxial deformation with $\lambda_1 = \lambda$ and various lateral contractions defined by $\lambda_2 = \lambda_2(\lambda)$. Three cases are considered: the isochoric case with $\lambda_2 = 1/\sqrt{\lambda}$, and two cases described by Eq. (6.12) with $v_0 = 0.5$ and exponent $\beta = 3$ and 6, respectively. In 3D, P_2 takes values between $-1/2$ and 1 in the limits corresponding to perfect alignment perpendicular and parallel to the reference (stretch) direction, respectively. Eq. (6.12) captures the rapid contraction of networks that strain stiffen exponentially, as discussed in Section 6.1.1.4.4. Compared with the isochoric case, the alignment is much more pronounced if λ_2 decreases fast with increasing $\lambda_1 = \lambda$. Therefore, most networks which are not embedded in a volume conserving matrix exhibit more alignment than the isochoric-deforming networks such as gels and rubber. This rapid fiber alignment leads to pronounced strain stiffening in dry networks.

5.3.2   Thermal Networks of Flexible Filaments

The classical theory of rubber elasticity was developed starting with the works of Kuhn (1934) and Guth and Mark (1934). In the affine model it is considered that elastomeric networks are constructed from filaments (molecules) of n "links," or Kuhn

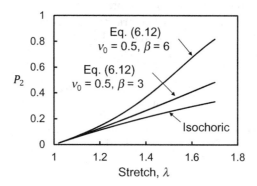

Figure 5.2 Fiber alignment parameter predicted by the affine model for uniaxial tension and assuming three functions describing the lateral contraction, $\lambda_2 = \lambda_2(\lambda)$. The alignment is evaluated relative to the stretch direction.

segments, each of length a. These are subjected to thermal fluctuations and are represented by the Gaussian model of Section 2.4.1.1. The end-to-end length of a filament is evaluated based on the mean position of the fluctuating crosslinks. The crosslinks are implicitly considered of pin-joint type and excluded volume interactions of the filaments are not considered.

Since a central force network deforms affinely (James and Guth, 1943), the end-to-end vector of a filament in the deformed configurations, $\mathbf{L}$, is related to the respective reference vector $\mathbf{L}_0$ as $\mathbf{L} = \mathbf{F}\mathbf{L}_0$. The force in the filament is given by Eq. (2.27):

$$\mathbf{f}_f = 3\frac{k_B T}{na^2}\mathbf{L}. \tag{5.21}$$

The Cauchy stress is computed with Eq. (5.14). We take this opportunity to slightly reformulate the derivation in Section 5.3.1 and rewrite Eq. (5.14) by replacing the integration over the distribution function in the current configuration with a sum over all N_f chains in the current volume V:

$$\boldsymbol{\sigma} = \frac{1}{V}\sum_{k=1}^{N_f}\mathbf{L}^{(k)}\otimes\mathbf{f}_f^{(k)} = 3\frac{k_B T}{na^2}\frac{1}{V}\sum_{k=1}^{N_f}\left(\mathbf{L}^{(k)}\otimes\mathbf{L}^{(k)}\right). \tag{5.22}$$

Using $\mathbf{L} = \mathbf{F}\mathbf{L}_0 = L_0\mathbf{F}\mathbf{N}$, and assuming that the end-to-end length is statistically independent from the orientation, $\mathbf{N}$, Eq. (5.22) becomes:

$$\sigma_{ij} = 3\frac{k_B T}{na^2}\frac{N_f}{V}\langle L_0^2\rangle F_{im}F_{jn}\langle N_m N_n\rangle. \tag{5.23}$$

As in Eq. (5.14), the averages of the end-to-end vector length and that of the unit vector orientation can be written separately since these two variables are statistically independent. The average $\langle N_m N_n\rangle$ is performed over the sphere of unit radius and takes the value $1/3\delta_{mn}$ for a uniform distribution of filament orientations. Hence, using the relation $\langle L_0^2\rangle = na^2$, Eq. (5.23) becomes:

$$\boldsymbol{\sigma} = \frac{N_f}{V}k_B T\mathbf{F}\mathbf{F}^T. \tag{5.24}$$

The network inherits the entropic nature of stress from the filament definition and, hence, $\sigma_{ij} \sim k_B T$.

Since in the Gaussian model the filaments have zero axial force only when their two ends are collocated, an energetic volume-dependent component needs to be added to preserve physical significance. Then, the stress reads:

$$\boldsymbol{\sigma} = \rho_\# k_B T\mathbf{B} - g(J)\mathbf{I}, \tag{5.25}$$

where $\rho_\# = N_f/V$ is the number density of filaments in the current configuration effectively engaged in the network (excluding dangling ends and free chains) and $J = \det\mathbf{F}$ is the Jacobian of the transformation.

Consider a uniaxial deformation in direction x_1 (the coordinate system is aligned with the eigenvectors of $\mathbf{C}$ and $\mathbf{R} = \mathbf{I}$), defined by

$$
\mathbf{F} = \begin{pmatrix} \lambda_1 & 0 & 0 \\ 0 & \lambda_2 & 0 \\ 0 & 0 & \lambda_3 \end{pmatrix}, \tag{5.26}
$$

with $\lambda_1, \lambda_2, \lambda_3$ being the stretch ratios in the three directions. The Cauchy stress reads $\sigma_{ii} = G_0 \lambda_i^2 - g(J)$, where $G_0 = \rho_\# k_B T$ is the "network modulus." If the network is isotropic, the lateral contraction is identical in the two directions orthogonal to the stretch direction and $\lambda_2 = \lambda_3$.

Imposing the traction free boundary condition in the x_2 and x_3 directions, $\sigma_{22} = \sigma_{33} = 0$, it results that $g(J) = G_0 \lambda_2^2 = G_0 J / \lambda_1$. Then, the Cauchy stress in the loading direction results:

$$
\sigma_{11} = G_0 \left(\lambda_1^2 - \frac{J}{\lambda_1} \right). \tag{5.27a}
$$

The PK1 and PK2 stresses can be computed with Eqs. (5.9) and (5.10) and result:

$$
S_{11} = G_0 J \left(\lambda_1 - \frac{J}{\lambda_1^2} \right), \tag{5.27b}
$$

$$
\Pi_{11} = G_0 J \left(1 - \frac{J}{\lambda_1^3} \right). \tag{5.27c}
$$

Note that $G_0 J = \rho_{\#0} k_B T$, which is the network modulus computed with the strand number density in the reference configuration. If the material is incompressible, $J = 1$, and $\rho_\#$, the number density of filaments in the current configuration, becomes equal to the number density in the reference configuration, $\rho_{\#0}$.

In biaxial deformation, $\mathbf{F}$ is given by Eq. (5.26). λ_1 and λ_2 are imposed, while λ_3 results from the condition of zero stress in the x_3 direction. The Cauchy, PK1 and PK2 stresses result:

$$
\sigma_{ii} = G_0 \left(\lambda_i^2 - \frac{J^2}{\lambda_1^2 \lambda_2^2} \right), \tag{5.28a}
$$

$$
S_{ii} = G_0 J \left(\lambda_i - \frac{J^2}{\lambda_i \lambda_1^2 \lambda_2^2} \right), \tag{5.28b}
$$

$$
\Pi_{ii} = G_0 J \left(1 - \frac{J^2}{\lambda_i^2 \lambda_1^2 \lambda_2^2} \right), \tag{5.28c}
$$

where $i = 1, 2$ and the normal stress component in direction x_3 vanishes.

These expressions apply to shear stress states as well, in principal directions. A pure shear deformation applied in the $x_1 - x_2$ plane, with γ being the engineering shear strain, has $\lambda_1 = 1 + \gamma$, $\lambda_2 = 1 - \gamma$, while λ_3 results from the condition that the normal stress in the x_3 direction vanishes. If the material is incompressible, superimposing a hydrostatic deformation $\mathbf{F}_h$, that is, applying the total deformation gradient $\mathbf{F}^* = F F_h$, makes no difference and the states characterized by $\mathbf{F}^*$ are equivalent with each other for any $\mathbf{F}_h$. Selecting $\mathbf{F}_h = [1/(1 - \gamma)]\mathbf{I}$, the deformation that has to be realized in the experiment is a uniaxial stretch in the x_1 direction with $\lambda_1 = (1 + \gamma)/(1 - \gamma)$, confined in direction x_2, $\lambda_2 = 1$. Hence a uniaxial tension confined in the transverse direction mimics a pure shear deformation in the incompressible case. Since it is straightforward to be implemented experimentally, such biaxial deformation was used extensively to characterize the shear response of elastomers (which undergo isochoric deformation). This procedure is known as the "strip-biaxial test."

Two of the assumptions made while deriving Eq. (5.24) require further discussion. It was assumed that all chains have n Kuhn segments, and $\langle L_0^2 \rangle = na^2$, that is, segments are ideal random walks. To examine the effect of these assumption while retaining the condition that chains are Gaussian, we rewrite Eq. (5.22) by allowing each segment to have its own end-to-end length and number of Kuhn segments: $\boldsymbol{\sigma} = (3k_B T/V)\sum_{k=1}^{N_f}\mathbf{L}^{(k)} \otimes \mathbf{L}^{(k)}/n^{(k)}a^2$. Then, Eq. (5.24) becomes $\boldsymbol{\sigma} = \rho_N k_B T\langle L_0^2/na^2 \rangle \mathbf{B}$, where the average is performed over all chains. Clearly, the difference relative to Eq. (5.24) is the presence of $\langle L_0^2/na^2 \rangle$, which is not necessarily 1.

If the chains are represented by random walks but are not of the same length, $\langle L_0^2/na^2 \rangle = 1$ and Eq. (5.24) is recovered. This is the case of vulcanizates which are crosslinked in the melt state (chains have ideal conformations) and the distance between crosslinks (segment length) is Poisson distributed. The conclusion that chain length polydispersity does not affect the shear modulus of Gaussian networks without defects is also obtained within the more accurate phantom chain network model (Section 6.1.1.3.5) in Graessley (1975) and Higgs and Ball (1988).

If all segments of the network have the same length, n, as in the case of end-linked gels (Sakai et al., 2010), $\langle L_0^2/na^2 \rangle = \langle L_0^2 \rangle/na^2$ and the value of the respective parameter depends on the relation between the mean end-to-end length of the actual network and the ideal chain end-to-end length. The actual end-to-end length may be controlled by swelling with a solvent. As the polymer swells, the undeformed end-to-end length varies linearly with the sample size, and hence inversely with $\rho_\#^{1/3}$. It results that in this case the modulus scales with the number density of active strands differently from the prediction of Eq. (5.24), that is, as $G_0 \sim \rho_\#^{1/3} k_B T/na^2$. This implies that, as the degree of swelling increases, G_0 decreases. This trend is observed experimentally in gels (Ruland et al., 2018). Note that additional mechanisms may contribute to the experimentally observed decrease of the modulus during swelling.

If the affine assumption is eliminated and the network is allowed to relax, structural heterogeneity leads to states in which the equilibrium undeformed chain length $L_0^{(k)}$ is unrelated to $n^{(k)}$ (obviously, $L_0^{(k)} < n^{(k)}a$ must be fulfilled). It results via numerical simulations (Gusev, 2019) and analytic considerations (Lin et al., 2019) that $\langle L_0^2/na^2 \rangle < 1$ and networks are softer than predicted by Eq. (5.24). Heterogeneity and nonaffinity lead to strong softening in athermal networks too, as discussed in Chapter 6.

The effect of chain length polydispersity is not obvious in networks of nonlinear chains (e.g., Langevin chains) forced to deform affinely. An analysis of this problem (Tehrani and Sarvestani, 2017) indicates that the prediction of G_0 with an affine model that accounts for the Poisson distribution of chain lengths is close to the prediction of the affine model in which all chains are of length equal to the mean of the distribution.

It is important to observe that the affine model is identical to the phenomenological neo-Hookean model (Section 7.3.1). The neo-Hookean model is defined by a free energy function given by Eq. (7.10) in which the only non-zero coefficient is α_1, that is, $\Psi = \alpha_1(I_1 - 3)$. Then, Eq. (7.11) predicts a Cauchy stress of the form (Eq. (5.24)), that is, $\boldsymbol{\sigma} = 2\rho\alpha_1\mathbf{B} + g(J)\mathbf{I}$. This result can be reached also by computing the total free energy of the ensemble of entropic filaments using the entropy given by Eq. (2.25). The free energy variation associated with a generic $\mathbf{F}$ and expressed in terms of the principal stretches results of neo-Hookean form $\Psi = \alpha_1(I_1 - 3)$. The detailed derivation in available in texts on polymer physics (e.g., Rubinstein and Colby, 2003); see also Rubinstein and Panyukov (2002) for additional details on rubber elasticity.

5.3.3   Thermal Networks of Semiflexible Filaments

The methods outlined in Section 5.3.2 can be used to infer the stiffness of networks of semi-flexible filaments deforming affinely. To keep the analysis simple, the filament deformation is limited to the linear regime. The expansion in series of the right side of Eq. (2.48) provides an approximation for the filament axial stiffness (equivalent to $3k_BT/na^2$ in Eq. (5.21)) which results in $90E_fI_fL_p/L_0^4$ for a filament of length L_0 and persistence length L_p. The axial stiffness of a network segment of length l_c becomes $90E_fI_fL_p/l_c^4$. Following the path leading to Eq. (5.38), the Young's and shear moduli of the network can be derived as:

$$E_0 = 15\rho k_BT\frac{L_p^2}{l_c^3}; \quad G_0 = 6\rho k_BT\frac{L_p^2}{l_c^3}. \tag{5.29}$$

However, the filament response is nonlinear, Eq. (2.48), and the stress–strain response of the network departs from linearity at a critical stress σ_c. The type of stiffening observed in experiments at stresses $\sigma > \sigma_c$ is discussed in Chapter 6 (see Section 6.1.1.4.5 for examples). It is possible to infer σ_c, the limit stress of the linear elastic segment of the stress–strain curve within the limitations of the affine model. The semiflexible filament force–stretch response of Eq. (2.48) departs from its linear approximation at an axial force $P \approx E_fI_f/L_0^2 = k_BTL_p/L_0^2$. This filament force corresponds to a stress of magnitude:

$$\sigma_c = \frac{1}{6}\rho k_BT\frac{L_p}{l_c^2}, \tag{5.30}$$

which also implies that $G_0l_c \sim \sigma_cL_p$. With G_0 and σ_c measured from experiments and l_c evaluated based on the density, this relation can be used to infer the filament persistence length directly from rheological measurements (Lin et al., 2010). It also helps estimate the strain at the transition from linear to nonlinear elasticity, which results as proportional to l_c/L_p (see also Section 6.1.1.4.10). This is a consequence of the affine assumption and of the requirement that filaments deform axially.

5.3.4 Athermal Networks of Axially-Deforming Fibers

The solution for athermal networks of axially deforming fibers can be obtained following the procedure used in Section 5.3.2. The axial stress in a fiber oriented along unit vector $\mathbf{n}$ in the deformed configuration is proportional to the Green–Lagrange strain experienced by the fiber, $\hat{E}$. The force in the fiber can be written as:

$$\mathbf{f}_f = E_f A_f \hat{E} \mathbf{n}. \tag{5.31}$$

The Cauchy stress can be computed using Eq. (5.22):

$$\boldsymbol{\sigma} = \frac{1}{V}\sum_{k=1}^{N_f} \mathbf{L}^{(k)} \otimes \mathbf{f}_f^{(k)} = \frac{E_f A_f}{V}\sum_{k=1}^{N_f} L^{(k)}\hat{E}^{(k)}(\mathbf{n}\otimes\mathbf{n}) = \rho E_f A_f \left\langle \hat{E}^{(k)}(\mathbf{n}\otimes\mathbf{n})\right\rangle, \tag{5.32}$$

where V is the volume of the current configuration and the fiber density results from $\rho = \langle L \rangle N_f / V$. The average is performed over the distribution of fiber orientations in the current configuration. Observing that $\hat{E}$ can be expressed in terms of the applied Green–Lagrange strain $\mathbf{E}$ as $\hat{E} = \mathbf{N}^T \mathbf{E}\mathbf{N} = \mathbf{n}^T \mathbf{F}^{-T}\mathbf{E}\mathbf{F}^{-1}\mathbf{n}$, the Cauchy stress can be written on components as:

$$\sigma_{ij} = \rho E_f A_f \left(\mathbf{F}^{-T}\mathbf{E}\mathbf{F}^{-1}\right)_{kp}\left\langle n_k n_p n_i n_j\right\rangle. \tag{5.33}$$

The term $\mathbf{e} = \mathbf{F}^{-T}\mathbf{E}\mathbf{F}^{-1} = (1/2)\left(\mathbf{I} - \mathbf{B}^{-1}\right)$, where $\mathbf{B} = \mathbf{F}\mathbf{F}^T$ is the Almansi strain, which is the large strain in the Eulerian description. In Eq. (5.33), the average is performed over the distribution of fiber orientations in the current configuration. It is more convenient to reformulate the average in terms of the distribution over the unit sphere in the reference configuration. To this end, one may write $\mathbf{n} = \mathbf{F}\mathbf{N}/|\mathbf{F}\mathbf{N}| = (1/\lambda(\mathbf{N}))\mathbf{F}\mathbf{N} = \mathbf{R}\mathbf{N}$, where $\mathbf{R}$ is the rotation included in the deformation gradient. With this, Eq. (5.33) becomes:

$$\sigma_{ij} = \rho E_f A_f e_{kp} R_{k\alpha} R_{p\beta} R_{i\gamma} R_{j\delta}\left\langle N_\alpha N_\beta N_\gamma N_\delta\right\rangle. \tag{5.34}$$

If the distribution of orientations is uniform over the unit sphere (random in 3D) in the reference configuration, the average can be evaluated and results in $\left\langle N_\alpha N_\beta N_\gamma N_\delta\right\rangle = 1/15\left(\delta_{\alpha\beta}\delta_{\gamma\delta} + \delta_{\alpha\gamma}\delta_{\beta\delta} + \delta_{\alpha\delta}\delta_{\beta\gamma}\right).$[3] Then, Eq. (5.34) leads to:

$$\boldsymbol{\sigma} = \frac{1}{15}\rho E_f A_f (\mathrm{tr}(\mathbf{e})\mathbf{I} + 2\mathbf{e}). \tag{5.35}$$

The PK1 and PK2 stresses result from using Eqs. (5.9) and (5.10), respectively.

---

[3] In two dimensions, this average becomes $\left\langle N_\alpha N_\beta N_\gamma N_\delta\right\rangle = 1/8\left(\delta_{\alpha\beta}\delta_{\gamma\delta} + \delta_{\alpha\gamma}\delta_{\beta\delta} + \delta_{\alpha\delta}\delta_{\beta\gamma}\right).$

As a particular case, consider a uniaxial deformation applied in direction x_1, with $\mathbf{F}$ given by Eq. (5.26) and $\lambda_2 = \lambda_3$. Requiring that the normal stress in the two directions orthogonal to the loading direction vanish, the Cauchy stress in the loading direction reads:

$$\sigma_{11} = \frac{1}{12}\rho E_f A_f \left(1 - \frac{1}{\lambda_1^2} \right). \tag{5.36}$$

In the small strain limit, the Green–Lagrange strain $\mathbf{E}$ is replaced with the equivalent small strain, ε, such that Eq. (5.36) becomes:

$$\sigma_{11} = \frac{1}{6}\rho E_f A_f \varepsilon_{11}, \tag{5.37}$$

and $\varepsilon_{22} = \varepsilon_{33} = -\varepsilon_{11}/4$. It results that the Young's and shear moduli of the affinely deforming athermal random network are:

$$E_0 = \frac{1}{6}\rho E_f A_f; \quad G_0 = \frac{1}{15}\rho E_f A_f, \tag{5.38}$$

and the effective Poisson ratio is $\nu = 1/4$.

In two dimensions, the Young's modulus results in $E = 1/3\rho E_f A_f$ and the shear modulus is $G = 1/8\rho E_f A_f$, while the Poisson ratio becomes $\nu = 1/3$. These results were obtained by Cox (1952), who also discusses the effect of the preferential fiber orientation on the elastic constants of the network.

The elastic constants are proportional to the network density and to $E_f A_f$, which is related to the axial rigidity of fibers. The scaling $E_0 \sim E_f A_f$ is expected since the only deformation mode available is the axial mode. The scaling $E_0 \sim \rho$ is a signature of the affine deformation. In fact, $E_0 \sim \rho$ is sufficiently characteristic for the affine deformation to allow one to infer that deformation is affine simply based on the observation of the $E_0 \sim \rho$ scaling in experiments. This issue is discussed further in Chapter 6.

A discussion of the assumptions made in this model is necessary. Since it was assumed that the affine deformation is applied to all points of the network, initially straight fibers remain straight during deformation. The bending deformation mode is not engaged for any macroscopic applied strain, which eventually leads to the scaling of the network modulus with the axial rigidity of fibers, $E_0 \sim E_f A_f$. However, the imposed affine deformation modifies the angles between fibers at crosslinks. If the network is pin-jointed, the angle variation does not lead to the variation of the strain energy. But if the crosslinks are welded joints, the change of the angles at crosslinks should lead to fiber bending, which is ignored in the affine model.

5.3.5   The Stress–Optical Law

Consider a situation in which the network free energy function depends on $\mathbf{C}$ through the scalar invariant $\mathbf{C} : \mathbf{A} = \langle \mathbf{CN} \cdot \mathbf{N} \rangle = \langle \lambda^2 (\mathbf{N}) \rangle$, that is, $\Psi = \Psi(\mathbf{C} : \mathbf{A})$, where $\mathbf{A}$ is the structure (or orientation) tensor defined in Section 5.1. Then, Eq. (5.8) indicates that the PK2 stress can be written as $\mathbf{\Pi} = 2\rho_0 \Psi' \mathbf{A}$, where Ψ' indicates the derivative of the free energy function with respect to its argument. If Ψ' is independent of $\mathbf{A}$, the stress is proportional to the orientation tensor – a relation known as the stress–optical law. The stress–optical law forms the theoretical basis of photoelasticity.

The free energy of thermal filaments is proportional to the square of the filament stretch (e.g., Eq. (2.25)), and hence it is proportional to $\mathbf{C} : \mathbf{A}^f$, where $\mathbf{A}^f = \mathbf{N} \otimes \mathbf{N}$ for a given fiber. Consider now an ensemble of thermal fibers subjected to an affine deformation characterized by $\mathbf{C}$. Their free energy is proportional to $\mathbf{C} : \langle \mathbf{A}^f \rangle$, where the average is computed over all fibers in the ensemble. However, $\mathbf{C} : \langle \mathbf{A}^f \rangle = \mathbf{C} : \mathbf{A}$, where $\mathbf{A}$ is the orientation tensor of the system and, hence, in this case, $\Psi \sim \mathbf{C} : \mathbf{A}$ and the stress–optical law applies.

In rubber elasticity, both the affine and phantom models indicate that the stress–optical law applies. It has been occasionally claimed that the stress–optical law also applies to collagen-based networks (Bancelin et al., 2015; Nesbitt et al., 2015). However, the evidence for this claim is scarce. In general, the validity of the stress–optical law cannot be extended to the broader range of network materials which do not deform affinely.

Closing this overview of the affine models, it is useful to summarize the main implications of the affine deformation assumption:

▶ A consequence of the affine deformation approximation is that the network inherits the mechanical behavior of individual fibers. If the network is composed from fibers of different properties, the network behavior is predicted as the weighted average of fiber contributions (similar to the rule of mixtures).
▶ The network moduli predicted by the affine models scale linearly with the network density, ρ, and are proportional to the axial rigidity of fibers.
▶ As described in Section 5.4, in athermal networks, if fibers are not straight in the undeformed configuration, the network moduli are not necessarily proportional to the axial fiber rigidity.

5.4 Strict Affine Deformation of Networks of Tortuous Fibers

Imposing an affine deformation does not always imply that the bending mode is not activated. An example is provided by the strict affine deformation of curved fibers.

Consider a fiber described by a curve in 3D which is defined by the reference position vector $\mathbf{W}(X_1, X_2, X_3)$. Let s be the contour length variable and t a parametrization of space $\mathbf{X}$. The tangent vector to the curve is $\mathbf{T} = d\mathbf{W}/ds = d\mathbf{W}/dt[1/(ds/dt)] = \mathbf{W}'/|\mathbf{W}'|$, where use was made of $ds/dt = \sqrt{X_1'^2 + X_2'^2 + X_3'^2} = |\mathbf{W}'|$, and a prime indicates derivative with respect to parameter t. The norm of the tangent vector is unity. The curvature is, by definition, given by

$$\kappa = \left| \frac{d\mathbf{T}}{ds} \right|, \qquad (5.39)$$

and can be evaluated in terms of $\mathbf{W}$ as:

$$\kappa = \frac{|\mathbf{W}' \times \mathbf{W}''|}{|\mathbf{W}'|^3}. \tag{5.40}$$

A strict affine deformation moves $\mathbf{W}$ into $\mathbf{w} = \mathbf{FW}$. We write $\mathbf{F}$ in principal coordinates as $F_{ii} = \lambda_i$ and $F_{ij} = 0, i \neq j$, with λ_i being the principal stretches. The curvature becomes:

$$\kappa^* = \frac{|\mathbf{w}' \times \mathbf{w}''|}{|\mathbf{w}'|^3}$$

$$= \frac{\left[\lambda_2^2\lambda_3^2\left(X_2'X_3'' - X_3'X_2''\right)^2 + \lambda_1^2\lambda_3^2\left(X_3'X_1'' - X_1'X_3''\right)^2 + \lambda_2^2\lambda_1^2\left(X_1'X_2'' - X_2'X_1''\right)^2\right]^{1/2}}{\left[\lambda_1^2 X_1'^2 + \lambda_2^2 X_2'^2 + \lambda_3^2 X_3'^2\right]^{3/2}}.$$

$$\tag{5.41}$$

It is clear that, for a generic deformation with all principal stretches independent of each other, $\kappa^* \neq \kappa$. Even for a dilatation strain, with $\lambda_1 = \lambda_2 = \lambda_3 = \lambda$, the fiber curvature changes from κ to $\kappa^* = \kappa/\lambda$.

Since the curvature changes, a bending moment of magnitude $E_f I_f(\kappa^* - \kappa)$ develops, and any infinitesimal fiber segment stores strain energy in the bending mode. If fibers are straight in the initial configuration $\kappa = \kappa^* = 0$ for any $\mathbf{F}$, that is, a fiber does not develop curvature if it is initially straight. Under these conditions, only the axial mode is enabled and hence the strict affine deformation applied to a system of straight fibers leads to axial strain energy storage and to the scaling of the effective network modulus with $E_f A_f$, as seen in Section 5.3.4. If fibers are tortuous in the reference configuration, both axial and bending modes are enabled by a strict affine deformation and the network stiffness depends on both $E_f A_f$ and $E_f I_f$. Models accounting for both axial and bending deformation of fibers have been developed by several authors (e.g., Komori and Itoh, 1991a, 1991b).

References

Bancelin, S., Lynch, B., Bonod-Bidaud, C., et al. (2015). Ex-vivo multiscale quantitation of skin biomechanics in wild-type and genetically-modified mice using multiphoton microscopy. *Sci. Rep.* **5**, 17635.

Bazant, Z. P. & Oh, B. H. (1985). Microplane model for progressive fracture of concrete and rock. *ASCE J. Eng. Mech.* **111**, 559–582.

Carol, I., Jirasek, M. & Bazant, Z. P. (2004). A framework for microplane models at large strain, with application to hyperelasticity. *Int. J. Sol. Struct.* **41**, 511–557.

Cox, H. L. (1952). The elasticity and strength of paper and other fibrous materials. *Br. J. Appl. Phys.* **3**, 72–81.

Gasser, T. C., Ogden, R. W. & Holzapfel, G. A. (2006). Hyperelastic modelling of atrial layers with distributed collagen fiber orientations. *J. Roy. Soc. Interfaces* **3**, 15–35.

Graessley, W. W. (1975). Elasticity and chain dimensions in Gaussian networks. *Macromolecules* **8**, 865–868.

Gusev, A. A. (2019). Numerical estimates of the topological effect in the elasticity of Gaussian polymer networks and their exact theoretical description. *Macromolecules* **52**, 3244–3251.

Guth, E. & Mark, H. (1934). Zur innermolekularen, Statistik, insbesondere bei Kettenmolekiilen I. *Monats. F. Chem.* **65**, 93–121.

Hatami-Marbini, H. & Picu, R. C. (2008). Scaling of non-affine deformation in random semiflexible fiber networks. *Phys. Rev. E* **77**, 062103.

Higgs, P. G. & Ball, R. C. (1988). Polydisperse polymer networks: elasticity, orientational properties, and small angle neutron scattering. *J. de Physique* **49**, 1785–1811.

James, H. M. & Guth, E. (1943). Theory of the elastic properties of rubber. *J. Chem. Phys.* **11**, 455–481.

Komori, T. & Itoh, M. (1991a). A new approach to the theory of the compression of fiber assemblies. *Textile Res. J.* **61**, 420–428.

Komori, T. & Itoh, M. (1991b). Theory of the general deformation of fiber assemblies. *Textile Res. J.* **61**, 588–594.

Kuhn, W. (1934). Uber die Gestalt fadenformiger Molekule in Losungen. *Kolloid-Z* **68**, 2–15.

Lee, D. H. & Carnaby, G. A. (1992). Compressional energy of random fiber assembly, part I: Theory. *Textile Res. J.* **62**, 185–191.

Lin, T. S., Wang, R., Johnson, J. A. & Olsen, B. D. (2019). Revisiting the elasticity theory of Gaussian phantom networks. *Macromolecules* **52**, 1685–1694.

Lin, Y. C., Yao, N. Y., Broedersz, C. P., et al. (2010). Origins of elasticity in intermediate filament networks. *Phys. Rev. Lett.* **104**, 058101.

Nesbitt, S., Scott, W., Macione, J. & Kotha, S. (2015). Collagen fibrils in skin orient in the direction of applied uniaxial load in proportion to stress while exhibiting differential strains around hair follicles. *Materials* **8**, 1841–1857.

Rubinstein, M. & Colby, R. H. (2003). *Polymer physics*. Oxford University Press, Oxford.

Rubinstein, M. & Panyukov, S. (2002). Elasticity of polymer networks. *Macromolecules* **35**, 6670–6686.

Ruland, A., Chen, X., Khansari, A., et al. (2018). A contactless approach for monitoring the mechanical properties of swollen hydrogels. *Soft Matt.* **14**, 7228–7236.

Sakai, T., Akagi, Y., Matsunaga, T., et al. (2010). Highly elastic and deformable hydrogel formed from tetraarm polymers. *Macromol. Rapid Commun.* **31**, 1954–1959.

Tehrani, M. & Sarvestani, A. (2017). Effect of chain length distribution on mechanical behavior of polymeric networks. *Eur. Poly. J.* **87**, 136–146.

6 Mechanical Behavior and Relation to Structural Parameters

Network materials are diverse and exhibit a broad range of behaviors. This chapter presents the common features of their mechanical response, that emerge from the mechanics of the underlying fiber network. Networks of thermal and athermal fibers are discussed separately. Likewise, the behavior of crosslinked and non-crosslinked networks is reviewed in separate sections. The mechanical response at small and large strains is linear and nonlinear, respectively, and is described in terms of parameters such as the small strain modulus (network stiffness), the Poisson ratio, the tangent stiffness of the nonlinear branch, etc. The relation between material behavior and parameters describing the network structure is defined. The analysis provides guidelines for the design of stochastic networks with prescribed material behaviors.

The mechanical response of materials is typically probed in uniaxial tension/compression and in shear. Network materials exhibit significantly different responses in tension and compression, while the behavior in shear is qualitatively similar to that observed in tension.

Situations in which the network preserves its structural integrity during loading are discussed in this chapter. Damage accumulation, rupture, and the relation between structural parameters and the strength and toughness of network materials are discussed in Chapter 8. The behavior described in this chapter is independent of the deformation rate; rate-dependent effects are presented in Chapter 9.

6.1 Mechanical Behavior of Crosslinked Networks

Many network materials have a crosslinked fiber network as their main structural component. Crosslinking increases the stiffness and strength of the network and reduces its deformability. It also reduces the residual strains upon unloading and the ability of the material to dissipate energy during cyclic loading.

This section focuses on the relationship between network parameters and network behavior. Therefore, it is appropriate to begin by summarizing the structural parameters that control the mechanical behavior of crosslinked networks.

The structural parameters of a generic network are discussed in Chapter 4. The key geometric parameters are: the network density, ρ, defined as the total fiber length per unit volume (for 3D networks) or area (for 2D networks), the crosslink density, ρ_b, defined as the number of crosslinks per unit volume or area, the mean segment length, l_c, the fiber

Table 6.1 Minimum set of independent non-dimensional groups of structural parameters used to describe the mechanical behavior of various types of network materials

Network Type	Non-dimensional Groups of Structural Parameters	Number of Relevant Groups
2D, athermal	$\langle z \rangle$, l_b/L_0, and either ρL_0 or l_c/L_0	3
2D, thermal	$\langle z \rangle$, L_p/l_c, and either ρL_0 or l_c/L_0	3
3D, cellular, athermal	ρl_b^2 or l_b/l_c	1
3D, fibrous, athermal	$\langle z \rangle$, l_b/L_0, and either ρL_0^2 or l_c/L_0	3
3D, fibrous, thermal	$\langle z \rangle$, L_p/l_c, and either ρL_0^2 or l_c/L_0	3

length, L_0, the fiber persistence length, L_p, which describes the tortuosity of fibers (Section 2.4.2), and the connectivity, $\langle z \rangle$, defined in Section 4.2.2.

The fiber material is described by the modulus, E_f, yield stress, σ_y^f, and the strain hardening rate. The fiber modulus may be used as the unit of stress and hence it is not part of the list of parameters. Networks of relatively low density and low crosslink density deform mainly elastically and the plastic parameters of fibers have little to no influence on their overall behavior. The situation is opposite in dense and densely crosslinked networks, which may deform plastically. These issues are discussed in Section 6.1.4.

The fiber cross-section is typically characterized by the diameter, d. The diameter enters the area, A_f, and the moment of inertia of the cross-section, I_f, which are proportional to the fiber axial and bending rigidities, respectively. Although they are associated with different deformation modes, A_f and I_f are not independent in the case of athermal fibers and a single parameter, that is, d, suffices. In networks composed from the same type of fiber, it is common to use for the purpose of characterizing the shape of the fiber cross-section, a parameter with units of length defined as $l_b = \sqrt{E_f I_f / E_f A_f}$. In the case of athermal fibers of circular cross-section, $l_b = d/4$. Parameter l_b also appears in the description of networks of semiflexible filaments (protein fibers of biological importance), but it is not present in the description of Gaussian and Langevin thermal networks, in which filament bending is not considered a viable mode of energy storage. Furthermore, in semiflexible molecular networks the axial and bending stiffnesses are decoupled, as they are not related to the diameter of the filament. In these cases, correlating the axial and bending rigidities with a single parameter is not possible.

This leaves us with a list of seven parameters that, with the exception of $\langle z \rangle$, have units of length or inverse length. The two parameters describing the plastic behavior of fibers are left out of this set, but may be added if athermal networks are discussed and fiber plasticity is important (Section 6.1.4). The Buckingham π analysis indicates that six non-dimensional groups can be formed. In addition, a relation exists between ρ and l_c (Section 4.2.3), and yet another relation links ρ, ρ_b, $\langle z \rangle$, and l_c (Eq. (4.21)). The remaining independent non-dimensional groups are shown in Table 6.1 for 2D and 3D networks of thermal and athermal fibers.

With the exception of the cellular networks whose geometry can be described in the average sense by only one non-dimensional group, a minimum of three groups characterize the structure of all other network types. In the cellular case, $\langle z \rangle = 4$ and the fiber length is identical to the segment length, since each fiber has only two crosslinks.

The groups shown in Table 6.1 represent a minimum set. In the presence of constituents other than the network, dangling ends or fibers not connected to the network, or of structure beyond that of an idealized random network, additional parameters are needed. For example, at least one more parameter is needed to characterize preferential fiber alignment, flocculation requires an additional descriptor of density fluctuations on scales much larger than the fiber length, while, in the presence of a matrix, additional parameters are needed to describe the properties of the embedding medium.

In the remainder of this chapter, the set of network parameters shown in Table 6.1 are related to the various aspects of the network mechanical behavior.

6.1.1   Tension and Shear

6.1.1.1   Generic Mechanical Behavior

Crosslinked network materials exhibit a broad range of behaviors when loaded in tension or shear. It is possible to divide the response in tension into three categories. Figure 6.1(a) shows schematically the stress–stretch curves representative for these categories.

Type A corresponds to a linear elastic behavior up to failure. The effective stiffness and the failure stress of materials in this class are usually large. This type of behavior is seen in densely crosslinked networks, such as paper tested in low humidity conditions, in which the fibers do not have the kinematic freedom to reorganize during loading.

Type B is a behavior with softening. The stress–stretch curve is linear elastic up to a stress beyond which softening is observed. The onset of softening is often referred to as a yield point. The analogy to the yielding phenomenon observed in ductile continua is not quite valid since softening may not be caused by the onset of plastic deformation of fibers. However, we will refer here to the change of slope as the "yield point," for

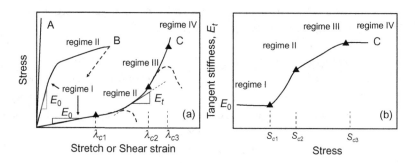

Figure 6.1 (a) Schematic representation of stress–stretch curves showing three types of network behavior labeled A, B, and C. (b) Tangent stiffness–stress representation of curve C in (a).

convenience. Regimes I and II defined in Figure 6.1(a) correspond to the deformation before and after the yield point, respectively. Network materials that exhibit this type of behavior can be divided into two categories based on the nature of unloading. In a small number of cases (e.g., elastomers), the material is hyperelastic and the unloading stress–stretch curve overlaps the loading curve. In most cases, if unloading is performed beyond the onset of softening, hysteresis and/or residual strain are observed. The unloading branch of the characteristic curve may have a slope smaller or equal to that of the loading branch (shown schematically in Figure 6.1(a) by the dashed curve). Failure may be catastrophic or gradual, cases in which the stress–stretch curve drops to zero stress rapidly or in a gradual manner, respectively. Type B is observed in spunbonded nonwovens, where softening is due to fiber yielding, in hydroentangled nonwovens, where softening is due to relative fiber sliding and disentanglement, in paper tested in humid environments, where fiber deformation and possibly relative sliding of fibers facilitate plastic deformation and creep, and in other network materials.

Type C is a behavior with stiffening. In the absence of damage, the characteristic curve exhibits four regimes (although in practical situations only a subset of these may be observed). The small strain response is linear elastic (regime I), followed by two stiffening regimes (regimes II and III), after which the stress–stretch curve becomes linear again in regime IV. This type of behavior is typical for various biological connective tissue and for protein networks such as actin and fibrin gels. Monotonic stiffening is made possible by the fact that fibers in such networks have sufficient kinematic freedom to reorganize during deformation. If damage occurs, the characteristic curve develops a peak, as shown schematically by the dashed lines in Figure 6.1(a). The number of regimes visible in an experiment depends on the magnitude of the stress at which damage sets in and leads to softening followed by the peak stress.

Transitions between regimes are marked by triangles in Figure 6.1(a) and correspond to critical stretches and corresponding stresses, denoted by λ_{ci} and $S_{ci}, i = 1 \ldots 3$. Often, it is more convenient to use the representation of Figure 6.1(b) in which the tangent stiffness, E_t, is plotted against the stress. This plot allows for easier identification of the various regimes. The sketch in Figure 6.1(b) corresponds to the type C curve in Figure 6.1(a) and clearly shows the two extreme linear regimes in which E_t is stress-independent, and the two stiffening regimes, regimes II and III, in-between.

A different representation, known as the Mooney–Rivlin plot, is widely used in the elastomers literature as an alternative representation of the uniaxial stress–stretch curve. This is discussed further in Section 6.1.1.4.5.

The three types of behavior denoted by A, B, and C in Figure 6.1 are exemplified next. Figure 6.2(a) shows stress–stretch curves for nonwovens of two types (Patel and Kothari, 2001). The first is made from spunbond polypropylene fibers which were subjected after spinning to a thermal pressing and bonding process (heat-sealing). Heat-sealing leads to the formation of inter-fiber crosslinks at most fiber-to-fiber contacts and a densely crosslinked network results. The strength of the crosslinks created in this manner is not controlled and has a broad distribution. Such networks

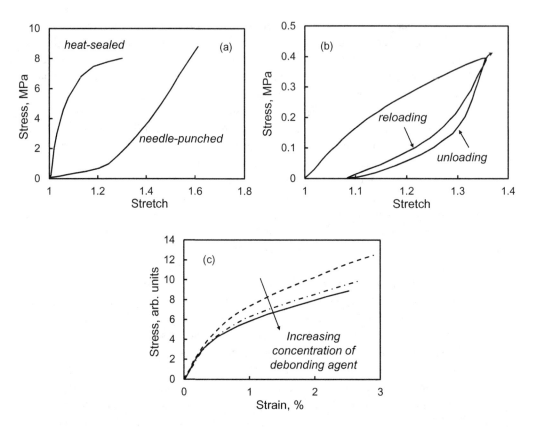

Figure 6.2 (a) Stress–stretch curves for heat-sealed and needle-punched nonwovens of similar areal density and sample dimensions (data from Patel and Kothari, 2001). (b) Stress–stretch curve for electrospun PCL network (republished with permission of ASME from Duling et al., 2008, permissions conveyed through Copyright Clearance Center Inc). (c) Stress–strain curves for paper chemically treated to reduce the strength of the inter-fiber bonds, at various concentrations of debonding agent (data from Seth and Page, 1981).

exhibit in uniaxial tension type B behavior: a linear elastic regime I followed by an apparent yield point and softening. In this case, softening is associated with crosslink rupture and/or fiber yielding. Which process controls the occurrence of the yield point depends on the relative magnitude of the crosslink strength and the yield stress of the fiber material. The typical mechanical response for nonwovens is of type B.

The second curve in Figure 6.2(a) shows the behavior of a needle-punched nonwoven of polyester fibers. The fiber diameter and areal density of this nonwoven are comparable to those of the heat-sealed nonwoven of stress–stretch curve shown in the same figure. Yet, the characteristic curve is quite different. Its convex shape is due to the much larger kinematic freedom of fibers, which are not crosslinked. Needle-punching is a process by which fibers on one side of the mat are pulled across the thickness with the help of a barbed needle. This is not a weaving process, but some degree of fiber entanglement results. The technique provides structural integrity and stiffness to nonwovens made from non-crosslinked fibers of finite length. The slack of fibers introduced by needle-punching leads to the J-shape stress–stretch curve shown in Figure 6.2(a). Hydroentangled non-woven mats exhibit similar behavior. It is important to note that the nonlinearity in both cases shown in this figure is associated with energy dissipation and the response of these materials beyond the initial (and short ranged) regime I is not elastic.

Figure 6.2(b) shows the behavior of another nonwoven subjected to tension: an electrospun network of polycaprolactone (PCL) fibers used for tissue engineering (Duling et al., 2008). This example is shown to highlight the change of the network behavior from the initial loading to the reloading section of the curve. The initial loading branch exhibits the usual concave shape with gradual softening which, in this case, is associated with crosslink rupture. Damage accumulation also accounts for the large hysteresis observed upon unloading. However, the reloading branch is convex, which indicates that the removal of crosslinks during the first loading provides sufficient kinematic freedom to the fibers to change the material response from type B to type C. Note the significant difference relative to a metallic material, for example, which would exhibit concave loading and concave reloading.

The behavior shown in Figure 6.2(b) is typical for networks exhibiting the Mullins effect. The effect refers to situations in which the stress–strain curve during reloading follows the previous unloading segment, as seen in Figure 6.2(b), up to the largest stress reached in the previous cycle. The reloading curve is convex as long as the stress remains below the previous cycle maximum. If the stress increases beyond the largest value previously reached, the stress–strain curve becomes concave and follows the curve that would have been obtained in monotonic loading. The Mullins effect is associated with damage accumulation. Damage leads to softening and a concave stress–strain curve, but new damage accumulates only provided the stress increases beyond the largest stress previously experienced by the sample. In the absence of damage accumulation, the material behavior is hyperelastic.

Figure 6.2(c) shows a particularly revealing dataset that supports the present discussion (Seth and Page, 1981). Several types of paper were made from commercial softwood pulp after adding various amounts of a chemical agent to the pulp, before sheet making, which reduces the strength of the inter-fiber bonds. The top curve in Figure 6.2(c) corresponds to untreated pulp. As the concentration of debonding agent increases and the bond strength is reduced, the apparent yield point decreases. This sheds light on the physical origin of the yield point in this case.

Figure 6.3 shows stress–stretch curves obtained with a mat of electrospun reconsti-tuted collagen nanofibers of 400 nm diameter (Meng et al., 2012). The inset shows the curve for a mat held at 33% relative humidity for several days (dry conditions). This material sustains large stresses and exhibits type B behavior: an initial linear response followed by softening. The stiff response is due to the high density of (primarily cohesive) contacts between nanofibers. The same network shows a very different behavior when hydrated. In this case, the stress decreases by almost two orders of magnitude, the ductility increase dramatically, while the overall shape of the curve changes from concave to convex. The J-shaped curve obtained in the hydrated state is due to the lower density of inter-fiber contacts (relative to the dry state), which provides more kinematic freedom for fiber reorganization during stretching.

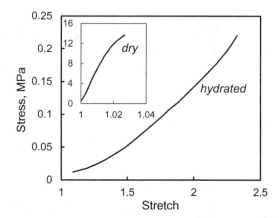

Figure 6.3 Stress–stretch curves for a mat of electrospun reconstituted collagen nanofibers in hydrated and dry (inset) conditions. In the hydrated state the behavior is of type C, while in dry conditions, it is of type B. Data from Meng et al. (2012)

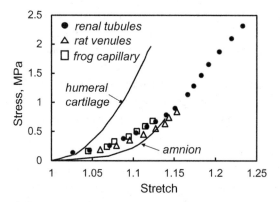

Figure 6.4 Stress–stretch curves for several biological tissues showing type C behavior. Data from Huang et al. (2005), Mauri et al. (2015), Welling et al. (1995), and Swane et al. (1989)

Figure 6.4 shows typical stress–stretch curves obtained in uniaxial tension for a number of biological tissues in which the mechanical response is primarily controlled by a network of collagen fibers. These include samples of the human glenohumeral cartilage (Huang et al., 2005), the human amnion (Mauri et al., 2015), rabbit renal tubules (Welling et al., 1995), rat venules, and frog capillaries subjected to inflation (Swane et al., 1989). All curves exhibit strain stiffening and the characteristic J-shape observed for most collagen-based connective tissue (type C in Figure 6.1). While the small strain stiffness differs from tissue to tissue, and often from sample to sample of the same tissue, the functional form of the stiffening behavior is similar (Section 6.1.1.4). It is important to observe that, under physiological conditions, the applied strains do not exceed ~15%, and hence the tissue functions in regime I and in the lower part of the strain stiffening regime II.

6.1.1.2   Mechanical Stability Condition: The Maxwell Criterion

Consider a structure made from trusses (segments that transmit only axial force and no moments) pin-jointed at all crosslinks. The stability of such constructs was studied in the early days of solid mechanics and a stability criterion was developed by Maxwell (1864). Maxwell's criterion is based on counting the total number of degrees of freedom of the structure and comparing with the total number of constraints imposed by connectors. If the structure has N_X crosslinks and N_e elements, the total number of degrees of freedom is DN_X, where D is the dimensionality of the space in which the structure is embedded, while the number of constraints is N_e. The total number of rigid body modes of the structure is $3(D - 1)$. Hence, one may write the relation $DN_X - N_e = 3(D - 1) + q$, where q represents the number of overconstrained deformation modes (if $q < 0$) or the number of floppy modes (if $q > 0$). If the system is overconstrained, it is stiff against any perturbation. If it has floppy modes, the structure has no stiffness when subjected to an infinitesimal deformation that matches the floppy eigenmodes.

This is made clear by an example. Consider the 4-bar system shown in Figure 6.5(a). It has $N_X = N_e = 4$ and is defined in 2D, that is, $D = 2$. Therefore, $q = 1$, that is, the frame has one floppy mode. Specifically, the frame is a mechanism if loaded in shear, but it is stiff if loaded in tension parallel to any of the bars. By adding a diagonal bar, as shown in Figure 6.5(b), $N_X = 4$, N_e increases to 5, and $q = 0$, that is, the frame is marginally stable. One more diagonal bar (Figure 6.5(c)) renders $q = -1$, which indicates that the frame is overconstrained. We refer to the floppy structure as being subisostatic, and a structure with no floppy modes as isostatic.

This reasoning can be applied to a truss network structure of any architecture. However, it is more convenient to evaluate the stability (or isostaticity) condition in terms of the mean connectivity index, $\langle z \rangle$, which avoids the need to count all crosslinks and segments of the structure. Observe first that $\langle z \rangle$ is related to N_X and N_e. Specifically, ignoring dangling segments, $N_e = \langle z \rangle N_X/2$. Replacing in Maxwell's criterion, one obtains $N_X(D - \langle z \rangle/2) = 3(D - 1) + q$. At the limit of stability, $q = 0$, and hence $D - \langle z \rangle_c/2 = 3(D - 1)/N_X$, which goes to zero as the number of crosslinks N_X increases. The stability condition results:

$$\langle z \rangle_c = 2D. \tag{6.1}$$

Interestingly, most networks of practical interest have $\langle z \rangle < \langle z \rangle_c$ and are sub-isostatic. In 2D, Mikado networks have $\langle z \rangle \le \langle z \rangle_c = 4$. If fibers of a Mikado structure are of infinite length, then $\langle z \rangle = 4$. However, realistic structures (e.g., paper) have fibers of

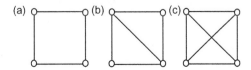

Figure 6.5 Examples of (a) sub-isostatic, (b) marginally stable, and (c) isostatic 2D structures.

finite length and this creates dangling segments which contribute little to the mechanics of the network. Hence, the actual mean connectivity of such quasi-2D networks is smaller than 4. Fibrous networks in 3D have similar mean connectivity, $\langle z \rangle \leq 4$, which is much smaller than the isostaticity value of $\langle z \rangle_c = 2D = 6$. Open cell foams have connectivity similar to that of 3D Voronoi tessellations, that is, $\langle z \rangle = 4$, which is also below the isostatic 3D limit. Networks made from bundles of fibers that split and merge at crosslinks (e.g., buckypaper and collagen networks) have $z = 3$, a value much smaller than $\langle z \rangle_c$.

Sub-isostatic networks may be stabilized in several ways. Enabling the bending deformation mode of fibers and transforming the crosslinks from pin-joints to welded joints that transmit both forces and moments (type 1 in Figure 3.1) renders the structure isostatic. This is the situation found in most network materials, since a connector that has absolutely zero bending stiffness and does not transmit moments is only a model, and not physical reality. Likewise, in most practical cases, crosslinks allow the transmission of moments between fibers.

Another interesting situation is that of a fibrous network with crosslinks of type 2 in Figure 3.1. In this case moments are transmitted along fibers, but only forces are transmitted between fibers. This situation is encountered in biological semiflexible networks. The Maxwell counting analysis presented in Huisman and Lubensky (2011) indicates that the network is isostatic if the number of crosslinks per fiber is larger than three in 2D and six in 3D.

Sub-isostatic networks may also be stabilized by the presence of self-equilibrating internal stress states (eigenstress). Tensegrity structures (Fuller, 1976) are an excellent demonstration of this principle. These are structures made from elements loaded in tension (e.g., ropes) and struts loaded in compression. Maxwell counting indicates that they are sub-isostatic and, hence, floppy. Nevertheless, the spatial arrangement of the components allows for the development of a self-equilibrated internal stress which, in turn, provides finite stiffness to the structure. To account for the stabilizing effect of internal stress, the Maxwell relation may be modified as (Calladine, 1978):

$$DN_X - N_e - s = 3(D - 1) + q, \tag{6.2}$$

where s is the number of independent self-equilibrated stress states in the structure.

Sub-isostatic structures without residual stress loaded in one of their zero stiffness modes acquire stiffness once a certain strain is reached during deformation. The transition from the floppy to the stiff deformation mode is a critical point (Sharma et al., 2016). The strain that needs to be applied in order to reach a regime of non-zero stiffness increases monotonically with the distance to the isostaticity point of the undeformed network, $\langle z \rangle_c - \langle z \rangle$.

6.1.1.3   Linear Elastic Behavior: The Small Strain Stiffness
6.1.1.3.1   *Generic Behavior and Relation to Network Parameters*
The relation between the network stiffness measured in regime I (Figure 6.1) and the set of parameters listed in Table 6.1 is discussed in this section. The relation derived

applies equally to the shear and Young's moduli and, hence, symbols E_0 and G_0 are used interchangeably throughout this section. Typically, nonwovens are tested in uniaxial tension and E_0 is reported. Elastomers and biological tissue are tested in both tension and shear, while semiflexible protein networks are probed primarily by rheometry and the storage modulus G' (or G_0) is reported. Since regime I is linear elastic, the shear and Young's moduli of isotropic networks are related, $G_0 = E_0/2(1 + \nu)$. Here ν is the usual, small strain Poisson ratio of linear elasticity. E_0 and G_0 exhibit a much larger variation with network parameters than the Poisson ratio.

Considering $\langle z \rangle$ to be constant (its effect is discussed in Section 6.1.1.3.5), the remaining two non-dimensional groups listed in Table 6.1 involve l_b, L_0, and l_c. l_c is equivalent to ρ through Eqs. (4.11), (4.13), and (4.14). We construct using these variables the structural parameter w defined for athermal fibrous networks as:

$$w = \left(\delta \rho L_0^{D-1}\right)^x \left(\frac{l_b}{L_0}\right)^2. \tag{6.3a}$$

In the case of cellular networks,

$$w = \left(\delta \rho l_b^{D-1}\right)^x, \tag{6.3b}$$

with D being the dimensionality of the embedding space and x a parameter. $\delta \rho = \rho - \rho_c$ is the density measured relative to the stiffness percolation threshold. To recall the results related to percolation discussed in Chapter 4, note first that, since we consider here welded crosslinks and fibers with non-zero bending stiffness, the transport and stiffness percolation thresholds are identical. Further, the density at percolation in 2D is given by Eq. (4.38) as $\rho_c L_0 = 5.637$, while in 3D it is given by Eq. (4.41), which can be rearranged as $\rho_c L_0^2 = 0.89 L_0/d$. It should be noted that the density entering Eq. (6.3) represents only fibers connected to the network. If fibers not connected to the network are present, it is assumed that they do not contribute to strain energy storage and are not counted when evaluating ρ.

For convenience, the network modulus is normalized by the affine prediction provided by Eq. (5.38), that is, it is written as $E_0/\rho E_f A_f$. Figure 6.6 shows the variation of the normalized elastic modulus with w. The main figure shows data for 2D fibrous Mikado mat-like structures (Shahsavari and Picu, 2013), while the inset shows equivalent results for 3D fibrous networks. In both cases, exponent x was adjusted to obtain the collapse of the data onto a single curve. The data points correspond to structures of different combinations of parameters, with L_0 varying by a factor of 8, l_c (and ρ) changing by one order of magnitude, and l_b/L_0 varying by six orders of magnitude. An excellent collapse is obtained for both 2D and 3D data and a curve with two well-defined regimes emerges. We refer to this representation as the "master plot."

All networks considered here are structurally isotropic, that is, the fiber orientation distribution is uniform over the unit sphere. Therefore, networks are also mechanically isotropic in the small strain limit. The master plots shown in Figure 6.6 are obtained by

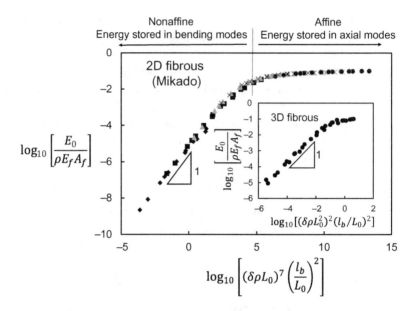

Figure 6.6 Dependence of the small strain stiffness on network parameters. The data points represent many combinations of network parameters. The main figure shows data for 2D fibrous Mikado networks (Shahsavari and Picu, 2013), while the inset shows data corresponding to 3D fibrous networks.

probing the network stiffness in uniaxial tension. One may also probe the network in shear, biaxial, or triaxial tension and measure the shear, biaxial, and bulk moduli, respectively. E_0 may be computed based on any of these moduli using the Poisson ratio evaluated in the same test, provided the applied deformation remains in the linear elastic regime. The resulting values overlap with those in the master plot of Figure 6.6, which is a consequence of the isotropy of the structure and the linearity of the response.

The master plot provides valuable information about network behavior. For large values of w, that is, large network densities and/or fibers relatively stiff in bending (large l_b), a plateau is obtained, which indicates that the normalized modulus is independent of w. In this regime $E_0 \sim \rho E_f A_f$, which suggests that, for the corresponding range of network parameters, the deformation is affine.

As w is reduced, a transition to a regime in which the normalized modulus scales linearly with w results. The modulus decreases fast with decreasing density in this regime, although all structures considered are significantly above the stiffness percolation threshold. Specifically, $E_0 \sim \rho \delta \rho^x$, $E_0 \sim L_0^{(D-1)x-2}$ and $E_0 \sim E_f A_f l_b^2 = E_f I_f$. Since the densities of most network materials are generally significantly larger than the percolation threshold, here and in the remainder of this chapter we approximate $\delta \rho \approx \rho$. Therefore, $E_0 \sim \rho^{x+1}$.

Exponent x depends on the network architecture. The data in Figure 6.6 indicates that for 2D fibrous structures, $x = 7$ and hence the stiffness is approximately proportional to ρ^8, while for the equivalent 3D fibrous network, $x = 2$ and the stiffness is proportional to ρ^3. In open cell foams, Voronoi networks and other 3D cellular networks, the stiffness is proportional to ρ^2 (Gibson and Ashby, 1988).

In periodic lattices, $E_0 \sim \rho^2$ follows directly from dimensional arguments based on the fact that the bending deformation mode of fibers controls the network behavior. Specifically, consider that the stored strain energy is associated with the bending of fibers of length l_c which have bending stiffness $E_f I_f / l_c^3$. The force bending a representative fiber can be computed as $E_f I_f \delta / l_c^3 \sim E_f I_f \varepsilon l_c / l_c^3$, where δ is the beam deflection, which may be approximated as the applied strain ε times l_c. On the other hand, the bending force scales as the applied stress divided by the area corresponding to one fiber in the plane normal to the direction of the load, that is, $\sim l_c^2$. Hence, the effective stiffness results in $E_0 \sim E_f I_f / l_c^4 \sim E_f I_f \rho^2$. Therefore, the scaling $E_0 \sim \rho^2$ is a trivial result emerging from the affine approximation applied to the bending modes. The occurrence of higher exponents x indicates that additional physics is at play and the nonaffinity of the deformation field is essential.

To gain insight into the physical aspects of this behavior, it is useful to evaluate the degree of nonaffinity of the deformation field as a function of w. Figure 6.7(a) shows the nonaffinity parameter of Eq. (5.11b) computed for the 2D fibrous networks used to produce the data in Figure 6.6. As w decreases, the nonaffinity increases monotonically, with the variation being more pronounced in the transition region between the two regimes of the master plot of Figure 6.6. The fact that the nonaffinity measure is small at large values of w agrees with the observation that the modulus follows the prediction of the affine model: $E_0 \sim \rho E_f A_f$. The large and small w regions of the master plot are denoted as affine and nonaffine, respectively. It should nevertheless be

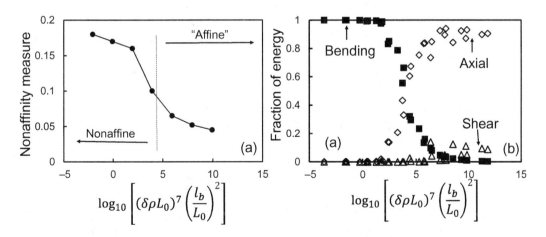

Figure 6.7 (a) Variation of the nonaffinity measure of Eq. (5.11b) with the structural parameter w (also used on the horizontal axis of Figure 6.6) for 2D Mikado networks. (b) Fraction of the total strain energy stored in the axial, bending, and shear deformation modes of fibers function of the structural parameter w. Similar results are obtained for other types of networks in 2D and 3D.

observed that the deformation of stochastic networks is never exactly affine, even at large w. This is in strong contrast with the deformation of a periodic network structure, which is affine on and above the scale of periodicity, as long as strain localization does not take place.

Further, it is useful to evaluate the partition of strain energy stored in the various deformation modes of fibers. Since the fibers of athermal networks behave as mechanical beams, strain energy may be stored in the axial, bending, torsion, and shear modes. Figure 6.7(b) shows the fractions of the total strain energy stored in the bending, axial, and shear modes of 2D fibrous networks used to generate the data shown in the main part of Figure 6.6. Structures with large w store most of the strain energy in the axial deformation mode of fibers, while those with small w store the largest fraction of the strain energy in the bending mode. It results that the transition from one regime of the master plot to the other corresponds to a bending to axial transition (Head et al., 2003; Wilhelm and Frey, 2003). The curves in Figure 6.7(b) representing the axial and bending modes cross at a value of w which corresponds to the affine–nonaffine transition of the master plot (Figure 6.6). The bending to axial transition is not associated with percolation, since the transition is observed at network densities much larger than those of percolation.

The fact that the axial mode dominates at large w is in agreement with the observation that deformation is approximately affine in the respective w range (Figure 6.7(a)) and with the fact that the network modulus is proportional to the axial rigidity of fibers, $E_0 \sim E_f A_f$. In contrast, at small w, $E_0 \sim E_f I_f$, the deformation is highly nonaffine and strain energy is stored primarily in the bending deformation mode of fibers. We call the large w regime "affine" and "axially dominated," and the low w regime, "nonaffine" and "bending dominated."

The shear mode does not store significant amounts of energy, but its contribution is noticeable in the affine regime, where l_c is small and a large fraction of fiber segments have a low aspect ratio. The torsion mode is engaged in 3D networks, but it does not store significant amounts of energy either, which is a remarkable and somewhat surprising result.

The strongly biased energy partition observed in Figure 6.7(b) suggests an interesting interpretation. The network behaves as if the two essential deformation modes, axial and bending, are coupled in series. The softest mode stores the largest fraction of the strain energy – a situation similar to that of two springs connected in series. Small w values result if l_b is small and l_c is large, which implies that the bending stiffness of fibers is much smaller than the axial stiffness. Hence, the bending mode stores the largest fraction of the strain energy. The opposite situation is encountered when w is large.

The results discussed here apply equally to thermal networks of semiflexible filaments. However, some details are different, as outlined by the following analysis. Consider a network composed of worm-like chains, as is adequate for semiflexible filaments. Based on the results in Figure 6.7(b), we represent the network by a set of three springs coupled in series. The three springs represent the axial, bending, and thermal modes of the WLC filaments. The stiffness of the axial spring is

$k_a = E_f A_f / l_c = E_f I_f / (l_c l_b^2)$, that of the spring representing the bending mode is $k_b = E_f I_f / l_c^3$, while the stiffness of the filament representing the thermal component is $k_t \sim E_f I_f L_p / l_c^4$ (Section 2.4.4). The effective stiffness of the set of three springs is computed as $1/k_{eff} = 1/k_a + 1/k_b + 1/k_t$. Considering that E_0 (or G_0) is proportional to k_{eff}/l_c, the modulus can be written as:

$$\frac{E_0}{\rho E_f A_f} \sim \frac{1}{1 + \left(\frac{l_c}{l_b}\right)^2 \left(\alpha_1 + \alpha_2 \frac{l_c}{L_p}\right)}, \tag{6.4}$$

where α_1 and α_2 are numerical constants.

In the athermal limit, $L_p \to \infty$. For $l_b \gg l_c$, $E_0 / \rho E_f A_f \to const$ and the axial and affine limit is recovered. For $l_b \ll l_c$, Eq. (6.4) leads to $E_0 / \rho E_f A_f \sim (l_b/l_c)^2$, which is equivalent to $E_0 \sim \rho^2 E_f I_f$, and the bending-dominated nonaffine limit is recovered. This simplified analysis leads to the trivial scaling of the modulus with the density. The nontrivial exponents emerge from the collective behavior of fibers in the nonaffinely deforming network.

If the thermal component dominates, that is, $L_p \to 0$, Eq. (6.4) indicates that $E_0 / \rho E_f A_f \sim (l_b/l_c)^2 (L_p/l_c)$. Therefore, $E_0 \sim \rho^2 E_f I_f L_p / l_c \sim \rho^2 k_B T L_p^2 / l_c$, which is identical to Eq. (5.29) representing the affine small strain stiffness of thermal networks.

Let us recall that for the affine model of classical rubber elasticity, in which the chains are loaded only in the axial mode, the network stiffness results proportional to the (number) density of elastically active chains $E_0 \sim G_0 \sim \rho_\# k_B T$ (Section 5.3.2). Neglecting dangling ends and strands not connected to the network, $\rho_\# \sim \rho$. The proportionality of E_0 and ρ emerges, once again, from the affine deformation assumption.

6.1.1.3.2 *Discussion of the Role of Randomness: Floppy Modes*

The results presented in Section 6.1.1.3.1 indicate that one of the key effects of nonaffinity is the modification of the stiffness–density relationship and the emergence of nontrivial scaling exponents. Two perspectives are summarized here to provide a deeper understanding of this observation.

The linear elastic deformation of periodic structures is affine on length scales equal and larger than that of periodicity. Therefore, their stiffness scales linearly or quadratically with the density when the deformation is dominated by the axial or bending deformation modes of fibers, respectively. One may inquire how fast this scaling changes (exponent x in $E_0 \sim \rho^x$ increases) when the degree of structural randomness increases. This issue was studied in the literature on foams and cellular materials (Zhu et al., 2000; Roberts and Garboczi, 2002; Schraad and Harlow, 2006), but the study reported in Heussinger and Frey (2007) is closer to the perspective taken here. A family of 2D periodic networks are considered in Heussinger and Frey (2007) and structural stochasticity is introduced by imposing random displacements of the crosslinks. The magnitude of the random displacements, in units of l_c, is sampled from a distribution of imposed variance. The scaling of the stiffness with network parameters in the bending-dominated regime is evaluated for multiple values of the structural perturbation distribution variance. It is concluded that the transition away from the trivial affine scaling is slow and convergence to the corresponding nontrivial exponent of a fully stochastic network takes place only once the standard deviation of the distribution becomes larger than approximately $l_c/2$.

A treatment aiming to quantify the relationship between the degree of nonaffinity and the nontrivial scaling exponents was developed in Heussinger and Frey (2006) and Heussinger et al. (2007). It starts with the observation that, in the limit of infinitesimal fiber diameter, $E_f I_f$ vanishes and the structure converges to a network of trusses. Since $\langle z \rangle$ is below the isostatic limit of Eq. (6.1), floppy modes exist. Their magnitude depends on the level of nonaffinity allowed locally by the fiber arrangement. It is assumed that the relative nonaffine displacements scale linearly with L_0. The bending stiffness of fibers stabilizes these floppy modes, rendering them non-floppy, that is, of finite stiffness. One infers that the strain energy stored in the bending mode of a segment of length l accommodating a relative nonaffine displacement $\delta_{na} \sim L_0$ should scale as $E_f I_f \delta_{na}^2/l^3 \sim E_f I_f L_0^2/l^3$. Integrating over the entire (Poisson) distribution of segment lengths and considering the number density of segments to be equal to one segment per l_c^2 area of the network (in 2D), an estimate of the total strain energy density is obtained. This allows computing a relation between the effective modulus of the network and the density which take the form $E_0 \sim \rho^x$. For a 2D Mikado network, x results equal to 7, which is close to the value found numerically (Figure 6.6).

The floppy modes analysis provides conceptual insights into the origin of the exceptionally strong dependence of the stiffness on density in the nonaffine regime, based on reasonable considerations about the mechanics of the fibrous assembly. It shows that nonaffinity leads to the increase of exponent x above the prediction of affine models.

6.1.1.3.3 Stiffness–Density Relation for Several Network Materials
Cellulose Networks

Paper is a dense and densely crosslinked network of cellulose fibers. There are many types of paper that vary greatly in density. The basis weight of paper is evaluated per unit area of the sheet as the density multiplied by the paper thickness. Common paper has thickness of about 100 μm, but special papers can be made as thin as 10 μm. The basis weight of lineboard is larger than 200 g/m^2, office (copier) paper is about 80 g/m^2, while the paper for newspapers is about 40 g/m^2. The density of paper at percolation can be estimated with Eq. (4.38) and results in the range 0.5–1 g/m^2 (Niskanen, 1998; Alava and Niskanen, 2006), function of the fiber length.

Given its high fiber density, the deformation of paper is approximately affine. Consequently, the stiffness is expected to be proportional to the density (both in the 2D projection and in 3D). This is verified experimentally for densities larger than about 500 Kg/m^3 ($\sim$50 g/m^2) (Niskanen, 2012). Figure 6.8 shows the Young's modulus of paper of different densities (Rigdahl and Hollmark, 1986) and two fiber lengths: 1.7 and 2.2 mm. The vertical axis is normalized by the affine estimate. It is seen that at densities above the threshold indicated by the arrow, the affine estimate approximates the data well, while below the respective threshold, the modulus depends nonlinearly on the density. Based on the limited density range available

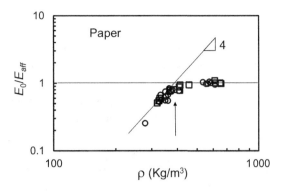

Figure 6.8 Variation of the elastic modulus of low-density paper with the sheet density. The vertical axis is normalized by the affine prediction ($E_{aff} \sim \rho$). Data for papers with fiber length 1.7 mm and 2.2 mm is included. Data from Rigdahl and Hollmark (1986)

below the nonaffine–affine transition, $E_0 \sim \rho^4$ for both fiber lengths considered. This data set indicates a transition density of about 400 Kg/m^3. The curve in Figure 6.8 has the general features of the master plot of Figure 6.6.

It is important to emphasize the distinction between the real structure, which is three-dimensional, and the 2D projection. As fibers are stacked on top of each other, the top fibers make contact only with several fibers immediately below them, and not with all fibers through the thickness. A thickness emerges beyond which the sheet develops two surface layers and a core. The structure of the core is independent of the sheet thickness. Therefore, the thickness increases linearly with the basis weight (or the density in projection, ρ_{2D}). Let t be the thickness of the sheet. The modulus evaluated in projection, E_0^{2D}, is related to the real modulus in 3D as $E_0^{2D} = t E_0^{3D}$, while the projected density is related to the real density as $\rho_{2D} = t\rho_{3D} = t\rho$. If $E_0^{2D} \sim \rho_{2D}$, it results that $E_0^{3D} \sim \rho$. If one measures the apparent stiffness, that is, the 2D stiffness normalized by the basis weight, ρ_{2D}, it is observed that it becomes independent of the basis weight at large area weights (or when t becomes large enough). Since $E_0^{3D}/\rho = E_0^{2D}/\rho_{2D}$, the same statement can be made about the normalized 3D stiffness, which is what is observed in Figure 6.8 at large ρ.

Collagen Networks

Figure 6.9 shows data for type I reconstituted collagen networks of different densities (Vader et al., 2009; Motte and Kaufman, 2012; Licup et al., 2015). The collagen concentration is varied from 0.5 to 4 mg/ml in all three works. The resulting gels are tested by rheometry at 37°C. The results of the three data sets are highly consistent with each other and demonstrate that the modulus (storage modulus, G', in this case) varies with the density as $G' \sim \rho^3$. This is in close agreement with the numerical data shown in the inset of Figure 6.6 for 3D athermal fibrous networks. It is important to observe that, for the entire range of physiological collagen concentrations, the gel deformation is in the nonaffine, bending-dominated regime of the master plot of Figure 6.6.

Actin

The rheology of semiflexible protein networks exhibits the thermal signature of their filaments. The arguments presented in MacKintosh et al. (1995) indicate that the stiffness (here, storage modulus) should scale with the density as $G' \sim \rho^{11/5}$. This relation follows from the assumption that deformation is affine and the filaments behave like entropic worm-like chains of axial stiffness proportional to $k_B T L_p^2 / l_c^4$ (Section 2.4.4). It is implicitly assumed that the filaments are loaded axially during network deformation. Experimental results from Janmey et al. (1991) obtained with actin solutions of various concentrations, shown in Figure 6.10, suggest that the modulus is a power law of network density with an exponent close to 2. Furthermore, rheology data for actin solutions crosslinked by the actin-binding protein scruin presented in Gardel et al. (2004) indicates a stiffness–density relation of the

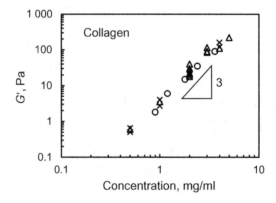

Figure 6.9 Variation of the storage modulus with the collagen concentration in reconstituted collagen gels tested at 37°C. Data from Vader et al. (2009), Motte and Kaufman (2012), and Licup et al. (2015)

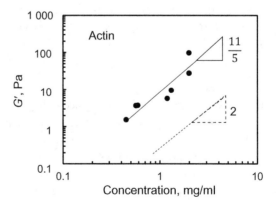

Figure 6.10 Variation of the storage modulus of actin gels with the actin concentration. Data from Janmey et al. (1991)

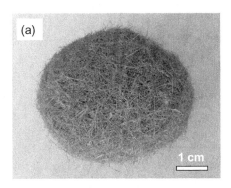
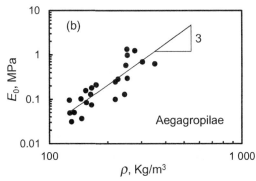

Figure 6.11 (a) An aegagropila cluster and (b) the dependence of its small strain stiffness with the fiber density. Adapted from Verhille et al. (2017)

form $G' \sim \rho^{2.5}$, a function of the degree of crosslinking and bundling of the actin filaments.[1] The trivial scaling $G' \sim \rho^2$ corresponding to the affine deformation of fibrous 3D networks (Eq. (6.4)) is also shown in Figure 6.10 for reference. This data set supports the conclusion that deformation is approximately affine, since the scaling exponent is close to 2. This implies that the signature of the axial deformation mode of individual filaments must be visible in the behavior of the network. This data is too noisy to support a doubtless conclusion regarding the value of the exponent.

Aegagropilae: Networks of Rods

An unusual example of a fibrous network is provided by the aegagropilae clusters encountered along the Mediterranean coast. These are clumps of fibers produced by the natural decay of the acicular leaves of the marine plant *Posidonia Oceanica*. The fibers aggregate forming clusters which are washed ashore. The clusters are ellipsoidal and have sizes varying from 1 to 10 cm (Figure 6.11(a)), while the fiber length ranges from 0.5 to 20 mm. Fibers are straight, densely packed, and somewhat entangled. This provides finite tensile and compressive stiffness to the cluster. While the resulting network is not of practical importance, it provides an example of fibrous network with a nontrivial scaling of the stiffness with the density. Figure 6.11(b) shows such data from Verhille et al. (2017), indicating that the stiffness measured in compression is a power function of density with $x = 3$, $E_0 \sim \rho^3$. This scaling is similar to that shown in the inset of Figure 6.6 for fibrous 3D network models and to that of Figure 6.9 obtained for reconstituted collagen.

Gels and Rubber

Gels are crosslinked in an aqueous solution at a predefined concentration. Upon crosslinking, the gel may swell or may be kept at the concentration at which it was crosslinked. We explore first a case in which the elasticity of a gel is probed without

---

[1] Scruin produces both crosslinking and bundling of actin filaments.

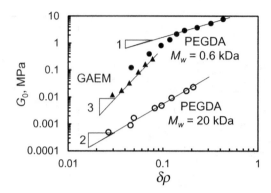

Figure 6.12 Scaling of the shear modulus of two types of gels with the density of the polymer relative to the percolation density, $\delta\rho = (\rho - \rho_c)$. Data for polyethylene glycol with acrylate end groups (PEGDA) of molecular weight 0.6 and 20 kDa (de Molina et al., 2015) and for a green algae extract methacrylate (GAEM) in the equilibrium swelled states (Ruland et al., 2018) is shown.

swelling, in the as-crosslinked state. Figure 6.12 shows the variation of the shear modulus with the network density of a gel of polyethylene glycol with polymerizable acrylate end groups (PEGDA). The actual density of the polymer is known and the percolation density, ρ_c, is estimated based on the strand molecular weight (de Molina et al., 2015). Additional discussion on gelation is presented in Section 4.5. Data for $\rho > \rho_c$ and two strand molecular weights is shown. In both cases $G_0 \sim (\rho - \rho_c)^x$, with $x > 1$. The smaller molecular weight samples show a transition to linear scaling $G_0 \sim (\rho - \rho_c)$ at larger densities. The behavior is qualitatively similar to that described in Figure 6.6. The authors of the cited work associate the emergence of exponents x larger than 1 with the heterogeneity of the gel. Indeed, as discussed in this section, heterogeneity may produce an increase of the degree of nonaffinity, which leads to an increase of x.

Along the same lines, data collected from the literature and presented in Oyen (2014) demonstrates scaling of the modulus with the polymer concentration with $x = 2$ for agar gels and with $x = 3$ for acrylamide gels. Agar is a physical gel, while acrylamide is a chemical gel. The exponents x reported for chemical networks are generally clustered in the range 2–2.5 (Candau et al., 1981; Vasiliev et al., 1985; Yohsuke et al., 2011), while a broader range is reported for physical gels, with values as high as 3.8 for swelled polyvinyl alcohol gels (Yohsuke et al., 2011).

The figure includes data from Ruland et al. (2018) for gels made from a green algae extract methacrylate (GAEM), which is crosslinked at a predefined concentration and then allowed to swell to equilibrium. To allow data interpretation, it is necessary to determine the expectation for the stiffness–equilibrium concentration scaling based on the usual affine deformation assumption. The first step in this process is to evaluate the gel concentration in the equilibrium state, after swelling.

The degree of swelling depends on the solvent–polymer affinity. In the case of good solvents, the network volume increases as the solvent is absorbed into the gel.

The energy penalty for this process is associated with stretching the molecular strands during network expansion. This is balanced by the free energy gain due to mixing.

In the Flory–Huggins theory of solutions, the free energy of the mixture includes a mixing term proportional to the product of the concentrations of the two species and a coefficient χ which quantifies the strength of the interaction between the two species. The equilibrium state can be evaluated from the balance of thermodynamic forces associated with mixing and network deformation. In Flory's theory of gels (Flory, 1971), the equilibrium concentration of the gel upon swelling is given by $\tilde{c} \sim n^{-4/5}\tilde{v}^{-3/5}a^{-6/5}$, where a is the monomer size, n is the number of monomers per strand, and $\tilde{v}$ is the excluded volume parameter given by $\tilde{v} = a^3(1 - 2\chi)$. This relation is supported by experimental data. To infer the stiffness of the gel, one may invoke the affine assumption, which leads to $G_0 \sim \rho_\# k_B T$ (Section 5.3.2), where $\rho_\# = \tilde{c}/n$. Eliminating the strand length, n, between Flory's expression for the concentration and G_0, one obtains (de Gennes, 1979):

$$G_0 \sim k_B T \tilde{c}^{2.25}(1 - 2\chi)^{3/4}. \tag{6.5}$$

This expression refers to equilibrium gels in the swelled state. The stiffness may be controlled either by varying $\tilde{c}$ or by varying the affinity, χ. In the first case, gels are made at various polymer concentrations and allowed to swell by using the same solvent. This is the case of the experiments leading to the data shown by triangles in Figure 6.12. The other possibility is to use different solvents such to control gel stiffness via χ. The exponent describing the dependence of the stiffness on network density (here, taken proportional to the concentration), that is, 2.25 (Eq. (6.5)), is frequently quoted as the reference value for gels in good solvents. When using θ-solvents,[2] the exponent increases to 3 (Zrinyi and Horkay, 1987).

Rubber elasticity models predict a linear dependence of the stiffness on the density of elastically active molecular strands effectively engaged in the network. Both the affine and phantom network models predict that the nominal stress of an isotropic network subjected to uniaxial deformation (e.g., applied in direction x_1) is given by $S_{11} = G_0(\lambda_1 - \lambda_2^2/\lambda_1)$, where G_0 is the network modulus, while λ_1 and λ_2 are the stretches in the loading and transverse directions, respectively (Section 5.3.2).

In the affine model, G_0 is proportional to the number density of elastically active strands (in the current notation and assuming all strands are connected to the network, $\rho_\# = \rho/l_c$), $G_0 = (\rho/l_c)k_B T$. In the phantom network model, $G_0 = (1 - 2/\langle z \rangle)(\rho/l_c)k_B T$ (see also Section 6.1.1.3.5). The constrained junction theory (Ronca and Allegra, 1975; Erman and Mark, 1997) predicts the modulus as

---

[2] Polymer conformations are described by the self-avoiding walk model which accounts for the excluded volume interaction of chain segments. These interactions force the polymer coil to expand relative to the prediction of the random walk model in which segment overlap is allowed. Solvents may modify the volume spanned by the coil. In a good solvent, polymer chains expand because interactions between polymer segments and solvent molecules are favored. If the solvent is poor, the chain contracts because interactions between polymer segments are favored. The θ-condition is reached when the solvent (called a θ-solvent) is sufficiently poor to allow the chain to take conformations compatible with the random walk model.

$G_0 = (1 - 2h/\langle z\rangle)(\rho/l_c)k_BT$, where factor h takes values between 0 and 1. Hence, the affine model prediction is larger than that of the phantom network model, while the constrained junction theory provides values between these two limits. Experimental data falls below the affine prediction.

The small strain stiffness E_0 is related to G_0. This relationship may be obtained by taking the derivative of $S_{11} = G_0(\lambda_1 - \lambda_2^2/\lambda_1)$ with respect to λ_1 and considering the limit in which both stretches are close to 1. It results that $E_0 = 2G_0(1 - d\lambda_2/d\lambda_1)$. In the isochoric case, which is appropriate for rubber, $J = \lambda_1\lambda_2^2 = 1$ and, hence, $\lambda_2 = 1/\sqrt{\lambda_1}$, while $E_0 = 3G_0 \sim \rho$. In the small strain limit, $d\lambda_2/d\lambda_1 = -\nu$, where ν is the Poisson ratio of the equivalent isotropic material, and the usual relation $E_0 = 2G_0(1 + \nu)$ results.

Although the stiffness is proportional to the density in rubber elasticity, which is usually an indication that deformation is affine, it should be noted again that this is only an approximation in stochastic networks. The deformation of rubbers is not exactly affine; direct evidence for crosslink fluctuations was obtained by neutron scattering (Oeser et al., 1988; Erman and Mark, 1997).

On the macroscopic scale, the degree of nonaffinity may be evaluated based on the ratio of the second and first coefficients in the Mooney–Rivlin model, α_2/α_1, Eq. (7.10). Coefficients α_1 and α_2 result by fitting the Mooney–Rivlin model to experimental data. If $\alpha_2 = 0$, the material behavior is neo-Hookean and affine. Non-zero values of α_2/α_1 indicate that deformation is nonaffine. Experiments confirm that reducing the molecular weight between crosslinks and increasing $\langle z\rangle$, which, according to the present discussion, should render deformation more affine, leads to a reduction of α_2/α_1 (Erman and Mark, 1997).

An additional consideration must be taken into account when discussing rubbers. In dry rubber, that is, rubber not exposed to a solvent, the volume not occupied by molecular strands (the free volume) is small. Hence, the density of excluded volume interactions, that is, interactions of nonbonded type at sites which are not crosslinks, is large. This mandates that, under small enough strains, the deformation of the strands is approximately affine. In some sense, increasing the degree of packing is equivalent to increasing the degree of crosslinking in a network without nonbonded interactions. This contrasts with the situation of high free volume networks, where strands interact only at crosslinks and affinity may be enforced only by rendering the fibers stiffer in bending. This distinction is important.

When a low free volume network, such as dry rubber, is subjected to swelling, the role of excluded volume interactions is gradually reduced. Large degrees of swelling (which may not be reachable in practice) are required for rubber to develop the high free volume state normally encountered in nonwovens and gels. Therefore, the scaling of the stiffness with the effective network density for rubber in various equilibrium swelled states is weaker than that shown in Figure 6.12 for gels. The exponents x obtained with swelled rubbers remain in the vicinity of 1 (e.g., Fontaine et al., 1989).

6.1.1.3.4 *Effect of the Crosslink Density*

The crosslink density is related to the density and the mean connectivity through Eq. (4.21): $\rho = \rho_b\langle z\rangle l_c/2$. This relation holds if ρ does not include contributions from dangling and free filaments, as assumed throughout this text. In this case, ρ_b is not an independent parameter. Further, the density and the mean segment length are related through equations of the type $l_c \sim \rho^{-y}$, where $y = 1$ or $y = 1/2$ function of the type of network considered – see also the discussion in Chapter 4, Eqs. (4.14) and (4.13). Therefore, the relation between the density and the crosslink density can be written as $\rho^{y+1} \sim \rho_b \langle z \rangle$. This allows reformulating of all results presented above in terms of ρ, as functions of ρ_b and $\langle z \rangle$.

Situations exist in which only the nominal density, ρ_{nom}, is known, while the real density of the network, ρ, is not known. For example, in molecular networks prepared by mixing a polymer in solution with a crosslinking agent, only the concentration of the two reacting components is known. In such situations it is useful to estimate ρ function of ρ_{nom} and the crosslinking agent density.

Consider a population of fibers of length L_0 and nominal density ρ_{nom}, which are crosslinked with crosslink density ρ_b. Assuming connectivity $\langle z \rangle = 4$, the mean segment length can be computed as $l_c = \rho/2\rho_b$.

We evaluate the probability that fibers have exactly k crosslinks. Since the fiber length is given and assuming that the segment length is Poisson distributed, it is possible to infer that the distribution of the number of segments per fiber is described by a discrete Poisson distribution given by:

$$p(k) = \frac{1}{k!} \left(\frac{L_0}{l_c}\right)^k \exp\left(-\frac{L_0}{l_c}\right), \qquad (6.6)$$

with the mean number of crosslinks per fiber being L_0/l_c and the necessary normalization condition $\sum_{k=0}^{\infty} p(k) = 1$. Fiber segments not contributing to network mechanics are the dangling ends of fibers with multiple crosslinks, and the fibers not connected to the network, or connected at one crosslink only, that is, fibers with $k = 0$ and $k = 1$. The number of fibers not connected to the network (per unit volume) is $(\rho_{nom}/L_0)(p(0) + p(1))$. Their total length per unit volume is given by $\rho_{nom}(1 + L_0/l_c) \exp(-L_0/l_c)$. The total length of dangling ends[3] per unit volume is given by $2\rho l_c/L_0$. Therefore, the network density can be expressed in terms of the nominal density as:

$$\rho = \rho_{nom} \left[1 - \left(1 + \frac{L_0}{l_c}\right) \exp\left(-\frac{L_0}{l_c}\right)\right] \Big/ \left(1 + 2\frac{l_c}{L_0}\right). \qquad (6.7)$$

Replacing $l_c = \rho/2\rho_b$ in Eq. (6.7) one obtains an implicit relation linking the network density, ρ, the nominal density, ρ_{nom}, and the crosslink number density, ρ_b. Figure 6.13(a) shows the prediction of Eq. (6.7). In highly crosslinked networks with small l_c/L_0, ρ is slightly smaller than ρ_{nom}, primarily due to the presence of dangling ends. As the crosslink density decreases and l_c/L_0 increases, the contribution of the chains not connected to the network becomes important.

---

[3] Considering two dangling ends per fiber of length L_0; note that this does not apply to continuous tortuous fibers, a case in which the contribution of dangling ends to this calculation can be neglected.

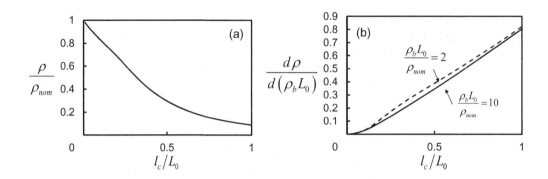

Figure 6.13 (a) Relation between network density, nominal density and the ratio l_c/L_0, Eq. (6.7). (b) Rate of change of the network density relative to the variation of the crosslink density, function of l_c/L_0. Curves for two values of $\rho_b L_0/\rho_{nom}$ are shown indicating weak sensitivity to this parameter.

To evaluate how the density of the network changes upon the variation of the crosslink density, one may express Eq. (6.7) in terms of ρ_b and take the derivative with respect to this variable. The result is shown in Figure 6.13(b) in terms of l_c/L_0. As expected, it is seen that when l_c is small relative to L_0, the network density does not change significantly as the crosslink density varies. However, as the crosslink density decreases (l_c increases), ρ changes approximately in proportion to ρ_b. The relationship has $\rho_b L_0/\rho_{nom}$ as a parameter which takes values larger than 1. Two curves corresponding to $\rho_b L_0/\rho_{nom} = 2$ and 10 are shown, indicating weak sensitivity to this parameter.

The analysis discussed in this section is purely geometric and assumes good mixing of filaments and crosslinking agent. In molecular networks, such as epoxy, the distribution of hardener may be uneven and crosslinking may stop due to increasingly reduced mobility of reacting species as the network forms. This mechanism results in conversion ratios smaller than 100%.

6.1.1.3.5 *Effect of Network Connectivity*

Most networks of practical importance are of fibrous type. In such structures, fibers are continuous through crosslink sites and the connectivity is $z = 4$. If fibers are of finite length, dangling ends are present. However, dangling ends are assumed not to contribute to network mechanics. This changes the connectivity of the crosslinks located at fiber ends from $z = 4$ to $z = 3$, or even $z = 2$, if two fibers are connected at their ends.

The mean connectivity of the network is smaller than 4 and may be computed as follows. The fraction of all crosslinks located at fiber ends is $p_{end} \approx 2/(L_0/l_c - 1)$,[4] while the fraction of crosslinks located in the middle segment of fibers is $1 - p_{end}$. Crosslinks in the central part of fibers may have either $z = 4$ or $z = 3$, while a fiber

---

[4] The approximate sign indicates the exclusion of (the few) cases in which fibers have no dangling ends. If fibers are connected such that no dangling ends are present in the entire network, $p_{end} = 2/(L_0/l_c + 1)$.

end crosslink may have $z = 3$ or $z = 2$. Expressing these conditions in terms of the fractions of crosslinks that belong to one of the three types, one may express the probability of occurrence of crosslinks with $z = 4$, $z = 3$, and $z = 2$ as $p_{z=4} = (1 - 2p_{end} + 2\bar{n}_{z=2})/(1 - p_{end} + 2\bar{n}_{z=2})$, $p_{z=3} = (p_{end} - 2\bar{n}_{z=2})/(1 - p_{end} + 2\bar{n}_{z=2})$, and $p_{z=2} = 2\bar{n}_{z=2}/(1 - p_{end} + 2\bar{n}_{z=2})$, respectively. These probabilities are expressed in terms of p_{end} and a parameter, $\bar{n}_{z=2} = p_{z=2}(1 - p_{end})/2(1 - p_{z=2})$, which is related to the probability of end-linked fibers, that is, of crosslinks with $z = 2$. Note that two parameters are needed, one accounting for the number of crosslinks per fiber, p_{end}, and the other representing the network architecture, $\bar{n}_{z=2}$. The mean connectivity can be computed as:

$$\langle z \rangle = 2p_{z=2} + 3p_{z=3} + 4p_{z=4} = \frac{4 - 5p_{end} + 6\bar{n}_{z=2}}{1 - p_{end} + 2\bar{n}_{z=2}}. \tag{6.8}$$

This expression allows computing the mean connectivity in terms of l_c/L_0. When l_c/L_0 is small, $p_{end} = p_{z=2} = \bar{n}_{z=2} \approx 0$ and $\langle z \rangle \approx 4$. In the other limit, at percolation, fibers forming the percolating backbone have on average two crosslinks per fiber and, hence, $p_{end} = p_{z=2} = 1, \bar{n}_{z=2} = 1/2$, and $\langle z \rangle \approx 2$. A statistical analysis of this issue is presented in Zagar et al. (2011), who suggest that the mean connectivity at percolation is $\langle z \rangle_c = 2$. The analysis in Broedersz et al. (2012) leads to a percolation threshold of $\langle z \rangle_c = 2.6$.

In fibrous networks in which fibers are bundled, the role of the crosslinks is played by sites where bundles split or merge. The network nodes connecting bundles have effective $\langle z \rangle = 3$ in 2D and $z = 4$ in 3D (Chapter 10). In molecular networks the connectivity is controlled by the functionality of the crosslinking molecules. Most molecules establish trivalent or quadrivalent crosslinks, while connectors of higher connectivity, such as for example polyhedral oligomeric silsesquioxane (POSS), may be used. Lattice structures of high $\langle z \rangle$ can be constructed by 3D printing or self-assembly. Therefore, it is of interest to investigate the dependence of the elastic modulus on $\langle z \rangle$.

In the affine limit, fibers deform in the axial mode and the axial stretch depends exclusively on the orientation of the respective segment (Chapter 5). The modulus depends on the density of elastically active strands and is independent of $\langle z \rangle$.

In the nonaffine regime, increasing $\langle z \rangle$ leads to the rapid increase of the modulus, as large values of $\langle z \rangle$ promote affinity. This is suggested by the schematic master plot shown in Figure 6.14. The curve shifts to the left as $\langle z \rangle$ increases, while the affine plateau remains unchanged. When probed in a narrow range of w, the modulus increases toward the affine limit as $\langle z \rangle$ increases, but the rate of increase is a function of w, as reported in Buxton and Clarke (2007) and Lindstrom et al. (2010).

Lindstrom et al. (2010) report an exponential dependence of the modulus on $\langle z \rangle$, of the type $E_0 \sim \exp\left[a(\langle z \rangle - \langle z \rangle_c)\right]$ for $\langle z \rangle$ in the range 3–3.8, while Zagar et al. (2011) report a stronger scaling $E_0 \sim (\langle z \rangle - \langle z \rangle_c)^5$ for a broader range of $\langle z \rangle$. $\langle z \rangle_c$ is the connectivity percolation threshold. Clearly, the issue of whether the modulus scales with $\langle z \rangle - \langle z \rangle_c$ exponentially or as a power function cannot be clarified by varying $\langle z \rangle$ in a narrow range. In fact, exploring a large range of variation of $\langle z \rangle$ is not possible since the affine limit is reached fast and the modulus becomes independent of $\langle z \rangle$ in that limit.

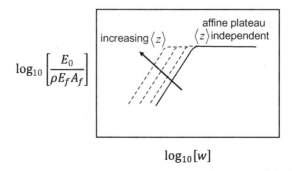

$$\log_{10}\left[\frac{E_0}{\rho E_f A_f}\right]$$

$$\log_{10}[w]$$

Figure 6.14 Schematic representation of the effect of $\langle z \rangle$ on the master plot of Figure 6.6.

A somewhat similar situation is encountered in thermal networks. The phantom network theory of rubber elasticity includes the effect of $\langle z \rangle$ on the network modulus. This theory was developed to account for crosslink fluctuations (i.e., for nonaffinity), so as to relax the strong affine deformation assumption of the classical theory. The model predicts a relation between network modulus and $\langle z \rangle$ of the form: $E_0 = E_0^{aff}(1 - 2/\langle z \rangle)$, where E_0^{aff} is the network stiffness in the affine limit (see also the discussion about gels and rubber in Section 6.1.1.3.3). This is a much weaker dependence on $\langle z \rangle$ than the exponential and power functions discussed in this section for athermal networks. However, it does capture the underlying physics by indicating that the deformation becomes more affine and the network becomes stiffer, as $\langle z \rangle$ increases. The general trend of stiffness reduction upon decreasing $\langle z \rangle$ was confirmed in experiments performed with networks produced by end-linking PDMS chains with either trifunctional, $z = 3$, or tetrafunctional, $z = 4$, crosslinking agents (Mark et al., 1979). As expected, this data indicates that increasing $\langle z \rangle$ leads to an increase of the effective stiffness, although the scaling $E_0/E_0^{aff} \sim 1 - 2/\langle z \rangle$ could not be fully validated based just on two values of $\langle z \rangle$.

To address this limitation, recent experiments of Nishi et al. (2017) compare reference ideal molecular networks having tetrafunctional crosslinking groups ($\langle z \rangle = z = 4$) and all segments of the same molecular weight, with similar networks in which defects are introduced in a controlled manner and the effective connectivity is reduced. The density of defects is represented by the probability p that a strand of the reference network is preserved in the defective network. In the reference network without defects, $p = 1$. The results indicate that, for given connectivity $\langle z \rangle$ of the reference nondefective network, the modulus depends on p as $G(p)/G(1) = (p - 2/\langle z \rangle)/(1 - 2/\langle z \rangle)$. $G(1)$ and $G(p)$ are moduli of the reference network and defective networks, respectively. This relationship applies away from the percolation threshold. The respective functional form is also predicted by the effective medium theory developed for linear spring networks in Feng et al. (1985). The experimental data in Nishi et al. (2017), which correspond to $\langle z \rangle = 4$, provide strong support for this expression for p values far above the threshold of the sol–gel transition. The effect of various types of defects (dangling ends, loops) on the modulus is evaluated analytically in Lang (2018) and Lin et al. (2019) and numerically in Gusev (2019), and expressions in terms of $\langle z \rangle$ are reported in Lin et al. (2019) for each type of defect. Experiments with end-linked networks having controlled defect populations are currently used to evaluate the validity of this functional form.

6.1.1.3.6 *Effect of Fiber Tortuosity in Athermal Networks*

Tortuosity changes drastically the effective axial and bending stiffnesses of a fiber. This is discussed in Section 2.3.5, where it is shown that the shape of the fiber determines its stiffness and nonlinear behavior. In a stochastic network in which each fiber has a different shape, the effect of tortuosity is amplified. The only general statement that can be made is that increasing the degree of tortuosity and the fraction of nonstraight fibers leads to the reduction of the effective network stiffness.

To make these ideas quantitative, it is necessary to consider a particular case. Ban et al. (2016a) studied the effect of tortuosity in 3D networks in which each fiber has sinusoidal shape of wavelength $2l$ and amplitude ςl, where l is the end-to-end length of the respective fiber. The tortuosity is computed with Eq. (2.9) and is expressed in terms of ς as $c = 2E_2(-\varsigma\pi)/\pi$, where $E_2()$ is the complete elliptic integral of the second kind. Figure 6.15 shows the variation of the network stiffness normalized by the stiffness of the network of straight fibers having the same geometry and connectivity, function of the tortuosity parameter, c, and of the fraction of tortuous fibers in the network. The results show that these two parameters affect the stiffness equally and the decrease of the stiffness with increasing c is rapid. This data set should be viewed as a generic guideline, since considering fibers of other shapes should change the position of the contours in Figure 6.15, although the general appearance of the map should not change.

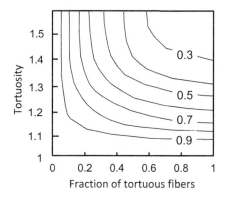

Figure 6.15 Map showing the stiffness of a network with tortuous fibers, E_0, normalized by the stiffness of the same network with straight fibers, function of the tortuosity parameter, c (Eq. (2.9)) and of the fraction of tortuous fibers in the network. Republished with permission of ASME from Ban et al. (2016a), permission conveyed through Copyright Clearance Center Inc.

6.1.1.3.7 Networks of Fibers with Noncircular Section

While most studies consider fibers of circular cross-section, situations in which the fiber cross-section has a different shape exist. Examples include paper, which is made from ribbon-like fibers, and collagen networks in which fibers are bundled and the bundles do not have a circular section. Therefore, it is of interest to inquire to what extent the master plot of Figure 6.6 changes when rendering the fiber cross-section noncircular.

Noncircular sections have two principal directions of inertia, corresponding to the principal moments $I_{f\,max}$ and $I_{f\,min}$. Fibers are more compliant in bending in the plane perpendicular to the axis corresponding to $I_{f\,min}$. Since the difference between the two principal moments of inertia may be large, the fiber bending compliance may be very different in the two principal planes. When bent in the stiff mode characterized by $I_{f\,max}$, lateral buckling takes place to allow bending to occur exclusively in the softer mode, that is, in the plane of $I_{f\,min}$. The energy penalty for this transformation is associated with fiber torsion. This mechanism is discussed in Section 2.3.4.

This detail of the fiber-scale mechanics is expected to influence the network-scale behavior. Since the network structure is stochastic, the orientation of the principal axes of fiber cross-sections relative to the global coordinate system is also stochastic. Based on the considerations in the previous paragraph it is expected that the nonaffine, bending-dominated branch of the master plot depends on both $I_{f\,max}$ and $I_{f\,min}$, while the affine, axially-dominated branch is expected to be independent of the shape of the fiber cross-section. Figure 6.16 shows the master plot for a Voronoi network of fibers of noncircular cross-section (Deogekar and Picu, 2017). In the case of Voronoi networks, the stiffness scales with the density as $E_0 \sim \rho \cdot \delta\rho$ or, if $\delta\rho \approx \rho$, as

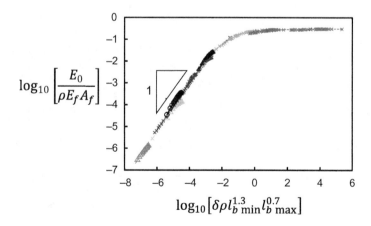

Figure 6.16 Master plot for 3D Voronoi networks of fibers with noncircular cross-section. The cross-section of fibers is characterized by two principal moments of inertia $I_{f\,min}$ and $I_{f\,max}$, which define $l_{b\,min}$ and $l_{b\,max}$ that appear on the horizontal axis. Reprinted with permission from Deogekar and Picu (2017), copyright (2017) by the American Physical Society

$E_0 \sim \rho^2$. If fibers have a circular section, and in the bending regime of the master plot, the modulus depends on the fiber cross-section parameters as $E_0/\rho E_f A_f \sim l_b^2$, that is, $E_0 \sim E_f I_f$. If fibers have a noncircular section, the scaling becomes $E_0 \sim E_f I_{f\,min}^{0.65} I_{f\,max}^{0.35}$. This is written in terms of the non-dimensional groups $l_{b\,min} = \sqrt{E_f I_{f\,min}/E_f A_f}$ and $l_{b\,max} = \sqrt{E_f I_{f\,max}/E_f A_f}$ which appear on the horizontal axis of the plot in Figure 6.16.

The relation $E_0 \sim E_f I_{f\,min}^{y} I_{f\,max}^{1-y}$ reflects the shared contribution of the two bending modes. The softer mode dominates when $y > 0.5$, as seen in Figure 6.16. However, y depends on the ability of fibers to undergo lateral buckling. If the difference between $I_{f\,min}$ and $I_{f\,max}$ is maintained, but the torsional stiffness is increased such as to prevent fiber rotation into the softer mode, the scaling becomes $E_0 \sim E_f I_{f\,min}^{0.5} I_{f\,max}^{0.5}$, which indicates that the two modes contribute equally to the overall network response. The results in Figure 6.16 are obtained by considering fiber cross-sections with $G_f J_f \ll E_f I_{min}$, which allows for lateral buckling. Figure 6.16 extends the master plot of Figure 6.6 to the more general case of networks of fibers with arbitrary cross-section.

6.1.1.3.8 *Effect of Preferential Fiber Alignment*

Preferential fiber alignment causes elastic anisotropy. The type and degree of anisotropy are functions of the type and degree of alignment. It was observed that, for given anisotropy, the master plot of Figure 6.6 remains unchanged, provided the vertical axis is normalized with the affine model prediction for the respective probing direction relative to the principal directions of anisotropy (Missel et al., 2010). This implies that the functional form of the dependence of stiffness on network parameters is independent of alignment. The anisotropic stiffness tensor can be written as the affine anisotropic stiffness multiplied by a function of density and l_b, which represents the master plot.

This observation is useful when predicting the stiffness of an anisotropic network in the nonaffine range. The affine model can be used to infer the affine plateau for any type and degree of fiber orientation (e.g., Section 5.2). Further, the master plot for isotropic networks (Figure 6.6) may be used to extrapolate the anisotropic affine prediction to the nonaffine case.

6.1.1.3.9 *Effect of the Variability of Fiber Properties*

Fibers forming a given network material are generally not identical. For example, paper is made from cellulose fibers whose length and cross-section shape and dimensions are stochastic variables. Counterexamples are nonwovens made from a single type of fiber and some molecular networks. Spunbonded fibers have identical diameters and properties, while molecular networks in which filaments do not bundle and do not form crystallites are also examples of networks made from filaments of identical properties.

Fiber bundling occurs in many biological networks. Assuming strong cohesion and/or internal crosslinking within bundles (as is the case with collagen), individual

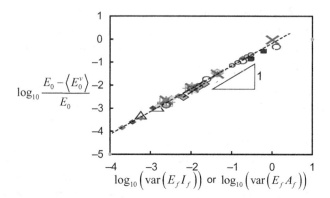

Figure 6.17 Effect of the variability of fiber properties ($E_f I_f$ or $E_f A_f$) on the mean network stiffness, $\langle E_0^v \rangle$, relative to the stiffness of the same network composed from identical fibers, E_0. Data for networks in the affine range, networks in the nonaffine range and various distributions of fiber properties is included. Reprinted from Ban et al. (2016b) with permission from Elsevier

bundles behave like effective fibers. Since the bundle size differs from bundle to bundle, the mechanical properties of the effective fibers representing the bundles are also different.

Therefore, it is of interest to inquire to what extent the variability of fiber properties within a given network influences the network stiffness. This problem was studied in Ban et al. (2016b) by considering Voronoi networks in which the properties of fibers are selected from distributions. In separate models, the mean of the distribution is kept constant, while the variance is gradually increased. For networks in the affine regime, $E_f A_f$ is varied, while $E_f I_f$ and ρ are kept constant. For nonaffine networks, $E_f A_f$ and ρ are kept constant and $E_f I_f$ is varied from fiber to fiber. It is observed that, in all cases, the mean network stiffness $\langle E_0^v \rangle$ decreases as the variance of the distribution from which the relevant fiber material property is selected increases. Specifically, it results that $(E_0 - \langle E_0^v \rangle)/E_0 \sim \mathrm{var}(E_f I_f)$ for networks in the nonaffine regime, where E_0 is the modulus of the network of same geometry without fiber properties variability. For networks in the affine regime $(E_0 - \langle E_0^v \rangle)/E_0 \sim \mathrm{var}(E_f A_f)$. Since E_0 and $\langle E_0^v \rangle$ depend on density in the same way, the normalized modulus difference is independent of density. These results are shown in Figure 6.17.

6.1.1.3.10 *Effect of the Crosslink Compliance*

Real crosslinks have finite stiffness. A typical example is provided by the intra-cellular actin network which is crosslinked by actin-binding proteins with relatively small stiffness. Some physical crosslinks in gels may be also considered elastically soft. Similarly, during the compression of fibrous wads, when high densities are reached and high contact forces develop, the finite stiffness of the contacts between fibers has a noticeable effect on the network-scale response (Section 6.2.1). The contribution of crosslink compliance is not considered in the previous sections of this chapter since the focus is placed on the contribution of fibers to the network stiffness. It is useful at this point to explore the conditions in which the crosslink compliance manifests itself in the mechanical behavior of the network.

It is easy to understand the extreme case of this problem: when the crosslinks become very soft, they store the largest fraction of the strain energy, while fibers move as rigid objects. In this case, the network stiffness scales linearly with the crosslink stiffness and with the crosslink number density.

This issue is studied in Zagar et al. (2015) for 3D random networks subjected to shear and in Hossain et al. (2019) in the context of compression of planar mats. The problem may be formulated in terms of a non-dimensional parameter that contrasts the stiffness of the crosslinks, k_X, with the bending stiffness of fiber segments: $k^* = k_X l_c^3 / E_f I_f$. Large values of this parameter correspond to the rigidly crosslinked network limit. As k^* decreases, the network stiffness becomes proportional to k_X. Note that k^* decreases when k_X decreases or when the density increases (reducing l_c). Figure 6.18 shows the variation of the modulus of networks with flexible crosslinks normalized by the modulus of the corresponding networks with rigid crosslinks, versus $k_X l_c^3 / E_f I_f$. The interesting result here is that the crossover from the rigid crosslinks limit (right side of the plot) to the regime in which the modulus scales linearly with k_X takes place for k^* in the range 10–100. The data in Hossain et al. (2019) indicates that the transition from fiber-dominated to crosslink-dominated energy storage takes place at values of k^* between 100 and 1 000, that is, roughly one order or magnitude larger than the regime reported in Zagar et al. (2015). Figure 6.18 includes data from Ovaska et al. (2017), who performed a similar study considering mat-like networks of planar geometry. Their results agree with those of Zagar et al. (2015).

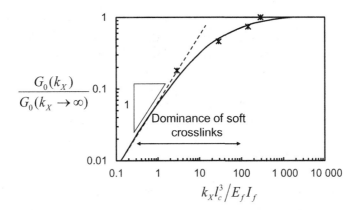

Figure 6.18 Effect of the crosslink stiffness on network stiffness. The normalized crosslink stiffness is represented by the non-dimensional parameter $k^* = k_X l_c^3 / E_f I_f$. The vertical axis is normalized by the stiffness of the same network with rigid crosslinks. The curve is plotted based on data from Zagar et al. (2015), while the data points are from Ovaska et al. (2017) and correspond to planar fibrous mats.

6.1.1.3.11 Size Effects

Size effects refer to the situation when a material property becomes dependent on the size of the sample used for its evaluation. Size effects are common in heterogeneous materials. In continuum composites with stochastic microstructure, as for example particulate composites with randomly distributed inclusions, the distribution of particles defines an internal length scale. If particles are randomly distributed, the relevant internal length scale is the average distance between particles. If particles are clustered, a mesoscale length scale roughly associated with the distance between clusters (and quantifiable using the two-point distribution function, see Section 4.3) can be defined. It is established that the sample size must be larger than about 10 times the largest of these internal length scales in order for the measured elastic constants to be sample size independent. If particles form a percolated structure, the characteristic sample size discussed here diverges.

The direct comparison of network materials and stochastic composites is difficult due to the fact that networks are structures made from discrete entities – the fibers. However, the deformation of a network observed on scales much larger than the mean segment length may be compared with the deformation of a heterogeneous continuum. A homogeneous continuum deforms affinely. Therefore, for networks with parameter w (Figure 6.6) in the affine range, the degree of nonaffinity is low (Figure 6.7) and the size effect is weak. The degree of nonaffinity increases as w decreases and networks in the nonaffine range should be compared with heterogeneous continua whose degree of heterogeneity increases as w decreases. These networks are expected to exhibit significant size effects.

In the presence of size effects, the predicted elastic constants depend on the imposed boundary conditions. The sensitivity to boundary conditions is eliminated once the sample is large enough to be size effect-free.

To substantiate this physical picture, Figure 6.19 presents data for 2D fibrous and 3D cellular networks. Figure 6.19(a) and (b) shows data for 2D Mikado networks: the effective stiffness, E_0, normalized by the stiffness of the same network of "infinite" size, versus the model size, L (Shahsavari and Picu, 2013). Results for two values of w are shown: Figure 6.19(a) corresponds to $w = -5.6$, which, according to the master plot of Figure 6.6, is a highly nonaffine case, while Figure 6.19(b) corresponds to an affine network with $w = 6.4$. The modulus is evaluated by performing uniaxial tension tests. Three boundary conditions are used for each w value: (i) free boundaries in the direction orthogonal to the loading direction, (ii) the same as in (i) except that the network structure is periodic, with periodicity L, and (iii) the boundaries in the direction orthogonal to the loading direction are restricted to remain straight and traction free in the average sense. The important observation is that the sensitivity to boundary conditions is large in the small w case, and is almost nonexistent in the large w case. The nonaffine networks do not reach the model size-independent asymptote even for model sizes as large as 20-times the fiber length, L_0. As opposed to this, network with large w produce model size-independent predictions for $L > 5L_0$. All boundary conditions used are of mixed type, but it results that condition (i) is close to the Voigt, upper limit, while condition (iii) is close to the Reuss, lower limit.

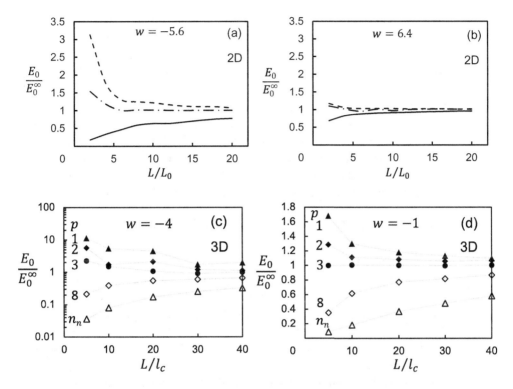

Figure 6.19 Sample size dependence of E_0 for (a) highly nonaffine ($w = -5.6$) and (b) affine ($w = 6.4$) 2D fibrous networks of Mikado type (adapted from Shahsavari and Picu (2013) with permission from Elsevier), and for (c) nonaffine ($w = -4$) and (d) close to affine ($w = -1$) 3D cellular networks of Voronoi type (adapted from Merson and Picu (2020) with permission from Elsevier). The dashed, dashed dot, and continuous curves in (a) and (b) correspond to the boundary conditions (i), (ii), and (iii) described in the text, respectively. The curves in (c) and (d) correspond to various values of parameter p defining the generalized boundary conditions applied in the 3D case, as described in the text and in Merson and Picu (2020).

Figure 6.19(c) and (d) shows similar results for 3D cellular networks of Voronoi type (Merson and Picu, 2020), for $w = -4$ and $w = -1$, which correspond to the nonaffine range and nonaffine-to-affine transition, respectively. Multiple curves are shown in each figure. In this work, the generalized boundary conditions introduced by Glüge (2013) are used to mitigate the size effect. This technique allows developing models that interpolate between the boundary conditions that lead to the Voigt and Reuss upper and lower bounds of the elastic constants by adjusting a parameter of the model denoted here by p. The review of this method goes beyond the scope of the current discussion; the interested reader is referred to the cited references for details. Here, we consider p to be an internal model parameter which leads to the Voigt upper limit when $p = 1$ and to the Reuss lower limit when p is equal to the number of intersections of fibers with the outer surface of the model, $p = n_n$.

Two important observations can be made. The first refers to the distance between the two bounds, which is much larger in the nonaffine case with $w = -4$ compared with the more affine case with $w = -1$. Note that the vertical axis for $w = -4$ is logarithmic. While in the 2D case, curves converge to the size independent asymptote (Figure 6.19(b)) for relatively small model sizes, the asymptote is never reached in the 3D case, even for models much larger than those used in 2D.

The second observation refers to the effect of p. The upper bound reached for $p = 1$ converges to the infinite model size limit from above, while the lower bound ($p = n_n$) converges from below. As p increases, the curvature of the function $E_0/E_0^\infty (L/l_c)$ changes sign. A value of p results for which the curvature is zero and the predicted $E_0(p)/E_0^\infty$ is model size independent. This value is $p = 3$. Models with this p predict the stiffness in the limit of infinite model size even though their size is rather small. The vertical axes in Figure 6.19(c) and (d) are normalized by this prediction. The major advantage of this method is that it allows predicting the size-independent material parameters using small models.

The results in Figure 6.19 suggest the possibility that the master plot of Figure 6.6 is subjected to size effects. The size effect is weak in the affine regime but may be pronounced for low w values. This issue was studied in Merson and Picu (2020), where it is shown that the size effect does not modify the scaling of the modulus with the network parameters, such that the general shape of the curve in Figure 6.6 is independent of the model size. However, it affects the vertical position of the low w branch of the curve and the value of w at the affine to nonaffine transition.

The size effect in network materials was also studied in Ostoja-Starzewski and Stahl (2000) using a model relevant for planar cellulose networks. They observe that the upper bound converges faster to the asymptote corresponding to the infinite model size than the lower bound. This trend is confirmed by the results shown in Figure 6.19(a).

A size effect is also observed in indentation, where it is related to the size of the indenter. In indentation, the elastic modulus can be evaluated by fitting the Hertzian contact model to the measured force–displacement curve (Johnson, 1985).[5] The modulus turns out to be dependent on the indenter diameter: It increases as the indenter size decreases and becomes independent of this parameter for large enough probing tips (Andrews et al., 2001). The indentation size effect is due to the presence of large strain gradients and network compaction under the indenter and is expected to be observed even in periodic networks whose behavior is not affected by the complexities associated with stochastic heterogeneous microstructures. While the usual size effect emerges from the mechanical heterogeneity of the microstructure, the size effect measured in indentation may be caused by the inherently nonlocal (micropolar) nature of fiber interactions in the network (Section 7.3.3) and by the formation of a low free volume, compacted network layer under the indenter. The formation of a compacted layer effectively modifies the size and shape of the indenter, which may modify the functional form of the indentation force-displacement relation. The interplay between

---

[5] Fitting the loading branch of the indentation force–displacement curve is acceptable for hyperelastic materials. When working with elastic–plastic materials, the Oliver–Pharr method (Oliver and Pharr, 2004) requires fitting the unloading branch of the indentation curve.

nonlocality and heterogeneity is a subject insufficiently explored to date in the context of network mechanics. Understanding the relation between the indentation size effect and the more common size effect evaluated with samples of different dimensions requires a solid theoretical understanding of this interplay.

6.1.1.4 Nonlinear Elastic Behavior

6.1.1.4.1 *Representation of Nonlinear Behavior*

The nonlinear behavior of network materials is described in general terms in Section 6.1.1.1 based on the stress–stretch and tangent stiffness–stress curves (Figure 6.1). Before diving into the details of the nonlinear behavior, it is necessary to inquire about the optimal way to describe and quantify this complex issue. While the linear elastic behavior of a material is fully described by the stiffness tensor, the nonlinear behavior is multifaceted and allows for multiple descriptions. In general terms, two aspects are of interest: the criterion defining the transition from the linear to the nonlinear regime and the type of subsequent nonlinearity. The transition is characterized by a critical stretch denoted in Figure 6.1 by λ_{c1}. The characterization of nonlinearity can be based, in principle, on a deformation-dependent tangent stiffness tensor similar in structure to the four-dimensional stiffness tensor of linear elasticity. Since the network becomes anisotropic during deformation, such tensorial representation of stiffening/softening would capture the mechanical effect of structural anisotropy. However, this type of characterization is not only impractical, but also nonunique since the tangent stiffness depends on the deformation path.

In practice, the material behavior is characterized based on several specific tests – for example, uniaxial tension/compression and shear. Several pairs of stress and strain measures can be used: the Cauchy stress, $\boldsymbol{\sigma}$, function of the deformation gradient $\mathbf{F}$ (or the principal stretches), the first Piola–Kirchoff stress (PK1 or nominal stress), $\mathbf{S}$, function of $\mathbf{F}$, and the second Piola–Kirchoff stress (PK2), $\boldsymbol{\Pi}$, function of the Green–Lagrange strain, $\mathbf{E}$. $\mathbf{S}$ is work conjugate with $\mathbf{F}$, and $\boldsymbol{\Pi}$ is work conjugate with $\mathbf{E}$ (Eq. (5.8)), which makes these pairs preferable. Experimental results are conveniently expressed in terms of $\mathbf{S}$ and $\mathbf{F}$, since $\mathbf{S}$ is proportional to the force applied, which is directly measured in any mechanical test.

Network materials are often characterized in uniaxial tension/compression. Stretch is imposed in a given direction, for example, x_1, and the associated stress, as well as the stretches in directions orthogonal to the loading direction, λ_2, λ_3, are measured. In this case $\mathbf{F}$ is diagonal and $F_{ii} = \lambda_i$. The only non-zero component of the stress is the normal stress in the loading direction. In this case, the nonlinear behavior is described by the function $S_{11}(\lambda_1)$ or, under small deformations, by $\sigma_{11}(\varepsilon_{11})$, $\varepsilon_{11} = \lambda_1 - 1$. The tangent stiffness E_t is computed either as $\partial S_{11}/\partial \lambda_1$ or $\partial \sigma_{11}/\partial \lambda_1$,[6] and E_t can be plotted versus the corresponding stress to obtain a representation similar to that in Figure 6.1 (b). While this is straightforward and rather commonplace, one should recall that it provides a limited description of the nonlinear behavior.

---

[6] This is equivalent to the tangent stiffness computed in terms of the small strain, that is, $\partial \sigma_{11}/\partial \varepsilon_{11}$.

A similar situation is encountered when performing rheological characterization. The deformation in the rheometer is of simple shear, constant volume type, which is significantly different from that imposed in the uniaxial test in which the variation of the material volume is not constrained. The shear modulus is measured in these tests in three ways: (i) by applying a monotonic shear strain, while recoding the shear stress, (ii) by applying a small amplitude oscillatory load about a non-zero mean and evaluating the storage modulus G', which is then associated with the mean stress of the oscillatory load (the "differential modulus"), and (iii) in a more standard way, by applying oscillatory load of zero mean and recording the frequency-dependent complex modulus, $G(\omega) = G'(\omega) + iG''(\omega)$, where G' and G'' are the storage and loss moduli, respectively. In the case of network materials, the loss modulus is typically smaller than the storage modulus, while G' exhibits a frequency-independent plateau (see Chapter 9). The value of the plateau is taken to be the shear modulus of the material. To obtain the representation in Figure 6.1(b), it is necessary to use the second method (ii) and to plot the differential modulus versus the mean stress.

The tangent stiffness versus stress curve also depends on the stress–strain pairs used. For example, consider that the nominal stress is measured in a uniaxial test. The nonlinear response may be represented as $\partial S_{11}/\partial\lambda_1$ versus S_{11}, or as $\partial\sigma_{11}/\partial\lambda_1$ versus σ_{11}. Considering that PK1 is related to $\boldsymbol{\sigma}$ as $\mathbf{S} = J\boldsymbol{\sigma}\mathbf{F}^{-T}$ and that $\mathbf{F}$ is diagonal, $F_{ii} = \lambda_i$, the tangent stiffness evaluated as $\partial S_{11}/\partial\lambda_1$ may be written

$$\frac{\partial S_{11}}{\partial\lambda_1} = \frac{\partial\left(\lambda_2^2\sigma_{11}\right)}{\partial\lambda_1} = \lambda_2^2\frac{\partial\sigma_{11}}{\partial\lambda_1} + 2\lambda_2\sigma_{11}\frac{d\lambda_2}{d\lambda_1}, \tag{6.9}$$

which can be re-arranged as:

$$\frac{dS_{11}/d\lambda_1}{S_{11}} = \frac{d\sigma_{11}/d\lambda_1}{\sigma_{11}} + \frac{2}{\lambda_2}\frac{d\lambda_2}{d\lambda_1}. \tag{6.10}$$

The second term on the right side of Eq. (6.10) may be large if the Poisson contraction is pronounced – which is the case with networks of large free volume not embedded in a volume-preserving matrix. This expression indicates that the dependence of the tangent stiffness on stress is different when different pairs of stress–strain measures are used.

It results that the nonlinear behavior of various materials can be compared only when obtained from the same type of test and when described in terms of the same pair of stress and kinematic variables. In this chapter, and throughout the book, we consistently use the nominal stress (PK1), $\mathbf{S}$, and the deformation gradient, $\mathbf{F}$ (or principal stretches), to describe the nonlinear behavior. In uniaxial tension, the tangent stiffness is computed as $E_t = \partial S_{11}/\partial\lambda_1$.

A specific type of representation of the nonlinear behavior, known as the Mooney–Rivlin plot, is used in the literature on rubber. This facilitates the comparison with the neo-Hookean and Mooney–Rivlin models described in Section 7.3.1. To this end, the nominal stress in uniaxial tension, S, is plotted as $S/(\lambda - 1/\lambda^2)$ versus $1/\lambda$. The neo-Hookean model predicts a straight horizontal line in this representation, as can be inferred from Eq. (5.28); note that, since rubber is incompressible, $J = \lambda_1\lambda_2^2 = 1$ in a uniaxial test. The Mooney–Rivlin model predicts a straight line whose intercept of the vertical axis is defined by coefficient α_1 in the free energy function of Eq. (7.10), and its slope is defined by α_2 of the same equation. For the many network materials that do not undergo deformations as large as those sustained by rubber, and whose behavior is not described accurately by the Mooney–Rivlin model, this representation is not particularly useful. Due to this reason we do not make extensive use of the Mooney–Rivlin plot in this chapter. The stiffness–stress and the Mooney–Rivlin representations are compared in Figures 6.35 and 6.36.

6.1.1.4.2 *Type B: Softening*

Networks that exhibit type B behavior (Figure 6.1(a)) can be divided into nonlinear elastic and inelastic. When characterized in uniaxial tension/compression and using the nominal stress–stretch relation, elastomers show nonlinear elastic behavior of type B for stretches below approximately 2. Nonwovens show inelastic type B behavior, with hysteresis and residual strains being observed upon unloading from stresses beyond the yield stress.

Softening of $S(\lambda)$ may be caused by multiple mechanisms:

(i) *Damage*. In densely crosslinked networks with a distribution of bond strengths, softening is associated with crosslink rupture. This leads to inelastic type B behavior, hysteresis, and the Mullins effect.

(ii) *Onset of plastic deformation of fibers*. An affinely deforming network yields at a strain equal to the yield strain of the fibers (see also Section 6.1.4), which leads to type B behavior.

(iii) *Friction and adhesion between fibers*. Inter-fiber friction is important in nonwovens and sparsely crosslinked athermal networks. Static friction is larger than dynamic friction. During loading, the network transitions from an initially locked structure to an evolving mode in which fibers slide relative to each other and rearrange dynamically; this transition entails softening and causes the occurrence of an apparent yield point. Adhesion between fibers is important in networks of nanofibers (Chapter 10). Loading–unloading cycles lead to dynamic bonding and debonding of adhesion-stabilized fiber bundles. This dissipative mechanism causes softening during monotonic tension (Figure 10.3(b)).

Networks of type B exhibit regimes I and II (Figure 6.1(a)). Regime I is characterized by the stiffness E_0, whose relation to the structural parameters of the network is discussed in Section 6.1.1.3. The stretch at the transition between regimes I and II, λ_{1c}, corresponds to the onset of operation of one of the inelastic mechanisms (i) to (iii) mentioned. If softening is produced by damage, regime II depends on the rate of damage accumulation. Diffuse damage causes a yield point and a regime II of approximately constant slope. An example of this type relevant for paper is shown in Figure 6.2(c). In this densely crosslinked network, the processes controlling regime II are crosslink rupture, plastic deformation of fibers, and the relative sliding of fibers at contacts. On the other hand, if damage accumulation is rapid and spatially localized, it leads to global failure and regime II is absent. A behavior of this type is shown in

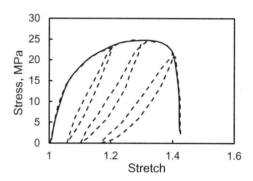

Figure 6.20 Stress–stretch curve of a polypropylene nonwoven. The unloading–reloading curves are shown with a dashed line, while the continuous line represents the envelope corresponding to monotonic loading. Adapted from Chen et al. (2016)

Figure 6.20 and is obtained with a nonwoven made from polypropylene fibers of area weight 220 g/m^2 (Chen et al., 2016). The stress–strain curve exhibits the linear elastic regime I followed by gradual softening leading to rupture, with no apparent regime II. The unloading and reloading curves shown with dashed lines indicate that the effective stiffness decreases and the residual strain increases as the material is stretched. The effective stiffness is defined by the slope of the unloading branch at the beginning of each unloading segment. This indicates that the continuous softening observed in the monotonic curve is due to damage accumulation.

Similar stress–stretch curves and similar phenomenology are presented in Mao et al. (2017) for carbon nanotube buckypaper and nanocellulose paper tested in uniaxial tension, and in Niskanen (2012) for regular cellulose paper.

6.1.1.4.3 *Type B to C: Softening Followed by Stiffening*
A relatively small number of network materials exhibit a mechanical response in-between types B and C. In such cases, the initial network stiffness is large, but softening begins at relatively small strains. The softening regime is followed by stiffening resembling the behavior of type C networks.

Figure 6.21(a) shows uniaxial nominal stress–stretch curves for a bisphenol A diglycidyl ether epoxy with various hardener concentrations. The curve labeled 100% corresponds to the epoxy with the stoichiometrically correct concentration of hardener, as prescribed by the manufacturer. The other curves correspond to epoxies with 80% and 70% of the nominal hardener concentration. The percolation threshold below which the material remains a liquid is reached in the vicinity of 60% hardener concentration. Epoxies are densely crosslinked networks and the strands between two successive crosslinks are relatively rigid. This conveys high strength but restricts the toughness and the ductility of the material. As can be seen in Figure 6.21(a), the epoxy with nominal hardener concentration (100%) fails at about 4% engineering strain, which is quite typical for these materials. The stress–stretch curve develops a peak, but the ductility is too low to support substantial post-critical deformation. Reducing the

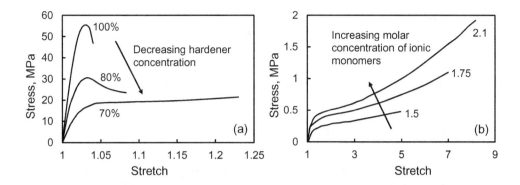

Figure 6.21 (a) Nominal stress–stretch curves for epoxy with stoichiometric (curve labeled 100%) and reduced hardener concentration (curves labeled 80% and 70%). Decreasing the crosslink density reduces the peak stress, increases the stretch at failure, and allows the emergence of a stiffening regime beyond the yield point. (b) Nominal stress–stretch curves for polyampholytes of different molar concentrations of ionic monomers. Adapted from Sun et al. (2013) with permission from Springer Nature, copyright (2013)

crosslink density decreases the strength and increases the ductility. However, the most interesting aspect in the context of the present discussion is the gradual emergence of a stiffening branch of the stress–stretch curve beyond the yield point, as the crosslink density is reduced. At the same time, the peak at yield disappears. This indicates that the sparsely crosslinked network of the epoxy with the lowest hardener concentration has sufficient kinematic freedom to avoid localization and to reorganize during loading such as to sustain strain stiffening. Therefore, these networks exhibit behaviors of types B and C at small and large stretches, respectively.

Inquiring about the physical origin of the yield point, it is of interest to observe that the yield strain is largely independent of the crosslink density. Yield in these networks is likely associated with the onset of relative sliding of polymeric chains. Given the large crosslink density and the dense packing of the chains, the deformation is approximately affine at small stretches. Hence, the yield strain measured macroscopically is equal to the strain causing irreversible local relative motion of the chains; the yield strain is a material parameter dependent only on packing. This issue is discussed further in Section 6.1.4 in relation to network yielding caused by the plastic deformation of fibers.

A behavior exhibiting both softening and stiffening is typical for interpenetrating network gels and elastomers (Section 11.3). Interpenetrating network gels (also known as double networks) are, as the name indicates, made from interpenetrating networks and derive remarkable properties from their synergetic interaction. The mechanical properties of single network gels made from one of the two components of a double network gel are significantly different and not particularly spectacular, exhibiting relatively small ductility and low strength in tension. However, when put together, the ensemble acquires exceptional toughness and ductility, and increased strength.

Figure 6.21(b) shows an example of a similar nature (Sun et al., 2013). Polyampholytes are polymers with randomly distributed anionic and cationic groups which assemble in gels stabilized by ionic interactions. The bonds formed have a wide distribution of strengths. The strong bonds stabilize the network, while the weak bonds break and reform providing an internal energy dissipation mechanism which greatly enhances the material toughness. The small strain stiffness of the material is large since in the initial state all bonds contribute to stress production. Softening takes place gradually as the weak bonds break randomly. With increasing density of such events, spatial correlations of bond failure develop leading to an apparent yield point characterized by rapid softening. Once a large enough fraction of the initial bonds is eliminated, the remaining network with strong bonds has the kinematic freedom to undergo large deformations and exhibits strain stiffening; note the very large ultimate stretch in Figure 6.21(b).

The stress at the apparent yield point depends on the network and bond densities. In the particular case shown in Figure 6.21(b), softening is not so dramatic to produce loss of stability. However, the deformation of some other double network gels may become localized at the apparent yield point (Na et al., 2006).

Behaviors of types A and B may be viewed as special cases of the response shown in Figure 6.21. This leads us to the discussion of type C networks. The type C behavior is viewed here as the genuine mechanical behavior of networks in the absence of internal dissipation mechanisms (friction, adhesion, plasticity) and damage accumulation.

6.1.1.4.4 Type C: Stiffening

Networks of type C exhibit multiple regimes of strain stiffening, as shown schematically in Figure 6.1(b). We discuss first the general features of the uniaxial response. Figure 6.22(a) shows the nominal stress $S_{11} = S$ versus stretch $\lambda_1 = \lambda$ for model 3D

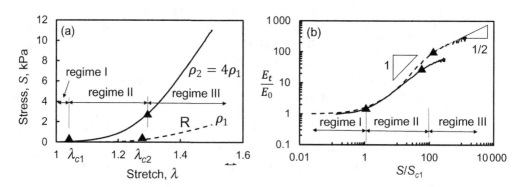

Figure 6.22 (a) Nominal stress–stretch curves for 3D athermal cellular networks of two densities. The transitions between the deformation regimes identified in the text are marked by triangles. (b) Data in (a) replotted as normalized tangent stiffness versus normalized stress. The vertical axis is normalized by the small strain modulus, E_0, and the horizontal axis is normalized by the stress at the transition between regimes I and II, S_{c1}.

cellular networks of two densities, ρ_1 and $\rho_2 = 4\rho_1$. The fibers have circular cross-sections and $l_b = 10^{-5}$. The linear elastic regime I is visible at small stretches, followed by two regimes of strain stiffening. The transition between regimes is marked in Figure 6.22 by triangles. The transition from regime I to regime II takes place at a stretch of $\lambda_{c1} = 1.03$, that is, 3% engineering strain. The transition stretch is largely independent of density and of l_b; see further discussion in Section 6.1.1.4.10. The regime I stiffness, E_0, differs between the two curves by a factor of 16, which follows the prediction of the equation $E_0 \sim \rho^2$ (with $\rho \gg \rho_c$) discussed in Section 6.1.1.3.

Figure 6.22(b) shows the data in Figure 6.22(a) plotted as tangent stiffness, $E_t = \partial S / \partial \lambda$, versus nominal stress, S. The stiffness is normalized by E_0 and the stress is normalized by the stress at the transition from regime I to II, S_{c1}. With this normalization, the two curves overlap in the linear elastic regime and the nonlinear regimes may be compared.

Regime II is characterized by a slope of 1, which indicates that $\partial S / \partial \lambda \sim S$, that is, the stress is an exponential function of the stretch: $S \sim \exp{(\alpha\lambda)}$. As discussed in Section 6.1.1.4.5, this type of stiffening is observed in many athermal networks tested in uniaxial tension and in shear. Regime III is characterized by a slope of $1/2$, which indicates that the stress varies quadratically with the stretch: $S / S_{c2} = (A(\lambda - \lambda_{c2}) + 1)^2$, where A is a constant. Regime IV, shown schematically in Figure 6.1(b), is not reached in this case.

Importantly, the nature of stiffening described by the slope of the curve in Figure 6.22(b) in regime II is independent of the network density and l_b (or parameter w). These parameters influence E_0 and the transition stress $S_{c1} = E_0(\lambda_{c1} - 1)$, but not the slope in Figure 6.22(b) which defines regime II stiffening. Similarly, the network connectivity, $\langle z \rangle$, has a strong effect on E_0 (Section 6.1.1.3.5), but has no effect on the functional form of strain stiffening (provided $\langle z \rangle$ is not as large such as to render the deformation affine). On the other hand, the stress range of regime II decreases as w increases and, for values of w in the affine range of the master plot, regime II disappears, and the network enters from the linear elastic regime I directly into regime III.

The free volume of these networks is large, on the order of 98%. Therefore, dynamic contact formation between fibers during the tension test is rare and excluded volume interactions make little to no contribution to the stress.

In the remainder of this subsection, we consider the lower density network of Figure 6.22 and analyze various aspects of its kinematics and strain energy storage. This network is denoted as the reference network R.

Nonaffinity

Figure 6.23 shows the nonaffinity parameter of Eq. (5.13) evaluated for network R. The nonaffinity is evaluated incrementally. At each increment of the global applied strain, the increment of the local strain is compared with the affine increment computed based on the network state in the previous step. This network is bending dominated at small strains and its deformation is highly nonaffine. The value of the nonaffinity parameter in regime I is comparable with that shown in Figure 6.7(a) at small w. The nonaffinity is constant during regime I but decreases fast in regime II.

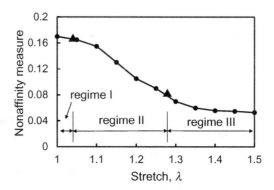

Figure 6.23 Variation of the nonaffinity parameter of Eq. (5.13) during the large strain deformation of the reference network R (lower density network in Figure 6.22).

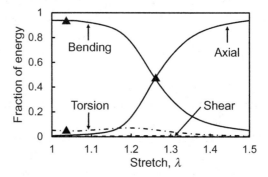

Figure 6.24 Fraction of the total strain energy stored in the axial, bending, torsion, and shear deformation modes of fibers during the uniaxial tensile deformation of the reference network R.

The transition between regimes is shown by triangles. The nonaffinity is small and approximately strain-independent in regime III, its values being comparable with those corresponding to the affine regime (large w) of Figure 6.7(a). Therefore, the data indicates that the degree of nonaffinity decreases in regime II and the network deforms approximately affinely in regime III. This conclusion is supported by direct experimental observations.

Energy Partition
The partition of strain energy between the various fiber deformation modes is shown as a function of stretch in Figure 6.24 for the reference athermal network R. Since the network is bending-dominated at small strains, most of the strain energy is stored in the bending mode in regime I. The role of the axial mode increases rapidly during regime II, at the expense of the bending mode. At the transition to regime III the axial mode becomes energetically dominant. This strain-controlled bending-to-axial energy transition correlates with the observation made in relation to Figure 6.23 that deformation becomes largely affine in

regime III. The bending-to-axial transition shown in Figure 6.7(b) is controlled by w and applies to regime I. Here we see a similar bending-to-axial transition occurring as an initially bending-dominated and nonaffine network is subjected to large strains.

Interestingly, the torsion and shear deformation modes do not store much strain energy and hence do not contribute significantly to stress production. While it is expected that the shear mode is not engaged if fibers are sufficiently slender, the observation that the torsion mode plays a minor role at all strains is somewhat of a surprise. This situation is similar to that discussed in relation to Figure 6.7(b).

The Poisson Effect
The Poisson effect is usually observed and evaluated in the uniaxial test. The Poisson ratio, v, of 3D random isotropic networks of athermal filaments loaded uniaxially in regime I is approximately 0.3.[7] The affine model predicts $v = 1/4$ for 3D random networks (Eq. (5.38)) and $v = 1/3$ for 2D networks. The Poisson ratio in regime I does not change significantly if the network is rendered nonaffine by reducing w.

The situation is quite different in regimes II and III. The Poisson ratio is defined in the linear elastic limit. An incremental version, v_i, can be used to characterize the lateral contraction/expansion under large strains:

$$v_i = -\frac{d \ln \lambda_2}{d \ln \lambda_1}. \tag{6.11}$$

Here λ_1 is the stretch in the loading direction, while λ_2 is the stretch in one of the transverse directions, or the average stretch in the transverse directions if the material remains transversely isotropic during deformation. v_i of Eq. (6.11) is computed based on the logarithmic (true) strains $\ln \lambda_1$ and $\ln \lambda_2$ in the longitudinal and transverse directions, respectively, and reduces to the usual Poisson ratio at small strains.

Figure 6.25 shows the variation of the incremental Poisson ratio during the uniaxial deformation of the reference network R. v_i is approximately 0.3 in regime I and increases fast in regime II, up to values as large as 6. In regime III, v_i levels off and decreases. The inset shows the variation of the stretch in the transverse direction, λ_2, with the applied stretch, λ. The ratio of the current to the reference volumes is given by the Jacobian and is equal to $\lambda_1 \lambda_2^2$. The volume decreases rapidly, reaching 65% of the initial volume at $\lambda_1 = \lambda = 1.3$.

The variation of the incremental Poisson ratio during the tensile test is weakly sensitive to the structural parameter w of Eq. (6.3) – see also Figure 8.6. This implies that denser networks of fibers stiffer in bending exhibit volume variations similar to those of the less dense networks made from fibers soft in bending, provided the network architecture is similar and the network is isotropic (fibers are randomly oriented). The network structure has strong influence on the Poisson ratio (see also Section 6.1.1.4.12) and hence the argument outlined in this paragraph applies to comparisons of networks of the same structure and with different w.

---

[7] We discuss here the case of networks with large free volume, not embedded in (solid or liquid) matrix.

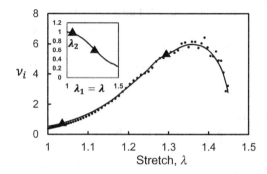

Figure 6.25 Variation of the incremental Poisson ratio of Eq. (6.11) during uniaxial tension for the reference network R (the low density network in Figure 6.22). The inset shows the variation of the stretch in the transverse direction with the applied axial stretch. The boundaries separating regimes I, II, and III are marked by triangles.

The incremental Poisson ratio can be approximated in regimes I and II by a power function of the form $v_i = v_0 \lambda^\beta$, where v_0 is the small strain Poisson ratio and β is an exponent which takes values larger than 5. This implies that the transverse stretch (inset to Figure 6.25) may be expressed in terms of the applied stretch, λ, as:

$$\lambda_2 = \lambda_3 = \exp\left[-\frac{v_0}{\beta}\left(\lambda^\beta - 1\right)\right]. \tag{6.12}$$

Exponent β changes little with w and, hence, Eq. (6.12) applies across the low w range of the master plot of Figure 6.6.

The Poisson ratio of isotropic linear elastic materials is bounded above by the value of 0.5. Dry networks exhibit spectacular Poisson contraction and drastic volume reduction during stretching, which appears to conflict with the thermodynamic considerations that impose restrictions on elastic constants and define the upper limit of v. Two arguments can be mentioned in support of the result in Figure 6.25. The first is that networks are not continua. Their kinematics is, in fact, closer to that of mechanisms. The second argument is based on the observation that networks develop preferential fiber alignment and become anisotropic at large strains. Ting and Chen (2005) showed that the upper limit of the Poisson ratio is eliminated in the case of anisotropic linear elastic continua, which potentially provides a theoretical basis for the large Poisson ratios observed in dry networks.

Lateral contraction is reduced in the presence of an embedding matrix. When the material is subjected to uniaxial tension, the network tends to contract in the direction transverse to the load much more than the matrix is able to contract. The network applies a compressive stress on the matrix which, in turn, resists the volumetric change (Figure 11.3(c)). In gels, for example, the aqueous embedding fluid imposes restrictions on lateral contraction. If there is no drainage, the effective Poisson ratio of the gel is 0.5. Drainage may take place if the hydrostatic stress wins over the network–solvent affinity. In such cases, lateral contraction is controlled by the energetic cost of expelling water from the gel. A detailed analysis of this situation is presented in Section 9.3.

The Poisson contraction is reduced in nonwovens (relative to Figure 6.25) due to the effect of excluded volume interactions (contacts) and inter-fiber friction. Excluded volume interactions also limit the Poisson effect in elastomers. In these materials the Poisson ratio is controlled by the short-range nonbonded interactions between molecular strands, and not by the behavior of the network. Since rubber has liquid-like density, it is incompressible and the Poisson ratio is equal to 0.5.

The Poisson effect is further limited by the presence of inclusions that may be embedded in the network. A prominent example is cartilage, which is a highly structured 3D collagen network embedding glycosaminoglycans (GAG) – large and bulky brush-like polysaccharide molecules. The role of the GAGs is to maintain the 3D structure of the network open, even under the action of large compressive stresses occurring in joints under normal physiological conditions. GAGs are charged molecules and retain water, preventing drainage when the tissue is compressed. This mechanism is essential in maintaining the proper function of the joints.

It is of interest to note that quasi-two-dimensional networks, such as nonwoven mats and biological collagen–elastin membranes, exhibit different Poisson effects in the in-plane and out-of-plane directions. Specifically, the out-of-plane effect is much more pronounced than the in-plane one, and this is somewhat independent of the thickness of the mat. Examples of such measurements are provided in Vader et al. (2009) for networks of reconstituted collagen, Mauri et al. (2015) and Perrini et al. (2015) for the human amnion, and Wong et al. (2019) for electrospun polymeric nanofiber mats. The difference between the in-plane and out-of-plane behaviors is caused by the preferential fiber alignment. In quasi-2D networks, fibers are randomly oriented in the plane of the mat.

The dependence of the Poisson effect on preferential alignment can be understood in simple terms by replacing the fibers with trusses, while otherwise retaining the network structure. Since the connectivity in such mats is typically $\langle z \rangle \approx 4$, the structure is sub-isostatic and, if one ignores the stabilizing effect of the fiber bending stiffness, the mat behaves like a mechanism. Consider the pantograph mechanism shown in Figure 6.26(a). The effective Poisson ratio of this structure depends on the angle θ between the pin-jointed bars and is given by $v_i = -1/\tan\theta$. v_i is large when θ is small and, hence, the Poisson effect in preferentially aligned networks loaded in the direction of alignment is much larger than in networks with random fiber orientation.

Negative Poisson ratios (i.e., auxetic behavior) are occasionally observed in athermal networks with a large density of contacts and small or zero crosslink density (Rodney et al., 2015; Wong et al., 2019). When evaluated in tension, the auxetic effect is due to the tendency of fibers to separate, such as to reduce the contact density at the expense of storing strain energy. This tendency is balanced by topological constraints associated with entanglements. This effect is described in Section 6.2.2 and discussed in detail in Negi and Picu (2021).

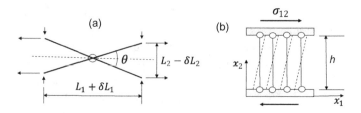

Figure 6.26 (a) Schematic showing the relation between fiber alignment and the Poisson effect. (b) Schematic suggesting the physical origin of the inverse Poynting effect in network materials.

The Poynting Effect

This effect was observed at the beginning of the twentieth century by Poynting (1909), who determined that the length of wires subjected to torsion increases. Equivalently, a sample subjected to shear tends to expand in the direction normal to the plane of the applied shear. If the expansion is constrained, as for example in simple shear, a compressive normal stress develops. To make ideas more precise, consider a generic hyperelastic material described by the free energy density function $\Psi(I_1, I_2, I_3)$, where I_1, I_2 and I_3 are the invariants of the right Cauchy–Green deformation tensor, $\mathbf{C}$ (Section 7.3.1), subjected to simple shear described by the deformation gradient tensor $\mathbf{F}$ with non-zero components $F_{12} = \gamma, F_{ii} = 1, i = 1 \dots 3$ and $\mathbf{C} = \mathbf{F}^\mathrm{T}\mathbf{F}$. The resulting shear stress is σ_{12}, and the direction normal to the plane of shear is x_2. The Cauchy stress can be computed with Eqs. (7.4) or (7.5) for the general and incompressible cases, respectively. Considering the incompressible case and assuming that the normal stresses σ_{11} and σ_{33} vanish, one obtains the relation between the normal stress σ_{22} and the applied shear stress as $\sigma_{22} = -\sigma_{12}\gamma$.[8] This implies that the normal stress is negative (compressive), which represents the usual Poynting effect, as reported in 1909. It also indicates that the Poynting effect is second order, in the sense that if the shear stress is a linear function of the shear strain, the normal stress is a quadratic function of γ.

Network materials often exhibit an inverse Poynting effect, that is, the normal stress σ_{22} associated with a simple shear experiment is tensile (positive). This is observed in hydrogels reinforced with collagen and fibrin (Janmey et al., 2007), collagen (Jansen et al., 2018), fibrin (Kang et al., 2009), and other protein fiber networks, and numerically in Licup et al. (2016).

The origin of the inverse Poynting effect may be understood qualitatively based on the schematic in Figure 6.26(b), which shows two horizontal plates connected by a set of vertical fibers. When shear is applied, while keeping the distance between plates constant, fibers are stretched, which leads to the development of normal and shear stresses. The normal stress σ_{22} is always positive (tensile). The two stress components are coupled, since both are related to the stretch of the fibers. It can be inferred based

---

[8] If the assumption that σ_{11} and σ_{33} vanish is relaxed, while requiring instead the pressure to be zero, the respective relation becomes: $\sigma_{22} - \sigma_{11} = -\sigma_{12}\gamma$.

on this schematic that the stiffness measured in shear, $G_t = \partial\sigma_{12}/\partial\gamma$, is a nonlinear function of σ_{22} which can be approximated for small σ_{22} as:

$$G_t \approx G_0 + A\sigma_{22}. \tag{6.13}$$

The simple configuration in Figure 6.26(b) has zero shear stiffness in the initial configuration (when γ is small) and, hence, $G_0 = 0$, while in this case $A = 3$.

This model indicates two other important aspects. If the distance h between the two plates in the x_2 direction is increased, A increases and the Poynting effect is enhanced. If the respective distance is reduced, the effective stiffness G_t decreases. This effect and the linear relation between shear stiffness and the normal stress, Eq. (6.13), are observed in network materials (e.g., Licup et al., 2015, 2016). The other aspect of importance is that the inverse Poynting effect may be controlled by adjusting the anisotropy of the material. If the angle of the fibers in Figure 6.26(b) in the reference configuration relative to axis x_2 increases, the normal stress σ_{22} and constant A in Eq. (6.13) decrease. In the limit when fibers are parallel to the x_1 axis, the effect vanishes. These issues are discussed in relation to the network architecture in Meng and Terentjev (2016), and based on general nonlinear hyperelasticity of anisotropic continua in Horgan and Murphy (2017), where it is shown that both normal and inverse Poynting effects may be obtained by adjusting the anisotropy of the hyperelastic material.

Preferential Fiber Orientation and the Stress–Optical Coefficient

Fibers become preferentially aligned in the stretch direction during tension, and the degree of alignment depends strongly on the lateral contraction. Figure 6.27 shows the variation of the orientation parameter P_2 (Eq. (4.3)) for the cellular reference network R. P_2 is evaluated based on the angle between the end-to-end vector of fibers and the stretch direction. Fibers are randomly oriented in the unloaded state and $P_2 = 0$ for $\lambda = 1$. The degree of alignment is close to zero in regime I and increases monotonically in regimes II and III. The figure also shows the slope of $P_2(\lambda)$, which exhibits a

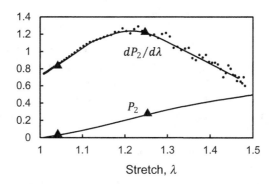

Figure 6.27 Variation of the orientation parameter $P_2(\lambda)$ and its derivative during the deformation of the reference network R. The transitions between the three deformation regimes are marked by triangles.

maximum at the transition from regime II to regime III. This indicates that the rate of alignment increases continuously in regime II and decreases in regime III. This behavior is consistent with the observation that Poisson contraction is pronounced during regime II and decreases in regime III (Figure 6.25).

The ratio between the deviatoric stress (proportional to the applied stress in the uniaxial test) and P_2 is known as the stress–optical coefficient (Section 5.3.5). This parameter is of interest since it establishes a relation between structural evolution and the network-scale stress. In some polymeric networks, the stress–optical coefficient is independent of stretch, which suggests that the global stress results as a sum of contributions of individual fibers and, hence, the structural anisotropy reflects directly in the anisotropy of the stress tensor. The affine model is a particular instantiation of this physical picture: fibers deform axially and, hence, the fiber-scale stress (the stress produced by a single fiber) has only one non-zero normal component in the direction of the fiber axis, while the network scale stress is the system average of these "stresslets." A more general form of this concept which applies to stress production in polymeric melts and networks is described in Gao and Weiner (1994) and Picu and Pavel (2003). The affine model applied to thermal networks (e.g., the affine rubber elasticity model) predicts that the Cauchy stress is proportional to P_2 and the stress–optical coefficient is a material constant. This is verified experimentally for elastomers at relatively small strains. However, the stress–optical rule does not apply to networks that stiffen exponentially (Figure 6.22). This is another indication that the stress production mechanism in exponentially stiffening nonaffine networks is different from that operating in affinely deforming networks.

To summarize the observations presented in Figures 6.22–6.27: regime I corresponds to a narrow range of strains (although, in networks with large E_0, it may correspond to a large range of stresses) and entails no structural evolution. Regime II corresponds to rapid strain stiffening described either by an exponential or a power stress–strain relation. Significant structural evolution takes place during regime II, including local instabilities and rapid fiber re-orientation supported by large nonaffinity (Figure 6.23) and causing a large Poisson effect (Figure 6.25). Since the networks discussed here are not embedded in a volume preserving matrix and excluded volume interactions between fibers are not essential (low fiber volume fraction situations), the network volume decreases significantly during uniaxial tensile loading. At the onset of regime III the rate of fiber alignment $dP_2/d\lambda$ reaches a maximum (Figure 6.27), the incremental Poisson ratio starts to decrease (Figure 6.25), and the energy storage switches from the bending mode to the axial mode (Figure 6.24). These trends continue into regime III, but the degree of incremental alignment and the lateral contraction gradually decrease. Deformation is approximately affine and fibers deform predominantly in their axial mode in regime III.

Damage accumulation and/or plasticity limit the number of the regimes that are actually observed in experiments. Denser networks, such as paper and some nonwovens, exhibit only regime I and, possibly, the onset of regime II. A behavior of type B is more likely for these systems than one of type C. Biological connective tissues deform in vivo within the range of regimes I and II. Some molecular networks of semiflexible protein filaments reach regime III. Regime IV (Figure 6.1(b)) may be visible in pre-aligned networks; an example of this type is given in Section 6.1.1.4.7.

The network parameters, $\rho, l_b, \langle z \rangle$, influence the nonlinear behavior primarily through the small strain stiffness, E_0. Specifically, increasing parameter w or the average connectivity $\langle z \rangle$ cause an increase of E_0 (Figure 6.6). In athermal networks, the stretch at the transition from regime I to II is largely independent of the network parameters. Therefore, the stress at this transition, $S_{c1} = E_0(\lambda_{c1} - 1)$, depends on structural parameters through E_0. The functional form of strain stiffening in regime II is independent of ρ and l_b, and is controlled by the network architecture.

The physical picture presented here for athermal networks also applies, in general terms, to thermal networks, although some important details are different. To outline these differences, several examples of both thermal and athermal network materials are discussed next.

6.1.1.4.5 Nonlinear Behavior of Several Network Materials
Collagen Networks and Biological Tissue

Collagen is deposited by cells in the form of fibrils that self-organize and form fibers and larger scale networks that play an essential structural role in the human and animal bodies. The elasticity of the skin, various membranes, and connective tissue, the mechanics of cartilage, and the elasticity of blood vessels are all related to the mechanics of collagen networks. Many pathologies are associated with defects of the multiscale collagen structure or with the accumulation of damage in collagen networks. These range from benign cosmetic issues, such as skin wrinkling and sagging, to serious, life threatening problems such as aneurisms. Given its importance, the mechanical behavior of collagenous tissue and of reconstituted collagen was studied extensively.

The dependence of the small strain stiffness of collagen networks on collagen concentration and crosslinking is discussed in Section 6.1.1.3.3 – see Figure 6.9. The nonlinear response of collagen networks of various types and tested in various conditions is shown in Figure 6.28.

Figure 6.28(a) presents a collection of data obtained from uniaxial tests performed on reconstituted type I collagen (Lake and Barocas, 2011), mouse dermis – the structural component of the skin (Bancelin et al., 2015), the human amnion – the structural component of the fetal membrane (Mauri et al., 2015), and the basement membrane of renal tubes (Welling et al., 1995). These networks are all athermal. The data is presented as tangent stiffness versus nominal stress, where the tangent stiffness is computed as $E_t = dS/d\lambda$, such that the results are directly comparable with those shown in Figure 6.22(b). Despite the significant difference between these structures, the data collapses on a line of slope 1, which indicates exponential strain stiffening. This is typical for many tissues, including cartilage, the liver capsule, tendon, etc. The basement membrane and the dermis samples are loaded to stress levels beyond those associated with damage initiation, and this leads to the softening visible in the upper part of the respective curves.

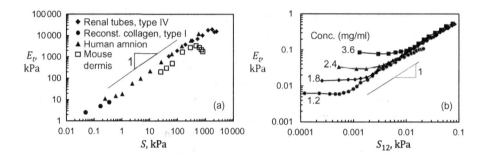

Figure 6.28 Tangent stiffness–stress plots for various collagen-based tissues and reconstituted collagen gels. The results in (a) are obtained in uniaxial tension and include data for renal tubules containing type IV collagen (Welling et al., 1995), human amnion (Mauri et al., 2015), reconstituted type I collagen (Lake and Barocas, 2011), and mouse dermis (Bancelin et al., 2015). The data in (b) are obtained with reconstituted type I collagen tested in shear. Adapted from Licup et al. (2015)

Figure 6.28(b) shows results from tests performed in simple shear on reconstituted type I collagen at $37°C$ (Licup et al., 2015).[9] Gels of four collagen concentrations are considered and the test is performed at constant volume. The tangent stiffness is computed in this case as the derivative of the shear stress to the engineering strain; note that, in this case, the Cauchy stress component σ_{12} is identical to the nominal stress component S_{12}. The reported tangent stiffness is $E_t = dS_{12}/d\gamma$, which is plotted against S_{12}.

Regimes I and II of the network response are clearly visible in Figure 6.28(b). The small strain stiffness, E_0, corresponding to the regime I plateau, increases with the concentration, as shown in Figure 6.9. All curves converge in regime II to a common line of slope 1 which indicates an exponential dependence of the shear stress on the engineering shear strain, in agreement with the behavior of collagen-based biological tissues shown in Figure 6.28(a). The fact that these curves merge in regime II also indicates that the strain at the transition between regimes I and II is independent of the collagen concentration (network density), an issue also discussed in Section 6.1.1.4.10.

The networks leading to the results in Figure 6.28(a) have collagen densities in the same range as those presented in Figure 6.28(b). However, the stiffness of the reconstituted collagen networks is orders of magnitude smaller than that of the tissues. This highlights an important problem with the current artificial collagen constructs, which are too soft to be used in tissue engineering. The discrepancy is due in part to the different fiber diameter in the two types of networks: reconstituted collagen networks have thinner fibers. It is likely that the network architecture is also different, although precise quantification is difficult and subjected to interpretation.

---

[9] The temperature has a large effect on the state of these networks. Lower temperatures may lead to bundling of fibers, which changes the network behavior (Jansen et al., 2018). This also applies to other protein fiber networks such as actin (Gardel et al., 2004) and fibrin (Kurniawan et al., 2017).

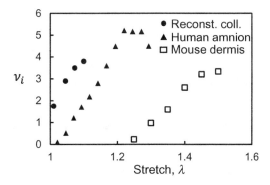

Figure 6.29 Variation of the incremental Poisson ratio during deformation for three of the collagen structures shown in Figure 6.28(a): The human amnion (Mauri et al., 2015), reconstituted type I collagen (Lake and Barocas, 2011), and mouse dermis (Bancelin et al., 2015). Adapted from Picu et al. (2018)

Figure 6.29 shows the incremental Poisson ratio of some of the collagen networks in Figure 28(a). The curves exhibit the features outlined in relation connection with Figure 6.25: the Poisson ratio is small at small strains, but increases rapidly as the network strain stiffens during regime II. Poisson ratios as large as 5 are measured. ν_i reaches a maximum and then decreases. The occurrence of a maximum in these curves is related to the onset of damage and not to the transition to regime III, as indicated in relation to Figure 6.25. This is supported by the observation that these networks do not enter regime III in the range of strains applied in these experiments, which is obvious from Figure 6.28(a) where only regime II is visible.

The strong Poisson effect exhibited by these structures can be described by Eq. (6.12). For example, the data in Mauri et al. (2015) can be fitted by selecting $\beta = 15$. An even stronger Poisson effect is reported in Vader et al. (2009), who worked with reconstituted collagen gels of 1 mg/ml treated with glutaraldehyde.

The curves in Figure 6.29 do not start from $\lambda = 1$. The horizontal shift is due to differences in the definition of the reference state of zero strain. Identifying the origin of the strain axis is a common problem when testing soft materials. In the initial stages of deformation, the force is below the resolution of the load cell and one is tempted to identify the state corresponding to zero strain as that in which the first force reading is obtained. This is misleading since the network may evolve at very small applied stress. In Mauri et al. (2015) it is shown that lateral contraction and fiber orientation may be measured below the force resolution limit of the respective instrument.

The strong Poisson effect observed in these networks suggests that fiber alignment should be pronounced. This is indeed observed in all studies that measure fiber orientation either by direct microscopic observations or by using birefringence. Examples of this include Mauri et al. (2015) and Vader et al. (2009), who work with human amnion and reconstituted collagen samples, respectively. Alignment is more pronounced during regime II of network deformation. However, the stress–optical

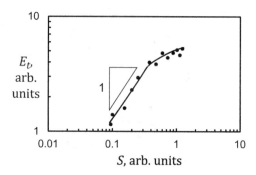

Figure 6.30 Tangent stiffness–stress for polyester felt samples. Arbitrary units are used for the two axes, which does not affect the scaling relation represented by the curve. Data from Kabla and Mahadevan (2006)

coefficient varies during deformation and the stress–optical rule does not apply to collagen structures. A counterexample related to the deformation of the mouse dermis is presented in Bancelin et al. (2015).

Felt

Felt is an athermal network obtained by pressing together fibers of finite length. It can be made from natural or artificial fibers. Pressing leads to negligible inter-fiber bonding. Although not crosslinked, fibers stay together due to entanglements and friction. Therefore, felt has mechanical properties somewhat similar to those of nonwovens, but generally it has lower strength. It exhibits all features of low density, sparsely crosslinked networks such as stress–strain curves of type C, and significant Poisson contraction.

Figure 6.30 shows the tangent stiffness versus stress curve of a felt made from polyester fibers of lengths between 4 and 20 cm and of diameter 30 μm (Kabla and Mahadevan, 2006). The test is performed in uniaxial tension and the tangent stiffness is computed based on the nominal stress, $E_t = dS/d\lambda$. Strain stiffening is similar to that observed in all biological and reconstituted collagen networks shown in Figure 6.28: $E_t \sim S$ in regime II, which indicates exponential stiffening. A short apparent regime III emerges at larger stress values. The network ruptures at larger stresses and this creates difficulties when interpreting the reduction of the stiffening rate at the end of regime II: it is unclear whether it is related to disentanglement and/or fiber rupture, or it is a genuine nonlinear elastic regime III of the type shown in Figure 6.22(b).

The incremental Poisson ratio of the same samples is shown in Figure 6.31. The inset shows the variation of the transverse stretch, λ_2, as a function of the applied stretch, $\lambda_1 = \lambda$. The incremental Poisson ratio reaches a maximum value of 2.8 within regime III of the curve shown in Figure 6.30. The rapid decrease of v_i beyond the maximum is likely due to damage accumulation. The inset shows the variation of the transverse stretch, which exhibits the characteristic behavior observed in Figure 6.25.

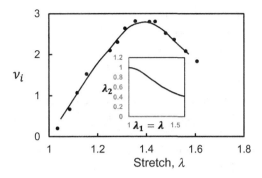

Figure 6.31 Variation of the incremental Poisson ratio during deformation for polyester felt samples. The inset shows the variation of the transverse stretch with the applied stretch, λ. Data from Kabla and Mahadevan (2006)

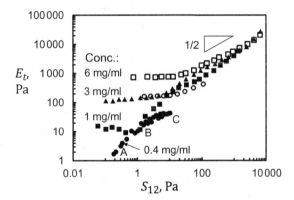

Figure 6.32 Tangent stiffness versus shear stress for fibrin networks of various concentrations. Data from Jawerth (2013) – filled circles, Kang et al. (2009) – open circles, and Kurniawan et al. (2017) – triangles and squares.

Fibrin Gels

Fibrin is a fibrous protein with a central role in blood clotting. Fibrin fibers form by the polymerization of fibrinogen and then aggregate to create a dense network. The in vivo fibrin networks embed platelets – a blood component – to create plugs that stop bleeding caused by wounds.

The mechanical behavior of fibrin networks with and without platelets was studied recently. Results for such networks without platelets obtained from shear tests are shown in Figure 6.32. The figure shows the tangent modulus computed as the derivative of the shear stress relative to the applied engineering shear strain versus the shear stress. Data from various sources are included, corresponding to concentrations of 0.4 mg/ml (Jawerth, 2013), 1 mg/ml (Kang et al., 2009; Kurniawan et al., 2017), 3 mg/ml, and 6 mg/ml (Kurniawan et al., 2017). The regime I plateau is visible at small stresses. The higher concentration gels stiffen in the second regime with a slope of approximately 1/2, that is, the exponential stiffening regime II is absent and the curve directly enters regime III from regime I. Samples with lower concentrations (0.4 and 1 mg/ml) exhibit both regimes II and III, characterized by slopes of 1 and 1/2, respectively.

The kinematics of fibrin fibers during tests on the low concentration gels have been evaluated by direct in situ microscopy observations in Jawerth (2013). It is reported that the nonaffinity parameter is large in regime II and decreases fast as the network enters regime III. This behavior is predicted by numerical models of athermal networks, as discussed earlier in this chapter.

Actin Gels

Actin is an important component of the cytoskeleton – the structural scaffold of the cell. It forms fibers (fibrillar actin) of diameters small enough for thermal fluctuations to play an essential role in their mechanics. Fibers have a persistence length larger than the mean segment length and, hence, actin fibers are considered thermal and semiflexible.

Networks of semiflexible thermal filaments present additional complexities since the strongly nonlinear behavior of their fibers reflects in the mechanics of the network. Fibers are stiff in bending on length scales comparable with, and smaller than, the persistence length, and their effective axial stiffness is proportional to the temperature, which is the signature of the entropic nature of their elasticity. They are generally represented with the worm-like chain model (Section 2.4.4) whose axial mechanical response is described by Eq. (2.48) and is shown graphically in Figure 2.20.

Figure 6.33 shows experimental data from Gardel et al. (2004) for actin of various concentrations crosslinked with scruin. The ratio between the crosslinker and actin concentrations in solution is 0.03. The tangent stiffness is computed as the derivative

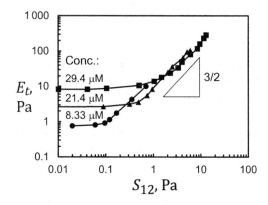

Figure 6.33 Tangent stiffness versus shear stress curves for actin networks of various concentrations crosslinked with scruin. Data from Gardel et al. (2004)

of the shear stress relative to the engineering shear strain, $E_t = d\sigma_{12}/d\gamma = dS_{12}/d\gamma$, and is shown in Figure 6.33 as a function of shear stress for several actin concentrations. Regimes I and II are well defined. The variation of the small strain stiffness (left plateau) with the actin concentration is in agreement with the data shown in Figure 6.10. In regime II, the stiffening rate is characterized by a slope of 3/2, that is, these networks stiffen faster than athermal networks. The larger slope is commonly associated with the contribution of the nonlinear behavior of individual filaments. Similar scalings have been observed in other networks of thermal semiflexible filaments such as vimentin (Lin et al., 2010) and neurofilament gels (Storm et al., 2005). Note that, in order for the signature of the axial deformation of fibers to be visible in the network-scale response, the network deformation has to be affine. Therefore, the interpretation that the 3/2 slope emerges from the fiber-scale mechanics is based on the tacit affine deformation assumption.

The cytoskeleton reconfigures dynamically as actin filaments inside the cell grow and depolymerize continuously. The crosslinks are not permanent either. They attach and detach dynamically at rates that depend on temperature and the applied stress. Such networks creep under constant load and are capable of self-healing (Chapter 9). Their response depends on the applied strain rate. In order to evaluate the behavior of the network without the contribution of the transient crosslinks, it is necessary to construct artificial gels in which actin filaments are crosslinked permanently. The effect of the crosslink behavior on network response was studied by several groups; some of these results are discussed in Section 6.1.1.4.9.

Agarose and Acrylamide Gels

Interpenetrating network (IPN) gels exhibit the complex behavior shown in Figure 6.21(b) characterized by large stiffness in regime I, softening, and the development of an apparent yield point, followed by strain stiffening (see also Section 11.3). These materials are exceptionally resilient as they deform to very large stretches before failure. The behavior after the yield point is interesting in the present context, as it is supposed to be controlled exclusively by the strongly crosslinked component network (referred to as network 2 in Section 11.3). Figure 6.34 shows the tangent stiffness versus nominal stress curves corresponding to the strain stiffening regime of two types of IPNs, beyond their softening regime that follows the yield point. The first example corresponds to an IPN in which the strongly bonded component is acrylamide-based (Na et al., 2006). The second example refers to a hydrophobic association hydrogel obtained by the micellar copolymerization of acrylamide and a small molar fraction of a hydrophobic monomer (Jiang et al., 2010).

Both types of gels exhibit exceptional ductility and strong stiffening with a slope of 3/2 in the tangent stiffness–nominal stress plot (Figure 6.34).[10] The fascinating aspect

---

[10] Note that this stiffening is similar to that of the actin gels and of gels of other semiflexible filaments such as vimentin, although the acrylamide molecular strands cannot be considered semiflexible filaments. This suggests the possibility that the 3/2 slope reported in Figure 6.33 is not entirely caused by the nonlinear elasticity of the semiflexible filaments, as commonly assumed in the literature on semiflexible networks.

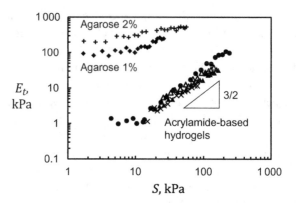

Figure 6.34 Tangent stiffness versus stress curves for agarose gels of 1% and 2% concentration (Kwon et al., 2010), the strain stiffening part of an acrylamide-based interpenetrating network deformed beyond the yield point (Na et al., 2006), and of a hydrophobic association acrylamide-based hydrogel at three concentrations (1%, 2%, and 4%) of the hydrophobic monomer (Jiang et al., 2010). All tests are performed in uniaxial tension.

of the interpenetrating network behavior is their exceptional ductility. The single network gel (corresponding to the strongly crosslinked component of the IPN) ruptures at about 1/3 of the failure stretch exhibited by the softened IPN network (which fails at a stretch of approximately 15!). The physical reason for the large ductility of double networks has not been clarified to date.

The hydrophobic association hydrogel exhibits stretches at failure larger than 10 and strong strain stiffening beyond the yield point, with $E_t \sim S^{3/2}$ for all concentrations considered.

The figure also includes data for two single network agarose gels of 1% and 2% concentration (Kwon et al., 2010). These gels rupture in tension at strains of about 15% and exhibit weak stiffening in this strain range.

Elastomers

Rubber is one of the most studied network materials. It stands apart from the other examples discussed in this section due to its low free volume and the strong non-bonded interactions of the molecular strands. The dense molecular packing enforces isochoric deformation.

Figure 6.35 shows results from uniaxial tests performed at room temperature with vulcanized natural rubber (Treloar, 1944). This is one of the reference data sets for rubber, used by many authors to validate constitutive descriptions. At stretches below 2, the nominal stress–stretch curve shows gradual softening, while continuous stiffening is observed for larger stretches (Figure 6.35(a)). The tangent stiffness–stress curve, $E_t \sim dS/d\lambda$ versus S (Figure 6.35(b)), makes these two regimes obvious.

Natural rubber crystallizes when strained. Strain-induced crystallization begins at threshold stretch values that decrease with decreasing temperature. At room

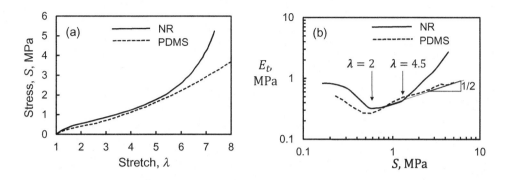

Figure 6.35 (a) Nominal stress–stretch of vulcanized natural rubber (NR) (data from Treloar, 1944) and of PDMS loaded in uniaxial tension at room temperature. (b) Data in (a) replotted as tangent stiffness versus stress.

temperature, crystallization begins at a stretch of approximately 4.5 (Rault et al., 2006). The range of the stiffness–stress curve bounded by $\lambda = 2$ and $\lambda = 4.5$ may be approximated as $E_t \sim S^{1/2}$. The stiffening rate increases rapidly for larger stretches due to the strain-induced crystallization process.

While a phenomenon of scientific and technological importance, crystallization adds complexity to the behavior of the network and obscures some of the features we wish to emphasize here. To outline the effect of crystallization, data for PDMS networks are included in Figure 6.35 (shown by the dashed line). PDMS does not undergo strain-induced crystallization and the stress–stretch curve of Figure 6.35(a) does not show the rapid upturn of the natural rubber curve at stretches above $\lambda = 4.5$. This is also seen in Figure 6.35(b), where the PDMS curve strain stiffens with a constant slope of approximately $1/2$ for the entire strain stiffening segment of the stress–stretch curve. The behavior of the PDMS network at stretches larger than $\lambda = 2$ is consistent with that of other networks discussed in this chapter. We consider the behavior in this stretch range to be representative for the response of the elastomeric network.

To place the representation in Figure 6.35(b) in relation with the more established way of reporting data in the rubber literature, Figure 6.36 shows the same natural rubber and PDMS curves rearranged as a Mooney–Rivlin plot, that is, as $S/\left(\lambda - 1/\lambda^2\right)$ versus $1/\lambda$. The curves are typical for such representation. The intercept of the vertical axis and the slope at $\lambda = 1$ correspond to the first and second coefficients of the Mooney–Rivlin model, α_1 and α_2, respectively (Eq. (7.10)). Both PDMS and natural rubber exhibit an initial decreasing trend as λ increases. This segment corresponds to the softening branch of the curves in Figure 6.35(b). Stiffening is observed for $\lambda > 2$ and the upturn is more pronounced in the natural rubber case.

To compare the response of elastomers with that of other network materials discussed in this section, one of the curves for biological collagen networks presented in Figure 6.28(a) (Mauri et al., 2015) is included in Figure 6.36. Collagen networks stiffen rapidly, at much smaller strains than rubber, and the Mooney–Rivlin softening

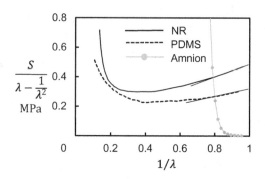

Figure 6.36 Mooney–Rivlin representation of the natural rubber and PDMS data in Figure 6.35. Data from tests on the human amnion (Mauri et al., 2015) from Figure 6.28(a) is included for reference; note that in this case S decreases to zero faster than the normalization parameter of the vertical axis as λ approaches 1 from above.

regime is not present. However, the type of stiffening observed in PDMS is similar to that of other affinely-deforming network materials (slope of 1/2 in Figure 6.35(b)).

Other elastomeric networks that do not undergo strain-induced crystallization show upturns in the Mooney–Rivlin plot. End-linked bimodal networks composed from long and short chains exhibit stiffening at decreasing strains as the fraction of the short chains increases (Erman and Mark, 1989). These upturns are independent of temperature, which indicates that they are not associated with strain-induced crystallization. This is in contrast with natural rubber for which the upturn of the Mooney–Rivlin curve can be eliminated or shifted to larger strains by increasing the temperature. Also, the stiffening regime begins at smaller strains when the network is swelled by a solvent. This observation is interpreted as a manifestation of the finite extensibility of the chains and a departure from the Gaussian behavior. Further discussion of these issues can be found in texts dedicated to rubber elasticity, such as Erman and Mark (1997) and Gent (2012).

6.1.1.4.6 *Effect of Fiber Crimp*

Fibers are tortuous in most network materials and examples are shown in Figures 2.12 and 2.15. Figure 2.12 shows that collagen fibers in blood vessels have a broad range of tortuosity (or crimp), and this appears to have an important functional role (see Section 2.3.5).

Figure 2.15 shows typical tortuosity of felt fibers, which also exhibit a range of slack. Fibers in nonwovens made by continuous spinning have similar geometric properties.

The presence of crimp causes a drastic reduction of the small strain stiffness, E_0, relative to the case of straight fibers (Section 6.1.1.3.6). Here we discuss the effect of crimp on the nonlinear behavior of the network.

Figure 6.37 shows numerical data for cellular networks with various degrees of fiber crimp. Crimp is quantified using the tortuosity parameter, c, of Eq. (2.9).

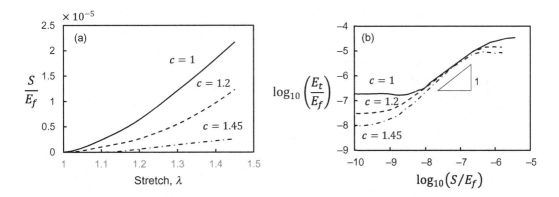

Figure 6.37 (a) Nominal stress–stretch curves for cellular networks of increasing fiber tortuosity, c. (b) Data in (a) replotted as tangent stiffness versus stress. Case $c = 1$ corresponds to straight fibers and is shown with a continuous line. In the other two cases, fibers have sinusoidal shape of amplitude equal to 0.3 of the end-to-end length in the $c = 1.2$ case, and 0.75 of the end-to-end length in the $c = 1.45$ case. In the given model, all fibers have the same c. Courtesy of E. Ban

Figure 6.37(a) shows nominal stress–stretch curves obtained in uniaxial tension with similar models. The stress is normalized by the fiber material modulus, E_f. The curves exhibit all typical features discussed in Section 6.1.1.4.4 for networks of straight fibers. Three regimes can be identified: regime I characterized by E_0, followed by the strain stiffening regime II and by regime III, which is characterized by a strain stiffening rate smaller than that observed in regime II. The effect of crimp is clearly visible in regime I, amounting to a substantial reduction of E_0 as the tortuosity parameter increases. No other obvious effects of crimp are observed in the representation of Figure 6.37(a).

Figure 6.37(b) shows the same data replotted as tangent stiffness versus stress and indicates that the nonlinear behavior is largely independent of fiber tortuosity. All networks with tortuous fibers ($c > 1$) stiffen exponentially, similar to the equivalent network with straight fibers ($c = 1$). Decreasing the density or decreasing l_b of a network of straight fibers leads to the same qualitative behavior as increasing the fiber tortuosity.

6.1.1.4.7 *Effect of Preferential Fiber Alignment*

Preferential fiber alignment modifies some aspects of the nonlinear mechanical behavior quantitatively, but not qualitatively; the essential features described in Sections 6.1.1.4.2 and 6.1.1.4.4 remain relevant.

Fiber alignment introduces anisotropy, the network becoming stiffer when probed in the alignment direction (Section 6.1.1.3.8). To gain an intuitive understanding of the effect of alignment on the nonlinear response, consider that a network with random fiber orientations in the undeformed state develops preferential alignment in the principal stretch direction when subjected to large deformations. Therefore, a network

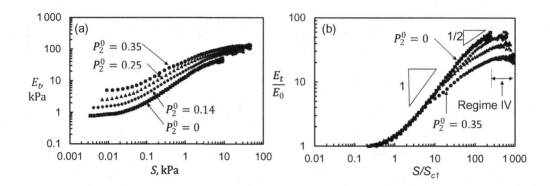

Figure 6.38 (a) Tangent stiffness versus stress curves for model cellular networks with increasing degrees of initial alignment characterized by P_2^0. (b) Curves in (a) replotted with the vertical axis normalized by the small strain modulus, E_0, such to overlap regime I.

with pre-aligned fibers loaded in tension in the direction of alignment should start with a larger value of the stiffness and follow roughly the same nonlinear path as the reference network without pre-aligned fibers, except that the strain range over which regime II is observed becomes smaller. Figure 6.38 demonstrates this type of behavior. Models with various degrees of initial alignment (characterized by parameter P_2^0) are generated and loaded in uniaxial tension in the direction of the initial alignment. Figure 6.38(a) shows the tangent stiffness, $E_t = dS/d\lambda$, versus nominal stress for several networks with increasing degrees of P_2^0. The curves shift up, reflecting the increase of the small strain stiffness, while the nonlinear response appears to be weakly dependent on P_2^0. However, differences are observed in Figure 6.38(b), where the curves shown in Figure 6.38(a) are shifted such as to overlap their regime I sections. It becomes obvious that, as P_2^0 increases, the range of regime II decreases, regime III becomes more pronounced, and, in the case of the largest P_2^0 value considered, regime IV appears; additional results are presented in Picu et al. (2018).

If probed in tension in a direction orthogonal to the direction of initial fiber alignment, the regime I stiffness decreases and the range of regime II increases with increasing degree of alignment. These trends are opposite to those shown in Figure 6.38.

Figure 6.39 shows the energy partition in pre-aligned networks. As the degree of initial alignment increases, the transition from the regime in which energy is stored predominantly in the bending mode to that in which the axial mode dominates takes place at smaller stretches. As discussed in Section 6.1.1.4.4 (Figure 6.24), this shift from bending to axially-dominated deformation coincides with the transition from regime II to regime III. The figure also shows that the bending-to-axial transition corresponds to the inflection point of the $P_2(\lambda)$ curve, which is also associated with the transition between regimes II and III, as shown in Figure 6.27. These observations reinforce the conclusion resulting from Figure 6.38(b) that increasing P_2^0 reduces the stress range of regime II and expands the range of regime III. If the network is probed

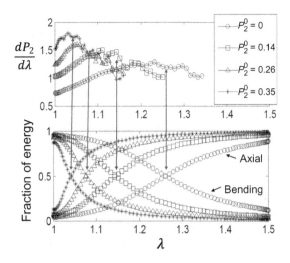

Figure 6.39 Energy partition in networks with increasing degree of initial fiber alignment. The transition from bending to axial energy dominance coincides with the transition from regime II to III, and corresponds to the inflection point of $P_2(\lambda)$ (maximum of $dP_2/d\lambda(\lambda)$). Republished with permission of ASME from Picu et al. (2018), permission conveyed through Copyright Clearance Center Inc.

in tension in a direction orthogonal to that of the initial preferential orientation, the opposite trends are observed.

6.1.1.4.8 *Effect of the Variability of Fiber Properties*

The situation in which fibers of a specific network have different properties is discussed in Section 6.1.1.3.9, with a focus on evaluating the effect of stochasticity on network stiffness. In the respective section, the degree of stochasticity of fiber properties is controlled by adjusting the variance of the distribution of the fiber bending/axial stiffness, while the mean of the distribution is kept constant. It is concluded that increasing stochasticity renders the network softer: the relative change of the network stiffness is proportional to the coefficient of variation of the distribution (Figure 6.17).

It is of interest to inquire how this type of stochasticity affects the nonlinear response. This problem was studied in Ban et al. (2016b) where it is shown that networks with very different degrees of stochasticity of fiber properties stiffen in the same way, that is, identical to the network of the same architecture in which all fibers have the same axial and bending stiffness. This allows us to apply the results discussed in Section 6.1.1.4.4 to networks with spatially varying fiber properties. This result is relevant for biological networks in which fibers are not identical. Collagen fibers are constructed by the organization of fibrils into bundles stabilized by internal enzymatic and nonenzymatic crosslinks. The size of bundles varies from one network segment to the other. Likewise, filament bundling occurs frequently in protein fiber networks, as a

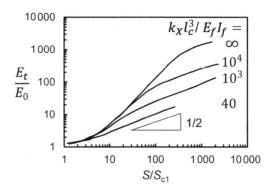

Figure 6.40 Tangent stiffness versus stress curves for networks with compliant crosslinks characterized by the non-dimensional parameter $k_X l_c^3 / E_f I_f$, where k_X is the normalized crosslink stiffness. $k_X l_c^3 / E_f I_f \to \infty$ corresponds to the rigidly crosslinked case. Data from Zagar et al. (2015)

function of the temperature and local chemistry, leading to the formation of networks of bundles with inherent bundle to bundle variability. The result discussed here helps rationalize the effect of such variability on the network behavior.

6.1.1.4.9 *Effect of the Crosslink Compliance*

In Section 6.1.1.3.10 it is shown that the crosslink compliance may have a significant effect on the small strain stiffness of the network. This effect is evaluated in terms of a non-dimensional parameter which compares the stiffness of the crosslinks, k_X, with the bending stiffness of fibers: $k^* = k_X l_c^3 / E_f I_f$. For values of k^* larger than $\sim 10^3$, the network behaves as if the crosslinks are infinitely rigid. As k^* decreases, E_0 decreases.

The crosslink compliance has a significant effect on the nonlinear network behavior as well. Figure 6.40 shows the tangent stiffness versus stress for several fibrous networks of similar architecture, with different values of $k_X l_c^3 / E_f I_f$. The values of k^* are selected in the transition region of the curve in Figure 6.18, from the rigid crosslinks limit to the soft crosslinks regime (Zagar et al., 2015). The vertical axis is normalized by E_0 and hence all curves overlap in regime I. It is observed that regime II is gradually eliminated as k_X decreases. When $k_X l_c^3 / E_f I_f$ enters the descending branch of the curve in Figure 6.18, regime II is entirely replaced by regime III characterized by $E_t \sim S^{1/2}$. A similar result is presented in Kasza et al. (2009).

The interpretation provided in Zagar et al. (2015) is that, as the crosslinks become sufficiently compliant, the fibers cease to deform and undergo rigid body displacements. It can be envisioned that, in this limit, the fibers become rigid connectors of the flexible crosslinks and the network becomes a network of deformable spring-like crosslinks connected by rigid fibers. The effective connectivity of this network (i.e., of the deforming component of the network, which are the soft crosslinks) is equal to the number of crosslinks per fiber and is much larger than 4. Hence, the network deforms affinely due to the large effective $\langle z \rangle$. The scaling $E_t \sim S^{1/2}$ observed in the limit of compliant crosslinks is a manifestation of this affine deformation mode (similar to regime III).

6.1.1.4.10 Onset of Nonlinear Deformation

The relevance of the linear elastic regime I depends on the magnitude of the stretch λ_{c1} marking the transition from regime I to regime II. Whether measured in uniaxial tension or in shear, λ_{c1} is generally small, below 1.05 (5% engineering strain). However, in many experiments λ_{c1} is too small to be measured, as the network stiffens once the stress becomes larger than the resolution of the testing instrument.

It is observed that in athermal networks that stiffen exponentially ($E_t \sim S$), the critical stretch is independent of parameter w, that is, of the network density and fiber properties, as long as the deformation is bending dominated. This was also observed in collagen (Vader et al., 2009; Piechocka et al., 2011; Motte and Kaufman, 2012; Licup et al., 2015) and fibrin gels (Jawerth, 2013). It is further reported that the threshold strain decreases as $\langle z \rangle$ increases (Zagar et al., 2011; Licup et al., 2015) and may depend on the network architecture.

In networks of thermal semiflexible filaments, it is observed that the threshold strain decreases as the mean segment length decreases or, equivalently, as the cross-link density increases (Tharmann et al., 2007; Broedersz et al., 2014). The relation $\gamma_{c1} \sim \rho_b^{-0.4}$ is proposed in Tharmann et al. (2007) for networks of semiflexible filaments tested in shear.[11]

An affine model for the deformation of networks of worm-like chains (MacKintosh et al., 1995) leads to the scaling $\gamma_{c1} \sim l_c/L_p$, where L_p is the filament persistence length. In this model, the filaments are assumed to move affinely and hence only the axial deformation mode is engaged. In this case, the axial constitutive behavior of filaments determines the network response. Since filaments are subjected to thermal fluctuations, their length in the unloaded state is smaller than the contour length, with the mean slack of the unloaded filament segment of length l_c being $l_c^2/6L_p$ (Section 2.4.4). The filament-scale axial strain required to eliminate the slack is $l_c/6L_p$. This theory assumes that regime II begins when the thermal undulations of filaments are pulled out and, hence, one expects $\gamma_{c1} \sim l_c/L_p$ since, according to the affine assumption, the local, fiber scale strain is identical to the global strain. It is implied that the nonlinear regime II is associated with fiber rotations, as captured by the affine model. Considering the relation between density and l_c ($l_c \sim \rho^{-0.5}$) and Eq. (4.21), one may rearrange the expression for the critical strain as $\gamma_{c1} \sim l_c/L_p \sim \rho_b^{-1/3}$, which is reasonably close to the result of Tharmann et al. (2007) discussed in the preceding paragraph. Clearly, this line of reasoning applies provided the multiple assumptions made about the deformation mechanisms are valid.

An alternative perspective on the onset of nonlinearity is based on the view that nonlinearity is primarily geometric at relatively small strains. Hence, it manifests itself

---

[11] The shear strain in a simple shear experiment is related to the principal stretches (eigenvalues of $\mathbf{U}$, with $\mathbf{C} = \mathbf{U}^2$) as $\lambda_{1,2}^2 = (1/2)\left(2 + \gamma^2 \pm \gamma\sqrt{4 + \gamma^2}\right)$. This expression relates γ_{c1} to the largest eigenvalue, λ_{c1}, for this type of loading.

once fibers are free to rotate in order to accommodate the imposed large deformations. For this to happen, the initial structure of the network must become unstable at a sufficient number of sites, such as to allow the rotation of a subset of fibers. As tensile stress is applied, fiber orientation in the stretch direction is resisted by compressive forces developing in fibers oriented more or less orthogonal to the stretch direction (Section 6.1.1.4.12). As long as these fibers undergo only small deformations, the overall deformation remains linear elastic. Softening of the transverse compressed fibers happens once the transverse fibers undergo large bending deformations. This takes place when the net force loading the transverse fibers reaches a threshold proportional to $E_f I_f / l_c^2$ (see Figure 2.8 and the discussion of buckling in Section 2.3.3). This is equivalent to a self-equilibrated stress in the transverse direction proportional to $E_f I_f / l_c^4$. Since the total, average stress in the transverse direction must be zero, as required by the boundary conditions of the uniaxial test, the transverse internal loads are necessarily self-equilibrated – a situation similar to that in tensegrity structures. This stress scales in proportion with the axial, applied stress which, at the onset of instability, should be $E_0 \varepsilon_{c1} = E_0 (\lambda_{c1} - 1)$. Hence, the strain at the transition scales as $\varepsilon_{c1} \sim E_f I_f / E_0 l_c^4$. For a bending-dominated network of athermal fibers, $E_0 \sim \rho^x E_f I_f$ (Section 6.1.1.3.1), and $\varepsilon_{c1} \sim l_c^{2x-4}$. For networks with $x = 2$ (e.g., cellular), ε_{c1} and λ_{c1} are independent of network density and fiber properties, as observed in experiments. If $x > 2$, ε_{c1} is independent of fiber properties, but is predicted to be a function of density, specifically $\varepsilon_{c1} \sim \rho^{2-x}$. Comparing networks of the same density, the transition strain is smaller if x is larger. These considerations do not apply to affinely deforming networks since in the affine model nonlinearity is entirely defined by the affine rotation of fibers (Section 5.3.5).

6.1.1.4.11 Size Effects

Size effects introduce errors in the evaluation of material parameters. Highly heterogeneous materials present strong size effects and, as the degree of heterogeneity increases, large samples must be used in order to predict sample size-independent properties. This issue is discussed in Section 6.1.1.3.11 in relation to the small strain stiffness, E_0. The size effect on the nonlinear response is briefly discussed here.

E_0 is affected by size effects to an extent which depends on the degree of nonaffinity and on the dimensionality of the embedding space (2D versus 3D networks). Size effects are stronger in 3D and increase in magnitude as the network becomes more nonaffine. It is shown in Merson and Picu (2020) that the size effect of the nonlinear response is correlated to the size effect of E_0. Stress–stretch curves corresponding to networks of different sizes and the same nominal parameters are very different. However, when plotted as tangent stiffness versus stress, while normalizing the stiffness by E_0, all curves overlap. This indicates that the functional form of strain stiffening is independent of the size effect. Hence, once samples large enough to eliminate the size effect of E_0 are used (see Section 6.1.1.3.11), the observed nonlinear response becomes size effect free. This conclusion reduces the discussion of size effects in the context of networks without damage to the analysis of the size effect of

E_0. As discussed in Section 8.1.8, stronger size effects are observed if damage accumulates and during fracture.

6.1.1.4.12 Mechanism of Strain Stiffening: The Emergence of Eigenstress

The central characteristic of type C network behavior is the pronounced strain stiffening which takes place in regimes II and III. The functional form of stiffening is defined by the slope of the tangent stiffness–stress curve, $E_t(S)$ (e.g., Figure 6.22(b)). In regime II, the slope is close to 1 for most network materials (Figures 6.28 and 6.30). A slope of 1 is equivalent to exponential stiffening, while a smaller slope corresponds to power law stiffening.

Affinely deforming networks of linear elastic athermal fibers stiffen sub-exponentially; if volume conservation is imposed, $E_t(S)$ may be approximated with a power function of exponent 1/2. In these networks, stiffening is due exclusively to the deformation-driven orientation of fibers in the principal tensile stretch direction. If the constitutive behavior of fibers is nonlinear, the network-scale response becomes a combination of the effect of alignment and fiber behavior. An example of this type is provided by networks of Langevin molecular strands. $E_t(S)$ of affinely deforming networks of semiflexible filaments is a power function of exponent 3/2 (Figure 6.33), and it is generally assumed that the exponent is a direct manifestation of the stiffening of axially loaded individual filaments, whose behavior is described by the worm-like chain model (Section 2.4.4).

The origin of the prevalent exponent of 1 of $E_t(S)$ in many network materials in regime II (virtually all collagen-based biological networks stiffen this way) was not fully explained to date. Several additional observations may assist in the development of an understanding of stiffening. The incremental Poisson ratio increases fast in regime II, and this implies rapid lateral contraction (Figures 6.25 and 6.29). The contraction is significantly larger than that predicted by the affine model in which volume conservation is imposed. The stiffening rate predicted by the affine model in nonisochoric conditions is larger than that predicted under the isochoric assumption but remains sub-exponential. As discussed in the preceding sections in this chapter (Figure 6.24), most of the strain energy is stored in the bending deformation mode of fibers throughout regime II, that is, during the entire regime of exponential strain stiffening, while in the affine model the energy is stored exclusively in the axial mode. In addition, the preferential alignment in regime II is not sufficiently different from that in regime III (Figure 6.27) to justify the observed change of stiffening behavior at the transition between the two regimes.

An important observation made by several authors (e.g., Onck et al., 2005) is the development of stress paths. These are chain-like structures of axially loaded fibers which transmit load across the network. Figure 6.41(a) shows a network stretched uniaxially to 20% strain, at the transition between regimes II and III, with the fibers which are most loaded in the axial mode (top 5% of the distribution of axial energies density) shown in black. These fibers form a stochastic sub-network that carries most of the applied load. The stress paths are less visible in regime II and stiffening is associated with their gradual organization throughout this regime. The structure of the

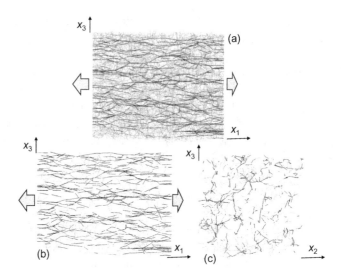

Figure 6.41 (a) Stress paths in a network at the transition between regimes II and III. Fibers are ranked in terms of their axial strain energy density, and the top 5% are shown in black. The other fibers are shown in gray. The same set of axially loaded fibers is reproduced (in black) in (b), while the top 5% of fibers with largest bending strain energy density are shown in gray. The other fibers of the network are not shown in (b). The axial projection of the figure in (b) is shown in (c).

stress paths does not evolve substantially during regime III. It is important to note that the overall degree of fiber orientation described by P_2 may remain modest even in the presence of stress paths, which implies that stress path organization does not require large global fiber alignment. The emergence of stress paths is also important since it leads to an increase of the spatial correlation of fields during network deformation; specifically, the correlation length asymptotes to the system size. As discussed in Chapter 7, this causes difficulties when attempting to define representative volume elements of the network for the purpose of performing numerical homogenization.

It is useful to inquire why the stress path structure does not collapse in directions x_2 and x_3. If this would happen, the load would be carried exclusively by a bundle of parallel fibers and the stiffness would be defined exclusively by the axial behavior of fibers. The answer to this question is provided graphically in Figure 6.41(b) and (c), where the stress paths in Figure 6.41(a) are shown along with the fibers which are most loaded in the bending mode (top 5% of the distribution of bending strain energy density). The fibers subjected to bending provide resistance to the collapse of the stress paths. This is visible in Figure 6.41(c) where the same network is viewed along the stretch direction, that is, along the stress paths. Therefore, a self-equilibrated stress field (eigenstress) develops, which is marginally engaged in load transmission, but controls lateral contraction and, therefore, strain stiffening. This is similar to the self-equilibrated stress field which provides stiffness to tensegrity structures. As the stress path structure develops during regime II, the eigenstress increases in magnitude. In

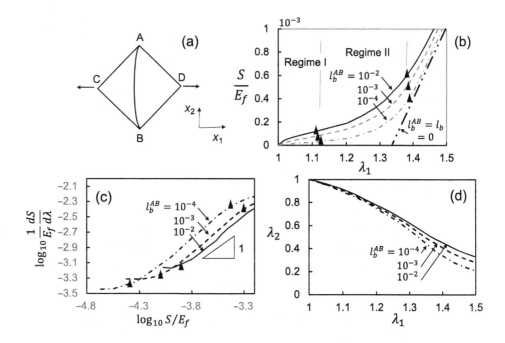

Figure 6.42 Mechanical behavior of a periodic 2D network of a repeat unit cell shown in (a), subjected to uniaxial extension in the x_1 direction. Three networks are considered, with all fibers of the same $l_b = 10^{-2}$, and with l_b^{AB} of segment AB reduced to 10^{-3} and 10^{-4}, respectively. (b–d) The nominal stress–stretch, tangent stiffness–stress, and transverse to axial stretch plots, respectively. Transitions between regimes I, II, and III are indicated by triangles in (b) and (c), and (b) includes the expected stress–stretch curve corresponding to a truss network with no reinforcement AB (labeled $l_b^{AB} = l_b = 0$).

regime III, the eigenstress remains approximately constant, which can be understood based on the nonlinear behavior of beams subjected to large deformations shown in Figures 2.8 and 2.9.

Figure 6.42 shows the behavior of the simplest structure which embodies these concepts. It is a periodic, two-dimensional network of beams with square architecture, as shown in Figure 6.42(a) (only the repeat unit is shown). Beam AB performs the function of the bending-dominated fibers in Figure 6.41(b) and prevents the rapid collapse of frame ABCD as the applied force increases. Figure 6.42(b) shows the nominal stress–stretch curve, Figure 6.42(c) shows the corresponding tangent stiffness–stress plot, and Figure 6.42(d) shows the transverse stretch versus the applied stretch. Three curves are shown in each plot, corresponding to structures with decreasing bending stiffness of fiber AB. In the reference case, all fibers have the same bending stiffness, with $l_b = 10^{-2}$. In the other two cases, l_b^{AB} of fiber AB is reduced to 10^{-3} and 10^{-4}, respectively.

This structure exhibits many of the features observed in athermal dry networks, including a short linear elastic regime I, followed by an exponential stiffening regime II. The energy is stored primarily in the bending mode in regimes I and II, and becomes axially dominated in regime III (see also Figure 6.24). The stiffening rate is controlled by the nonlinear behavior of AB and is independent of the effective bending compliance of this fiber (l_b^{AB}). The degree of lateral contraction is also independent of l_b^{AB}. The eigenstress is associated in this schematic with the force in the x_2 direction produced by AB. This force is balanced by the x_2 components of forces in the frame, such as to fulfill the zero traction boundary condition in the x_2 direction. This illustrates in a conceptual way the mechanism controlling the lateral contraction and stiffening rate of complex networks of the type shown in Figure 6.41.

6.1.2   Compression

6.1.2.1   Generic Mechanical Behavior in Compression

Similar to most engineering materials, the behavior of network materials in compression is different from that in tension or shear. This mandates a separate discussion. The focus of Section 6.1.2 is on crosslinked networks, while the mechanical behavior of non-crosslinked networks is presented in Section 6.2.

Uniaxial compression reduces the volume of the material. Network materials with large free volume, such as most nonwovens, collagen constructs, cellular materials (open cell foams), aerogels, etc., present little resistance to volumetric changes. Materials in which the network is embedded in a fluid, such as gels, swelled rubber, and biological connective tissue, may change volume only if drainage occurs. This imposes an energy penalty which, in turn, increases the effective bulk modulus. In other network materials in which filaments are densely packed, such as dry rubber and dense nonwovens, volume variation is restricted by nonbonded interactions (excluded volume) and such materials are, in general, weakly compressible or incompressible. To separate the effect of the embedding medium from the contribution of the network, we focus here on the response of networks without a matrix. This view, which may appear initially limited, represents the physical reality in a number of applications which are presented as examples in this section.

The generic response of a crosslinked network to compressive loads is shown in Figure 6.43 (Islam and Picu, 2018). These curves are obtained with models of athermal fibrous 3D networks similar to those used to investigate the behavior in tension discussed in Section 6.1.1.3.1. The crosslinks are considered rigid and transmit both forces and moments between fibers.

Figure 6.43(a) shows nominal stress–stretch curves for networks of three different densities. Three deformation regimes can be identified. Regime I is linear elastic and is characterized by the effective small strain stiffness, E_0. This regime ends at the threshold stretch and stress, λ_{c1} and S_{c1}, beyond which softening is observed. Regime II corresponds to a segment of the stress–stretch curve of approximately constant slope denoted by $E_{II} = dS/d\lambda$. This regime is bounded by the transition stretches λ_{c1} and λ_{c2} at the left and right ends, respectively. Regime III occurs for $\lambda > \lambda_{c2}$ and corresponds to rapid stiffening. The transition stretch between regimes I and II, λ_{c1}, is independent of the network density. λ_{c2} decreases as density increases, as shown schematically in Figure 6.43(a) by the curved dashed line.

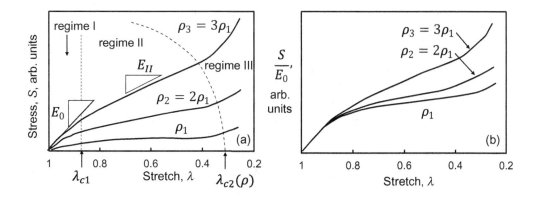

Figure 6.43 (a) Typical nominal stress–stretch curves for crosslinked networks loaded in compression. Curves corresponding to three densities are shown and the boundaries of the three regimes discussed in the text are shown by dashed lines. (b) Data in (a) replotted after normalizing the vertical axis by the small strain modulus.

While the response in regime I is elastic, regime II may be either elastic or inelastic. The characteristic feature typically observed in regime II is strain localization. At the transition stress S_{c1}, diffuse localization bands form. Continued loading in regime II leads to the formation of new bands and the slight growth of the previously nucleated bands. Strain localization implies partial or full compaction of the network in the respective region.

This process is well characterized in cellular materials with a periodic microstructure (Gibson and Ashby, 1988). If all cells are identical, localization corresponds to the collapse of a row of cells orthogonal to the applied loads. As loading continues, the material deforms at essentially constant stress, $S = S_{c1}$ and $E_{II} \approx 0$, by the propagation of a dominant localization band across the sample gauge. At the end of this process, the cells in the entire sample have collapsed. Cells being identical, localization takes place at a well-defined stress level, which depends on the geometry of the cell and the elastic properties of the struts and cell walls. If cells are irregular, the slope of the stress–stretch curve in regime II becomes non-zero.

The structure of random networks has no periodicity or regularity, and localization bands that span the entire sample do not form. Instead, localization is confined to softer regions and, once formed, the bands remain of limited spatial extent. It is energetically favorable to initiate a localization event at some other site in the sample, rather than to propagate an existing band whose growth is trapped by structural heterogeneity. Due to this reason, $E_{II} > 0$ in stochastic network materials.

Regime III corresponds to the densification of the material and the formation of contacts between fibers at sites other than the crosslinks, which entails rapid stiffening. Contacts form in regime II as well, but their rate of formation increases rapidly in regime III. The phenomenology described in Section 6.2.1 for non-crosslinked networks subjected to compression applies to regime III.

Figure 6.43(b) shows the curves in Figure 6.43(a) upon the normalization of the vertical axis with the small strain modulus. Regimes II and III do not collapse upon this normalization, which indicates that the density dependence of the nonlinear behavior is different from that of the linear behavior prevalent in regime I.

In order to determine whether the deformation is elastic, unloading has to be performed. It is conceivable that if fibers remain elastic during localization, the structure may fully rebound upon unloading, such that the overall response is elastic irrespective of the stretch at which unloading is performed. Such a situation is theoretically conceivable but is rarely encountered in practice. The common situation is the occurrence of residual strains upon unloading, even when unloading is performed at relatively small strains, for example, at the beginning of regime II. Inelastic behavior is caused by several mechanisms, including the plastic deformation of fibers, inter-fiber cohesion, and friction.

The three regimes are clearly defined in the tangent stiffness–stress representation. Figure 6.44 shows E_t versus S for network models representing reconstituted collagen of various concentrations (Picu et al., 2018). The effective normalized density (proportional to the fiber volume fraction) for each curve is indicated in Figure 6.44. Regime I appears as a plateau at $E_t = E_0$. The transient softening between regimes I and II is well-defined in this representation, while the broad regime II segment of constant E_{II} appears as a short horizontal plateau. E_{II} increases as the density of the network increases. The densification regime III is clearly defined at stresses larger than S_{c2}.

The deformation is entirely controlled by the bending deformation mode of fibers. Figure 6.45 shows the fractions of the total strain energy stored in the axial, bending, torsion, and shear modes for one of the curves in Figure 6.44. The bending energy dominates at all strains, while the fraction of the axial energy stored is very small. The torsion mode stores about 10% of the total energy. This situation is very different from

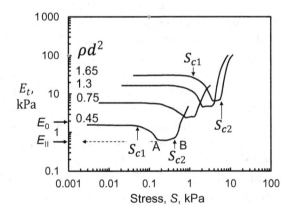

Figure 6.44 Tangent stiffness–stress representation of the behavior in compression of fibrous networks of different volume fractions without matrix. Segment AB corresponds to regime II. Both E_0 and E_{II} (indicated for the lowest curve with $\rho d^2 = 0.45$) increase with increasing density, but not in proportion to each other.

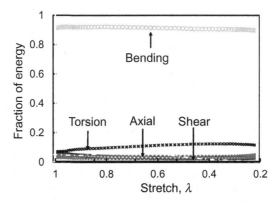

Figure 6.45 Partition of energy between the deformation modes of fibers during compression of crosslinked networks without matrix.

that in tension (Figure 6.24), where a transition from bending to axial mode dominance is observed. The dominance of the bending mode in compression is intuitive since local instabilities and densification must entail large bending deformations. The strain energy of the highly bent fibers is significantly larger than that of the other, moderately deformed fibers. Furthermore, this observation indicates that the affine models which are based on the assumption that fibers deform axially cannot be expected to provide accurate predictions in compression.

This analysis does not take into account the strain energy stored locally, at contacts between fibers. This becomes important in regime III if fibers have low contact stiffness, as for example in the case of some natural fibers (cellulose, cotton, mycelium) which are hollow and the fiber wall deforms easily under the action of contact forces. Another situation in which the contact energy becomes important is that of quasi-planar mats of fibers compressed in the direction perpendicular to the plane of the mat. In this case, the density of contacts is large and deformation at contacts becomes the softest deformation mode available (Hossain et al., 2019).

Figure 6.46(a) shows the incremental Poisson ratio (Eq. (6.11)) for the three networks of different density shown in Figure 6.43. At small strains, the Poisson ratio is approximately 0.3. A weak increase of the incremental Poisson ratio is observed during regime II, with the maximum remaining well below 1. This weak lateral expansion during compression is also observed in Figure 6.46(b) where images of a network compressed to different degrees are shown. This is in stark contrast with the behavior in tension (Figure 6.25), where the incremental Poisson ratio is seen to increase rapidly in regime II to very large values. The limited Poisson effect in uniaxial compression is a feature observed in many networks which are not embedded in matrix.

Compression leads to preferential fiber alignment in the plane perpendicular to the loading direction (Figure 6.47). The orientation parameter P_2 (Eq. (4.3)) is computed based on the fiber end-to-end vector and relative to the direction of compression. The

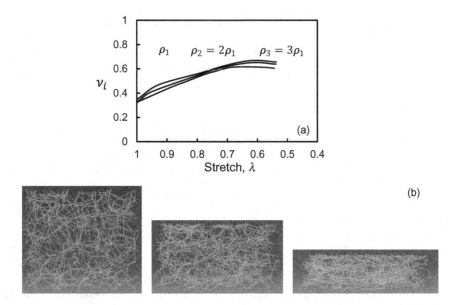

Figure 6.46 (a) Variation of the incremental Poisson ratio during compression for the three networks of different densities whose stress–stretch curves are shown in Figure 6.43. (b) Images of a network at different stages of compression demonstrating weak lateral expansion during deformation.

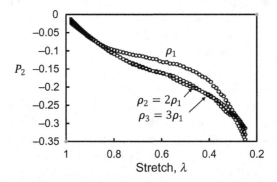

Figure 6.47 Orientation parameter P_2 computed based on the fiber end-to-end vector versus stretch, indicating alignment in the plane perpendicular to the direction of compression. The three curves correspond to cases shown in Figure 6.43.

density has little effect on the degree of alignment, which is also the case in tension. However, the alignment is less pronounced in compression and is generally compatible with the expectation based on the affine model. The weaker alignment in compression correlates with the weaker Poisson effect (Figure 6.46). Overall, fiber kinematics during compression is more constrained than in tension and the central structural evolution phenomenon is not alignment and structural reorganization as in

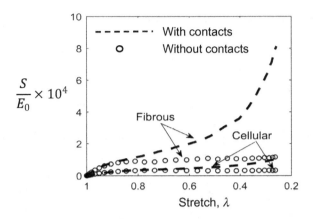

Figure 6.48 Stress–stretch curves for fibrous and cellular networks obtained with models in which contacts between fibers are (dashed lines) and are not (open circles) taken into account. Republished with permission of ASME from Islam and Picu (2018), permission conveyed through Copyright Clearance Center Inc.

tension, but rather the collapse and densification of network sub-domains observed during regime II.

Inter-fiber contacts are established during compression, which leads to continuous stiffening in regimes II and III. To demonstrate the role of contacts, Figure 6.48 shows stress–stretch curves for a fibrous and a cellular network of the same density and fiber properties. Two curves are shown for each case: one is evaluated by taking into account all contacts that form dynamically during compression – shown by dashed lines – and the other corresponds to cases in which contacts are ignored (phantom network) – shown by open circles (Islam and Picu, 2018). The curves corresponding to the actual and phantom networks overlap in regime I and at the beginning of regime II. In this range of strains, the mechanical response is controlled by the crosslinks. The curves depart at larger degrees of compression. The phantom networks develop a plateau with no stiffening, while the networks with contacts exhibit the regime II stiffening seen in Figure 6.43. This indicates the essential role played by contacts in the mechanics of networks subjected to compression.

It is of interest to enquire to what extent the concepts developed for non-crosslinked networks subjected to compression – a subject discussed in Section 6.2 – apply to the compression of crosslinked networks. The answer is somewhat obvious from Figure 6.48. For compressions up to $\lambda \approx 0.8$, the network behavior is controlled by the crosslinks, and the density of newly formed contacts is small. At large deformations, the number density of contacts becomes larger than that of crosslinks. In this regime the crosslinked network response is similar to that of a non-crosslinked ensemble of fibers of the same density. In the case of sparsely crosslinked networks of long tortuous fibers, in which the contact density is large even in the unloaded state, the crosslinks contribute little to the response in compression, which hence becomes similar to that of packings of non-crosslinked fibers.

6.1.2.2   Relation to Network Parameters

Three parameters describe the response of networks subjected to uniaxial compression: the small strain stiffness, E_0, the stress/stretch at the transition to regime II, S_{c1}, λ_{c1}, and the slope of the curve in regime II, E_{II}. It is desirable to establish the relation between these parameters and the network structural parameters (nominal network density in the unloaded, reference state, ρ, and fiber properties).

The stress–stretch curve has no discontinuity at the transition between tension and compression, that is, in the vicinity of the state corresponding to the unloaded network. Therefore, the tangent is unique and E_0 is identical in tension and compression. This allows us to apply all the results presented in Section 6.1.1.3 in relation to the behavior in tension, to regime I of the compression branch of the characteristic curve.

As discussed in Section 6.1.1.3.1, $E_0 \sim \rho^x$ with $x \geq 3$ for fibrous networks in 3D, while $E_0 \sim \rho^2$ for cellular networks. A large number of data support the scaling $E_0 \sim \rho^2$ for cellular materials, as described, for example, in Gibson and Ashby (1988), Andrews et al. (1999), and Roberts and Garboczi (2002). Furthermore, since the deformation is bending-dominated, in both cellular and fibrous cases $E_0 \sim E_f I_f$. The results related to the effect of the mean connectivity $\langle z \rangle$, fiber tortuosity, stochasticity of fiber properties, and fiber alignment discussed in Section 6.1.1.3 apply to the compression regime I.

Reports in the literature indicate that the regime I stiffness measured in compression is smaller than that measured in tension. For the most part, these works refer to soft materials. Examples include Harley et al. (2007), who work with reconstituted collagen networks embedding proteoglycans and indicate that the stiffness in compression is one order of magnitude smaller than that in tension, Puleo et al. (2013), who test acrylamide hydrogels and report that E_0 in compression is half of that in tension, and multiple studies on cartilage (hydrated collagen network) which also report significantly lower E_0 in compression relative to the corresponding tension value. These differences may be due to various reasons. The primary argument is related to the general shape of the tension and compression curves: continuous stiffening is observed in tension, while in compression the initial short linear regime I is followed by softening. Therefore, if one measures the secant stiffness, as opposed to the tangent stiffness at zero strain, the conclusion would unavoidably be that the material is softer in compression. In experiments it is hard to evaluate correctly the tangent stiffness at the origin, primarily considering that regimes I in tension and compression are limited to small strains. This requires consistently defining the reference unloaded state and using the same reference for the tension and compression tests. It also requires high resolution data acquisition in the small strain regime and the ability to measure small stresses, which is not always possible when working with soft materials.

An additional argument is related to drainage in networks with liquid embedding. The Poisson effect and the volume variation of a dry network in tension are much larger than those in compression (Figures 6.25 and 6.46). Therefore, in the presence of a liquid embedding, the driving force for liquid expulsion is larger in tension than in compression. Preventing drainage promotes stiffening in tension. Hence, the fluid–network interaction may introduce tension–compression asymmetry.

Occasionally, it has been suggested that the tension–compression asymmetry may be due to the fact that fibers are able to carry tensile loads, but bend under compression and carry only small compressive loads. This argument is rooted in the thought that fibers deform axially, as suggested by the affine model. However, deformation is bending dominated in most 3D networks. In the limit of small global and local deformations, the response in bending is symmetric (relative to the reversal of the loading direction). Since structural evolution in regime I is negligible, geometric nonlinearity is absent in the small strain limit and the response must be symmetric.

The transition between regimes I and II is defined by S_{c1} and λ_{c1}. For cellular networks it can be shown that $S_{1c}/E_f \sim \rho^2$ (Gibson and Ashby, 1988). Since $E_0/E_f \sim \rho^2$ in these structures, it results that the strain at the transition is independent of density. This is shown schematically in Figure 6.43(a), where the dashed line corresponding to λ_{c1}, and marking the transition between regimes I and II, is vertical. The value of λ_{c1} may be inferred from the constants of proportionality of the stress–density and stiffness–density relations and results in approximately $\lambda_{c1} = 0.95$ (Gibson and Ashby, 1988). In fibrous materials, $0.95 < \lambda_{c1} \leq 1$. In some situations (see the examples presented in Section 6.1.2.3) regime I is hardly distinguishable, and $\lambda_{c1} \approx 1$.

The slope of the characteristic curve in regime II, E_{II}, did not receive much attention in the literature. This is primarily due to the fact that in periodic cellular materials $E_{\text{II}} \approx 0$, which does not require additional discussion. In random cellular networks (open cell foams) and in fibrous networks, E_{II} is larger than zero due to the stochasticity of the microstructure.

Figure 6.49 shows results from models of fibrous networks of various densities and fiber diameters (Islam and Picu, 2018). The data is plotted in a manner similar to the

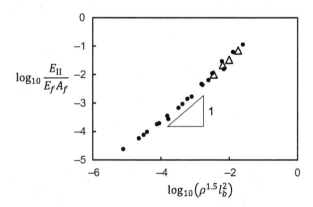

Figure 6.49 The variation of the slope of the stress–stretch curve in regime II, E_{II}, with the two network parameters, ρ and l_b (data from Islam and Picu, 2018). The exponent x_{II} leading to data collapse in this case is 1.5. Triangles represent experimental data from Harley et al. (2007). ρ represents the density in the unloaded state.

master plot of Figure 6.6. The vertical axis is normalized by $E_f A_f$ and the variable on the horizontal axis is $w_{\mathrm{II}} = \log_{10}(\rho^{x_{\mathrm{II}}} l_b^2)$, which is equivalent to the variable w on the horizontal axis of Figure 6.6. ρ represents the nominal density in the initial, unloaded state. The exponent x_{II} is adjusted to obtain the collapse of the data on a line of slope 1 and a value of $x_{\mathrm{II}} = 1.5$ results, that is, $E_{\mathrm{II}} \sim E_f I_f \rho^{1.5}$. The fact that $E_{\mathrm{II}} \sim E_f I_f$ is expected, since the deformation is bending dominated throughout regime II and into regime III. The value of the exponent x_{II} is expected to depend on the network architecture, but no systematic investigation of this issue is reported in the literature at this time. Results for several fibrous networks are presented in Section 6.1.2.3.

Figure 6.49 includes experimental data from Harley et al. (2007) pertaining to reconstituted collagen gels of different densities. To make them comparable with the numerical results, the values are normalized to align the lowest density data point to the collapsed data set. This normalization does not affect the scaling of E_{II} with ρ, which results very close to that obtained numerically.

6.1.2.3 Nonlinear Behavior of Several Network Materials in Compression

6.1.2.3.1 Crosslinked Felts of Glass, Carbon, and Steel Fibers

Fibrous networks of glass fibers are typically used for sound and thermal insulation. These are felts of non-crosslinked fibers clumped and partially entangled. Interest exists in developing crosslinked versions of these felts which have higher stiffness but preserve the energy dissipation capability. An experimental data set of this type is presented in Mezeix et al. (2009). These authors work with glass, stainless steel, and carbon fibers of approximately 10 μm diameter, which are tested in the original non-crosslinked state and upon crosslinking. Figure 6.50(a) compares the stress–stretch curves of non-crosslinked and crosslinked glass fiber felts. The crosslinked networks exhibit the three regimes described in Section 6.1.2.1, while the non-crosslinked felt has a characteristic curve that stiffens continuously during compression. The curves are approximately parallel beyond the initial part of regime II of the crosslinked case,

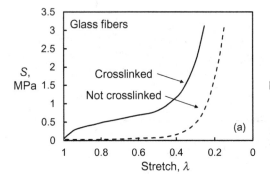
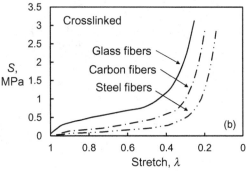

Figure 6.50 Nominal stress–stretch curves for (a) crosslinked and un-crosslinked felts of glass fibers, and (b) crosslinked felts of glass, carbon, and stainless steel fibers. Data from Mezeix et al. (2009)

which supports the discussion of Section 6.1.2.1 indicating that the effect of the crosslinks becomes less important as the density of contacts between fibers increases during compression.

Figure 6.50(b) shows stress–stretch curves for crosslinked felts of stainless steel, carbon, and glass fibers. All these materials have the same initial mass density of 150 Kg/m^3, but different ρ and different fiber properties. The three networks exhibit different E_0 and E_{II}. When plotting these three data points against $w_{\mathrm{II}} = \log_{10}\left(\rho^{x_{\mathrm{II}}} l_b^2\right)$ in a manner similar to Figure 6.49, a scaling $E_{\mathrm{II}} \sim E_f I_f \rho^{0.5}$ results. The dependence $E_{\mathrm{II}} \sim E_f I_f$ is expected since the dominant fiber deformation mode is bending. The exponent x_{II} is smaller than that emerging from the data in Figure 6.49, but it is also less accurate due to the small number of data points used for its evaluation.

6.1.2.3.2 *Mycelium*

Mycelium is the vegetative part of a fungus. Mycelium grows by the apical tip extension of mycelium filaments, or hyphae, which originate from spores. After an initial growth phase, hypha initiate random branching, forming tree-like colonies. Colonies interconnect randomly through hyphal fusion (anastomosis) to form a random network structure. Figure 6.51 shows an example of such a network.

Hyphae are hollow fibers whose role is to transport enzymes from the organism to the source of food and to carry nutrients in the reverse direction. Their wall has a complex layered structure, with the primary component being chitin nanofibers.

Mycelium growth is driven by the search for nutrients and oxygen and the branching density and network topology are largely controlled by the environmental conditions. Starting from this observation, interest has developed over the last decade to use mycelium as a biodegradable fibrous material, for applications such as packaging, artificial leather, and food products. From the point of view of the present discussion, mycelium is an interesting network material that may exist in the dry form, or as a network embedded in a matrix.

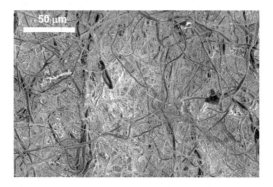

Figure 6.51 Scanning electron micrograph of the mycelium network.

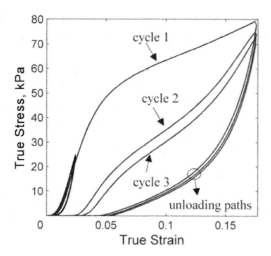

Figure 6.52 Experimental behavior of dry mycelium subjected to uniaxial compression. Reprinted with permission from Islam et al. (2017)

Mycelium networks in the dry, not-embedded state are characterized by standard mechanical testing in Islam et al. (2017). The material behaves in compression like a cellular network, exhibiting stress–stretch curves similar to those in Figure 6.43. It shows a well-defined regime I, with E_0 being proportional to ρ^2, followed by softening, and regime II, characterized by $E_{\mathrm{II}} > 0$. Digital image correlation observations of the deformation field demonstrate a phenomenology similar to that described in Section 6.1.2.1: densified regions appear stochastically at the onset of regime II and remain localized; continued loading does not necessarily lead to the extension of the localization bands, but rather to the increase of their number. The sample undergoes densification in a more or less homogeneous fashion during regime III.

Figure 6.52 demonstrates another feature of network deformation which is encountered in many network materials: regime II is usually associated with plasticity and internal dissipation. In the case of dry mycelium, plasticity is due to the rearrangement of fibers and the occasional rupture of hyphae, while dissipation happens primarily due to inter-fiber friction. The figure shows regime I and the early part of regime II, along with several unloading–reloading cycles. Unloading performed in regime I leads to no hysteresis, while large hysteresis loops and residual strain are observed when unloading from regime II.

6.1.3 Multiaxial

While uniaxial and shear loading are convenient for characterization purposes, network materials are subjected to multiaxial loading when in service. Therefore, it is of importance to characterize their behavior under such deformations and to compare with the uniaxial tension response.

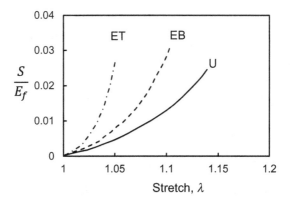

Figure 6.53 Nominal stress–stretch curves for uniaxial (U), equi-biaxial (EB), and equi-triaxial (ET) tensile loadings predicted by cellular network models.

In the principal directions of the right Cauchy–Green tensor, $\mathbf{C} = \mathbf{F}^T\mathbf{F}$ (where $\mathbf{F}$ is the deformation gradient, Section 5.1), multiaxial deformation is characterized by the three principal stretches, $\lambda_i, i = 1, 2, 3$. If the network is isotropic, the principal directions of stress coincide with the principal directions of strain, and the stress tensor is also diagonal in the same (principal) coordinate system.

For the purpose of the present discussion it is sufficient to consider extreme situations, that is, the equi-biaxial (EB) and equi-triaxial (ET) cases, which are to be contrasted with the uniaxial (U) loading discussed in the preceding sections of this chapter. In the EB case, stretch is imposed in two directions, $\lambda_1 = \lambda_2 = \lambda$, while the stress in the third direction vanishes. For an isotropic material and by symmetry, this implies $S_1 = S_2 = S$ and $S_3 = 0$. In the ET case, equal stretch is imposed in all three directions $\lambda_1 = \lambda_2 = \lambda_3 = \lambda$ and a hydrostatic stress state results: $S_1 = S_2 = S_3 = S$.

As discussed in Section 6.1.1.4.4, one of the important characteristics of the uniaxial response in tension is the rapid lateral contraction during regime II. This is described by the incremental Poisson ratio which is seen in Figure 6.25 to increase rapidly from a value below 0.5 to large values. It is intuitive that in the presence of such a large Poisson effect, strong coupling should exist between the deformation applied in multiple directions, and this should lead to more rapid stiffening in multiaxial loading compared with the uniaxial case.

Figure 6.53 shows the typical response to U, EB, and ET tensile loadings of a cellular network in which damage does not occur. Stress increases more rapidly in EB and ET relative to U in all deformation regimes. This type of behavior is observed in all multiaxial tests performed with network materials; a comparison with experimental data is presented in this subsection. To understand the details, it is necessary to discuss each deformation regime separately.

Regime I is linear elastic and is characterized by the effective stiffness $E_0 = S/(\lambda - 1)$. In the uniaxial case, E_0^U is identical to E_0 discussed in Section 6.1.1.3. In the EB case, $E_0^{EB} = E_0^U/(1 - v)$, where v is the small strain value of the

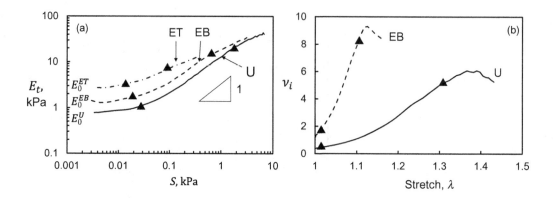

Figure 6.54 (a) Tangent stiffness versus nominal stress curves for uniaxial (U), equi-biaxial (EB), and equi-triaxial (ET), loading modes, and (b) incremental Poisson ratio versus stretch for the U and EB cases. The transition between regimes I and II and between regimes II and III are marked by triangles in both panels.

incremental Poisson ratio. The effective modulus in the ET case is $E_0^{ET} = E_0^U/(1 - 2\nu)$, as predicted by Hooke's law. Under small strains, ν of isostatic and isotropic networks has values below 0.5 and, hence, E_0^{EB} and E_0^{ET} are properly defined and are larger than E_0^U.

Regime II is better described in the tangent stiffness–stress representation. Figure 6.54(a) shows E_t versus S for the U, EB, and ET cases obtained with a network model of low w, which deforms in a nonaffine way. The three deformation regimes are visible for each of the curves. As in Figure 6.22, regime I corresponds to the horizontal plateau at small stress values. E_0^U, E_0^{EB}, and E_0^{ET} are related through the linear elasticity expressions indicated in the previous paragraph, with a Poisson ratio $\nu = 0.35$.

The strain stiffening regime II is visible in all three curves (Picu et al., 2018). This regime is bounded below by the transition stress S_{c1} (and stretch λ_{c1}) and above by the transition stress to regime III, S_{c2} (and stretch λ_{c2}). The stress range of regime II decreases as the deformation becomes more constrained, that is, going from U to ET. However, the slope of the E_t versus S curves is independent of the loading mode and equal to 1. Hence, one may write in regime II:

$$E_t = \frac{dS}{d\lambda} \sim S_0 S, \tag{6.14}$$

where S_0 depends on the loading mode. To clarify this dependence, impose the conditions that at the end of regime I the value of the stress must be S_{c1} and the tangent stiffness must be E_0 for the specific type of load considered. It results that $S_0 \sim 1/(\lambda_{c1} - 1)$ and, since λ_{c1} decreases when going from U to ET, S_0 must increase, that is, $S_0^U < S_0^{EB} < S_0^{ET}$.

This discussion indicates that the E_t versus S curves are simply shifted up in the logarithmic plot of Figure 6.54(a), which provides a unitary description of the nonlinear elastic response under multiaxial and uniaxial loadings. While this applies to the description in terms of the nominal stress, it is not expected to apply identically if the Cauchy stress is used instead. To see this difference, compare the U and ET cases: with $\lambda_2 = \lambda_3 = \lambda^*$ being the stretch in the direction transverse to the loading direction, $S^U = \lambda^{*2}\sigma^U$ and $S^{ET} = \lambda^2\sigma^{ET}$. Deriving the tangent stiffnesses in terms of S and σ, that is, E_{t-S} and $E_{t-\sigma}$, one obtains (Eq. (6.10)):

$$\frac{E_{t-S}^U}{S} = \frac{E_{t-\sigma}^U}{\sigma} + 2\frac{d\lambda^*/d\lambda}{\lambda^*}; \quad \frac{E_{t-S}^{ET}}{S} = \frac{E_{t-\sigma}^{ET}}{\sigma} + \frac{2}{\lambda}. \tag{6.15}$$

These expressions indicate that the functional form of the E_{t-S} versus S curves in the U and ET regimes II is generally expected to be different when using the nominal stress and the Cauchy stress descriptions.

Regime III is visible in Figure 6.54(a) and is characterized by the ½ slope, which is also reported in Figure 6.22(b). As in the uniaxial case, the degree of nonaffinity decreases in regime III. In fact, networks which are initially in the affine range of the master curve of Figure 6.6 and are loaded uniaxially enter directly regime III from regime I. Since the stress range of regime II decreases as the degree of triaxiality increases, regime III becomes dominant. This is important when analyzing experimental results. As shown later in this section, a slope of ½ is observed in the tangent stiffness–nominal stress representation in some experiments. To determine whether this indicates that the network is intrinsically affine (has large w) or the slope is merely a manifestation of the high degree of triaxiality, the $E_t - S$ curves obtained in the multiaxial experiment must be compared with the equivalent uniaxial curve. Figure 6.54 indicates that if regime II appears in the uniaxial curve, the slope of ½ observed in the multiaxial case is likely caused by triaxiality.

Figure 6.54(b) shows the variation of the incremental Poisson ratio during deformation for the U and EB cases (compare with Figure 6.25). The two curves show similar features. A maximum is reached after the transition from regime II to regime III. In regime III the ability of the network to reorganize to accommodate the stretch is reduced, which implies that the rate of fiber alignment and the rate of lateral contraction decrease. The transition between regimes is marked by triangles in Figure 6.54(a) and (b).

It is instructive to compare this behavior with that observed experimentally. An interesting data set is provided by Yohsuke et al. (2011), who work with soft chemically crosslinked polyacrylamide (PAAm) gels and with physically crosslinked polyvinyl alcohol (PVA) gels. The gels are allowed to swell to equilibrium in a good solvent before testing, and their mechanical behavior is probed uniaxially and equibiaxially. The results are shown in Figure 6.55 in a tangent stiffness versus nominal stress representation.

The behavior of the two gels is different. The physical gel stiffens with a well-defined regime II of slope 1 (exponential stiffening). The chemical gel exhibits a behavior similar to that of PDMS (see Figure 6.35(b)): regime I is followed by softening, which leads to stiffening at larger strains, with the slope of the large strain stiffening branch being ½. The biaxial curves run parallel to the uniaxial ones in both

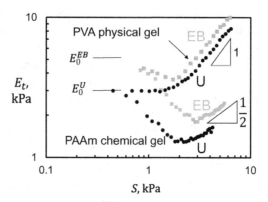

Figure 6.55 Tangent stiffness–stress curves for PAAm and PVA gels under uniaxial (U) and equi-biaxial (EB) deformations. Data from Yohsuke et al. (2011)

cases, which is in agreement with the results in Figure 6.54(a). The authors indicate that strain-induced crystallization is not a possible reason for the difference between the two gels. The small strain modulus of both gels is a power law of the polymer concentration in the equilibrium state. Power law relations of this type are common in gels (Figure 6.12). However, the exponent of the physical gel is almost twice as large as that of the chemical gel. Based on the discussion in Section 6.1.1.3, one may infer that the physical gel is more nonaffine than the chemical gel. This indicates that the different stiffening observed may be associated with the different degrees of non-affinity in the two networks. The larger nonaffinity of the physical gel allows faster reorganization during deformation which, in turn, leads to exponential stiffening. The chemical gel is an example of a more affine network which exhibits a broad regime III response.

Two data sets for PDMS rubber, including U, EB, and shear loading, are presented in Meunier et al. (2008) and Bernardi et al. (2017). Shear loading is implemented via a strip biaxial test, as often done with incompressible materials. This test consists in loading uniaxially a strip of material of large width, which is restricted from contract-ing in the lateral direction, that is, $\lambda_1 = \lambda$ and $\lambda_2 \approx 0$. The curves corresponding to the uniaxial, strip biaxial, and equi-biaxial tests are parallel, with the strip biaxial curve being located between U and EB. The curves resemble those in Figure 6.55, and have a slope of approximately 1/2 in the strain stiffening regime.

Figure 6.56 shows uniaxial and equi-biaxial results from tests performed with the Glisson capsule – a collagen-based membrane that covers the liver (Bircher et al., 2016). The membrane, which is normally attached to the liver, was separated for the purpose of these tests. Once again, the curves corresponding to U and EB are parallel, which confirms that the functional form of stiffening observed in the uniaxial test remains unchanged under multiaxial conditions. Stiffening is described by a power law in this case, and the slope in Figure 6.56 is 1/2.

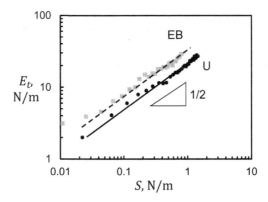

Figure 6.56 Tangent stiffness–stress curves for uniaxial and equi-biaxial tests performed with the Glisson capsule. Data from Bircher et al. (2016)

To summarize, the central features emerging from these experiments are:

(i) The effective small strain stiffness increases with increasing triaxiality, the increase being commensurate with expectations from linear elasticity, and

(ii) The multiaxial and uniaxial curves are parallel in the tangent stiffness–nominal stress representation, which indicates that the functional form of strain stiffening is identical under uniaxial and multiaxial loadings.

These observations provide valuable input for the development of constitutive descriptions valid in the nonlinear range of multiaxial loading.

6.1.4   Elastic versus Elastic–Plastic Athermal Fibers

Many athermal networks of biological and engineering importance are made from elastic–plastic fibers. Therefore, it is of interest to enquire to what extent plastic deformation of fibers may influence the global mechanical behavior of the network. The basic premises of this discussion are presented in Section 2.3.6. We aim to establish the relation between the yield stress of individual fibers, σ_y^f, and the yield stress of the network, σ_y^n. The discussion is restricted to permanently crosslinked networks without matrix and in which inter-fiber contact formation is infrequent. This restriction excludes situations in which the network-scale yield point is caused by mechanisms other than fiber plasticity.

Yielding on the network scale requires that a sufficiently large number of fibers enter the plastic regime and this set of fibers forms a percolated sub-domain. If this is not the case, the plastically-deforming fibers are fully embedded in an elastically deforming network, and the overall response remains elastic.

Understanding yielding in affinely deforming networks is straightforward. For large w values (Figure 6.6), fibers deform mainly in their axial mode. An axially deforming fiber segment of length l_c yields when $F_y^f/A_f = \sigma_y^f$, where F_y^f is the axial force at yielding. The mean force per fiber scales with the applied network-scale stress as $F^f \sim \sigma l_c^2$ and, at network yielding, $F_y^f \sim \sigma_y^n l_c^2$. This mean field argument allows computing the ratio of the network yield stress to the fiber yield stress as:

$$\frac{\sigma_y^n}{\sigma_y^f} \sim \left(\frac{d}{l_c}\right)^2. \tag{6.16}$$

Further, in the affine deformation regime, the small strain modulus is a linear function of density, $E_0 \sim \rho E_f A_f$ (Section 6.1.1.3.1), and, therefore:

$$E_0/E_f \sim \rho d^2 \sim (d/l_c)^2. \tag{6.17}$$

On the other hand, the ratio of the yield stresses can be written $\sigma_y^n/\sigma_y^f = E_0 \varepsilon_y^n/E_f \varepsilon_y^f$. Combining this equation and Eq. (6.16), and using Eq. (6.17), it results that for affinely deforming networks $\varepsilon_y^n/\varepsilon_y^f = const$. It is expected based on the affine assumption that the local strain is equal to the global, applied strain, which implies that the constant is actually equal to 1. Therefore, the network yields at a strain equal to the yield strain of individual fibers:

$$\varepsilon_y^n = \varepsilon_y^f. \tag{6.18}$$

This simple result was obtained numerically in explicit network simulations of 2D (Mikado) and planar mats (Zhang et al., 2019; Domaschke et al., 2020). For densely crosslinked paper which deforms almost affinely, Räisänen et al. (1996) conclude that the nonlinear response of individual fibers reflects identically in the nonlinear response of the network. The yield surface of networks with preferential in-plane fiber orientation and of a range of densities is evaluated in Ma et al. (2018). They observe that the in-plane yield stress scales with the modulus, which implies that ε_y^n is independent of ρ, which further suggests that Eq. (6.18) applies in the respective case.

The nonaffine network case can be understood based on a similar argument. Nonaffine networks have low w values (Figure 6.6), corresponding to low densities, ρ, and/or small fiber diameters, d. In this case fibers deform mainly in their bending mode. Cylindrical fibers subjected to bending begin deforming plastically when $M_y^f d/2I_f = \sigma_y^f$, where M_y^f is the bending moment which causes yielding. The mean fiber moment scales with the mean stress as $M^f \sim \sigma l_c^3$. This relation may be inferred by considering that $M^f \sim F^f l_c$ and relating the force to the mean stress as $F^f \sim \sigma l_c^2$. Therefore, at yielding one has $M_y^f \sim \sigma_y^n l_c^3$. It results that:

$$\frac{\sigma_y^n}{\sigma_y^f} \sim \left(\frac{d}{l_c}\right)^3. \tag{6.19}$$

This expression implies that, as the fiber diameter or the density decrease (the aspect ratio of fiber segments, l_c/d, increases), the network yield stress decreases fast relative to the fiber yield stress.

However, this does not mean that nonaffine networks yield easily. In fact, the opposite conclusion should be drawn. To infer it, we evaluate the ratio of the network and fiber yield strains, $\varepsilon_y^n/\varepsilon_y^f$. The ratio of the network small strain modulus and the fiber modulus can be written as $E_0/E_f \sim \rho^2 I_f \sim (d/l_c)^4$. Here, the relation $E_0 \sim \rho^x E_f I_f$ from Section 6.1.1.3.1 is considered, with $x = 2$, which is appropriate for cellular networks in 3D. Following the arguments outlined in the above paragraphs for the affine case, and using the ratio of the moduli and Eq. (6.19), the ratio of the network and fiber yield strains results:

$$\frac{\varepsilon_y^n}{\varepsilon_y^f} \sim \frac{l_c}{d}. \tag{6.20}$$

Equation (6.20) implies that, as the fiber diameter decreases (or the mean segment length increases), the yield strain of the network increases rapidly relative to the yield strain of fibers. In many cases, such as in collagen-based connective tissue and most nonwovens, the aspect ratio of fiber segments is at least 10. In the case of networks of nanofibers, the aspect ratio may be close to 100 or larger. This implies that ε_y^n is 10- to 100-times larger than the yield strain of fibers, that is, such networks are expected to never exhibit plasticity associated with the plastic deformation of fibers.

This analysis indicates that affine networks are expected to yield at a strain comparable with the yield strain of fibers. If fibers are elastic–plastic, it is unlikely for affine networks to exhibit hyperelastic behavior of the type shown in Figure 6.22, and a softening response compatible with yielding is expected instead (type B behavior). Nonaffine networks of thin fibers have the opposite behavior. They exhibit soft hyperelastic behavior and do not deform plastically. A combined situation may be observed in the transition regime from affine to nonaffine, amounting to an initial elastic branch including regimes I and II of the hyperelastic response, followed by yielding and softening which emerge once a large enough fraction of the fibers deforms plastically. Yielding and damage accumulation may occur concurrently in real networks.

Before closing this section, it should be noted that cyclic loading in the plastic range leads to the development of residual stresses. Unloading from a state in which a significant number of fibers have deformed plastically leads to residual strains in each fiber. Since the structure is stochastic, the residual strains that develop in different fibers are not equal, which necessarily leads to a complex residual stress state in the unloaded state of the network. However, networks with residual self-equilibrated stress are stiffer than those without residual stress (Huisman and Lubensky, 2011; Broedersz and MacKintosh, 2012). Therefore, the effective network modulus measured in cyclic loading with increasing cycle amplitudes should increase as the amplitude increases into the plastic range. This trend was observed in numerical models using approximately affine 2D networks in Zhang et al. (2019).

6.2 Mechanical Behavior of Non-crosslinked Networks

A large number of network materials important in many applications are not cross-linked. Athermal fibers clumped together forming fiber wads are used for thermal and sound insulation in a variety of applications ranging from construction to clothing. In these cases, fibers are not embedded in matrix and hold together due to inter-fiber friction and cohesion. Dense packings of athermal fibers dispersed in fluids form suspensions, which are encountered in paper making, food processing, and cosmetics.

This section is devoted to the mechanical behavior of athermal fiber wads subjected to compression, and to the mechanics of non-crosslinked fibrous mats loaded in tension. A vast amount of literature is available on the mechanical behavior of thermoplastics, above and below the glass transition temperature. The interested reader should consult relevant texts, such as Flory (1971) and Rubinstein and Colby (2003), for the polymer physics aspects, including the relationship between material structure and properties, and Ward and Sweeney (2013) and Larson (1988) for the engineering and constitutive modeling aspects. The rheology of suspensions and liquid crystalline complex fluids (dense suspensions of rod-like polymers) are summarized in Larson (1999).

6.2.1   Compression of Non-crosslinked Athermal Fiber Wads

This section refers to low density non-crosslinked fiber masses, such as glass, polymeric and cellulose fiber wads, felts made from natural fibers, and packings of metallic and carbon fibers. These materials are cheap, lightweight, and are able to efficiently dissipate energy due to the friction between fibers. This makes them useful in the textile industry as fillers, as the base material for various nonwovens, insulation material in buildings and automobiles, filler material with damping properties for open cell structures used in the aerospace industry, etc.

Interest in their mechanical properties arose during the growth of the textile industry in the late nineteenth and early twentieth centuries. For example, in 1938, Schofield observed that the stress required to compress a wad of wool scales proportionally to the third power of the wool volume fraction (Schofield, 1938).

The first micromechanical model describing the uniaxial compression of fiber assemblies is due to van Wyk, who derived an expression that relates the microstructural parameters of the network to the observed cubic relation between stress and fiber volume fraction (van Wyk, 1946). This model is based on several assumptions including the fact that fibers do not slide relative to each other at contacts, friction is absent, the bending deformation mode of fibers dominates, and deformation is affine on the scale of the mean segment length of the network.

Some of these assumptions have been relaxed in subsequent works. Carnaby and Pan (1989) accounted for relative fiber sliding at contacts, which allowed them to account, to some extent, for the experimentally observed hysteresis. Komori and Itoh (1991) considered that the fiber segment length (distance between contacts) is polydisperse, which is different from van Wyk's assumption that all stress-producing beam segments have the same length. In a later work, the same group considered the beams to be curved, which takes into account fiber crimp (Komori et al., 1992). Neckar (1997) extended van Wyk's model to represent biaxial compression of fiber wads with random fiber orientations in the plane of compression. Toll developed a generalization of van Wyk's equation for cases in

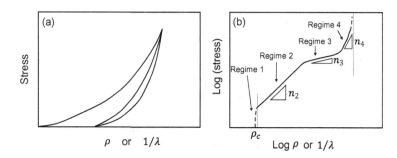

Figure 6.57 Schematic representation of the stress–density curve during uniaxial compression of a non-crosslinked fiber assembly. The curve in (a) shows loading, unloading, and reloading branches. The curve in (b) is the double logarithmic representation of the loading branch in (a). Compressive stress is shown to be positive.

which fibers are preferentially oriented in the plane normal to the compression direction before the mechanical test (Toll and Manson, 1995; Toll, 1998).

Figure 6.57 shows schematically the mechanical behavior of fiber wads in compression. The stress is shown in Figure 6.57(a) as a function of the fiber volume fraction (or the inverse of stretch). Consider a material volume of height h_0 and h (measured in the x_3 direction) in the undeformed and deformed configurations, respectively, and of unit area in the $x_1 - x_2$ plane, subjected to compression in the x_3 direction. As h decreases, the density, ρ, and the fiber volume fraction, $f = \pi d^2 \rho / 4$, increase. Since the Poisson effect in fiber assemblies subjected to compression is close to zero, the applied nominal stress, S, is approximately equal to the Cauchy stress, while the volume fraction and the density are related to the sample height and the stretch ($\lambda = h/h_0$) as $\rho/\rho_0 = f/f_0 = 1/\lambda$. ρ_0 and f_0 are the initial density and initial fiber volume fraction, corresponding to the sample height h_0. The key features of the mechanical response emerging from Figure 6.57(a) are:

- $S(\rho)$ increases monotonically;
- The network rearranges during compression and this causes hysteresis and residual strain during the first loading-unloading cycle; and
- The hysteresis decreases in subsequent cycles, provided the cycle amplitude remains constant.

Figure 6.57(b) shows a double logarithmic representation of the loading branch of the curve in Figure 6.57(a), which outlines several deformation regimes. Specifically:

- Stress is zero for densities below the percolation threshold, ρ_c (Section 4.5);
- Regime 1 is defined in the vicinity of the percolation threshold;
- Regimes 2, 3, and 4 are described by power laws and appear as straight lines in this representation. The exponent of regime 3 is smaller than that of regime 2, while the exponent of regime 4 is larger than that of regime 2. It is unusual to observe all these regimes in a given experiment due to experimental difficulties associated with varying the volume fraction in such a broad range; and

• Stress diverges fast beyond regime 4 due to the full compaction of the sample when high densities are reached.

Regime 1: This regime corresponds to the vicinity of stiffness percolation when, at a critical density, ρ_c, the network stiffness becomes non-zero. The density at percolation depends on the fiber aspect ratio, as discussed in Section 4.5.

Percolation is a critical phenomenon and, in the vicinity of the critical point, the characteristic parameters of the system – here, the stiffness – increase as a power function of the control parameter – here, the density. In many practical situations, the density is significantly larger than ρ_c and the behavior is not controlled by the proximity to the critical point. In these cases, the network response depends on the difference $\delta\rho = \rho - \rho_c$.

Regimes 2 and 3. These two regimes are described by power functions. It is instructive to derive the stress–density relation based on simple concepts. This discussion follows to some extent the derivation of van Wyk (1946) but departs from it in several important ways.

To this end, consider fibers packed in a volume of dimensions $1 \times 1 \times h$, as shown in Figure 6.58(a). The network consists of tortuous fibers that establish contacts at sites separated by the characteristic length ζ, that is, the mesh size. Another characteristic length scale is the mean segment length, l_c, or the distance between contact points measured along the fiber. If contacts are dense, that is, l_c is comparable or smaller than the persistence length of fibers, fiber segments between contacts become approximately straight and $l_c \approx \zeta$.

The stress applied at the upper boundary of the system is balanced at the microstructural scale by contact forces, P. Consider an intermediate state in which the system is loaded by the macroscopic nominal stress S and an increment dS is applied. The stress increment is related to the mean contact force increment as:

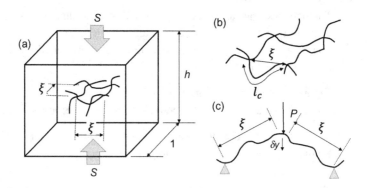

Figure 6.58 Schematic representation of packing of non-crosslinked fibers and notation used in the derivation of the stress–density relation.

$$dS = \frac{dP}{\zeta^2}. \tag{6.21}$$

This load increment leads to an increment of elastic deformation, dy. To evaluate dy, we approximate the actual complex loading of fibers by the bending of a simply supported curved beam of contour length $2l_c$ and end-to-end length 2ζ, loaded at the mid-point by the incremental force dP (Figure 6.58(c)). Considering small incremental deformations,

$$dP \sim \frac{E_f I_f}{\zeta^2 l_c} dy. \tag{6.22}$$

This expression is similar to that derived in the simple Euler–Bernoulli straight beam theory, except that the denominator accounts for the presence of two length scales, ζ and l_c. The derivation of this relation uses results from Section 2.3.5. The force–deformation relation for the curved beam can be written $dP = k_{bd}^{crv} dy$, where k_{bd}^{crv} is the effective bending stiffness. Figure 2.13(c) indicates that k_{bd}^{crv} can be expressed in terms of the bending stiffness of the equivalent straight beam, that is, $k_{bd}^{str} \sim E_f I_f/\zeta^3$, by observing that $k_{bd}^{crv}/k_{bd}^{str} \sim c$ and using the definition $c = \zeta/l_c$. Hence, $k_{bd}^{crv} \sim E_f I_f/\zeta^2 l_c$, which leads to Eq. (6.22). The term $E_f I_f$ indicates that only the bending deformation mode is considered. Curved beams loaded as shown in Figure 6.58(c) are subjected to axial, bending, torsional, and shear modes. However, the deformation is controlled primarily by bending and torsion, both leading to scalings similar to that in Eq. (6.22), with $E_f I_f$ being replaced by $G_f J_f$ in the case of torsion.

To make progress with the analytic derivation, we impose now the affine assumption which requires that the local strain associated with dy is identical to the global strain applied macroscopically:

$$\frac{dy}{\zeta} = \frac{dh}{h}. \tag{6.23}$$

Since the network density ρ is the total fiber length divided by the volume, $1 \times 1 \times h$, it results that $dh/h = -d\rho/\rho$.

Further, we make use of the definition of the contact density in 3D, $\rho_X \sim 1/\zeta^3$, and its relation to the fiber density, $\rho_X \sim \rho/l_c$, to obtain:

$$\zeta \sim \left(\frac{l_c}{\rho}\right)^{1/3}. \tag{6.24}$$

Combining Eqs. (6.21) to (6.24), the stress increment results:

$$dS \sim -\frac{E_f I_f}{l_c^2} d\rho. \tag{6.25}$$

To render Eq. (6.25) useful, the relation between the density and the mean segment length has to be specified. Predicting this relation is a long-standing issue in the literature. van Wyk (1946), and essentially all theoretical works based on his model, used the assumption that the number of contacts per fiber scales linearly with the fiber volume fraction, which is equivalent to:

$$\rho l_c d = const. \tag{6.26}$$

This relation is similar to that obtained from packing arguments in Section 4.2.3.2, Eq. (4.13).[12] It corresponds to a newly constructed system formed by the aggregation of fibers subjected to the topological constraint imposed by the excluded volume condition (Section 4.2.3.2). In the initial stages of loading, this scaling is expected to remain valid since the system structure does not depart significantly from the initial one. Based on this assumption, Eq. (6.25) leads to $dS \sim -E_f I_f \rho^2 d\rho$, which can be integrated to obtain the stress–density relation:

$$S(\rho) \sim E_f \left(\rho^3 - \rho_0^3 \right). \tag{6.27}$$

Density ρ_0 corresponds to a reference state in which $S(\rho_0) = 0$, that is, ρ_0 can be taken to be equal to the stiffness percolation density, ρ_c. Equation (6.27) was derived by van Wyk and is often taken as the refence theoretical prediction when interpreting experimental data.

It is useful to consider another limit case, in which the contact density has increased so much during compression that segments between successive contacts along fibers are approximately straight. In this case, there is only one characteristic length in the problem, since $l_c \approx \zeta$. Therefore, based on dimensional considerations, the relation between density and l_c is of the form:

$$\rho l_c^2 = const. \tag{6.28}$$

With this relation, Eq. (6.25) becomes $dS \sim -E_f I_f \rho \, d\rho$, which gives after integration the stress–density relation:

$$S(\rho) \sim E_f \left(\rho^2 - \rho_0^2 \right). \tag{6.29}$$

If one assumes that the two length scales ζ and l_c become proportional during deformation (but are not necessarily equal), Eq. (6.24), which is of geometric nature and is not subjected to interpretation, mandates the scaling of Eq. (6.28) and the result of Eq. (6.29) follows.

Equations (6.27) and (6.29) indicate that the slopes of the stress–density plot in Figure 6.57(b) in regimes 2 and 3 are $n_2 = 3$ and $n_3 = 2$, respectively.

Regime 4. At large enough deformations, fibers become preferentially aligned in the plane perpendicular to the loading direction and the initially isotropic 3D network

---

[12] Equation (6.26) actually indicates that $\rho l_c = const$, since the diameter d does not change during deformation. d is added to Eq. (6.26) to render the group non-dimensional. Any other deformation-independent length can be used for this purpose (e.g., L_0).

becomes transversely isotropic. The system is expected to become asymptotically similar to the case of a pre-aligned, mat-like fiber network which is compressed in the direction perpendicular to the plane of the mat. This problem was studied by Toll (Toll and Manson, 1995; Toll, 1998), who derived the stress–density relation:

$$S(\rho) \sim E_f\left(\rho^5 - \rho_0^5\right). \tag{6.30}$$

This implies that at large enough volume fractions, the characteristic curve of Figure 6.57(b) should develop a slope $n_4 = 5$.

It is necessary to compare the predictions of this theory with experimental and numerical results. Figure 6.59(a) shows experimental data from Baudequin et al. (1999) and Mezeix et al. (2009) obtained from compression tests performed with glass wools of different initial densities, carbon fiber wool, and wools of glass fiber yarn. The plot shows the nominal stress versus ρ/ρ_0, where ρ_0 is the initial density of the wad, after carding and before testing. Note that $\rho_0 > \rho_c$ corresponds to the unloaded state of the network. The initial mass density of each sample, ρ_{m0}, is indicated in the figure. The mass density is related to the volume fraction, f, as $\rho_m = \rho_f f$, where ρ_f is the density of the fiber material, and is related to ρ as $\rho_m = \pi d^2 \rho_f \rho/4$. ρ_0 is different for each data set shown.

The curves exhibit regimes 2 and 3, with the response in regime 2 being described by Eq. (6.27), with exponent $n_2 = 3$, as predicted by the theory. A transition to regime 3 is observed at larger densities. Samples of glass wool of the lowest density (10 Kg/m^3) exhibit the transition at lower ρ/ρ_0 and regime 3 is characterized by an exponent close to 2.

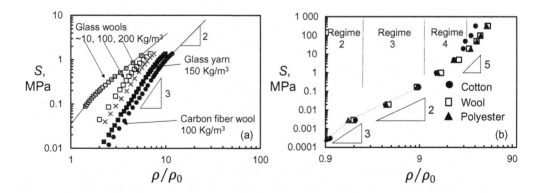

Figure 6.59 (a) Stress–density plots for glass fiber, glass yarn, and carbon fiber wools subjected to compression (data from Baudequin et al. (1999) and Mezeix et al. (2009)). Compressive stress is shown as positive. The initial mass density, ρ_{m0}, is indicated for each data set. (b) Stress–density plots for cotton, wool, and polyester fiber wads (Baljasov, 1976), based on data collected in Neckar and Das (2012). The same value of ρ_0 is used for all three curves. All regimes shown schematically in Figure 6.57(b) are visible in this data set.

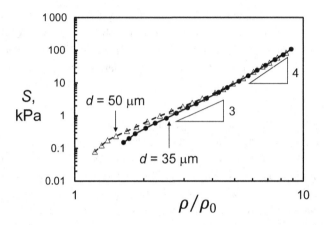

Figure 6.60 Stress–density plots for wads of polyamide fibers (data from Alkhagen and Toll, 2007). The two curves correspond to fibers of two diameters. Compressive stress is shown to be positive.

In the tests leading to the data in Figure 6.59(a), samples were compressed to moderate densities and a transition to regime 4 was not observed. In the experiments reported in Alkhagen and Toll (2007), carded polyamide fiber masses are subjected to compression to volume fractions as high as 20%. Figure 6.60 shows stress-relative density curves for fibers of two diameters. In both cases, regime 2 is followed by an increase of the slope, indicating a transition to regime 4, while regime 3 is not observed. The authors develop a theory supporting these experimental results, which is applicable to low and high densities and to multiaxial loading. It is also of interest to observe that the fiber diameter has little influence on the shape of the characteristic curve, especially at densities much larger than the percolation threshold, when the contact density is large.

Similar results are presented in Alkhagen and Toll (2009) for wools composed from mixtures of fibers of different diameters. The presence of fibers of different properties leaves the characteristic curve largely unchanged.

If fibers are pre-aligned in the plane perpendicular to the loading direction, the prediction of Eq. (6.30) is expected to apply in regime 2. In experiments it is observed that exponent n_2 takes values from 3 to 5 as the level of pre-alignment increases. A data set for steel fiber wool demonstrating this trend is presented in Masse et al. (2006).

A particularly interesting data set which exhibits multiple regimes and extends to large densities is provided by Baljasov (1976). Nominal stress-relative density curves for wool, cotton, and polyester fibers are shown in Figure 6.59(b). The data shown is adapted from Neckar and Das (2012). The curves for the three materials overlap (the horizontal axis is normalized with the same threshold density, ρ_0) and exhibit regimes 2, 3, 4, and the rapid rise beyond regime 4, as indicated schematically in Figure 6.57(b). These results, along with those in Figure 6.59(a), demonstrate the presence of the softer

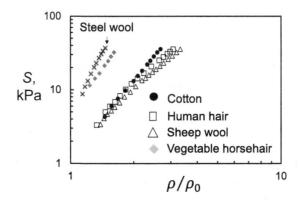

Figure 6.61 Stress–density plots for wools of various types. Compressive stress is shown as positive. Data from Poquillon et al. (2005)

regime 3 characterized by Eq. (6.29) and in which the scaling of the density with the mean segment length of Eq. (6.28) applies.

Power functions relating the stress and the density often emerge from experimental data. However, the exponents are not always as much in agreement with the theoretical considerations presented in this section as the data in Figures 6.59 and 6.60. Figure 6.61 presents a collection of stress–density curves for various wools organized based on data from Poquillon et al. (2005). Samples of dry human hair, sheep wool, steel wool, vegetable horsehair, and cotton are tested in confined uniaxial compression. The samples are not pre-conditioned by applying a preliminary compression cycle. Curves of slopes ranging from 2.8 for sheep wool to 5.2 for vegetable horsehair are obtained. In view of this variability, it is necessary to inquire about the physical or procedural causes that may cause the apparent departure from theoretical predictions. Several considerations along these lines are:

- It is important to subject samples to a preliminary compression cycle in order to diminish or remove the effect of carding or other sample preparation steps. As seen in Figure 6.57(a), the initial loading curve is quite different from the reloading curve. However, subsequent loading–unloading cycles produce consistent stress–stretch curves which are therefore representative for the pre-conditioned material state.
- Fiber pre-alignment in the plane perpendicular to the loading direction leads to regime 2 exponents in the range 3–5, in agreement with model predictions described in this section. Although most industrially produced samples are anisotropic, with fibers often being significantly aligned in the machine direction, the degree of alignment is not always observed and reported.
- Contacts between fibers have finite stiffness and this leads to the storage of some amount of energy at contact sites. If the contact density is relatively low, l_c is much larger than the fiber diameter and most strain energy is stored in the fiber bending

mode. In the opposite case, when contacts are compliant and the contact number density is high, the contact energy may become dominant. This is expected to be the case in networks of hollow fibers, such as cellulose fibers and mycelium filaments (hyphae), at high contact densities. Hossain et al. (2019) developed models that explore this range of system parameters and observed significant dependence of exponent n_2 on the inter-fiber contact stiffness.

Simulations of compression of fiber masses (Beil and Roberts, 2002; Durville, 2005; Barbier et al., 2009; Subramanian and Picu, 2011; El-Rahman and Tucker, 2013) clarified additional details of this processes. Durville (2005) performed compression of non-crosslinked tortuous fiber assemblies and observed regimes 2, 3, and 4, with slopes close to those predicted by the theoretical considerations discussed in this section, particularly at intermediate levels of tortuosity. The transition from regime 2 to regime 3 was also observed in Subramanian and Picu (2011) in systems of initially straight fibers subjected to triaxial compression. This work also underlines the importance of pre-conditioning when analyzing the mechanical behavior. The stress–density curve exhibits only one slope, between 2 and 3, in the first loading cycle, while the loading branch of subsequent loading–unloading cycles exhibits the expected transition between regimes 2 and 3. It was further observed that introducing inter-fiber friction at contacts has little effect on the mechanical behavior – a conclusion supported by the results of Barbier et al. (2009) and Subramanian and Picu (2011), but not supported by the data in El-Rahman and Tucker (2013).

6.2.2   Tension of Non-crosslinked Networks: Entanglements in Athermal Networks

The mechanical behavior of non-crosslinked networks subjected to tension is much less intuitive and less studied than the material response in compression. This section presents considerations leading to a qualitative understanding of this problem. To make concepts accessible, the discussion is focused on quasi-two-dimensional networks, that is, mats of fibers randomly oriented in the plane of the mat. This does not reduce the generality of the ideas presented.

It is necessary to underline the central importance of nonaffinity in defining the mechanical behavior of these networks. Stress is produced by fiber interactions which take place at contacts. If deformation is affine, the relative fiber positions remain unchanged since all inter-fiber distances vary in proportion. In such situations, the number of contacts remains constant during deformation. However, the relative fiber positions change if deformation is nonaffine. Nonaffinity is an intrinsic property of stochastic networks and is more pronounced if fibers are tortuous. Therefore, contacts are expected to form and open dynamically during any type of deformation of a stochastic fiber network, including when subjected to tension.

The mechanisms leading to the formation of entanglements in athermal networks are discussed here in the (idealized) limit of no inter-fiber friction. Figure 6.62(a) shows a set of three fibers belonging to a much larger mat of continuous ("infinite")

Figure 6.62 Schematic representation of (a) a segment of a stochastic network of fibers forming a mat and (b) a planar woven fibrous structure.

fibers which is loaded such that, at the scale of these particular fibers, AB is pulled out in the x_1 direction. Several situations may be envisioned:

(i)   If fibers have no bending stiffness, AB can be pulled out with no resistance. Such a network has no mechanism of stress production.
(ii) If fibers have finite bending stiffness and are straight in the zero energy configuration, the state of Figure 6.62(a) may exist only under external constraints. If allowed to relax, fibers would become straight, most contacts would be lost, and the network would come apart. This argument applies equally to the periodic network of woven fibers shown in Figure 6.62(b). No stress is required to deform a layered structure of straight fibers without friction.
(iii) If fibers have finite bending stiffness and their relaxed, minimum energy configuration is tortuous, pulling out fiber AB creates forces at contacts with fibers CD and EF. If the ends of CD and EF are not fixed, they would move in the direction perpendicular to the plane of the mat, normal to x_1, which creates an auxetic effect. If such motion of CD and EF is constrained, the pull out of AB requires the deformation of all fibers involved and a non-zero network-scale stress emerges even in the absence of frictional interactions.

This analysis indicates that, in the zero (or low) friction case, the minimum requirements for the network to develop macroscopic stiffness are: (a) fibers must have finite bending stiffness, (b) fibers must be tortuous in the zero strain energy configuration, and (c) some type of constraints should exist leading to the development of a self-equilibrated internal stress state which is induced by the externally applied load. The periodic, woven structure of Figure 6.62(b) is a typical example of this type. Contacts are engaged when the network is loaded biaxially in the plane of the mat and a self-equilibrated stress produced by contact forces acting in the direction normal to the plane of the mat emerges.

The essential requirement for the emergence of network-scale stiffness is that of the presence of constraints. Two mechanisms produce constraints in thermal and athermal networks:

(i) inter-fiber cohesion, and
(ii) fiber entanglements.

Cohesion takes place at contacts and holds fibers together. If strong enough, it may lead to the re-organization of the network and to the formation of fiber bundles

(Chapter 10). In dense molecular networks, such as polymeric melts, cohesive inter-actions ensure that fibers are packed at liquid-like densities. This, in turn, enhances the topological interactions between filaments. Cohesion is not a central component of the physics of athermal networks of fibers of diameter larger than several microns.

Entanglements are a more effective mechanism for imposing constraints on the relative motion of fibers in fibrous assemblies with and without inter-fiber friction. The concept of entanglements is well-established in polymer physics. It is envisioned that, in melts, the dense inter-chain interactions effectively confine the representative chain to a (tortuous) tube. The chain is free to move along the tube axis but is prevented from escaping from it in the transverse direction. The tube diameter is a characteristic length that emerges from the dynamics in this confined geometry and is known as the entanglement length, L_e. The tube is viewed as a random walk in 3D at scales larger than L_e. Chain motion is not constrained on scales smaller than L_e. This physical picture is not particularly useful for athermal fiber networks, where the fiber density and the density of contacts are much smaller and thermal fluctuations are absent.

Entanglements are also present in solutions of polymeric chains. The effective tube diameter is a function of the polymer volume fraction: $L_e \sim f^{-2/3}$ for θ-conditions and $L_e \sim f^{-3/4}$ for the athermal solvent case (see Section 4.2.3.4.2 for the origin of these exponents). This relation quantifies the rate at which topological confinement of a representative chain decreases as the filament volume fraction in the solution is reduced. These considerations and the underlying concepts are described in detail in polymer physics texts and are not discussed further here.

Entanglements are notoriously difficult to be identified based on the network geometry alone. Whether a polymeric system is entangled or not is inferred from its response to mechanical stimuli and not by the direct inspection of the structure. This is due to the fact that topological confinement is statistical in nature, particularly so in the case of filaments subjected to thermal fluctuations.

To make progress in the direction of defining entanglements in stochastic filamentary systems, consider as a reference case the woven periodic network of Figure 6.62(b). Consider that the network is loaded in tension in the x_1 direction and the deformation in the x_2 direction is restricted. Contact forces in the direction normal to the plane of the network, $x_1 - x_2$, develop at each inter-fiber contact. Forces in the x_3 direction acting along the length of each fiber balance exactly. In addition, it is possible to identify closed loops formed by "interwoven" segments of the network. Specifically, consider a path that advances along a given fiber to a contact formed with another fiber which is positioned above the current fiber (in the positive x_3 direction); the path continues on this new fiber up to another contact with a fiber which is similarly positioned above the current fiber; the procedure repeats until the path forms a closed loop along which the subsequent fiber is always positioned above the previous fiber at all contacts, and the path returns to its starting point. An infinite number of such loops can be found in the periodic network of Figure 6.62(b). However, the relevant loop is the shortest (which is also the stiffest and hence the most constraining). In the periodic network, it corresponds to the repeat unit shown by the gray area in Figure 6.62(b).

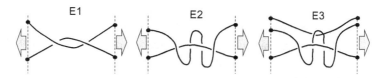

Figure 6.63 Three types of entangled configurations in stochastic fiber networks.

To extrapolate these ideas to stochastic network architectures, it is helpful to consider several specific fiber arrangements and to analyze the emergence of forces transverse to the loading direction. Three such situations are shown in Figure 6.63. In all these cases, fibers run continuously left to right and are considered fixed at both ends, such that their relative positions cannot change during deformation. Configuration E1 represents a braided structure. Transverse forces emerge at contacts when this structure is loaded in tension in the horizontal direction. E2 represents a somewhat different situation, in which the two fibers involved exchange positions only over the span of the central segment. The two are initially in contact and constrain each other in the direction orthogonal to the loading direction but would eventually disentangle and become independent at large strains. Case E3 is similar to E2, except that a third fiber stabilizes the loop and prevents fiber separation. E1 and E3 represent essential topological constraints, while E2 is a temporary constraint. The mechanism described here, leading to the development of an internal stress state, operates in all three cases.

A note on the likelihood of occurrence of these configurations in real fiber networks is required. Configuration E1 is similar to structures obtained by weaving. Nonwoven production methods generally do not lead to this type of fiber arrangement, particularly when continuous fibers are used. Spinning produces a layered structure and E2 type configurations occur frequently. Hydroentangling and needle-punching also lead primarily to configurations of E2 type. Such structures provide weak, temporary constraints and lead to relatively small network-scale stresses. This is the reason for which networks of spun (including electrospun) polymeric fibers have relatively low flow stress and need to be crosslinked before they become useful in engineering applications. To develop knotted configurations such as E3 while using continuous fibers, some type of weaving process is necessary. Braiding is applied in the production of yarn and ropes but is not typically performed with continuous fiber nonwovens.

Nonwovens of fibers with finite length, such as felt, may have configurations of both E1 and E2 types. All configurations shown in Figure 6.63 may exist in thermal networks of finite length filaments, such as polymeric solutions and melts. This is due to the ability of thermal filaments to diffuse, a process which allows continuous exchange of filament positions and the development of complex configurations. The probability to form knots of a geometry more complex than E3 is very low, even in thermal networks of polymeric chains.

Efforts have been made in polymer physics to determine the chain primitive path and the entanglement length based on geometric considerations. A computational procedure was devised (Rubinstein and Helfand, 1985; Foteinopoulou et al., 2006), which starts with a polymeric melt equilibrated at the proper density and determines the number of braided fiber pairs. To this end, the ends of the chains at the surface of the model are fixed and the chain length is gradually reduced by allowing the bonds along the chain to shrink. The procedure ends when the chains are pulled tight and hold on each other, as in E1. The entanglement length is defined as the mean distance between contact points in the resulting network. This procedure leads to estimates of the entanglement length in agreement with the values inferred from the variation of the mechanical properties of the melt with the chain length. For example, the relaxation time of the polymeric system scales with the chain length to the second power as long as the length is below L_e (Rouse relaxation regime), and with the chain length to a power close to 3 if the chain length is larger than L_e (reptation regime). L_e may be obtained by determining the chain length at which the scaling of the relaxation time changes.

Entanglements in athermal networks have been studied much less. A measure of the entanglement length based on geometric considerations is proposed in Negi and Picu (2021). In the spirit of the topological procedure described in relation to the periodic structure in Figure 6.62(b), the minimum length closed loop of interwoven segments starting from a generic contact is identified. The procedure is repeated for all contacts and the average minimum length loop is computed. Its contour length is taken as the entanglement length, L_e. While in the periodic woven network of Figure 6.62(b) the minimum length loops are composed from exactly four network segments, in stochastic networks the loops are composed from an arbitrary number of segments. Note that the shortest loop provides the largest effective constraint. This characteristic length is related to the density of E1 and E2 configurations in the network.

As in the periodic woven network of Figure 6.62(b), entanglements lead to the development of a self-equilibrated stress state in the network, which is enabled by the applied boundary conditions. This is somewhat similar to the situation of a tensegrity structure subjected to external loading. The tensegrity structure has a pre-existing internal stress state that stabilizes the sub-isostatic network and provides non-zero stiffness. The applied loads create a stress that superimposes to this internal stress state. In the case of an entangled fibrous structure, the self-equilibrated internal stress is produced by the far field and has a similar stabilizing role; see also the discussion related to Figure 6.41.

Another topological measure of entanglements in athermal networks was proposed in Grishanov et al. (2012) based on the number of position exchanges of fiber pairs required to transform a given network structure into a layered arrangement of fibers. This measure takes into account only essential entanglements (E1 and E3 in Figure 6.63). Unfortunately, this entanglement measure was not related to the network mechanical behavior.

The following discussion demonstrates the relation between the entanglement measure based on the minimum length loop and network mechanics. To make the discussion specific, we consider a network of tortuous (nonfractal) continuous fibers of persistence length L_p generated in a 3D volume and representing a fibrous mat. If

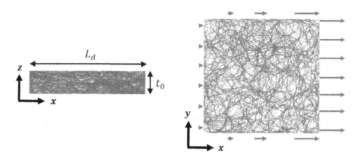

Figure 6.64 Side (left) and top (right) views of a mat of continuous non-crosslinked tortuous fibers of specified persistence length. The arrows show the boundary conditions applied when probing the mechanical response.

fibers are generated independently in 3D and then stacked to form the mat, the resulting structure is layered and only configurations of type E2 may be present. The other possibility is to generate fibers as random walks of given persistence length within a pre-defined problem domain, a case in which some degree of interweaving results and configurations of type E1 and E2 are created. Figure 6.64 shows the side and top views of such a network.

These networks are loaded in uniaxial tension, while enabling the formation of contacts between fibers. Figure 6.65(a) shows the normalized nominal stress, $\hat{S} = S/\rho E_f d^2$, versus stretch in the loading direction for networks of different values of the entanglement parameter L_e. The figure includes a curve representing the response of the phantom network which has the same structure, but in which fibers are not allowed to form contacts. The stress of the phantom network is due exclusively to the deformation of individual fibers and is taken as reference, $\hat{S}_{ref}$. Figure 6.65(b) shows the difference stress $\Delta\hat{S} = \hat{S} - \hat{S}_{ref}$, which is produced exclusively by excluded volume fiber interactions, including entanglements.

It is seen that the difference stress increases with increasing the degree of entanglement. Figure 6.65(c) shows the relation between L_e and the slope of the difference stress in Figure 6.65(b) (tangent stiffness), that is, $\partial(\Delta\hat{S})/\partial\lambda$, for networks of various volume fractions. A well-defined relation emerges, indicating that $\partial(\Delta\hat{S})/\partial\lambda \sim (L_e/d)^{-2}$. As the entanglement density decreases, L_e increases, and $\partial(\Delta\hat{S})/\partial\lambda$ decreases quadratically. This establishes the causal relationship between L_e and the stiffness of the entangled network.

The entanglement length results proportional to the mean segment length l_c (but is always larger than it) and is a function of the network volume fraction: $L_e \sim f^{-1/2}$. This scaling may be obtained with the overlap concentration procedure of polymer physics (see the discussion on thermal networks in Section 4.2.3.4.2) if the chains are replaced with rectilinear (nonfractal) curves in 3D.

Weakly entangled networks corresponding to large L_e exhibit a strong auxetic effect. Entanglements reduce the auxetic effect since they constrain the network from expanding in the mat normal direction. This is a manifestation of the self-equilibrated stress state discussed earlier in this section.

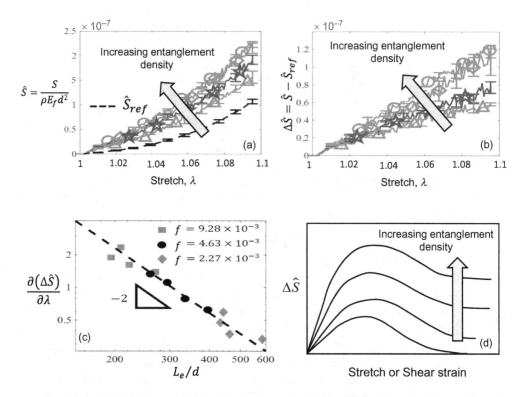

Figure 6.65 (a) Stress–stretch relation in networks of given fiber volume fraction and increasing entanglement density (decreasing L_e). $\hat{S}_{ref}$ is the normalized stress corresponding to the equivalent phantom network with no inter-fiber interactions. (b) Data in (a) replotted as difference stress versus stretch. (c) Relation between the slope of the curves in (b) and L_e for networks of various volume fractions, f (adapted from Negi and Picu, 2021), (d) Schematic representation of the expected difference stress–stretch relation in (b) for a broader stretch range.

The physical picture outlined in Figure 6.65(c) applies at relatively small strains. Both entanglements of type E1 and E2 contribute to stress production in this regime. Configurations E2 disappear gradually at larger strains. Therefore, the stress–strain curve is expected to exhibit a maximum, after which the stress reaches a L_e-dependent plateau. This is shown schematically in Figure 6.65(d). The stress peak and subsequent softening may be eliminated as the entanglement density increases. In the unentangled case in which configurations E1 are not present, the large strain plateau in Figure 6.65(d) drops to zero. This type of behavior is found experimentally in needle-punched and hydroentangled nonwovens, as for example in Moyo et al. (2016).

Since the mechanical response of non-crosslinked networks is controlled by fiber interactions at contacts, friction has a strong effect on the behavior. This is indeed observed in models similar to those used to produce the data in Figure 6.65, in which friction is enabled at contacts (Negi and Picu, 2021). Interestingly, friction does not change the physical picture of the effect of entanglements on network mechanics, and all conclusions discussed in this section remain qualitatively valid. However, increasing the friction coefficient leads to a rapid increase of the flow stress. Sliding at contacts takes place even for friction coefficients as high as 0.5. Therefore, fiber rearrangements occur during deformation in all cases. Friction forces lead to pronounced hysteresis and the emergence of large residual strains upon unloading. As friction increases, a large fraction of the work performed by the applied tractions is dissipated.

This brief account of the effect of the non-crosslinked network topology on the mechanical behavior is by no means exhaustive. In fact, key concepts are still to be established in this area. The goal of the present discussion is limited to setting up the stage for future developments.

References

Alava, M. & Niskanen, K. (2006). The physics of paper. *Rep. Prog. Phys.* **69**, 669–723.

Alkhagen, M. & Toll, S. (2007). Micromechanics of a compressed fiber mass. *J. Appl. Mech.* **74**, 723–731.

Alkhagen, M. & Toll, S. (2009). The effect of fiber diameter distribution on the elasticity of a fiber mass. *J. Appl. Mech.* **76**, 041014.

Andrews, E. W., Gioux, G., Onck, P., & Gibson, L. J. (2001). Size effects in ductile cellular solids. Part 2: Experimental results. *Int. J. Mech. Sci.* **43**, 701–713.

Andrews, E. W., Sanders, W. & Gibson, L. J. (1999). Compressive and tensile behavior of aluminum foams. *Mat. Sci. Eng. A* **270**, 113–124.

Baljasov, P. D. (1976). *Compression of textile fibers in mass and technology of textile manufacture*. Legkaya promyshlennost, Moscow.

Ban, E., Barocas, V. B., Shephard, M. S. & Picu, R. C. (2016a). Effect of fiber crimp on the elasticity of random fiber networks with and without embedding matrices. *J. Appl. Mech.* **83**, 041008.

Ban, E., Barocas, V. B., Shephard, M. S. & Picu, R. C. (2016b). Softening in random networks of non-identical beams. *J. Mech. Phys. Sol.* **87**, 38–50.

Bancelin, S., Lynch, B., Bonod-Bidaud, C., et al. (2015). Ex-vivo multiscale quantitation of skin biomechanics in wild-type and genetically-modified mice using multiphoton microscopy. *Sci. Rep.* **5**, 17635.

Barbier, C., Dendievel, R. & Rodney, D. (2009). Role of friction in the mechanics of nonbonded fibrous materials. *Phys. Rev. E* **80**, 016115.

Baudequin, M., Ryschenkow, G. & Roux, S. (1999). Non-linear elastic behavior of light fibrous materials. *Eur. Phys. J. B* **12**, 157–162.

Beil, N. B. & Roberts, W. W. (2002). Modeling and computer simulation of the compressional behavior of fiber assemblies. *Textile Res. J.* **72**, 341–351.

Bernardi, L., Hopf, R., Ferrari, A., Ehret, A. E. & Mazza, E. (2017). On the large strain deformation behavior of silicone-based elastomers for biomedical applications. *Poly. Testing* **58**, 189–198.

Bircher, K., Ehret, A. E. & Mazza, E. (2016). Mechanical characteristic of bovine Glisson's capsule as a model tissue for soft collagenous membranes. *J. Biomech. Eng.* **138**, 081005.

Broedersz, C. P. & MacKintosh, F. C. (2012). Molecular motors stiffen non-affine semiflexible polymer networks. *Soft Matter* **7**, 3186–3191.

Broedersz, C. P., Sheinman, M. & MacKintosh, F. C. (2012). Filament-length-controlled elasticity in 3D fiber networks. *Phys. Rev. Lett.* **108**, 078102.

Broedersz, C. P., Sheinman, M. & MacKintosh, F. C. (2014). Modeling semiflexible polymer networks. *Rev. Mod. Phys.* **86**, 995–1036.

Buxton, G. A. & Clarke, N. (2007). Bending to stretching transition in disordered networks. *Phys. Rev. Lett.* **98**, 238103.

Calladine, C. R. (1978). Buckminster Fuller's tensegrity structures and Clerk Maxwell's rule for the construction of stiff frames. *Int. J. Sol. Struct.* **15**, 161–172.

Candau, S., Peters, A. & Herz, J. (1981). Experimental evidence for trapped chain entanglements: Their influence on macroscopic behavior of networks. *Polymer* **22**, 1504–1510.

Carnaby, G. A. & Pan, N. (1989). Theory of the compression hysteresis of fibrous assemblies. *Text. Res. J.* **59**, 275–284.

Chen, N., Koker, M. K. A., Uzun, S. & Silberstein, M. N. (2016). In-situ X-ray study of the deformation mechanisms of non-woven polypropylene. *Int. J. Sol. Struct.* **97–98**, 200–208.

Deogekar, S. & Picu, R. C. (2017). Structure–properties relation for random networks of fibers with noncircular cross-section. *Phys. Rev. E* **95**, 033001.

Domaschke, S., Morel, A., Kaufmann, R., et al. (2020) Predicting the macroscopic response of electrospun membranes based on microstructure and single fiber properties. *J. Mech. Beh. Biomed. Mater.* **104**, 103634.

Duling, R. R., Dupaix, R. B., Katsube, N. & Lannutti, J. (2008). Mechanical characterization of electrospun polycaprolactone (PCL): A potential scaffold for tissue engineering. *J. Biomech. Eng.* **130**, 011006.

Durville, D. (2005). Numerical simulation of entangled materials mechanical properties. *J. Mater. Sci.* **40**, 5941–5948.

El-Rahman, A. I. A. & Tucker III, C. L. (2013). Mechanics of random discontinuous long-fiber thermoplastics. Part II: Direct simulation of uniaxial compression. *J. Rheol.* **57**, 1463–1489.

Erman, B. & Mark, J. E. (1989). Rubber-like elasticity. *Ann. Rev. Phys. Chem.* **40**, 351–374.

Erman, B. & Mark, J. E. (1997). *Structure and properties of rubberlike networks*. Oxford University Press, New York.

Feng, S., Thorpe, M. F. & Garboczi, E. (1985). Effective medium theory of percolation on central force elastic networks. *Phys. Rev. B* **31**, 276–280.

Flory, P. (1971). *Principles of polymer chemistry*. Cornell University Press, Ithaca, NY.

Fontaine, F., Noel, C., Monnerie, L. & Erman, B. (1989). Stress–strain–swelling behavior of amorphous polymeric networks: Comparison of experimental data with theory. *Macromolecules* **22**, 3352–3355.

Foteinopoulou, K., Karayannis, N. C., Mavrantzas, V. G. & Kroger, M. (2006). Primitive path identification and entanglement statistics in polymer melts: Results from direct topological analysis on atomistic polyethylene models. *Macromolecules* **39**, 4207–4216.

Fuller, R. B. (1976). *Synergetics: Explorations in the geometry of thinking*. Macmillan, New York.

Gao, J. & Weiner, J. H. (1994). Nature of stress on the atomic level in dense polymer systems. *Science* **266**, 748–752.

Gardel, M. L., Shin, J. H., MacKintosh, F. C., et al. (2004). Elastic behavior of crosslinked and bundled actin networks. *Science* **304**, 1301–1305.

de Gennes, P. G. (1979). *Scaling concepts in polymer physics*. Cornell University Press, Ithaca, NY.

Gent, A. N. (ed.) (2012). *Engineering with rubber. How to design rubber components*, 3rd ed. Hanser Publishers, Munich.

Gibson, L. J. & Ashby, M. F. (1988). *Cellular solids: Structure and properties*. Pergamon Press, Tarrytown, NY.

Glüge, R. (2013). Generalized boundary conditions on representative volume elements and their use in determining the effective material properties. *Comput. Mat. Sci.* **79**, 408–416.

Grishanov, S., Tausif, M. & Russell, S. J. (2012). Characterization of fiber entanglement in nonwoven fabrics based on knot theory. *Comp. Sci. Technol.* **72**, 1331–1337.

Gusev, A. A. (2019). Numerical estimates of the topological effect in the elasticity of Gaussian polymer networks and their exact theoretical description. *Macromolecules* **52**, 3244–3251.

Harley, B. A., Leung, J. H., Silva, E. C. C. M. & Gibson, L. J. (2007). Mechanical characterization of collagen-glycosaminoglycan scaffolds. *Acta Biomater.* **3**, 463–474.

Head, D. A., Levine, A. J. & MacKintosh, F. C. (2003). Distinct regimes of elastic response and deformation modes of crosslinked cytoskeletal and semiflexible polymer networks. *Phys. Rev. E* **68**, 061907.

Heussinger, C. & Frey, E. (2006). Floppy modes and nonaffine deformations in random fiber networks. *Phys. Rev. Lett.* **97**, 105501.

Heussinger, C. & Frey, E. (2007). Role of architecture in the elastic response of semiflexible polymer and fiber networks. *Phys. Rev. E* **75**, 011917.

Heussinger, C., Schaeter, B. & Frey, E. (2007). Nonaffine rubber elasticity for stiff polymer networks. *Phys. Rev. E* **76**, 031906.

Horgan, C. O. & Murphy, J. G. (2017). Poynting and reverse Poynting effects in soft materials. *Soft Matter* **13**, 4916–4923.

Hossain, M. S., Bergstrom, P. & Uesaka, T. (2019). Uniaxial compression of 3D entangled fiber networks: Impacts of contact interactions. *Model. Simul. Mater. Sci. Eng.* **27**, 015006.

Huang, C. Y., Stankiewicz, A., Ateshian, G. A. & Mow, V. C. (2005). Anisotropy, inhomogeneity and tension–compression nonlinearity of human glenohumeral cartilage in finite deformation. *J. Biomech.* **38**, 799–809.

Huisman, E. M. & Lubensky, T. C. (2011). Internal stresses, normal modes and nonaffinity in three-dimensional biopolymer networks. *Phys. Rev. Lett.* **106**, 088301.

Islam, M. R. & Picu, R. C. (2018). Effect of network architecture on the mechanical behavior of random fiber networks. *J. Appl. Mech.* **85**, 081011.

Islam, M. R., Tudryn, G., Bucinell, R., Schadler, L. & Picu, R. C. (2017). Morphology and mechanics of fungal mycelium. *Sci. Rep.* **7**, 13070.

Janmey, P. A., Euteneuer, U., Traub, P. & Schliwa, M. (1991). Viscoelastic properties of vimentin compared with other filamentous biopolymer networks. *J. Cell. Biol.*, **113**, 155–160.

Janmey, P. A., McCormick, M. E., Rammensee, S., et al. (2007). Negative normal stress in semiflexible biopolymer gels, *Nat. Mater.* **6**, 48–51.

Jansen, K. A., Licup, A. J., Sharma, A., et al. (2018). The role of network architecture in collagen mechanics. *Biophys. J.* **114**, 2665–2678.

Jawerth, L. M. (2013). The mechanics of fibrin networks and their alterations by platelets. PhD thesis, Harvard University.

Jiang, G., Liu, C., Liu, X., et al. (2010). Network structure and compositional effects on the tensile mechanical properties of hydrophobic association hydrogels with high mechanical strength. *Polymer* **51**, 1507–1515.

Johnson, K. L. (1985). *Contact mechanics*. Cambridge University Press, Cambridge.

Kabla, A. & Mahadevan, L. (2006). Nonlinear mechanics of soft fibrous networks. *J. R. Soc. Interface* **4**, 99–106.

Kang, H., Wen, Q., Janmey, P. A., et al. (2009). Nonlinear elasticity of stiff filament networks: Strain stiffening, negative normal stress and filament alignment in fibrin gels. *J. Phys. Chem. B* **113**, 3799–3805.

Kasza, K. E., Koenderink, G. H., Lin, Y. C., et al. (2009). Nonlinear elasticity of stiff biopolymers connected by flexible linkers. *Phys. Rev. E* **79**, 041928.

Komori, T. & Itoh, M. (1991). A new approach to the theory of the compression of fiber assemblies. *Text. Res. J.* **61**, 420–428.

Komori, T., Itoh, M. & Takaku, A. (1992). Model analysis of the compressibility of fiber assemblies. *Text. Res. J.* **62**, 567–574.

Kurniawan, N. A., van Kempen, T. H. S., Sonneveld, S., et al. (2017). Buffers strongly modulate fibrin self-assembly into fibrous networks. *Langmuir* **33**, 6342–6352.

Kwon, H. J., Rogalsky, A. D., Kovalchick, C. & Ravichandran, G. (2010). Application of digital image correlation method to biogel. *Poly. Eng. Sci.* **50**, 1585–1593.

Lake, S. P. & Barocas, V. H. (2011). Mechanical and structural contribution of non-fibrillar matrix in uniaxial tension: A collagen–agarose co-gel model. *Ann. Biomed. Eng.*, **39**, 1891–1903.

Lang, M. (2018). Elasticity of phantom model networks with cyclic defects. *ACS Macro Lett.* **7**, 536–539.

Larson, R. G. (1988). *Constitutive equations for polymer melts and solutions.* Butterworths, Stoneham, MA.

Larson, R. G. (1999). *The structure and rheology of complex fluids.* Oxford University Press, Oxford.

Licup, A. J., Munster, S., Sharma, A., et al. (2015). Stress controls the mechanics of collagen networks. *Proc. Nat. Acad. Sci.* **112**, 9573–9578.

Licup, A. J., Sharma, A. & MacKintosh, F. C. (2016). Elastic regimes of subisostatic athermal fiber networks. *Phys. Rev. E* **93**, 021407.

Lin, T. S., Wang, R., Johnson, J. A. & Olsen, B. D. (2019). Revisiting the elasticity theory of Gaussian phantom networks. *Macromolecules* **52**, 1685–1694.

Lin, Y. C., Yao, N. Y., Broedersz, C. P., et al. (2010). Origins of elasticity in intermediate filament networks. *Phys. Rev. Lett.* **104**, 058101.

Lindstrom, S. B., Vader, D. A., Kulachenko, A. & Weitz, D. A. (2010). Biopolymer network geometries: Characterization, regeneration and elastic properties. *Phys. Rev. E* **82**, 051905.

Ma, Y. H., Zhu, H. X., Su, B., Hu, G. K. & Perks, R. (2018). The elasto-plastic behavior of 3D stochastic fiber networks with crosslinkers. *J. Mech. Phys. Sol.* **110**, 155–172.

MacKintosh, F. C., Kas, J. & Janmey, P. A. (1995). Elasticity of semiflexible biopolymer networks. *Phys. Rev. Lett.* **24**, 4425–4428.

Mao, R., Goutianos, S., Tu, W., et al. (2017). Comparison of fracture properties of cellulose nanopaper, printing paper and buckypaper. *J. Mater. Sci.* **52**, 9508–9519.

Mark, J. E., Rahalkar, R. R. & Sullivan, J. L. (1979). Model networks of end-linked PDMS chains. III. Effect of the functionality of the cross links. *J. Chem. Phys.* **70**, 1794–1797.

Masse, J. P., Salvo, L., Rodney, D., Brechet, Y. & Bouaziz, O. (2006). Influence of relative density on the architecture and mechanical behavior of a steel metallic wool. *Scripta Mater.* **54**, 1379–1383.

Mauri, A., Ehret, A. E., Perrini, M., et al. (2015). Deformation mechanisms of human amnion: Quantitative studies based on second harmonic generation microscopy. *J. Biomech.* **48**, 1606–1613.

Maxwell, J. C. (1864). On the calculation of the equilibrium and stiffness of frames. *Phil. Mag.* **27**, 294–299.

Meng, F. & Terentjev, E. M. (2016). Nonlinear elasticity of semiflexible filament networks. *Soft Matter* **12**, 6749–6756.

Meng, L., Arnoult, O., Smith, M. & Wnek, G. E. (2012). Electrospinning of in-situ crosslinked collagen nanofibers. *J. Mater. Chem.* **22**, 19412.

Merson, J. & Picu, R. C. (2020). Size effects in random fiber networks controlled by the use of generalized boundary conditions. *Int. J. Sol. Struct.* **206**, 314–321.

Meunier, L., Chagnon, G., Favier, D., Orgeas, L. & Vacher, P. (2008). Mechanical experimental characterization and numerical modelling of an unfilled rubber. *Poly. Testing* **27**, 765–777.

Mezeix, L., Bouvet, C., Huez, J. & Poquillon, D. (2009). Mechanical behavior of entangled fibers and entangled cross-linked fibers during compression. *J. Mater. Sci.* **44**, 3652–3661.

Missel, A. R., Bai, M., Klug, W. S. & Levine, A. J. (2010). Affine–nonaffine transition in networks of nematically ordered semiflexible polymers. *Phys. Rev. E* **82**, 041907.

de Molina, P. M., Lad, S. & Helgeson, M. E. (2015). Heterogeneity and its influence on the properties of difunctional poly(ethylene glycol) hydrogels: Structure and mechanics. *Macromolecules* **48**, 5402–5411.

Motte, S. & Kaufman, L. J. (2012). Strain stiffening in collagen I networks. *Biopolymers* **99**, 35–46.

Moyo, D., Anandjiwala, R. D. & Patnaik, A. (2016). Micromechanics of hydroentangled nonwoven fabrics. *Textile Res. J.* **87**, 135–146.

Na, Y. H., Tanaka, Y., Kawauchi, Y., et al. (2006). Necking phenomenon in double network gels. *Macromolecules* **39**, 4641–4645.

Neckar, B. (1997). Compression and packing density of fibrous assemblies. *Textile Res. J.* **67**, 123–130.

Neckar, B. & Das, D. (2012). *Theory of structure and mechanics of fibrous assemblies.* Woodhead Publishing, New Delhi.

Negi, V. & Picu, R. C. (2021). Tensile behavior of non-crosslinked networks of athermal fibers in the presence of entanglements and friction. *Soft Matter* **17**, 10186–10197.

Nishi, K., Fujii, K., Chung, U. I., Shibayama, M. & Sakai, T. (2017). Experimental observation of two features unexpected from the classical theories of rubber elasticity. *Phys. Rev. Lett.* **119**, 267801.

Niskanen, K. (ed.) (1998). *Paper physics.* Tappi Press, Helsinki.

Niskanen, K. (ed.) (2012). *Mechanics of paper products.* De Gruyter, Berlin.

Oliver, W. C. and Pharr, G. M. (2004). Measurement of hardness and elastic modulus by instrumented indentation: advances in understanding and refinements to methodology. *J. Mater. Res.* **19**, 3–20.

Oeser, R., Ewen, B., Richter, D. & Fargo. B. (1988). Dynamic fluctuation of crosslinks in a rubber: A neutron-spin-echo study. *Phys. Rev. Lett.* **60**, 1041–1044.

Onck, P. R., Koeman, T., van Dillen, T. & van der Giessen, E. (2005). Alternative explanation of stiffening in cross-linked semiflexible networks. *Phys. Rev. Lett.* **95**, 178102.

Ostoja-Starzewski, M. & Stahl, D. C. (2000). Random fiber networks and special elastic orthotropy of paper. *J. Elast.* **60**, 131–149.

Ovaska, M., Bertalan, Z., Miksic, A., et al. (2017). Deformation and fracture of echinoderm collagen networks. *J. Mech. Beh. Biomed. Mater.* **65**, 42–52.

Oyen, M. L. (2014). Mechanical characterization of hydrogel materials. *Int. Mater. Rev.* **59**, 44–59.

Patel, P. C. & Kothari, V. K. (2001). Effect of specimen size and strain rate on the tensile properties of heat-sealed and needlepunched nonwoven fabrics. *Indian J. Fibr. Text. Res.* **26**, 409–413.

Perrini, M., Mauri, A., Ehret, A. E., et al. (2015). Mechanical and microstructural investigation of the cyclic behavior of human amnion. *J. Biomech. Eng.* **137**, 061010.

Picu, R. C., Deogekar, S. & Islam, M. R. (2018). Poisson's contraction and fiber kinematics in tissue: Insight from collagen network simulations. *J. Biomech. Eng.* **140**, 021002.

Picu, R. C. & Pavel, M. C. (2003). Scale invariance of the stress production mechanism in polymeric systems. *Macromolecules* **36**, 9205–9215.

Piechocka, I. K., van Oosten, A. S. G., Breuls, R. G. M. & Koenderink, G. H. (2011). Rheology of heterotypic collagen networks. *Biomacromolecules* **12**, 2797–2805.

Poquillon, D., Viguier, B. & Andrieu, E. (2005). Experimental data about mechanical behavior during compression tests for various matter fibers. *J. Mater. Sci.* **40**, 5963–5970.

Poynting, J. H. (1909). On pressure perpendicular to the shear planes in finite pure shears, and on the lengthening of loaded wires when twisted. *Proc. R. Soc. London A* **82**, 546–559.

Puleo, G. L., Zulli, F., Piovanelli, M., et al. (2013). Mechanical and rheological behavior of pNIPAAM crosslinked macrohydrogel. *React. Funct. Polym.* **73**, 1306–1318.

Räisänen, V. I., Alava, M. J., Nieminen, R. M. and Niskanen, K. J. (1996). Elastic–plastic behaviour in fibre networks. *Nordic Pulp Paper Res. J.* **11**, 243–248.

Rault, J., Marchal, J., Judeinstein, P. & Albouy, P. A. (2006). Stress-induced crystallization and reinforcement in filled natural rubbers: ^{2}H NMR study. *Macromolecules* **39**, 8356–8368.

Rigdahl, M. & Hollmark, H. (1986). Network mechanics. In *Paper structure and properties*, J. A. Bristow & P. Kolseth, eds. Marcel Dekker, New York, pp. 241–266.

Roberts, A. P. & Garboczi, E. J. (2002). Elastic properties of model random 3D open-cell solids. *J. Mech. Phys. Sol.* **50**, 33–55.

Rodney, D., Gadot, B., Martinez, O. R., Rolland du Roscoat, S. & Orgeas, L. (2015). Reversible dilatancy in entangled single-wire materials. *Nature Mat.* **15**, 72–77.

Ronca, G. & Allegra, G. (1975). An approach to rubber elasticity with internal constraints. *J. Chem. Phys.* **63**, 4990–4998.

Rubinstein, M. & Colby, R. H. (2003). *Polymer physics.* Oxford University Press, Oxford.

Rubinstein, M. & Helfand, E. (1985). Statistics of the entanglement of polymers: Concentration effects. *J. Chem. Phys.* **82**, 2477–2483.

Ruland, A., Chen, X., Khansari, A., et al. (2018). A contactless approach for monitoring the mechanical properties of swollen hydrogels. *Soft Matt.* **14**, 7228–7236.

Schofield, J. (1938). Research on wool felting. *J. Textile Inst.* **29**, T239–T252.

Schraad, M. W. & Harlow, F. H. (2006). A stochastic constitutive model for disordered cellular materials: Finite strain uniaxial compression. *Int. J. Sol. Struct.* **43**, 3542–3568.

Seth, R. S. & Page, D. H. (1981). The stress–strain curve of paper. In *The role of fundamental research in paper making: Transactions of the symposium held at Cambridge, September 1981*, 2nd ed. (January 1, 1983). Mechanical Engineering Publications Limited, London, pp. 421–452.

Shahsavari, A. S. & Picu, R. C. (2013). Size effect on mechanical behavior of random fiber networks. *Int. J. Sol. Struct.* **50**, 3332–3338.

Sharma, A., Licup, A. J., Rens, R., et al. (2016). Strain-driven criticality underlines non-linear mechanics of fibrous networks. *Phys. Rev. E* **94**, 042407.

Storm, C., Pastore, J. J., MacKintosh, F. C., Lubensky, T. C. & Janmey, P. A. (2005). Nonlinear elasticity in biological gels. *Nature* **435**, 191–194.

Subramanian, G. & Picu, R. C. (2011). Mechanics of three-dimensional, nonbonded random fiber networks. *Phys. Rev. E* **83**, 056120.

Sun, T. L., Kurokawa, T., Kuroda, S., et al. (2013). Physical hydrogels composed of polyampholytes demonstrate high toughness and viscoelasticity. *Nature Mater.* **12**, 932–937.

Swane, G. T. G., Smaje, L. H. & Bergel, D. H. (1989). Distensibility of single capillaries and venules in the rat and frog mesentery. *Int. J. Microcirc. Clin. Exp.* **8**, 25–42.

Tharmann, R., Claessens, M. M. A. E. & Bausch, A. R. (2007). Viscoelasticity of isotropically cross-linked actin networks. *Phys. Rev. Lett.* **98**, 088103.

Ting, T. C. T. & Chen, T. (2005). Poisson's ratio for anisotropic elastic materials can have no bounds. *Q. J. Mech. Appl. Math.* **58**, 73–82.

Toll, S. (1998). Packing mechanics of fiber reinforcements. *Polym. Eng. Sci.* **38**, 1337–1350.

Toll, S. & Manson, J. A. E. (1995). The elastic compression of a fiber network. *J. Appl. Mech.* **62**, 223–228.

Treloar, L. R. G. (1944). Stress–strain data for vulcanized rubber under various types of deformation. *Trans. Faraday Soc.* **40**, 59–70.

Vader, D., Kabla, A., Weitz, D. & Mahadevan L. (2009). Strain-induced alignment in collagen gels. *PLoS ONE* **4**, e5902.

Vasiliev, V. G., Rogovina, L. Z. & Slonimsky, G. L. (1985). Dependence of properties of swollen and dry polymer networks on the conditions of their formation in solution. *Polymer* **26**, 1667–1676.

Verhille, G., Moulinet, S., Vandenberghe, N., Adda-Bedia, M., & LeGal, P. (2017). Structure and mechanics of aegagropilae fiber network. *Proc. Nat. Acad. Sci.* **114**, 4607–4612.

Ward, I. M. & Sweeney, J. (2013). *Mechanical properties of solid polymers*, 3rd ed. Wiley, Chichester, UK.

Welling, L. W., Zupka, M. T. & Welling, D. J. (1995). Mechanical properties of basement membrane. *Physiology* **10**, 30–35.

Wilhelm, J. & Frey, E. (2003). Elasticity of stiff polymer networks. *Phys. Rev. Lett.* **91**, 108103.

Wong, D., Andriyana, A., Ang, B. C., et al. (2019). Poisson's ratio and volume change accompanying deformation of randomly oriented electrospun nanofibrous membranes. *Plast. Rubber Comp.* **48**, 456–465.

van Wyk, C. M. (1946). Note on the compressibility of wool. *J. Textile Inst.* **37**, T285–T292.

Yohsuke, B., Urayama, K., Takigawa, T. & Ito, K. (2011). Biaxial strain testing of extremely soft polymer gels. *Soft Matter* **7**, 2632–2638.

Zagar, G., Onck, P. R., & van der Giessen, E. (2011). Elasticity of rigidly cross-linked networks of athermal filaments. *Macromolecules* **44**, 7026–7033.

Zagar, G., Onck, P. R. & van der Giessen, E. (2015). Two fundamental mechanisms govern the stiffening of crosslinked networks. *Biophys. J.* **108**, 1470–1479.

Zhang, M., Chen, Y., Chiang, F. P., Gouma, P. I. & Wang, L. (2019). Modeling the large deformation and microstructure evolution of nonwoven polymer fiber networks. *J. Appl. Mech.* **86**, 011010.

Zhu, H. X., Hobdell, J. R. & Windle, A. H. (2000). Effects of cell irregularity on the elastic properties of open-cell foams. *Acta Mater.* **48**, 4893–4900.

Zrinyi, M. & Horkay, F. (1987). On the elastic modulus of swollen gels. *Polymer* **28**, 1139–1143.

7 Constitutive Formulations for Network Materials

7.1 Are Continuum Models Appropriate for Network Materials?

Constitutive models define the relation between stress and strain or, in a more general sense, between thermodynamic forces and fluxes. They provide the closure of the system of equations required for the solution of boundary value problems. Historically, the development of constitutive models for various engineering materials made possible the prediction of the behavior of components and systems, as needed in design.

Network materials are composed from distinct fibers and are not continua. Therefore, it is appropriate to ask whether a continuum model is adequate for such structures.

Obviously, a metallic polycrystal is not a continuum either, but continuum models are adequate and can be used for such materials. This is possible due to the large difference between the scale at which the discrete nature of the material manifests itself (the atomic scale, in the case of a metallic polycrystal) and the scale at which continuum fields are defined. It is necessary to discuss the limitations of the applicability of continuum models to fiber networks, including the issue of scale separation:

(1) Stochastic networks are highly heterogeneous assemblies of fibers. The degree of heterogeneity is discussed in Section 4.3, where it is shown, as an example, that the auto-correlation function of the local density of a specific type of paper is a power function (Figure 4.6). This indicates that the respective network is a fractal object over a range of scales bounded above by a limit that corresponds to the cut-off of the auto-correlation function. This cut-off is of the order of several fiber lengths.

Consider a 2D Mikado network constructed by depositing in a 2D domain fibers of length L_0, with random center of mass positions and random orientations. The resulting structure is imaginarily divided using a grid of square elements and the density is evaluated within each element. The density in this model is spatially power law correlated and Figure 4.6 shows that similar density correlations are obtained experimentally. Further, it is possible to evaluate the local effective small strain stiffness of each square element of the grid and compute the auto-correlation function of the resulting stiffness map. Figure 7.1 shows that, after normalization, the stiffness and the density auto-correlation functions overlap

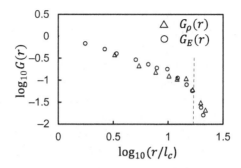

Figure 7.1 Two-point auto-correlation function of the density, $G_\rho(r)$, and of the elastic modulus, $G_E(r)$, for a 2D Mikado network probed with a square grid of size $2.2l_c$. The vertical dashed line marks the fiber length, L_0/l_c. Data from Hatami-Marbini and Picu (2009)

(Hatami-Marbini and Picu, 2009). This result is obtained by using a probing length scale (size of the square elements of the probing grid) approximately twice the mean segment length of the network. The one-to-one correspondence between local density and local stiffness indicates that structural fluctuations produce fluctuations of the mechanical fields. This correlation between deformation fields and structural heterogeneity was established experimentally for paper (e.g., Ostoja-Starzewski and Castro, 2003). A fractal-heterogeneous continuum would be the adequate representation of these networks.

(2) The characteristic length scales of the microstructure evolve during loading. The discussion at (1) suggests that a continuum homogenized representation may be adequate on scales larger than the cut-off of the auto-correlation functions shown in Figure 7.1, or larger than any length scale of heterogeneity associated with flocculation. This defines the size of the "representative volume element" (RVE) of the network. Obviously, the RVE size must be much smaller than the size of the continuum problem domain for the constitutive formulation to be useful. This situation is similar to that encountered in stochastic composites.

The significant difference between network materials and stochastic continuum composites is that the microstructure of the network changes significantly during deformation. Fiber alignment in the principal stretch direction begins at relatively small strains and continues until the formation of a system-spanning and load-bearing subnetwork of aligned fibers (Section 6.1.1.4.12). While structural correlations in the undeformed state have limited range, of one or several L_0 (several L_p in the case of tortuous fibers), the range of correlations in the deforming network increases rapidly as the load-carrying subnetwork emerges and becomes comparable with the sample size. The structure of this subnetwork depends on the architecture of the undeformed network and the specific boundary conditions applied.

This structural evolution has been identified early in the research on networks as the cause of long-range effects leading to the failure of the Saint-Venant principle (Thirlwell and Treloar, 1965). More recently it was observed that cells seeded on nonwoven fiber networks interact at large distances via the mechanical fields produced by their contractile action (Ma et al., 2013). Cell contractility leads to the formation of a load-carrying subnetwork, which extends the spatial correlations of the stiffness and kinematic fields way beyond the limit shown in Figure 7.1. The emergence of such long-range structural and field correlations is difficult to be predictively captured by a continuum model.

(3) Local instabilities are often observed during the macroscopically stable deformation regime of sparsely crosslinked networks. Such instabilities may produce local rearrangement of fibers. The occurrence of local instabilities is favored by the significant geometric nonlinearity of the network. Constitutive models generally homogenize over localization processes to predict a smooth macroscopic deformation field.

(4) Networks exhibit a pronounced Poisson effect (Section 6.1.1.4.4). An incremental Poisson ratio much larger than 0.5 is commonly observed in tension if the network is not embedded in a solid matrix (e.g., Picu et al., 2018). Auxetic behavior is also occasionally reported, primarily in sparsely crosslinked or non-crosslinked nonwovens (e.g., Wong et al., 2019a). Thermodynamic considerations impose limits on the values the Poisson ratio may take in isotropic materials and, hence, an isotropic elastic continuum with an arbitrary Poisson ratio is not conceivable. The upper limit of the allowable range of the Poisson ratio is eliminated in the presence of anisotropy (Ting and Chen, 2005). Large transverse contractions are associated with preferential fiber alignment which leads to structural anisotropy. However, the level of elastic anisotropy that develops is too low to justify Poisson ratios as large as those observed in experiments with some network materials. The representation of this aspect of the network mechanical behavior using continuum models is problematic.

Most of these issues can be understood based on the physical picture described in Section 6.1.1. In regime II of network deformation (Figure 6.1(a)), which corresponds to rapid strain stiffening, the network can be viewed as a mechanism. The deformation in regimes I and III is more amenable to a continuum description.

Continuum models for polymeric networks above the glass transition temperature have been developed over the last century. These networks (e.g., rubber, epoxy) do not have large free volume and molecules are densely packed. Strong excluded volume interactions restrict the Poisson ratio to the usual range $\nu \leq 0.5$. Hence, the dense nature of the network reduces to some extent the concerns outlined in this section. This justifies why models developed for molecular networks are generally much better representations of the observed physical behavior than their counterparts developed for nonwovens and other networks with large free volume.

7.2   Classification of the Current Constitutive Models for Network Materials

The behavior of common network materials presented in Chapter 6 is diverse. Molecular crosslinked networks, such as rubber and gels, exhibit hyperelastic behavior. Damage accumulation under large loads leads to the Mullins effect commonly seen in rubber composites and some athermal networks. Collagen networks in connective tissue and blood vessels also exhibit hyperelastic response, with little hysteresis and virtually no damage accumulation when loaded at physiological strain levels. Sparsely crosslinked nonwovens exhibit hyperelastic–plastic-type behavior, with strain stiffening during continuous loading, residual strains upon unloading, and hysteresis during cyclic loading. Densely crosslinked nonwovens and cellulose networks exhibit linear elastic behavior up to a stress level beyond which softening due to damage accumulation and rupture are observed.

Several scientific communities contributed to the development of constitutive models for network materials. The first works focused on rubber elasticity (thermal networks) (Wang and Guth, 1952) and on paper (athermal networks) (Cox, 1952) and led to hyperelastic and linear elastic models, respectively. Later work was dedicated to cellulose networks, nonwovens, and, more recently, biological tissues.

The available constitutive models are either phenomenological or mechanism based. Phenomenological models have thermodynamically-informed functional forms and are inspired by and fitted to experimental data (Humphrey et al., 1990). These models do not represent the microstructure or the deformation mechanisms. The mechanism-based models attempt to capture key aspects of the deformation mechanisms and generally include information about the microstructure and its evolution.

Phenomenological hyperelastic models predict the stress tensor using Eq. (5.8) based on a free energy function expressed in terms of the invariants of tensor $\mathbf{C}$. A large variety of such models exist, with slightly different versions being developed by different scientific communities. Several of the commonly used models of this type are reviewed in Section 7.3.

The current mechanism-based models for network materials associate stress with the deformation of individual fibers. Most models assume that only fiber stretch takes place during deformation. A small subset of models account for other modes of fiber deformation such as bending and shear. Of the various parameters describing the network architecture, the fiber orientation is considered the most important microstructural parameter. A number of models account for other microstructural parameters, such as the distribution of fiber crimp and the distribution of fiber length.

Despite ample evidence of the importance of nonaffine deformation in network mechanics, the majority of mechanism-based models make the assumption that fibers move and deform affinely with the macro deformation. The generic structure of such models is presented in Section 5.3. The application of this formulation to specific network materials is presented in Section 7.4. Several attempts have been made to account for some of the mechanisms causing nonaffinity within the general structure of affine models. This led to another class of models which are denoted here as quasi-affine and which are reviewed in Section 7.5. Only a handful of constitutive formulations consider nonaffinity as the central aspect of network mechanics and attempt to capture this behavior without solving the entire discrete network problem. These are outlined in Section 7.6.

7.3   Phenomenological Models

7.3.1   Hyperelastic Models

Phenomenological constitutive models are developed within continuum mechanics considering restrictions on the allowable functional forms imposed by thermodynamics and material symmetry considerations. Eq. (5.8) allows for the evaluation of stress if the free energy density is defined as a function of $\mathbf{C}$ or its invariants. The invariants can be written in terms of the principal stretches λ_i:

$$
\begin{aligned}
I_1^C &= \text{tr}(\mathbf{C}) = \lambda_1^2 + \lambda_2^2 + \lambda_3^2, \\
I_2^C &= \frac{1}{2}\left[(\text{tr}(\mathbf{C}))^2 - \text{tr}(\mathbf{C}^2)\right] = \lambda_1^2\lambda_2^2 + \lambda_1^2\lambda_3^2 + \lambda_2^2\lambda_3^2, \\
I_3^C &= \det(\mathbf{C}) = \det(\mathbf{F}^T\mathbf{F}) = J^2 = \lambda_1^2\lambda_2^2\lambda_3^2.
\end{aligned}
\tag{7.1}
$$

Since $\mathbf{C}$ and $\mathbf{B}$ have identical eigenvalues, they also have identical invariants. This allows dropping of the superscript C of I in Eq. (7.1) to simplify the invariant notation to I_1, I_2, I_3. Therefore, the free energy density per unit mass is written as $\Psi(I_1, I_2, I_3, T)$. The PK2 stress can be computed using Eq. (5.8):

$$
\mathbf{\Pi} = 2\rho_0\left[\left(\frac{\partial\Psi}{\partial I_1} + I_1\frac{\partial\Psi}{\partial I_2}\right)\mathbf{I} - \frac{\partial\Psi}{\partial I_2}\mathbf{C} + \frac{\partial\Psi}{\partial I_3}I_3\mathbf{C}^{-1}\right].
\tag{7.2}
$$

To derive this expression from Eq. (5.8), the following identities were used:

$$
\frac{\partial I_1}{\partial \mathbf{C}} = \mathbf{I}; \quad \frac{\partial I_2}{\partial \mathbf{C}} = I_1\mathbf{I} - \mathbf{C}; \quad \frac{\partial I_3}{\partial \mathbf{C}} = I_3\mathbf{C}^{-1}.
\tag{7.3}
$$

The Cauchy stress is computed starting from $\mathbf{\Pi}$ given by Eq. (7.2) and using Eq. (5.10) and results:

$$
\boldsymbol{\sigma} = 2\rho\left[\left(\frac{\partial\Psi}{\partial I_1} + I_1\frac{\partial\Psi}{\partial I_2}\right)\mathbf{B} - \frac{\partial\Psi}{\partial I_2}\mathbf{B}^2 + \frac{\partial\Psi}{\partial I_3}I_3\mathbf{I}\right].
\tag{7.4}
$$

In the incompressible case $J = I_3 = 1$ the free energy is a function of the first two invariants only, $\Psi(I_1, I_2, T)$, and only the first two terms remain in Eq. (7.4). The Cauchy stress can be rearranged using the Cayley–Hamilton equation: $\mathbf{B}^3 - I_1\mathbf{B}^2 + I_2\mathbf{B} - I_3\mathbf{I} = 0$ which, after multiplication with $\mathbf{B}^{-1}$ yields: $\mathbf{B}^2 = I_1\mathbf{B} + I_3\mathbf{B}^{-1} - I_2\mathbf{I}$. Replacing this expression in Eq. (7.4) and ignoring the hydrostatic term, $\boldsymbol{\sigma}$ results:

$$
\boldsymbol{\sigma} = 2\rho\left[\frac{\partial\Psi}{\partial I_1}\mathbf{B} - \frac{\partial\Psi}{\partial I_2}\mathbf{B}^{-1}\right] - P\mathbf{I}.
\tag{7.5}
$$

The hydrostatic term $P\mathbf{I}$ does not result from energetic considerations, rather it represents a constraint enforcing the isochoric deformation restriction.

Some network materials are approximately incompressible. A special formulation was developed for these cases in which the volumetric and isochoric deformations are derived from separate potentials. It is convenient to assume that the free energy can be written as a superposition of a component dependent on J (or I_3; $I_3 = J^2$) and a component associated with the isochoric deformation defined in terms of the deviatoric component of $\mathbf{C}$, $\overline{\mathbf{C}}$:

$$\Psi = \Psi_{iso}\left(\overline{\mathbf{C}}, T\right) + \Psi_v(J, T). \tag{7.6}$$

$\overline{\mathbf{C}}$ is related to $\mathbf{C}$ as $\mathbf{C} = J^{2/3}\overline{\mathbf{C}}$, Eq. (5.7).

The PK2 stress results:

$$\mathbf{\Pi} = 2\rho_0\left[\frac{\partial\Psi_v}{\partial J}\frac{\partial J}{\partial\mathbf{C}} + \frac{\partial\Psi_{iso}}{\partial\overline{\mathbf{C}}} : \frac{\partial\overline{\mathbf{C}}}{\partial\mathbf{C}}\right] = \rho_0 J\frac{\partial\Psi_v}{\partial J}\mathbf{C}^{-1}$$
$$+ 2\rho_0 J^{-2/3}\frac{\partial\Psi_{iso}}{\partial\overline{\mathbf{C}}} - \frac{2}{3}\rho_0 J^{-2/3}\left(\overline{\mathbf{C}} : \frac{\partial\Psi_{iso}}{\partial\overline{\mathbf{C}}}\right)\overline{\mathbf{C}}^{-1}. \tag{7.7}$$

and the Cauchy stress can be computed using Eq. (5.10):

$$\boldsymbol{\sigma} = \rho J\frac{\partial\Psi_v}{\partial J}\mathbf{I} + 2\rho\left[\overline{\mathbf{F}}\frac{\partial\Psi_{iso}}{\partial\overline{\mathbf{C}}}\overline{\mathbf{F}}^T - \frac{1}{3}\left(\overline{\mathbf{C}} : \frac{\partial\Psi_{iso}}{\partial\overline{\mathbf{C}}}\right)\mathbf{I}\right], \tag{7.8}$$

where the first term represents the energetic hydrostatic stress, while the second is the stress associated with the isochoric deformation.

These expressions allow for the evaluation of the stress tensor once a functional form is selected for the free energy. Rivlin (1948) introduced a general invariant-based energy function:

$$\Psi = \sum_{i,j}a_{ij}(I_1 - 3)^i(I_2 - 3)^j, \tag{7.9}$$

where a_{ij} are phenomenological coefficients, which may be temperature-dependent, and which are to be determined by fitting the constitutive model to experimental data.

A particular form of Eq. (7.9) is the Mooney–Rivlin model (Mooney, 1940; Rivlin, 1948) which has the energy function:

$$\Psi = \alpha_1(I_1 - 3) + \alpha_2(I_2 - 3). \tag{7.10}$$

The Cauchy stress can be computed with Eq. (7.4) as:

$$\boldsymbol{\sigma} = 2\rho\left[(\alpha_1 + \alpha_2 I_1)\mathbf{B} - \alpha_2\mathbf{B}^2\right]. \tag{7.11}$$

The small strain shear modulus results as $2(\alpha_1 + \alpha_2)$.

A particular form of the Mooney–Rivlin model known as the neo-Hookean model is obtained from Eqs. (7.10) and (7.11) by making $\alpha_2 = 0$. The neo-Hookean model is equivalent to the strict affine model described in Section 5.3. It is customary in rubber elasticity to evaluate an experimental data set in terms of these two simple constitutive equations. Since the neo-Hookean model corresponds to the affine deformation, the second term in the Mooney–Rivlin model is interpreted as an indicator of the degree of nonaffinity of the deformation. Hence, data sets can be compared based on the ratio of the second to the first coefficients of the Mooney–Rivlin model, that is, α_2/α_1 (Eq. (7.10)).

The Yeoh model (Yeoh, 1990, 1993) was developed for carbon black-reinforced rubber and was used subsequently to model the mechanical behavior of various tissues such as breast tissue and the lung parenchyma. It depends exclusively on the first invariant and the free energy for the compressible version reads:

$$\Psi = \alpha_1(I_1 - 3) + \alpha_2(I_1 - 3)^2 + \alpha_3(I_1 - 3)^3. \tag{7.12}$$

The Cauchy stress results:

$$\boldsymbol{\sigma} = 2\rho\left[\left(\alpha_1 + 2\alpha_2(I_1 - 3) + 3\alpha_3(I_1 - 3)^2\right)\mathbf{B}\right]. \tag{7.13}$$

Other models in which the free energy function has forms which are not polynomials of the invariants have been proposed. Of these, we mention only the model proposed in Delfino et al. (1997), which is used in biomechanics and is defined by the free energy expression:

$$\Psi = \frac{\alpha_1}{\alpha_2}\left\{\exp\left[\frac{\alpha_2}{2}\left(\overline{I}_1 - 3\right)\right] - 1\right\}, \tag{7.14}$$

and the Ogden model (Ogden, 1984), which is also broadly used, and has an energy function of the form:

$$\Psi = \sum_i \frac{\beta_i}{\alpha_i}\left(\lambda_1^{\alpha_i} + \lambda_2^{\alpha_i} + \lambda_3^{\alpha_i} - 3\right). \tag{7.15}$$

Retaining only the first term in the sum of Eq. (7.15), with $\alpha_1 = 2$, one recovers the incompressible neo-Hookean model. Taking two terms with $\alpha_1 = 2$ and $\alpha_2 = -2$, the incompressible Mooney–Rivlin model results.

Many models have been developed to represent the complexities of the nonlinear elastic deformation of rubber, some incorporating physical concepts and some being purely phenomenological. Along the line of the discussion in this section, we also mention the "localization model" (Gaylord and Douglas, 1990) and the "constrained chain" model developed based on the earlier junction fluctuation model (Flory and Erman, 1982) by assuming that constraints affect the entire representative chain rather than just the crosslinks. These models provide predictions in agreement with experimental results in tension and compression – see Han et al. (1999) for a comparison of various models. The constrained junction model is reviewed in Section 7.6.1. The localization model is based on a free energy function of the form $\Psi = \alpha_1(I_1 - 3) + \alpha_2(\lambda_1 + \lambda_2 + \lambda_3 - 3)$ in which the first term is neo-Hookean, while the second term represents the confining effect due to chain entanglements.

This brief account is not exhaustive and does not represent all aspects of phenomenological constitutive models developed for hyperelastic materials. The interested reader may consult reviews such as Holzapfel et al. (2000), which is focused on biomechanics applications, and Boyce and Arruda (2000), which discusses applications to rubber, as well as more general purpose monographs dedicated to nonlinear mechanics, such as Ogden (1984), Holzapfel (2000), and Gurtin et al. (2010).

Furthermore, the models reviewed here do not capture internal dissipation and damage. They are generally applicable to a broad class of network materials such as rubber, gels, and biological tissue at levels of stress that do not cause extensive damage but are not expected to represent accurately the constitutive behavior of textiles, nonwovens, and nanofiber mats.

7.3.2 Elastic–Plastic Models

Networks exhibit inelastic deformation, which is primarily associated with damage and fiber plasticity. Damage appears as crosslink failure, sliding of fibers at crosslinks without detachment, and fiber failure, with the first mechanism being dominant in the majority of networks. In non-crosslinked fiber assemblies (e.g., needle-punched non-wovens), inelastic deformation is associated with the rearrangement of fibers and sliding at fiber–fiber nonbonded contacts. This causes residual strains and hysteresis, and generally nonlinear behavior on the network scale, which increases the nonlinearity associated with the evolution of the fibrous structure.

This phenomenology is also observed in metals, polymers, and composites. Well-established phenomenological models developed for these other engineering materials have been used to represent the inelastic deformation of network materials. Reviews of continuum mechanics constitutive models abound in the literature. The interested reader should consult Gurtin et al. (2010) for a detailed account of elastic–plastic phenomenological models.

In relation to network materials, phenomenological models that account for inelasticity are used more often in the paper and cellulose products literature. In the presence of humidity, paper creeps under constant stress and deforms plastically in monotonic tests. Nonwovens and paper exhibit hysteresis in cyclic loading (Figure 6.2). Consequently, phenomenological models of elastic–plastic (Xia et al., 2002; Castro and Ostoja-Starzewski, 2003; Mäkelä and Östlund, 2003) and visco-plastic (Coffin, 2008) types have been used for this class of materials.

7.3.3 Micropolar and Nonlocal Models

Continuum theories generally consider that stress at a point depends exclusively on the strain at the respective point. Alternative theories in which internal rotations are added to the description of the kinematics have been developed. These are associated with internal higher order stresses known as couple stresses. The couple stress concept was introduced by Voigt (1887) and was extended by the Cosserat brothers (Cosserat and Cosserat, 1909) and later by many other researchers; reviews are available in Schaefer (1967) and Nowacki (1986).

In linear elasticity, the usual Cauchy stress, σ, is related to the strain, ε $\left(\varepsilon_{ij} = 1/2\left(u_{i,j} + u_{j,i}\right)\right)$, through the linear Hooke's law. The rotation ω $\left(\omega_{ij} = 1/2\left(u_{i,j} - u_{j,i}\right)\right)$ is not associated with stress. In the Voigt version of the couple stress theory, the gradient of the rotation is associated with the couple stress. The balance of linear momentum leads to the usual equilibrium equation for σ. The balance of angular momentum leads to an equivalent equation for the couple stress.[1] In the micropolar theory of Cosserat (Cosserat and Cosserat, 1909) and Eringen (1966), the internal rotation is entirely decoupled from the displacement field and is considered an independent field.

In another class of models, the locality assumption of the classical formulation is replaced with the assumption that stress at a given material point depends on the deformation (strain) in an entire neighborhood of the respective point (Eringen, 1972). These are known as nonlocal models. Specifically, $\sigma(\mathbf{x}) = \int \tilde{\mathbf{C}}_{nl}(\mathbf{y} - \mathbf{x}) : \varepsilon(\mathbf{y}) d\mathbf{y}$, where the integral is evaluated over a vicinity of point $\mathbf{x}$ and $\tilde{\mathbf{C}}_{nl}$ is a stiffness kernel which decays to zero for large values of the argument. The kernel defines the range of nonlocality and may be considered, for example, an exponential function.

A special case of nonlocality is provided by the gradient models (Altan and Aifantis, 1992). In this representation, both the strain, ε, and its gradient, $\varepsilon \otimes \nabla$, appear as independent kinematic descriptors. The Cauchy stress becomes a function of ε and of the strain gradient. A hyper-stress is introduced which is associated with the strain gradient. The balances of linear and angular momenta are expressed in terms of the Cauchy stress and the hyper-stress. A certain type of gradient model results as a particular case of Eringen's integral form of nonlocality. To obtain the gradient version, one may take the Fourier transform of the integral expression of $\sigma(\mathbf{x})$, expand the Fourier transform of kernel $\tilde{\mathbf{C}}_{nl}$ in series and retain only the first two terms, then transform back to obtain the corresponding gradient version of the elastic constitutive equation.

In both micropolar and gradient formulations, stress is related linearly to the internal rotation or the strain gradient fields. The strain gradient has units of inverse length. Therefore, the higher order elastic constants must have units of stress times length. It results that the formulation includes an internal length scale which, ideally, should be related to length scales of the material microstructure.

Micropolar models account for the existence of moments in the microstructure. Consider a volume element artificially separated from the material. The usual view in the Cauchy continuum is that only forces act on the faces of the volume element and they are directly related to the Cauchy stress. If the material microstructure is such that moments are also present at the volume element boundaries, these lead to a higher order stress of the type introduced in micropolar theories. Clearly, network materials are of this type.

An example is shown in Figure 7.2. Figure 7.2(a) shows a schematic of a network loaded in uniaxial tension. The problem domain is divided in volume elements and one of these is shown on the right side of the figure. Forces and moments transmitted by all fibers crossing the boundaries of the volume element sum up to tractions of finite value. A continuum model aiming to represent this situation must be micropolar.

As the size of the probing volume (δ) increases, the unbalanced moments, M_δ, acting along the boundaries of the element decrease and the magnitude of the force per

---

[1] Compare with the usual local theory in which conservation of angular momentum leads to the symmetry of the stress tensor.

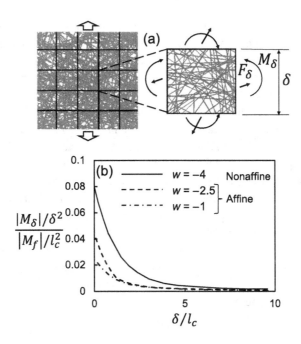

Figure 7.2 (a) Schematic of a volume element separated from a network. Forces and moments act along the boundaries of the volume element. (b) Variation with the size of the volume element, δ, of $\left(|M_\delta|/\delta^2\right)/\left(|M_f|/l_c^2\right)$ representing the mean specific moment acting on one of the faces of the volume element normalized by the mean moment carried by individual fibers.

unit area of the volume element boundary asymptotes to the far field Cauchy stress. Figure 7.2(b) shows the variation with the probing length δ, of the specific moment, $|M_\delta|/\delta^2$, acting along an edge of a volume element of size δ. The specific moment is normalized by the mean moment carried by individual fibers, $|M_f|/l_c^2$. The analysis is performed using 3D Voronoi networks of different $w = \log_{10}\left(\rho l_b^2\right)$ subjected to uniaxial far field tension. The network with $w = -4$ is strongly nonaffine, while network deformation is approximately affine for $w \geq -2.5$. The specific moment becomes negligible once $\delta \approx 5l_c$. This threshold δ provides an indirect indication of the magnitude of the internal length scale a continuum micropolar model representative for the mechanics of this network should have.

Note that, under uniaxial tension, that is, in the absence of imposed field gradients, there is no expectation of micropolar effects being activated in a homogeneous (micropolar) continuum. In order to trigger internal micropolar effects under such far field loading (i.e., a field with no nominal strain gradients), it is necessary to consider a heterogeneous continuum with micropolar constitutive description. Therefore, the nonlocality and heterogeneity problems are coupled in the case of stochastic networks.

Micropolar and gradient models have been used to represent the small strain response of fiber networks in Berkache et al. (2019a, 2019b) and Tyznik and Notbohm (2019). These works infer values of the internal length scales for the respective continuum models ranging from 1 to 10 fiber segment lengths, which is consistent with the expectation based on the results shown in Figure 7.2(b). The internal length scale depends on whether the network behavior is controlled by the axial or the bending deformation modes of fibers, and takes larger values when the network is softer, bending dominated, and more nonaffine. Micropolar and nonlocal models adequate for the large deformation of fibrous materials are not available at this time.

Micropolar effects become important in the presence of applied fields with nonzero gradients when the inverse of the strain gradient (which has units of length) is comparable with the internal length scale of the micropolar formulation. The small strain behavior of the network in the vicinity of crack tips, notches, point forces, and other stress concentrators cannot be captured with a classical continuum model and a micropolar description is required. An application to network materials (paper) within the limits of small strain theory is described in Isaksson and Hagglund (2009). However, further conceptual development is needed before generalized continuum models can be used to represent the mechanics of networks (including the large deformations regime) on scales comparable with the fiber length, fiber persistence length, or the mean segment length.

7.4   Affine Models

7.4.1   Strict Affine Models

Under the strict affine deformation assumption all points of the network move as described by the macroscale deformation gradient $\mathbf{F}$. A formulation based on this model is derived in Section 5.3. Here we review the affine models used in various contexts.

The majority of the available constitutive models for network materials are based on the affine deformation assumption. They follow the lead of the first models developed for such materials: the athermal model developed by Cox for cellulose networks (Cox, 1952) and that developed for thermal networks (rubber) by Wang and Guth (1952).[2] Cox's model assumes a 2D distribution of infinite fibers with random orientations, each fiber experiencing the far field strain and contributing to the stress independently from the other fibers. In Cox's model fibers are considered linear elastic and are loaded axially. Wang and Guth reduce the stochastic network of rubber to a set of three "representative" filaments oriented orthogonal to each other and forming the edges of a cubic unit cell; this is the so-called 3-chain model. The filaments are loaded in tension and their behavior is of Langevin type (Eq. (2.32)).

Details were gradually added to these pioneering models. For example, Cox's model was developed to account for generic fiber orientation distribution in 3D (Narter et al., 1999) and for fiber crimp (Hearle and Stevenson, 1964).

---

[2] For a review of the early work on rubber elasticity, see Treloar (1975).

The key physical assumptions made in these models are reproduced in all subsequent affine models: (i) fibers are loaded axially; the bending, shear, and torsion deformation modes of fibers are not active. As discussed in Section 5.3.1, this is mandated by the strict affine deformation assumption which implies that a straight segment before deformation remains straight after deformation.[3] (ii) Fibers deform independently and contribute individually to the overall stress. Since all fibers follow the imposed macro-deformation, no cooperative relaxation is possible. (iii) Fiber orientation becomes the central microstructural parameter. The fiber orientation distribution evolves as prescribed by the macro-deformation characterized by tensor $\mathbf{C}$.

These assumptions contravene the observations that (a) deformation is generally non-affine, which implies that the bending deformation mode of fibers is essential, (b) fibers move and deform cooperatively such as to reduce the total strain energy, while (c) the evolution of the orientation distribution function is generally different from the affine prediction. Furthermore, no correlation exists between the axial strain in fibers and their orientation relative to the principal stretch axes (Chandran and Barocas, 2006; Picu, 2011).

Affine models are formulated for various types of nonwovens. Examples include the formulation developed in Planas et al. (2007), which was used in Wong et al. (2019b) for generic nonwovens and in Ridruejo et al. (2012) and Martinez-Hergueta et al. (2016) for needle-punched nonwovens.

Affine models are used to describe the mechanics of biological materials (Lanir, 1983; Holzapfel et al., 2000; Gasser et al., 2006; Cacho et al., 2007). These works describe tissue microstructure in terms of families of fibers with preferential orientation, each family being represented by a structure tensor. Since fibers are assumed to deform independently, the strain energy and stress are written as sums of contributions from each fiber family.

In the polymer literature, a full chain incompressible model of the affine type was developed for rubber elasticity in Wu and van der Giessen (1993). Although it considers the entire distribution of chain orientations, this model provides predictions comparable with the 3-chain model of Wang and Guth (1952). Figure 7.3(a) shows experimental data for natural rubber from Treloar (1944) compared with predictions of two affine models: the affine model with Gaussian chains, and the 3-chain model with Langevin chains. Both models are fitted to the uniaxial data set and then used to predict the biaxial data. The Gaussian chain model fails to properly predict stiffening at large stretches and hence only the small and moderate stretch range can be fitted in the uniaxial case. Under biaxial deformation the response is much softer than observed in experiments. The affine model with Langevin chains properly fits the uniaxial stress–stretch curve, even at large deformations, but it fails to predict the biaxial curve. The Wu and van der Giessen model is somewhat closer to the biaxial data set than the 3-chain model with Langevin chains, but it is still softer than the experimental response. Note that fitting this experimental data set (which is often referenced in the literature) is challenging due to the fact that natural rubber undergoes strain-induced crystallization at large stretches (see

---

[3] Fibers curved in the initial configuration subjected to a strict affine deformation are loaded in a combination of modes; see Section 5.4.

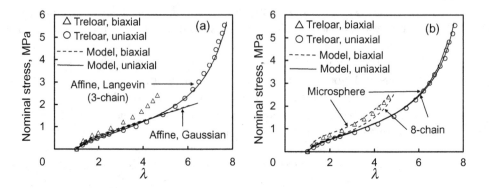

Figure 7.3 Stress–stretch curves for rubber subjected to uniaxial and biaxial loads. The symbols represent experimental data from Treloar (1944). This data set is compared with predictions of (a) affine models with Gaussian chains and the 3-chain model with Langevin chains, and (b) the 8-chain model and the nonaffine microsphere model. The figure includes data from Boyce and Arruda (2000) and Marckmann and Verron (2006)

Figure 6.35). If one takes the phenomenological view, as done here, this microstructural detail is irrelevant. But since multiple deformation mechanisms contribute to defining the characteristic curve, fitting models supposed to represent only one deformation mechanism is expected to present difficulties.

Affine models following the footsteps of Cox's model have been used for a long time in the textiles literature to represent the behavior of athermal fiberwebs subjected to various types of load. Examples include works by Hearle and collaborators (e.g., Hearle and Stevenson, 1964), Carnaby and collaborators (Carnaby and Pan, 1989a, 1989b; Lee and Carnaby, 1992), and Komori and collaborators (Komori et al., 1980, 1992; Komori and Itoh, 1991a, 1991b).

In general, the strict affine models provide predictions which are too stiff in the small strain regime (since the affine assumption over-predicts the small strain modulus of the network, see Section 6.1.1.3) and too soft at large strains. These discrepancies are less pronounced if incompressibility is enforced, as in the case of rubber, biological tissue, and gels, but are prominent in the case of networks which are not embedded in a volume preserving matrix, such as nonwovens. If the model is fitted to the uniaxial data, it typically underpredicts the biaxial response (Figure 7.3(a)). As already mentioned, the affine model provides better predictions for densely cross-linked networks. However, these networks usually do not reach large deformations without damage. If damage occurs, deformation becomes nonaffine and the predictions of a strict affine mode without damage ceases to be accurate.

7.4.2   Relaxing the Strict Affine Deformation Assumption

A number of affine models have been developed in which the strict affine deformation assumption is partly relaxed. The situations accounted for are shown in Figure 7.4.

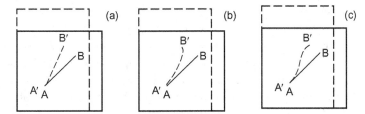

Figure 7.4 Three models used to describe the deformation of a fiber: strict affine deformation in (a), while in (b) and (c) affine displacements are applied only at fiber ends. In (b) the fiber ends are allowed to rotate to relax the moments (angles at fiber ends are unconstrained), while in (c) the fiber end orientation in the undeformed configuration is maintained during the deformation. A'B' represents the deformed shape of fiber AB and is shown here translated such as to overlap points A and A'.

Figure 7.4(a) shows the strict affine case in which all points of each fiber move affinely. This case is discussed in Sections 5.3 and 7.4.1.

An alternative is to apply affine deformation only on the crosslink scale, while fiber segments away from the crosslinks are allowed to seek configurations that minimize the potential energy. In this type of deformation crosslinks are moved with displacements $\mathbf{u}_{aff}$ computed based on the far-field strain (Eq (5.12)). Two types of models that differ in terms of the assumption made about the rotation of fibers at crosslinks may be developed. In the first such model, rotations are left free, which is equivalent to requiring the bending moment to vanish at the crosslink sites (Figure 7.4(b)). As in the strict affine model, fibers are not moving cooperatively. The crosslinks behave as rotating pin joints (type 3 in Figure 3.1). In principle, straight fibers should remain straight under such deformation conditions, but nonstraight fibers would deform primarily in bending. In general, the overall network modulus does not scale with $E_f A_f$ as in the strict affine model. This case was studied under both small and large deformations in Komori and Itoh (1991a, 1991b).

Another possibility is to restrict the angles at crosslinks to remain identical to those in the undeformed configuration (Figure 7.4(c)). This implies that all deformation modes of fibers are engaged, which provides a stiffer response compared to that of models in which the rotation at crosslinks is left free. Models of this type were developed in Astrom et al. (2000) and Wu and Dzenis (2005). In such models, the energy storage becomes biased toward the bending mode as the fiber bending stiffness increases (e.g., by increasing the fiber diameter) and the bending mode becomes stiffer. This behavior is opposite to what is generally observed in networks in which the affinity constraint is not imposed (Chapter 6): the softer deformation mode stores more energy and dominates the overall network behavior (Figure 6.7(b)).

7.4.3   A Note on Stability

The stability of deformation depends on the positive definiteness of the potential energy of the structure. In strict affine models, the energy functional of the network depends on the fiber arrangement and the axial constitutive behavior of fibers. If the fiber material is elastic and of identical stiffness in tension and compression, affinely deforming networks are unconditionally stable. If fibers store strain energy only in tension (are assumed to be floppy in compression), stability can be ensured only for uniform spatial distributions of fiber orientations, while proofs of stability are not available for nonuniform orientation distributions (Gasser et al., 2006).

Fibers do not move or deform cooperatively in any of the models discussed in Section 7.4. Under such conditions, network scale instabilities may result only if the fiber behavior is unstable. Structural instabilities in real networks always require the cooperative motion of groups of fibers. Since they do not capture this aspect of the mechanics, the affine models are not only stiffer, but also much more stable than real network structures.

7.5   Quasi-Affine Models

Several models that relax certain assumptions of the affine model are available. These are classified here in four categories: models that relax the assumption that fiber orientation evolves affinely, models accounting for fiber tortuosity within the affine framework, models accounting for damage within the affine framework, and unit cell models.

7.5.1   Models Relaxing the Assumption of Affine Fiber Orientation Evolution

All current mechanism-based constitutive models consider the deformation-driven preferential alignment of fibers as the key microstructural aspect of network evolution. Fiber orientation leads to anisotropy and strain stiffening, which are the dominant characteristics of the network mechanical response in the absence of damage.

Fiber alignment in the current configuration is described by the distribution function $p^*(\mathbf{n})$. Its relation with the corresponding distribution function in the reference configuration, $p(\mathbf{N})$, is defined by Eqs. (5.16) and (5.19). The degree of alignment predicted by the affine model in a uniaxial test depends on the lateral confinement. For a quasi-2D fibrous mat, no alignment results under biaxial tension with $\lambda_1 = \lambda_2 = \lambda$, while under uniaxial tension with confined lateral contraction $(\lambda_1 = \lambda; \lambda_2 = 1)$, $P_2 = (\lambda - 1)/(\lambda + 1)$, which increases almost linearly with λ for $1 < \lambda < 1.2$. However, under 2D isochoric uniaxial tension, that is, $\lambda_1 = \lambda; \lambda_2 = 1/\lambda$, the increase of P_2 with λ is twice as fast as in the confined case. Networks without an embedding matrix exhibit large Poisson contraction when loaded in uniaxial tension, with the ratio of the transverse and longitudinal stretches decreasing from 1 to values as low as $\lambda_2/\lambda_1 \approx 0.3$ at imposed stretches below $\lambda_1 = 1.4$ (e.g., Figures 6.29 and 6.31). In these cases, the affine reorientation of fibers predicts strong alignment, with P_2 increasing from 1 to ≈ 0.5.

Experimental observations of alignment are strongly system specific. For example, Vader et al. (2009) report experiments with reconstituted collagen networks of 1 mg/ml collagen concentration and observe stronger than affine alignment of fibrils. The alignment of polymer fiber nonwovens subjected to uniaxial stretching is measured in Chen et al. (2016) using X-ray diffraction, and the results indicate weaker than affine alignment.

Considering these complexities, it became desirable to predict the degree of fiber alignment based on considerations other than those associated with the affine model. For example, in Khansari et al. (2012), the evolution with strain of the orientation distribution function (in 2D) is predicted with a Fokker–Planck-type equation of the form:

$$\frac{\partial p^*}{\partial t} = \dot{\varepsilon} \frac{\partial}{\partial \theta} (p^* \sin 2\theta), \tag{7.16}$$

where $\dot{\varepsilon}$ represents the imposed strain rate and t is time. This equation can be integrated and the orientation distribution results as a function of strain and θ. The orientation distribution is then used to compute the Cauchy stress.

In Raina and Linder (2014), the mapping from unit vector $\mathbf{N}$ of the reference configuration to the corresponding unit vector $\mathbf{n}$ in the deformed configuration is performed with a function different from the affine $\mathbf{n} = \mathbf{FN}/|\mathbf{FN}|$. Specifically, it is assumed that in a frame aligned with the principal directions of $\mathbf{C}$, angle θ^* of a fiber measured from the eigenvector of $\mathbf{C}$ corresponding to the imposed stretch λ evolves with the stretch as:

$$\frac{d\theta^*}{d\lambda} = -\Gamma \sin 2\theta^*, \tag{7.17}$$

where Γ is a parameter used to control the degree of preferential orientation. This equation is solved analytically and provides a closed form solution for fiber orientation mapping.

These modifications of the affine model are phenomenological and are introduced to provide additional flexibility to the model and to represent a broader range of experimental behaviors.

7.5.2 Models Accounting for Fiber Tortuosity and Damage

Fibers are not straight in the majority of real networks and fiber tortuosity has a strong influence on network mechanics (Sections 6.1.1.3.6 and 6.1.1.4.6). Since in stochastic structures fibers are of different length and have different degrees of tortuosity, they respond differently to loading. This causes nonaffinity.

In mechanism-based constitutive models, fiber tortuosity is introduced primarily in order to model the soft network response at small strains, which is often observed experimentally (e.g., in biological tissue and needle-punched nonwovens). Since models generally consider that fibers are straight and are loaded axially, the tortuosity is accounted for by a modification of the fiber axial behavior. The bi-linear form shown in Figure 7.5 is often used for this purpose (Lanir, 1983; Cacho et al., 2007; Raina and Linder, 2014, 2015; Martinez-Hergueta et al., 2016). The fiber axial

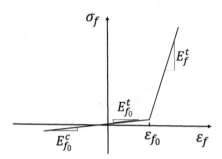

Figure 7.5 Axial constitutive response of fibers in models aimed to represent the mechanics of networks of tortuous fibers using straight fibers. The small effective stiffness for $\varepsilon_f < \varepsilon_{f0}$ accounts indirectly for the pull-out of fiber crimp and for the soft bending-dominated response in compression. The real axial constitutive response of fibers is recovered at strains larger than ε_{f0}.

response in tension is characterized by a small modulus E_{f0}^t up to strain ε_{f0}, and by a much larger modulus E_f^t at larger tensile strains. E_{f0}^t represents the pulling out of fiber crimp in the initial stages of stretch, while beyond ε_{f0} the fiber is straight and E_f^t is equal to the modulus of the fiber material, $E_f^t = E_f$. If the fiber behavior is nonlinear elastic or elastic–plastic, the nonlinearity or inelasticity manifest at strains larger than ε_{f0} (Silberstein et al., 2012). The axial response in compression is also soft, with effective modulus E_{f0}^c. This small compressive axial stiffness is supposed to account indirectly for fiber bending. In some models it is assumed that both E_{f0}^c and E_{f0}^t vanish, which may render the network floppy at small strains.

This representation ignores energy storage in bending and the proper behavior of fibers in compression.

Damage appears in networks at large strains either in the form of detachment of fibers from crosslinks or as fiber rupture. Crosslink failure is more often observed in network materials than fiber failure. In both cases, damage causes nonaffine local fiber kinematics, the redistribution of load from the failed filaments to the surrounding ones, and generally leads to gradual softening of the network response.

Since most current models do not account for fiber–fiber interactions and correlated kinematics, they cannot represent the actual load redistribution associated with the rupture of fibers or crosslinks. Nevertheless, in order to represent failure, the constitutive behavior of individual fibers in tension is arbitrarily modified by including a damage parameter which gradually reduces the axial stiffness of fibers (Ridruejo et al., 2012; Martinez-Hergueta et al., 2016) or by specifying the failure stress beyond which the fiber loses its load-carrying capacity.

Similarly, friction between fibers is occasionally represented indirectly by modifying the fiber behavior by the addition of a dissipative rheological element to the constitutive description of individual fibers (Planas et al., 2007; Wong et al., 2019b). Such models do not represent physical reality but may be tuned to fit experimental data.

7.5.3   Unit Cell Models

Several constitutive models have been developed for molecular networks with permanent crosslinks (e.g., rubber) by replacing the stochastic structure of the network
with a periodic lattice of unit cells. Due to the periodicity and assuming no global
instabilities, the behavior of one cell is representative for the behavior of the network.
The cells have polyhedral shapes such as tetrahedral, cubic, prismatic, or octahedral,
and contain a number of filaments equal to the number of vertices of the respective
cell. The filaments are connected at one end to the vertices and at the other end to a
node located in the center of the cell volume. Figure 7.6 shows the 4-chain tetrahedral
unit cell (Flory and Rehner, 1943; Treloar, 1946) and two versions of the 8-chain
model, with cubic (Arruda and Boyce, 1993) and prismatic (Bischoff et al., 2002)
unit cells.

Affine deformation is imposed on the cell scale, by moving the vertices with
displacements defined by the far field deformation gradient $\mathbf{F}$. The central node is
free and assumes positions that minimize the free energy of the cell. This introduces
some level of nonaffinity in the model. Note that, according to the phantom network
model of rubber elasticity (see for example, Rubinstein and Colby (2003) for a
derivation), the degree of nonaffinity decreases with increasing the connectivity
number $\langle z \rangle$ (Section 6.1.1.3.5). According to this model, in networks of high connectivity, for example, with $\langle z \rangle \geq 8$, the deformation of the unit cell is almost affine.

The constitutive behavior of filaments in these unit cell models may be thermal
(Gaussian or Langevin), as used in most models dedicated to rubber elasticity cited
here, or athermal (Diani et al., 2004; Silberstein et al., 2012).

Unit cell models are loaded in two ways: (i) the principal directions of the unit cell
(axes x_1, x_2, x_3 in Figure 7.6) are aligned with the reference frame or with the principal
axes of material anisotropy, or (ii) x_1, x_2, x_3 are aligned with the eigenvectors of $\mathbf{C}$.
Note that all these cells are anisotropic. The 8-chain model in Figure 7.6(b) has cubic
anisotropy, while the prismatic model in Figure 7.6(c) has anisotropy adjustable from
cubic to generic orthorhombic. This allows representing the actual material symmetry
in anisotropic network materials.

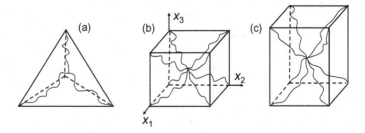

Figure 7.6 (a) A 4-chain model with tetrahedral unit cell, (b) an 8-chain cubic unit cell, and (c) an
8-chain prismatic unit cell. Octahedral and dodecahedral unit cells were proposed in Diani et al.
(2004).

An overly stiff response results if the deformation is such that an eigenvector of $\mathbf{C}$ aligns with one of the diagonals of the unit cell. A convenient remedy of this situation is to require that the principal directions of the cell (x_1, x_2, x_3) are aligned with the eigenvectors of $\mathbf{C}$ instead. This point of view was taken in Arruda and Boyce (1993) and led to better agreement with the experimental behavior of rubber under large strains. The model in Arruda and Boyce (1993) further assumes that all chains in the unit cell deform by a unique stretch which is postulated to be equal to the root mean square of the far field principal stretches, that is, $\lambda = \sqrt{\text{tr}(\mathbf{C})/3}$. This implies that the central node of the model remains centrally located throughout the deformation, which eliminates the nonaffinity. The response of the 8-chain model in uniaxial and biaxial loading is compared with Treloar's (1944) data in Figure 7.3(b). The model is fitted to the uniaxial data set and is used to predict the biaxial data. While the model is able to fit properly the uniaxial response, even at large stretches, it underpredicts the biaxial data. However, the error relative to the biaxial experimental data is smaller than in the case of the affine models – compare with Figure 7.3(a). This somewhat surprising result is due to the fact that none of the filaments of the unit cell is aligned with the eigenvector of $\mathbf{C}$. Hence, the filaments are stretched more gradually than in the affine model. The evolution of filament orientation during deformation is also different from that in the affine model since only chains oriented at roughly $45°$ relative to the principal directions of $\mathbf{C}$ are present in the model.

The unit cell models discussed in this section are compared in Boyce and Arruda (2000) and Marckmann and Verron (2006).

7.6   Nonaffine Models

7.6.1   Constraint Junction Models of Rubber Elasticity

The constraint junction model proposed by Ronca and Allegra (1975) and further developed by Flory (1976, 1985) and collaborators replaces the affine model with a representation in which the network crosslinks are tied to an affinely deforming background through a set of springs (virtual chains). The virtual chains allow some degree of nonaffinity (crosslink fluctuation).

If the virtual chains are very stiff, the crosslinks follow the affine deformation. If they are soft, the constraint imposed on the crosslinks is weak and the phantom network model is recovered. Hence, the constraint junction model interpolates between the affine and phantom models – see the discussion on gels and rubber in Section 6.1.1.3.3. The stiffness of the virtual chains is kept as a parameter in this model.

The condition that the virtual chains should not contribute to stress and strain energy storage mandates that their stiffness changes during deformation. Specifically, the stiffness decreases as $1/\lambda^2$, where λ is the stretch of the respective virtual chain (which depends on the nonaffine displacement of the crosslink to which the virtual chain is connected). Since in rubber elasticity chains are represented with the Gaussian or Langevin models, this implies that the effective number of segments of the virtual chains increases proportional to λ^2 during the deformation.

This concept was applied to networks of entangled polymers (Rubinstein and Panyukov, 2002). If the chain segments between crosslink points are long enough, entanglements are established. This leads to a reduction of the effective mesh size. The characteristic length of the network decreases as the entanglement density increases, from l_c, which represents the contour length distance between crosslinks, to $l_e < l_c$, which represents the distance between entanglements. One may postulate that entanglements behave just as another set of crosslinks. The constraint junction concept may be applied to both crosslinks and entanglements.

7.6.2   The Microsphere Model

The model proposed in Miehe et al. (2004), referred to as the "microsphere model," addresses specifically the problem of the representation of nonaffinity in a constitutive model, while accounting, to some degree, for the network microstructure. This model goes beyond the assumptions and limitations of the models presented in Section 7.5 as it replaces the ad-hoc homogenization performed in the affine model with a more formal procedure inspired by methods developed for composites with a continuum matrix.

The network is assumed to be composed from filaments loaded axially, in line with the physical view prevalent in the thermal networks literature. Since the model was developed for rubber, the filaments are either Gaussian or Langevin chains. The stretch of individual filaments, $\lambda(\mathbf{N})$, results from a free energy minimization procedure. This captures to some extent the actual cooperative kinematics of fibers. A field of fluctuations of the microstretch $f(\mathbf{N})$ is defined over the unit sphere of chain orientations such that $\lambda(\mathbf{N}) = f(\mathbf{N})\lambda^\infty(\mathbf{N})$, where $\lambda^\infty(\mathbf{N})$ is the macroscale stretch associated with $\mathbf{C}$ in the direction of the reference unit vector $\mathbf{N}$. The nonaffine microstretch field is assumed to follow the macrostretch in the p-average sense, that is, $\langle \lambda \rangle_p = \langle \lambda^\infty \rangle_p$, where $\langle\ \rangle_p$ represents the p-root averaging operator defined by $\langle y \rangle_p = \langle y^p \rangle^{1/p}$. The free energy of the system results from the minimization of the sum (integral over the unit sphere over all chain orientations) of the free energies of all filaments subjected to $\lambda = f\lambda^\infty$ and to the constraint expressed by the p-root averaging condition on λ. This minimization provides the nonaffine stretch fluctuation field $f(\mathbf{N})$ and the total free energy of the network. The value of p is kept as a parameter. The model also accounts for the lateral constraints imposed by the surrounding chains on the representative chain (the "tube") and for the nonaffine evolution of this constraint during deformation.

The trivial solution of the minimization problem is $f(\mathbf{N}) = 1$, a case in which $\lambda(\mathbf{N}) = \lambda^\infty(\mathbf{N})$ and the affine deformation is recovered. Another particular solution is obtained by assuming that $\lambda(\mathbf{N}) = \lambda$ is a constant, independent of the chain orientation. In this case all chains deform by the same stretch, which is taken equal to the p-root average of the macroscopic principal stretches, that is, $\lambda = \langle \lambda^\infty \rangle_p$. If $p = 2$, the 8-chain model of Arruda and Boyce (1993) is recovered.

This model provides sufficient flexibility to capture a broad range of responses, as demonstrated in the original article (Miehe et al., 2004) and in subsequent works (Kroon, 2010; Tkachuk and Linder, 2012) and provides a good compromise between computational efficiency and the representation of the microscale mechanics of stochastic networks in a continuum mechanics sense. Figure 7.3(b) shows its prediction of the uniaxial and biaxial tension responses of vulcanized rubber against experimental data from Treloar (1944). The model is fitted to the uniaxial experimental data and predicts the biaxial data.

7.6.3   Discrete-Continuum Network Models

Full representations of the network, in which each fiber is explicitly modeled as a beam or thermal/athermal truss, capture all details of the kinematics and provide a full description of nonaffinity. Such models are accurate, but computationally expensive, and do not provide close form solutions describing the constitutive behavior.

An intermediate solution between full network models and the analytic or semi-analytic models presented in this chapter is provided by discrete-continuum representations that belong to the class of sequential multiscale models. In such representations, a finite element model of the boundary value problem is defined on the continuum scale. The constitutive behavior is not defined analytically, rather, it is provided by subscale representative volume elements (RVE) that each contain a sub-domain of the network represented in full detail (Chan et al., 2019). Figure 7.7 shows an example of such a model. Each element of the continuum mesh is associated with one microscale RVE. The deformation gradient experienced by the element is applied as boundary conditions to the RVE, which is then solved and returns the stress tensor to the continuum scale. This procedure is iterated on both macro and micro scales until full convergence results. The RVEs are evolved throughout the entire deformation history.

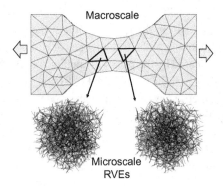

Figure 7.7 Multiscale discrete-continuum model of a network. A continuum finite element model is defined on the macroscale. The constitutive response of each element is provided by a microscale RVE containing a network representative for the actual material.

This model requires scale separation between the scale represented by the continuum model and that represented by RVEs. The RVE size must be much smaller than the size of the domain covered by the element. At the same time, the RVE must be large enough to remain representative, that is, its response should be free of size effects and its effective material properties should be independent of the boundary conditions applied (Sections 6.1.1.3.11 and 6.1.1.4.11). These restrictive conditions make the use of such models in the presence of large field gradients as encountered at crack tips and other concentrators difficult. However, they are quite effective in problems with weaker gradients, such as, for example, when studying networks with flocculation. Further, the solution of multiscale models is computationally expensive and requires massive parallel processing. Nevertheless, this type of representation provides a reasonable compromise between accuracy and efficiency, particularly when very large problem domains are considered.

7.7   Limitations of Current Mechanism-Based Models and Outlook

The current constitutive models for athermal network materials fail to account for several important aspects of the mechanics as follows:

▶ The nonaffine nature of the deformation is either ignored or accounted for heuristically.
▶ Fibers are assumed to evolve independently, despite the fact that fiber kinematics is actually collective, as it leads to lower system energies.
▶ Most current analytic or semi-analytic models assume that fibers deform axially, which ignores the fact that athermal fibers deform primarily in their bending mode.
▶ Fiber alignment is considered as the most representative descriptor of the microstructural evolution during deformation. Alignment is important. However, the dominant feature at large deformations is the development of a load-bearing sub-network, as described in Section 6.1.1.4.12. The structure of this sub-network is a result of collective deformation modes and cannot be easily predicted by homogenized models.
▶ Current models represent damage and inter-fiber friction by conveniently modifying the constitutive description of fibers. This indirect way of accounting for processes that involve both crosslink and fiber rupture and emerge from the collective kinematics of fibers can only be approximate, at best.
▶ If the network is embedded in a matrix, the two components are generally considered to contribute to the global stress independently, as two materials coupled in parallel. Since in reality network deformation is nonaffine, strong interactions between the matrix and fibers are expected.

The discussion in this chapter and Chapter 6 indicates that the affine assumption, which is made in almost all existing mechanism-based athermal network models, is physically inaccurate. On the other hand, properly capturing nonaffinity and the collective kinematics of fibers requires complex and computationally expensive models. To mitigate this situation, one is forced to make assumptions and include parameters which can be tuned to reproduce experimental data. In view of these current limitations of mechanism-based models, the phenomenological models remain an easier to use, easier to calibrate, and potentially more reliable alternative.

More works are dedicated to thermal networks and rubber elasticity than to athermal network materials. Thermal filaments are considered either Gaussian or Langevin and store free energy only in the axial mode. Significant attention was devoted to accounting for nonaffinity and the cooperative motion of the chains, both in unentangled and entangled networks. The present discussion does not represent fully the diversity of those opinions.

References

Altan, S. B. & Aifantis, E. C. (1992). On the structure of mode III crack tip in gradient elasticity. *Scripta Metall. Mater.* **26**, 319–324.

Arruda, M. & Boyce, M. C. (1993). A 3D constitutive model for the large stretch behavior of rubber elastic materials. *J. Mech. Phys. Solids* **41**, 389–412.

Astrom, J. A., Makinen, J. P., Hirvonen, H. & Timonen, J. (2000). Stiffness of compressed fiber mats. *J. Appl. Phys.* **88**, 5056–5061.

Berkache, K., Deogekar, S., Goda, I., Picu, R. C. & Ganghoffer, J. F. (2019a). Identification of equivalent couple-stress continuum models for planar random fibrous media. *Cont. Mech. Thermodyn.*, **31**, 1035–1050.

Berkache, K., Deogekar, S., Goda, I., Picu, R. C. & Ganghoffer, J. F. (2019b). Homogenized elastic response of random fiber networks based on strain gradient continuum models. *Mathem. Mech. Sol.* **24**, 3880–3896.

Bischoff, J. E., Arruda, E. M. & Grosh, K. (2002). A microstructurally-based orthotropic hyperelastic constitutive law. *J. Appl. Mech.* **69**, 570–579.

Boyce, M. C. & Arruda, E. M. (2000). Constitutive models of rubber elasticity: A review. *Rubber Chem. Techno.* **73**, 504–523.

Cacho, F., Elbischger, P. J., Rodriguez, J. F., Doblare, M. & Holzapfel, G. A. (2007). A constitutive model for fibrous tissues considering collagen fiber crimp. *Int. J. Non-linear Mech.* **42**, 391–402.

Carnaby, G. A. & Pan, N. (1989a). Theory of the compression hysteresis of fibrous assemblies. *Textile Res. J.* **59**, 275–284.

Carnaby, G. A. & Pan, N. (1989b). Theory of the shear deformation of fibrous assemblies. *Textile Res. J.* **59**, 285–292.

Castro, J. & Ostoja-Starzewski, M. (2003). Elasto-plasticity of paper. *Int. J. Plast.* **19**, 2083–2098.

Chan, V. W. L., Tobin, W. R., Zhang, S., et al. (2019). Image-based multiscale mechanical analysis of strain amplification in neurons embedded in collagen gel. *Comput. Meth. Biomech. Biomed. Eng.* **22**, 113–129.

Chandran, P. L. & Barocas, V. H. (2006). Affine versus non-affine fibril kinematics in collagen networks: Theoretical studies of network behavior. *J. Biomech. Eng.* **128**, 259–270.

Chen, N., Koker, M. K. A., Uzun, S. & Silberstein, M. N. (2016). In-situ X-ray study of the deformation mechanisms in non-woven polypropylene. *Int. J. Sol. Struct.* **97–98**, 200–208.

Coffin, D. (2008). Developing constitutive equations for paper that are valid for multi-time scales and large stresses. In: *Proceedings of progress in paper physics seminar*, Otaniemi, pp. 17–20.

Cosserat, E. & Cosserat, F (1909). *Theorie des corps deformables*. A. Hermann et Fils, Paris.

Cox, H. L. (1952). The elasticity and strength of paper and other fibrous materials. *British J. Appl. Phys.* **3**, 72–81.

Delfino, A., Stergiopulos, N., Moore, J. E. & Meister, J. J. (1997). Residual strain effects on the stress field in a thick wall finite element model of the human carotid bifurcation. *J. Biomech.* **30**, 777–786.

Diani, J., Brieu, M., Vacherand, J. M. & Rezgui, A. (2004). Directional model for isotropic and anisotropic hyperelastic rubberlike materials. *Mech. Mater.* **36**, 313–321.

Eringen, A. C. (1966). Linear theory of micropolar elasticity. *J. Math. Mech.* **15**, 909–923.

Eringen A. C. (1972) On nonlocal elasticity. *Int. J. Eng. Sci.* **10**, 233–248.

Flory, P. J. (1976). Statistical thermodynamics of random networks. *Proc. R. Soc. Lond. A* **351**,351–380.

Flory, P. J. (1985). Molecular theory of rubber elasticity. *Polym. J.* **17**, 1–12.

Flory, P. J. & Erman, B. (1982). The theory of elasticity of polymer networks. *Macromolecules* **15**, 800–806.

Flory, P. J. & Rehner Jr., J. (1943). Statistical mechanics of cross-linked polymer networks: I. Rubberlike elasticity. *J. Chem. Phys.* **11**, 512–520.

Gasser, T. C., Ogden, R. W. & Holzapfel, G. A. (2006). Hyperelastic modelling of artrial layers with distributed collagen fiber orientations. *J. Roy. Soc. Interfaces* **3**, 15–35.

Gaylord, R. J. & Douglas, J. F. (1990). The localization model of rubber elasticity. *Polym. Bull.* **23**, 529–533.

Gurtin, M. E., Fried, E. & Anand, L. (2010). *The mechanics and thermodynamics of continua.* Cambridge University Press, New York.

Han, W. H., Horkay, F. & McKenna, G. (1999). Mechanical and swelling behaviors of rubber: A comparison of some molecular models with experiments. *Math. Mech. Sol.* **4**, 139–167.

Hatami-Marbini, H. & Picu, R. C. (2009). Heterogeneous long-range correlated deformation in semiflexible random fiber networks. *Phys. Rev. E* **80**, 046703.

Hearle, J. W. S. & Stevenson, P. J. (1964). Studies in nonwoven fabrics: Prediction of tensile properties. *Textile Res J.* **34**, 181–191.

Holzapfel, G. A. (2000) *Nonlinear solid mechanics: A continuum approach for engineers.* Chichester, UK: Wiley.

Holzapfel, G. A., Gasser, T. C. & Ogden, R. W. (2000). A new constitutive framework for arterial wall mechanics and a comparative study of material models. *J. Elast.* **61**, 1–48.

Humphrey, J. D., Strumpf, R. K. & Yin, F. C. P. (1990). Determination of a constitutive relation for passive myocardium: A new functional form. *J. Biomech. Eng.* **112**, 333–339.

Isaksson, P. & Hagglund, R. (2009). Structural effects on deformation and fracture of random fiber networks and consequences on continuum models. *Int. J. Sol. Struct.* **46**, 2320–2329.

Khansari, S., Sinha-Ray, S., Yarin, A. L. & Pourdeyhimi, B. (2012). Stress–strain dependence for soy-protein nanofiber mats. *J. Appl. Phys.* **111**, 044906.

Komori, T. & Itoh, M. (1991a). A new approach to the theory of the compression of fiber assemblies. *Textile Res. J.* **61**, 420–428.

Komori, T. & Itoh, M. (1991b). Theory of the general deformation of fiber assemblies. *Textile Res. J.* **61**, 588–594.

Komori, T., Itoh, M. & Takaku, A. (1992). A model analysis of the compressibility of fiber assemblies. *Textile Res. J.* **62**, 567–574.

Komori, T., Makishima, K. & Itoh, M. (1980). Mechanics of large deformation of twisted-filament yarns. *Textile Res. J.* **50**, 548–555.

Kroon, M. (2010). A constitutive model for strain crystallizing rubber-like materials. *Mech. Mater.* **42**, 873–885.

Lanir, Y. (1983). Constitutive equations for fibrous connective tissue. *J. Biomech.* **16**, 1–12.

Lee, D. H. & Carnaby, G. A. (1992). Compressional energy of the random fiber assembly. I. Theory. *Textile Res. J.* **62**, 185–191.

Ma, X., Schickel, M. E., Stevenson, M. D., et al. (2013). Fibers in the extracellular matrix enable long-range stress transmission between cells. *Biophys. J.* **104**, 1410–1418.

Mäkelä, P. & Östlund, S. (2003). Orthotropic elastic-plastic material model for paper materials. *Int. J. Sol. Struct.* **40**, 5599–5620.

Marckmann, G. & Verron, E. (2006). Comparison of hyperelastic models for rubberlike materials. *Rubber Chem. Technol.* **79**, 835–858.

Martinez-Hergueta, F., Ridruejo, A., Gonzalez, C. & Llorca, J. (2016). A multiscale micro-mechanical model of needlepunched nonwoven fabrics. *Int. J. Sol. Struct.* **96**, 81–91.

Miehe, C., Goktepe, S. & Lulei, F. (2004). A micro-macro approach to rubber-like materials – Part I: The non-affine micro-sphere model of rubber elasticity. *J. Mech. Phys. Sol.* **52**, 2617–2660.

Mooney, M. (1940). A theory of large elastic deformation. *J. Appl. Phys.* **11**, 582–592.

Narter, M. A., Batra, S. K. & Buchanan, D. R. (1999). Micromechanics of the 3D fiberwebs: Constitutive equations. *Proc. R. Soc. Lond.* **455**, 3543–3563.

Nowacki, W. (1986). *Theory of asymmetric elasticity*. Pergamon Press, New York.

Ogden, R. W. (1984). *Nonlinear elastic deformations*. Wiley, New York.

Ostoja-Starzewski, M. & Castro, J. (2003). Random formation, inelastic response and scale effects in paper. *Proc. R. Soc. London A* **361**, 965–985.

Picu, R. C. (2011). Mechanics of random fiber networks: A review. *Soft. Matt.* **7**, 6768–6785.

Picu, R. C., Deogekar, S. & Islam, M. R. (2018). Poisson's contraction and fiber kinematics in tissue: Insight from collagen network simulations. *J. Biomech. Eng.* **140**, 021002.

Planas, J., Guinea, G. V. & Elices, M. (2007). Constitutive model for fiber-reinforced materials with deformable matrices. *Phys. Rev. E* **76**, 041903.

Raina, A. & Linder, C. (2014). A homogenization approach for nonwoven materials based on fiber undulations and reorientation. *J. Mech. Phys. Sol.* **65**, 12–34.

Raina, A. & Linder, C. (2015). A micromechanical model with strong discontinuities for failure in nonwovens at finite deformations. *Int. J. Sol. Struct.* **75–76**, 247–259.

Ridruejo, A., Gonzalez, C. & Llorca, J. (2012). A constitutive model for the in-plane mechanical behavior of nonwoven fabrics. *Int. J. Sol. Struct.* **49**, 2215–2229.

Rivlin, R. S. (1948). Large elastic deformations of isotropic materials. IV Further development of the general theory. *Phil. Trans. R. Soc. Lond. A* **241**, 379–397.

Ronca, G. & Allegra, G. (1975). An approach to rubber elasticity with internal constraints. *J. Chem. Phys.* **63**, 4990–4997.

Rubinstein, M. & Colby, R. H. (2003). *Polymer physics*. Oxford University Press, Oxford.

Rubinstein, M. & Panyukov, S. (2002). Elasticity of polymer networks. *Macromolecules* **35**, 6670–6686.

Schaefer, H (1967). Das Cosserat kontinuum. *ZAMM* **47**, 485–498.

Silberstein, M. N., Pai, C. L., Rutledge, G. C. & Boyce, M. C. (2012). Elastic–plastic behavior of non-woven fibrous mats. *J. Mech. Phys. Sol.* **60**, 295–318.

Thirlwell, B. E. & Treloar, L. R. G. (1965). Nonwoven fabrics. Part VI: Dimensional and mechanical anisotropy. *Text. Res. J.* **35**, 827–835.

Ting, T. C. T. & Chen, T. (2005). Poisson's ratio for anisotropic elastic materials can have no bounds. *Quart. J. Mech. Appl. Math.* **58**, 73–82.

Tkachuk, M. & Linder, C. (2012). The maximal advance path constraint for the homogenization of materials with random microstructure. *Phil. Mag.* **92**, 2779–2808.

Treloar, L. R. G. (1944). Stress–strain data for vulcanized rubber under various types of deformation. *Trans. Faraday Soc.* **40**, 59–70.

Treloar, L. R. G. (1946). The elasticity of a network of long-chain molecules. III. *Trans. Faraday Soc.* **42**, 83–94.

Treloar, L. R. G. (1975). *The physics of rubber elasticity*. Oxford University Press, Oxford.

Tyznik, S. & Notbohm, J. (2019). Length scale dependent elasticity in random three-dimensional fiber networks. *Mech. Mater.* **138**, 103155.

Vader, D., Kablea, A., Weitz, D. & Mahadevan, L. (2009). Strain-induced alignment in collagen gels. *PLoS ONE* **4**: e5902.

Voigt, W (1887). Theoretische studien uber die elastizitatsverhaltnisse der krystalle. *Abhandlungen der Mathematischen Classe der Koniglichen Gesellschaft der Wissenschaften zu Gottingen* **34**, 3–51.

Wang, M. & Guth, E. (1952). Statistical theory of networks of non-Gaussian flexible chains. *J. Chem. Phys.* **20**, 1144–1157.

Wong, D., Andriyana, A., Ang, B. C., et al. (2019a). Poisson's ratio and volume change accompanying deformation of randomly oriented electrospun nanofibrous membranes. *Plastic Rubber Comp.* **48**, 456–465.

Wong, D., Verron, E., Andriyana, A. & Ang, B. C. (2019b). Constitutive modeling of randomly oriented electrospun nanofibrous membranes. *Cont. Mech. Thermodyn.* **31**, 317–329.

Wu, P. D. & van der Giessen, E. (1993). On improved network models for rubber elasticity and their applications to orientation hardening in glassy polymers. *J. Mech. Phys. Sol.* **41**, 427–456.

Wu, W. F. & Dzenis, Y. A. (2005). Elasticity of planar fiber networks. *J. Appl. Phys.* **98**, 093501.

Xia, Q. S., Boyce, M. C. & Parks, D. M. (2002). A constitutive model for the anisotropic elastic–plastic deformation of paper and paperboard. *Int. J. Sol. Struct.* **39**, 4053–4071.

Yeoh, O. H. (1990). Characterization of elastic properties of carbon black filled rubber vulcanizates. *Rubber Chem. Technol.* **63**, 792–805.

Yeoh, O. H. (1993). Some forms of the strain energy function for rubber. *Rubber Chem. Technol.* **66**, 754–771.

8 Strength and Toughness of Network Materials

Understanding and predicting material failure is of obvious importance in applications. Network materials make no exception. Failure of engineering network materials is a key aspect in design and life prognosis, while rupture of biological networks leads to tissue damage and pain, and possibly to the catastrophic rupture of the respective organs and connective tissue components.

Fracture mechanics applies generally to continua. It provides a theoretical framework and design tools that predict the onset of crack growth. It also distinguishes between rapid growth leading to catastrophic failure and slow growth under small loads and specific environmental conditions. An important characteristic of Fracture mechanics is that it always assumes pre-existing cracks; the theory addresses crack growth but does not capture crack nucleation. Posing the nucleation problem within Fracture mechanics requires focusing on the microstructure and accounting for microscopic stress concentrators, an endeavor which is inherently subjected to uncertainties.

To circumvent some of these difficulties, damage mechanics was developed within continuum mechanics. This perspective is phenomenological and not specific with respect to what is considered damage in a specific application. It allows developing constitutive descriptions that account for the gradual degradation of stiffness and predict life reduction associated with the emergence and growth of damage. Damage is viewed as an internal field which evolves during macroscopic loading, as described by phenomenological laws. Since the physical nature of damage is not defined, it is not necessary to consider that cracks pre-exist the deformation of interest. This is an advantage relative to the micromechanics-based Fracture mechanics perspective. Nonetheless, the description remains phenomenological, does not represent explicitly the underlying physical processes, and requires experimental calibration.

Situations in which failure of network materials is associated with the growth of a pre-existing crack-like feature are not encountered frequently in applications. In most cases, the material is free of apparent macroscopic defects in the initial state and develops damage as it is loaded. The objective of theories in this field is to predict the strength and the toughness of virgin, un-notched samples. Strength is defined as the peak of the stress–strain curve, while toughness is the total area under the stress–strain curve up to failure, that is, the energy absorbed to failure per unit volume of material.

The discussion in this chapter is divided into two parts: results pertaining to the prediction of strength of networks without pre-existing defects are presented first, while the second part addresses the problem of crack growth in network materials.

Before going into the details, it is important to outline the central importance of network heterogeneity in the analysis of damage accumulation and failure. As discussed in Chapters 4 and 6, network materials are intrinsically heterogeneous. Mechanical heterogeneity – meaning fluctuations of the stress and strain fields for boundary conditions that should lead to spatially uniform fields in a homogeneous material – are introduced by multiple sources. In the simplest case of stochastic networks composed from fibers of the same type, heterogeneity is controlled by parameter w (Section 6.1.1.3.1). In the presence of inclusions or fibers of different properties the degree of heterogeneity increases. Mesoscale fluctuations of density associated with flocculation are another source of heterogeneity. Hence, in general, networks are heterogeneous on multiple scales and this controls the development of damage. Damage may be localized and may develop into a dominant crack-like feature which causes brittle failure, or may remain diffuse leading to gradual, ductile failure.

Some network materials are embedded in solid or liquid matrices, as for example in connective tissue, biological membranes, blood vessel walls, and gels. A fluid embedding without drainage imposes the condition that the total volume of the sample conserves during deformation. The isochoric condition restricts the natural tendency of the network to undergo large Poisson contraction when subjected to tension. Equivalently, one may view this situation as a multiaxial deformation imposed on the network by the combined action of the far field and of the matrix. On the other hand, the dynamic redistribution of fluid within the network during deformation applies minimal loads on fibers. Hence, the primary effect of a fluidic matrix on fracture is to modify the effective boundary conditions acting on the network. This observation aids in addressing such problems using the knowledge base developed for networks without a matrix, a case which is discussed in this chapter.

The situation is significantly different in the presence of a solid matrix, or of a viscoelastic matrix subjected to strain rates larger than the characteristic relaxation rate of the respective viscoelastic embedding. In these cases, the network and matrix interact strongly on the scale of individual fibers. This interaction changes the network kinematics relative to the reference pure network case. The two components – network and matrix – cannot be regarded as responding independently to the applied loads. In addition, the occurrence of damage in one of the components does not necessarily lead to global failure if the other component remains damage-free. This situation is similar to that of common fiber-reinforced composites. However, the problem of a stochastic network embedded in a solid matrix is somewhat different due to the pronounced heterogeneity of the network. A discussion of the effect of a solid matrix on network behavior is presented in Section 11.4.

Based on these considerations, this chapter addresses exclusively the rupture of pure networks with no embedding matrix.

8.1   Networks without Pre-existing Cracks

8.1.1   Failure Modes and Mechanisms

Phenomenological aspects of failure in networks without pre-existing defects are discussed in this section. Three types of mechanical response in uniaxial tension are described in Section 6.1.1.1, and are shown schematically in Figure 6.1(a). Type A is a linear elastic response. Type B has two regimes separated by a transition similar to yielding in elastic–plastic continua, while type C behavior is hyperelastic and presents several regimes. Figure 8.1 shows schematic representations of this behavior in the presence of damage accumulation and failure.

Type A materials fracture catastrophically, which leads to the rapid drop of the stress from a peak value to zero. The peak stress denoted by S_r is the ultimate tensile strength (UTS).

Type C materials rupture by developing a broad peak followed by either brittle-like failure or a gradual decrease of stress that may extend to large stretches. The stretch corresponding to the peak stress is denoted by λ_r, while the stretch at which the material entirely loses its load carrying capacity is denoted by λ_f.

In general, networks follow the stress–stretch curve corresponding to the case without damage up to close to peak stress. Increasing the strength of crosslinks and fibers moves the position of the peak up the "no damage" curve. Failure may take place in regime I if crosslinks are weak or may take place in regimes II and III if the crosslink strength is larger.

The behavior in Figure 8.1(b) is also frequently encountered. This represents the type B response of Figure 6.1(a), which consists in two regimes, of which the first is linear elastic. The well-defined reduction of the tangent stiffness at the end of regime I is usually associated with yielding. This implies plastic deformation of the network

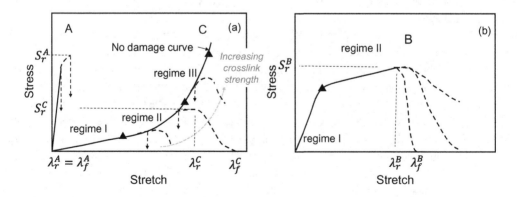

Figure 8.1 Schematic representation of stress–stretch curves corresponding to (a) type A and C, and (b) type B behaviors shown in Figure 6.1. Several possible post-peak behaviors are shown in each panel with dashed lines. Dashed lines with and without arrows indicate brittle and ductile failure, respectively.

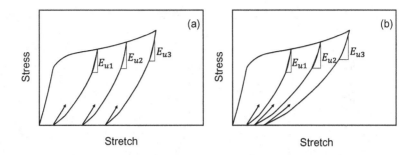

Figure 8.2 Schematic representation of stress–stretch curves of type B in which the yield point is associated with (a) plastic deformation of fibers and/or crosslinks and (b) damage accumulation.

due to the plastic deformation of fibers and crosslinks. However, the presence of an apparent yield point may also indicate damage initiation. The post-peak response may be either brittle or ductile. Three post-peak curves are shown in Figure 8.1(b) to suggest that either brittle or ductile behaviors are possible. The difference between the stretch at peak stress and the stretch at failure, $\lambda_f - \lambda_r$, quantifies the degree of brittleness.

When a stress–stretch curve of type B is obtained in experiments, one may enquire whether the yield point observed is associated with plastic deformation or with damage accumulation. A conclusion can be reached if unloading is performed from stress states beyond the yield point. Figure 8.2 shows two possibilities. If the branches of the stress–stretch curve obtained upon unloading from various stress values are approximately parallel to each other ($E_{u1} \approx E_{u2} \approx E_{u3}$), the yield point is most likely associated with plasticity. If the respective unloading branches have decreasing slopes ($E_{u1} > E_{u2} > E_{u3}$), as shown schematically in Figure 8.2(b), the yield point is associated with the onset of damage.

The discussion in Section 6.1.4 indicates the conditions under which plastic deformation is to be expected in crosslinked networks. In general, network plasticity is expected in dense and densely crosslinked networks which deform mostly affinely. In this case, the yield strain of the network is equal to the yield strain of fibers. Plastic deformation of fibers is improbable in nonaffine networks composed from fibers of small diameter.

Inter-fiber friction may also cause the emergence of a type B yield point. This happens in sparsely crosslinked nonwovens with a large fiber volume fraction. In these cases, the density of inter-fiber contacts is large and the yield point is associated with overcoming the effect of inter-fiber friction. Once this happens, fibers are free to move independently.

Inter-fiber friction leads to pronounced hysteresis. Strong hysteresis also emerges when fibers interact cohesively. On the other hand, damage in an elastic network produces no hysteresis. However, structural reorganization favored by damage accumulation in the presence of inter-fiber friction may cause some degree of hysteresis. These observations may be used to interpret experimental results.

Rupture may be classified as brittle or quasi-ductile. In general, brittle rupture implies the unstable propagation of a major crack. In highly heterogeneous network materials, mesoscale cracks are rough, as they result by the coalescence of localized damage. The mean plane of the localized damage region is approximately normal to the maximum principal tensile stress. The localization of damage in the form of a dominant crack occurs very close to the peak stress. Observations of paper deformation by digital image correlation (Borodulina et al., 2012) indicate large fluctuations of the strain field. As the deformation proceeds, these fluctuations grow in amplitude. Eventually they merge, forming a strain localization band at stresses close to the peak stress. Catastrophic failure is observed beyond the peak stress.

Ductile failure implies that localized damage regions remain localized and the network tends to separate in sub-domains that remain anchored to each other by a relatively small number of fibers. These sub-domains undergo relative rotations and translations as the connecting fibers rotate and deform. The stress carried by the structure decreases gradually as the number of connecting filaments between network sub-domains decreases. The resulting decreasing branch of the stress–stretch curve may extend to large stretches.

Ductile failure is possible when network heterogeneity is pronounced. Network toughness depends on the ability of the structure to reorganize as damage occurs. To understand this, consider the opposite limit of a "locked" network which tends to preserve its structure when loaded. Such a network remains in regime I, since geometric nonlinearity is prevented. Locking may be caused by dense crosslinking and/or the effect of friction and cohesion at inter-fiber contact points. Locked networks undergo brittle failure. On the other hand, a mechanically heterogeneous network may reorganize during deformation, which enables nonlinear behavior and delays damage localization. In such cases, the resulting post-peak behavior is quasi-ductile.

Figure 8.3 shows examples of stress–stretch curves that demonstrate the behavior described above. Figure 8.3(a) shows two stress–stretch curves obtained by testing commercial nonwovens made from polypropylene fibers of diameter $\sim$30 μm in uniaxial tension (Chen et al., 2016a). Samples are obtained by melt spinning followed by heat sealing, which leads to the formation of crosslinks at inter-fiber contacts. The two samples have weights per unit area of 68 g/m^2 and 220 g/m^2. The observed behavior is of type B, but the failure mode depends on network density. The strength, S_r, increases approximately in proportion to the density, while the stretch at the peak increases at a slower rate. The most significant difference between the two nonwovens is seen in the post-peak region of the curves. While the high-density material ruptures in a brittle way, the low-density network fails gradually and deforms up to stretches as large as 2. The difference $\lambda_f - \lambda_r$ is $\sim$0.1 and $\sim$0.9 in the high- and low-density cases, respectively.

Figure 8.3(b) shows the failure behavior of collagen networks extracted from starfish (Ovaska et al., 2017). To obtain these samples, the biological tissue was decellularized and then placed in a disintegrating solution of mercapto-ethanol to produce a collagen suspension. The solution is then filtered and subjected to

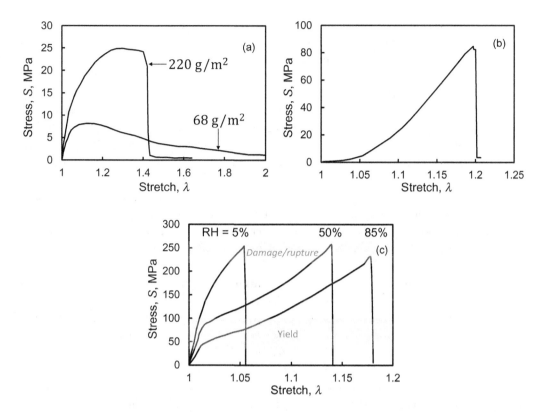

Figure 8.3 Examples of uniaxial tension tests performed up to failure with various network materials. (a) Polypropylene nonwovens of different base weight (data from Chen et al., 2016a), (b) reconstituted echinoderm collagen (data from Ovaska et al., 2017), and (c) nanofibrilated cellulose nanopaper tested at different relative humidity levels (RH) (adapted from Malho et al., 2015).

centrifugation, followed by treatment with a collagen crosslinking agent. The mechanical behavior is of type C and is typical for collagen networks. Rupture is brittle and occurs in regime III of network deformation. Significant variability is observed from sample to sample (not shown in Figure 8.3(b)), but rupture is always brittle.

Figure 8.3(c) shows stress–stretch curves for nanofibrilated cellulose nanopaper (Malho et al., 2015). These samples are composed from cellulose nanofibers of diameter ~20 nm and of several microns in length, which were not treated with additional bonding agents. The figure shows curves for three values of relative humidity. As humidity increases, paper absorbs water and fibers become elastic–plastic. Consequently, the nature of the response changes from type A to a combination of types B and C. Specifically, a yield point is observed at high humidity, followed by a type C response with stiffening. If unloading is performed from the region of the curve beyond the yield point, residual strains are observed, which is indicative of plastic deformation. Therefore, the yield point is associated with plasticity. Failure is brittle in all cases due to the high crosslink density of these materials. Interestingly, the strength is not affected by humidity. Therefore, the plastic response causing the yield point and damage accumulation leading to rupture are separate processes in these samples.

Network materials rupture due to fiber and/or crosslink failure, crosslink failure being prevalent. Paper and cellulose networks and heat-sealed nonwovens of polymeric fibers rupture due to crosslink failure. Thermally-bonded nonwovens rupture by fiber failure at the fiber insertion in thermal bonds (Figure 1.1(b)). This mechanism is due to the fact that the polymer in the thermal bond region is rendered brittle by the rapid melting and resolidification during bond formation. The mechanism of network rupture in collagen and collagen–elastin biological networks is less well understood, but it is to be expected that the molecular bridges between fibers, which play the role of crosslinks, fail before fibers rupture. The situation is also less clear in dense and densely crosslinked networks of nanofibers, such as nanopapers of cellulose, carbon nanofibers, boron nanofibers, carbon nanotubes, etc. In these networks, nanofibers interact cohesively and hence network rupture is a more complex process than in the case of noncohesive fibers.

Fiber failure is generally caused by axial deformation. Fiber failure in bending is rare and applicable only to brittle fibers, as encountered in some ceramic foams (Shi et al., 2020). However, these applications are of less practical importance than those of the other network materials discussed here. Therefore, fiber failure is characterized here by their maximum nominal stress in uniaxial tension, S_{fc}.

Crosslink failure is more complex and depends on the type of crosslink considered. This issue is discussed in Section 3.2.2, where relations that can be used to evaluate the loading conditions under which a given crosslink may fail are presented, Eqs. (3.1) to (3.3). To summarize: rotating crosslinks (Table 3.1) between molecular filaments fail in tension at a critical force, f_{Xc}. An example is provided in Figure 3.3, and Eq. (3.2) provides the critical condition for crosslink rupture. Welded crosslinks (Table 3.1) formed either by the interpenetration of two athermal fibers or by a distribution of nanoscale filaments connecting two fibers, such as in Figure 3.2(b), fail in a complex way due to the combination of forces and moments that tend to separate and to rotate the two fibers in contact. Failure criteria used in this case must be expressed in terms of all components of the force and moment transmitted by the crosslink, in a local coordinate system tied to the two fibers at the crosslink site. Equation (3.1) provides a failure criterion for such cases in terms of an equivalent force, which becomes equal to f_{Xc} at failure.

A description of crosslink rupture based on a cohesive element-type representation of individual crosslinks has also been used (Borodulina et al., 2012; Goutianos et al., 2018). This model has two parameters, of which one is the strength (equivalent to f_{Xc}), and the other is the total energy required to rupture the crosslink (the work of separation). Whether a force-based criterion or a cohesive model provide the best representations of rupture depends to a large extent on the crosslink compliance. The behavior of relatively stiff crosslinks which do not deform much up to network rupture may be described in terms of a force-based criterion, Eq. (3.2). A cohesive description is better suited for the case of compliant crosslinks. The work of separation becomes important when the large deformation of the crosslinks at rupture couples with the large deformation of the surrounding fibers. Numerical results available in the current literature indicate that the work of separation has a limited effect on network strength (Borodulina et al., 2012; Goutianos et al., 2018), with the caveat that these models consider relatively stiff crosslinks. Based on these observations, we consider here a description of crosslink rupture in terms of the critical force f_{Xc}.

To rationalize whether network rupture is due to fiber or crosslink failure, one may compare two non-dimensional parameters: the stress in fibers normalized by the fiber strength, S_f/S_{fc}, and the equivalent force transmitted by crosslinks (Eq. (3.1)) normalized by the crosslink strength, F_X/f_{Xc}. It is relatively easy to evaluate these parameters in an affine model. In this case, fibers always deform axially and the fiber stress can be written $S_f/E_f \sim \mathbf{N}^T \mathbf{EN}$, where $\mathbf{N}$ is the unit vector of the fiber orientation and $\mathbf{E}$ is the Green–Lagrange strain (Eq. (5.31)). Considering $z = 4$, that is, four fiber segments merging at each crosslink, the effective force transmitted by the crosslink is anywhere between one and two times the force carried by a fiber. The exact value depends on the relative orientation of the respective fibers. Denote this fraction as $\varsigma = F_X/f_f$. If $F_X/f_{Xc} > S_f/S_{fc}$, the probability of crosslink failure is larger than that of fiber failure. This condition can be rewritten as:

$$f_{Xc} < \varsigma S_{fc} A_f, \tag{8.1}$$

where ς ranges from 1 to 2. If this condition is fulfilled, network rupture is expected to occur predominantly by crosslink failure.

8.1.2 Relation between Network Strength and Structural Parameters

Establishing a relation between strength and material parameters controllable experimentally is essential for material development and design. The results presented in this section are obtained from models in which various material parameters are controlled independently, to establish their effect on network strength. The results are compared with the observed behavior of several network materials in Section 8.1.3.

The parameters describing the stochastic network structure are discussed in Chapter 6 (Table 6.1). The small strain stiffness and the nonlinear material behavior may be understood in terms of the nondimensional parameter w (Section 6.1.1.3.1 and Eq. (6.3)), which combines information about the network density, fiber aspect ratio, and fiber elastic properties. Therefore, it is of interest to evaluate the failure mode of networks with different w.

Figure 8.4(a) shows stress–stretch curves obtained with cellular network models of various w (Deogekar and Picu, 2018). The rupture of these networks is caused by crosslink failure ($S_{fc} \to \infty$), and the crosslink strength, f_{Xc}, is identical in all cases shown in Figure 8.4(a). The four networks labeled N1–N4 correspond to the points marked by stars on the master plot for the particular network architecture shown in Figure 8.4(b). Networks labeled N1, N2, and N3 are nonaffine, while network N4 is in the nonaffine to affine transition region.

The curves in Figure 8.4(a) indicate that, as w decreases, the elastic modulus E_0 decreases (as shown in Figure 8.4(b)), the peak stress decreases, and the stretch

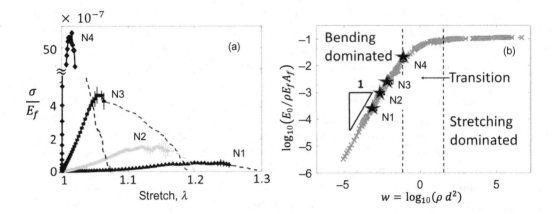

Figure 8.4 (a) Stress–stretch curves for cellular networks of various w. The four networks considered correspond to points on the master plot shown in (b) indicated by stars. The master plot shows the dependence of the small strain modulus, E_0, on various network parameters (see also Figure 6.6). The network density ρ is much larger than the percolation threshold, which allows replacing $\delta\rho$ used in the master plot of Figure 6.6, with ρ. Reprinted from Deogekar and Picu (2018) with permission from Elsevier

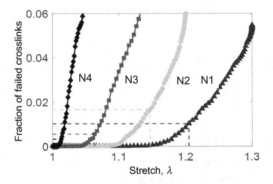

Figure 8.5 Fraction of failed crosslinks versus stretch for the four types of networks shown in Figure 8.4. Reprinted from Deogekar and Picu (2018) with permission from Elsevier

corresponding to the peak stress increases. As w decreases, the post-peak regime becomes broader. These observations indicate that, at given crosslink strength, network failure is brittle for large w and becomes gradually more ductile as w decreases.

Each of the curves in Figure 8.4(a) follow the stress–stretch curve that would be obtained if crosslink failure would be prevented (the "no damage curve" in Figure 8.1(a)), until very close to the peak stress. This indicates that the peak develops soon after the onset of damage accumulation. Figure 8.5 shows the variation of the fraction of failed crosslinks (out of the total number of crosslinks of the structure) during deformation for networks N1–N4.

The dashed lines in Figure 8.5 indicate the position of the peak stress. It results that less than 2% of the crosslinks fail before peak stress and the onset of crosslink failure

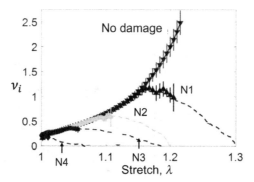

Figure 8.6 Variation of the incremental Poisson ratio during uniaxial deformation for networks N1, N2, N3, and N4 (Figure 8.4). Reprinted from Deogekar and Picu (2018) with permission from Elsevier

occurs at strains larger than 80% of the strain corresponding to peak stress.[1] The fraction of failed crosslinks increases fast beyond peak stress.

It is remarkable that the fraction of crosslinks that rupture before the peak stress is largely independent of w, particularly considering that the failure mode depends strongly on this parameter. Networks of large w are more brittle (Figure 8.4(a)) and fail due to the localization of damage leading to a crack. In samples of small w, damage is diffuse and more or less randomly distributed since the effective heterogeneity of these samples is more pronounced.

Damage has a strong effect on the overall network kinematics. The Poisson effect is the only macroscopic kinematic parameter in uniaxial tension, since the stretch in the loading direction is controlled. Figure 8.6 shows the variation of the incremental Poisson ratio, v_i (Eq. (6.11)), during stretch for the four networks discussed here. The curve corresponding to a network with damage is indistinguishable from that of the same network in which damage is inhibited up to very close to the peak stress. The distinguishing feature of the curves with damage is the fact that $v_i(\lambda)$ develops a peak at the stretch at which the stress develops a maximum, and beyond the peak stress v_i decreases. This indicates that damage relaxes the internal stress that forces the network to reorganize during the nonlinear deformation of regime II. Note that this peak occurs during regime II and its origin is different from that of the peak of the incremental Poisson ratio in networks without damage (Fig. 6.25), which emerges at the transition between regimes II and III. In the presence of crosslink and/or fiber failure, the fibers are less kinematically constrained and become less oriented in the stretch direction which, in turn, results in a markedly lower Poisson effect. The degree of nonaffinity of the network increases as damage accumulates.

The energy partition during the nonlinear deformation of athermal networks is introduced in Section 6.1.1.4.4 and discussed throughout Chapter 6. For athermal

---

[1] These systems are far from stiffness percolation and hence the low density of failed crosslinks at the peak stress is not caused by the proximity to the critical point and the expected sparse connectivity of the network in the respective conditions.

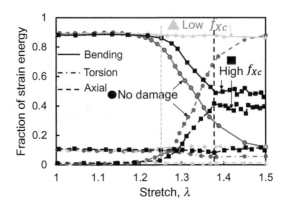

Figure 8.7 Fraction of the total strain energy stored in the axial, bending, and torsion deformation modes of fibers during uniaxial tension of network N2 (Figure 8.4), without and with damage accumulation. Two cases with damage are shown, with two values of crosslink strength, f_{Xc}, denoted here as "high" and "low." The shear energy accounts for about 1% of the total energy in all cases and is not shown. The left and right dashed vertical lines indicate the position of the peak stress in the two cases with low and high f_{Xc}, respectively.

nonaffine networks (low w) of relatively stiff crosslinks, the largest fraction of the work performed by the boundary tractions is stored in the bending deformation mode of fibers. The dominance of the bending mode persists throughout regimes I and II, while a transition from bending to axial dominance is observed to coincide with the transition to regime III (see Figure 6.24). This is associated with the formation of preferentially aligned force paths which, at stretches within regime III and beyond, carry most of the applied tensile load.

Figure 8.7 shows the energy partition of network N2 for the case in which damage is inhibited and in the presence of damage. Two cases are considered, with two different values of the crosslink strength, f_{Xc}. Increasing f_{Xc} moves the peak stress to higher values and postpones the onset of damage accumulation. Hence, this parameter may be used to control the regime of network deformation in which the peak stress is located.

The energy partition in the network without damage resembles that in Figure 6.24. The bending and axial modes exchange roles and become sequentially energetically dominant as the stretch increases. The curves corresponding to cases in which damage accumulation is enabled are different. The network with low f_{Xc} reaches the peak stress in regime II. Before the peak, the strain energy is stored predominantly in the bending mode. The bending mode remains dominant after the peak stress (marked by the left vertical dashed line in Figure 8.7). The network with high f_{Xc} reaches the peak stress at the beginning of regime III (marked by the right vertical dashed line). Before the peak, the fraction of the bending energy decreases and the fraction of the axial energy increases. Both bending and axial energy fractions level off beyond peak stress. The physical origins of these trends are identical to those causing the reduction of the incremental Poisson ratio observed in Figure 8.6 beyond peak stress.

Damage relaxes the internal constraints that lead to the formation of preferentially aligned force paths and hence inhibits the mechanism promoting the dominance of the axial mode.

The observations presented in Figures 8.6 and 8.7 provide insight into the effect of damage on network kinematics. To summarize, damage accumulation reduces the structural constraints that develop within the network during the large deformation regime, which, in turn, leads to reduced Poisson contraction, reduced fiber alignment, and more pronounced nonaffinity.

In the small strain regime I, the partition of strain energy is controlled by parameter w. The bending mode dominates at small values of w, while the axial mode dominates at large w (Figure 6.7(b)). If crosslinks are not perfectly rigid, strain energy is also stored in the inter-fiber bonds. The crosslinks are loaded in bending, axial, torsion, and shear modes. It is observed that the partition of the energy stored in crosslinks parallels the partition of the energy stored in fibers (Bergstrom et al., 2019). Specifically, at small w, when fibers are loaded mostly in the bending mode, the crosslink energy is also stored primarily in the bending mode. At large w, the fiber energy is stored predominantly in the axial mode, while the crosslink energy is stored in the shear mode.

To identify the relation between network strength and structural parameters, it is necessary to perform a parametric study by varying network parameters independently, while evaluating the strength. Before focusing on results from numerical models, it is desirable to perform a simple, "back-of-the-envelope" evaluation of the relationship of interest.

The stochastic nature of the network mandates that various quantities, such as the segment length, the strain energy stored in fibers, and the strain energy stored in crosslinks, are Poisson distributed. This is confirmed by numerical models that evaluate these quantities for each component of the network. Assuming that the effective force in crosslinks (F_{Xeq} of Eq. (3.1)) is Poisson distributed with mean $\overline{F}_{Xeq}$, it is possible to compute the fraction of crosslinks that, under given macroscopic loading, carry equivalent forces larger than their strength, f_{Xc}. This fraction results as the integral of the Poisson distribution from f_{Xc} to infinity, which has the form $\exp\left(-f_{Xc}/\overline{F}_{Xeq}\right)$. According to the results shown in Figure 8.5, the peak stress is reached when this fraction becomes equal to a constant, $\beta \approx 0.02$. Hence, when the peak stress is reached, the relation $\overline{F}_{Xeq\_r} = f_{Xc}/(-\ln\beta)$ must hold. $\overline{F}_{Xeq\_r}$ is the mean force carried by crosslinks at peak stress.

To relate the mean equivalent force in crosslinks $\overline{F}_{Xeq}$ to the applied stress, one may identify a surface which passes through crosslinks, while not cutting any fiber, and separates the body in two parts, and which is oriented on average orthogonal to the direction of the applied load. The applied stress can be written as the mean crosslink force divided by the area of this surface corresponding to a crosslink in the projection on the plane orthogonal to the external load. It results that $\overline{F}_{Xeq} \approx S/\rho_b l_c$. With the expression for $\overline{F}_{Xeq\_r}$ established in the previous paragraph, one may write the network strength as $S_r \sim \rho_b f_{Xc} l_c$.

This simple analysis establishes that if network rupture is associated with crosslink failure, the strength is independent of the fiber properties and is proportional to the

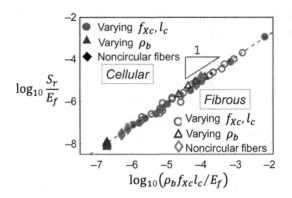

Figure 8.8 Variation of network strength with the non-dimensional group $\rho_b f_{Xc} l_c$ for cellular and fibrous networks with a broad range of network parameters. Reprinted from Deogekar et al. (2019) with permission from Elsevier

crosslink density and the crosslink strength. As discussed in the remainder of this section, this conclusion is broadly supported by numerical simulations and experimental results obtained with diverse network materials.

We turn now to numerical results obtained from large scale simulations of network failure. In these models, Eq. (8.1) is fulfilled and network rupture occurs through crosslink failure. If fiber (tensile) failure would be the controlling process, the results reported here are expected to apply identically, except that the critical parameter representing the crosslink strength, f_{Xc}, would be replaced by the fiber strength, $S_{fc} A_f$.

Figure 8.8 presents data collected from a large number of simulations with networks of different architecture (cellular and fibrous), different crosslink strengths, f_{Xc}, various densities, ρ, various crosslink number densities, ρ_b, and with circular and noncircular fiber cross-sections (Deogekar et al., 2019). The data corresponds to situations in which network rupture takes place in regime I, and to cases with larger f_{Xc}, in which the network undergoes large deformations before failure and ruptures in regimes II or III.

The results support the conclusion emerging from the simple estimate presented in the preceding paragraphs, which indicates that the strength is proportional to the crosslink density and the crosslink strength:

$$S_r \sim \rho_b f_{Xc} l_c. \tag{8.2}$$

It should be noted that the data in Figure 8.8 is expressed in terms of the nominal stress, while in Deogekar et al. (2019) it is presented in terms of the Cauchy stress. In uniaxial tension, the Cauchy and the nominal strengths are related as $S_r = \sigma_r \lambda_{2r}^2$, where λ_{2r} is the stretch in the direction transverse to the applied load, at peak stress. For small f_{Xc}, λ_{2r} is close to 1 and $S_r \approx \sigma_r$. For large values of f_{Xc} and small ρ_b, the lateral contraction at peak stress may be as large as $\lambda_{2r} \approx 0.7$, which implies a factor of ~ 0.5 between S_r and σ_r. On the logarithmic vertical axis of the plot in Figure 8.8, this

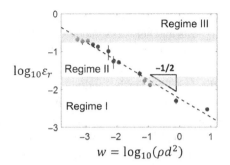

Figure 8.9 Variation of the true strain at peak stress with w for cellular networks of same crosslink strength. The shaded regions represent transitions between regimes I and II and regimes II and III. The data covers a broad range of network parameters. Reprinted from Deogekar and Picu (2018) with permission from Elsevier

difference corresponds to a variation of the vertical position of the data points of magnitude similar to the symbol size.

The result in Figure 8.8 is generally supported by other modeling works. Simulations of 3D fibrous networks by Heyden (2000) indicate that the strength scales linearly with the network density, provided the density is much larger than the stiffness percolation density. In these models, deformations are small and fracture takes place in regime I. An approximately linear dependence of the tensile strength with the crosslink density is reported in Borodulina et al. (2012) for networks aimed to represent paper. The strength of paper usually scales with the basis weight, as observed in tension (Borodulina et al., 2018) and in compression (Brandberg and Kulachenko, 2020). The linear scaling with the basis weight suggested by the models reported in Brandberg and Kulachenko (2020) is supported by experimental data presented in Popil (2017). A somewhat sublinear scaling of the strength with the crosslink density is reported in Goutianos et al. (2018).

The stretch at the peak stress, λ_r, is generally observed in experiments to decrease with increasing network density. The exact functional dependence is not established experimentally. However, Deogekar and Picu (2018) report a relation between the true strain at peak stress, $\varepsilon_r = \ln \lambda_r$, and w of the form:

$$\varepsilon_r \sim w^{-1/2}. \tag{8.3}$$

The data supporting Eq. (8.3) is shown in Figure 8.9 and corresponds to a broad range of ρ, and l_b, values, including the four types of networks N1–N4 referred to in Figures 8.4–8.6. All networks used to produce the data in Figure 8.9 have the same crosslink strength.

The response becomes more brittle as the fiber and crosslink density increase and the network deforms more affinely. The same trend is observed in simulations of spring lattices with disorder (Driscoll et al., 2016). The failure strain also decreases with increasing the mean connectivity of the network, $\langle z \rangle$. This is in line with the concept that promoting the affine deformation renders the behavior more brittle.

8.1.2.1   Brittle to Ductile Transition

Figure 8.9 suggests that networks undergo a ductile to brittle transition as w increases. In this process, the strength increases, while the network microstructure loses its ability to reorganize during deformation. The transition takes place upon increasing the density, the crosslink density, and/or the fiber diameter. In general terms, the affine networks are more brittle than the nonaffine ones. Therefore, any structural modification that decreases the degree of nonaffinity also leads to a reduction of the stretch at peak stress and, overall, to more brittle behavior. For example, increasing the connectivity, $\langle z \rangle$, leads to a reduction of the stretch at peak stress in reconstituted collagen networks (Burla et al., 2020).

The networks discussed here are far above the stiffness percolation threshold. However, if the network is in the proximity of this critical point, the distance to the isostaticity threshold may be used to control the fracture behavior. The deformability of sub-isostatic networks whose $\langle z \rangle$ is gradually increased past the isostaticity threshold is reduced in this process and a ductile to brittle transition is observed (Berthier et al., 2019).

The ductility may also be controlled by adjusting the degree of confinement which, in turn, modifies the degree of nonaffinity of the deformation field. Increasing confinement may be achieved either by increasing the network size or by applying restrictive boundary conditions. A size-induced ductile to brittle transition is reported in Dussi et al. (2020), fracture becoming more brittle as the network size increases; see also the discussion of size effects in Section 8.1.8. If confinement is introduced by the boundary conditions (e.g., compare uniaxial and triaxial loading conditions), a nominally nonaffine network may be forced to deform more affinely if the network size is sufficiently small ($L/l_c < 20$) (see also Section 6.1.1.3.11).

In molecular networks, the network architecture may be modified to impart high toughness and exceptional stretchability to the material. The relationship between the strength and toughness of molecular networks is discussed further in Sections 8.3 and 11.3.

8.1.3   Strength–Structure Relation for Several Network Materials

8.1.3.1   Gels

Normand et al. (2000) present data for two types of agarose gels termed "high viscosity" and "low viscosity." The low viscosity version was obtained by enzymatic degradation. Gels of different concentrations were produced and tested in uniaxial tension and compression. Figure 8.10(a) shows the variation of the strength measured in tension with the gel concentration. Both low and high viscosity gels exhibit an approximately linear variation with the gel concentration (or network density). The response is linear up to rupture and failure is catastrophic (brittle behavior).

The compressive strength also scales approximately linear with the concentration, but the strength in compression is about twice that in tension for the same gels.

Figure 8.10(b) shows the variation of the strain at peak stress with the concentration for the same agarose gels. Since failure takes place at peak stress, this also represents the rupture strain. It is observed that the rupture strain decreases with increasing gel

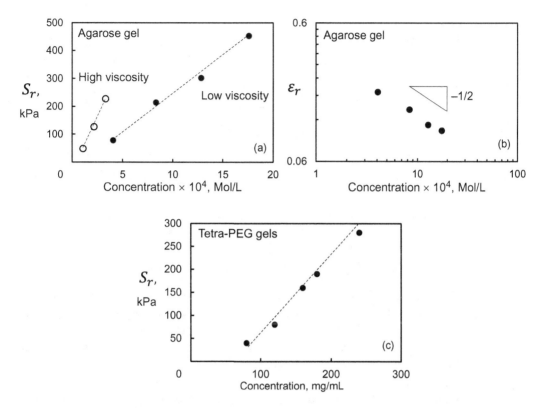

Figure 8.10 Variation of the (a) tensile strength and (b) strain at peak stress measured in uniaxial tension for agarose gels (data from Normand et al., 2000). (c) Strength versus gel concentration for tetra-PEG gels (data from Fukasawa et al., 2010).

concentration following a power law decay of exponent close to –0.5, in agreement with the numerical data shown in Figure 8.9.

The variation of the strength of a set of tetra-PEG gels with the concentration is shown in Figure 8.10(c) (Fukasawa et al., 2010). Tetra-PEG gels are made from molecular strands of the same length and are less heterogeneous than regular gels with polydisperse segment length distribution. Once again, the strength scales linearly with the gel concentration and, hence, with the network and crosslink densities.

A linear relation between strength and the crosslink density is also reported in Fang and Li (2012), who worked with protein gels in which the crosslink density is explicitly controlled. Crosslinking is photochemically initiated and is controlled by adjusting the density of modified lysine residues of the network proteins. Gels produced with these modified proteins also exhibit Young's modulus proportional to the crosslink density.

8.1.3.2   Cellulose Networks

Systematic data on the dependence of the strength on network parameters is generally scarce in the literature. However, strength is of particular importance in the paper

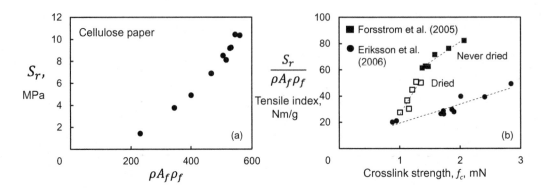

Figure 8.11 Variation of the strength of paper with (a) the network density (data from Eriksson et al., 2006), and (b) with the mean crosslink strength (data from Eriksson et al. (2006) and Forsstrom et al., 2005). The data indicates approximately linear relations of the strength with both these network parameters.

industry and this led to multiple studies of the strength–structure relation in this type of network material. Nevertheless, even in this intensely studied sub-field, very few researchers measured both the strength of crosslinks and the strength of paper, to be able to determine the relationship between f_{Xc} and S_r. The present discussion is based on data from two reports, by Eriksson et al. (2006) and Forsstrom et al. (2005).

Ericksson et al. measured the strength of sheets made of bleached, unbeaten kraft pulp fibers pressed to various densities. Figure 8.11(a) shows the variation of the paper strength with $\rho A_f \rho_m = \rho_m f$, where ρ_m is the density of the fiber material and f is the volume fraction. The strength increases almost linearly with network density.

Figure 8.11(b) shows the dependence of the tensile index of paper on the strength of the crosslinks of the same material. Measuring the strength of crosslinks is a complex experiment and has been performed in these studies with in-house developed instrumentation and methods. The data from Forsstrom et al. (2005) is split into two groups referring to never dried pulp and pulp dried once or multiple times. It is seem that, in all cases, the paper strength increases linearly with f_{Xc}, in agreement with the numerical results presented in Figure 8.8. This conclusion is somewhat expected considering that it is by now established that paper ruptures by the failure of crosslinks and not by fiber failure. Nevertheless, these data sets are unique in that no such direct measurements of both the crosslink and network strengths seem to be available in the literature for any other network material.

8.1.3.3   Nonwovens

Nonwovens made from melt spun polymeric fibers are used in many applications and their mechanical behavior was repeatedly evaluated. Here we present data demonstrating an approximately linear variation of the network strength with the network volume fraction.

The data in Figure 8.12 is collected from Choi et al. (1988), Patel and Kothari (2001), and Chen et al. (2016a) who work with polypropylene nonwovens. The

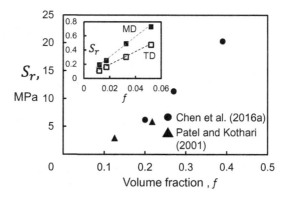

Figure 8.12 Strength–fiber volume fraction relation in polypropylene nonwovens. The main figure is based on data from Patel and Kothari (2001) and Chen et al. (2016a), and the inset presents data from Choi et al. (1988). A web thickness of 1 mm was assumed to convert the data in Choi et al. (1988) and Patel and Kothari (2001).

network strength is plotted versus the network volume fraction, $f = \rho A_f$. To evaluate the volume fraction from the area weight (equal to $f\rho_m t$, where t is the mat thickness), a thickness of 1 mm was assumed when replotting the data from Choi et al. (1988) and Patel and Kothari (2001). The results in Patel and Kothari (2001) and Chen et al. (2016a) refer to heat sealed nonwovens, that is, nonwovens in which crosslinks are created at fiber contacts during an intermediate step of heating under pressure. The main figure shows that S_r is a linear function of the volume fraction. The inset shows similar results for an as-deposited nonwoven tested in the machine (MD) and transverse (TD) directions and demonstrate similar scaling. However, the stress level is significantly reduced relative to the data in the main figure due to the different nature of fiber–fiber interactions.

8.1.4 Effect of Fiber Tortuosity

Most network materials are made from nonstraight fibers with a contour length larger than the end-to-end length. The degree of tortuosity is measured with the crimp parameter of Eq. (2.9), c, which is the ratio of the end-to-end and contour lengths.

It is observed that network strength is independent of c. Figure 8.13 shows numerical data from Deogekar and Picu (2018) obtained with cellular networks of constant crosslink density and increasing crimp. Two data sets are shown, corresponding to low and high crosslink strengths, f_{Xc}, and in both cases S_r is independent of c. In these networks, the fiber density increases as crimp increases, while the crosslink density and the average connectivity are kept constant. The figure further indicates that strength is independent of fiber density, in agreement with Eq. (8.2). The stretch at peak stress increases significantly as the tortuosity increases (c decreases) since the network compliance is very sensitive to c (Section 6.1.1.3.6).

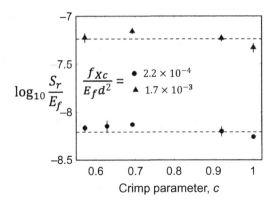

Figure 8.13 Normalized network strength versus the crimp parameter, c, for two values of the crosslink strength, f_{Xc}. Reprinted from Deogekar and Picu (2018) with permission from Elsevier

8.1.5   Effect of Fiber Aspect Ratio and Fiber Compliance

Some experimental and numerical studies suggest that increasing the fiber length, L_0, causes an increase of network strength, particularly if fibers are compliant. In the 3D simulations of fibrous networks presented in Heyden (2000), the strength increases linearly with the fiber length at constant fiber diameter. At the same time, the strain at peak stress is reported to decrease with increasing L_0. The strength also increases with increasing fiber length in the simulations of mat-like 3D networks reported in Goutianos et al. (2018). In the simulations presented in Borodulina et al. (2018) networks are created with fibers of unequal length. The distribution of fiber lengths had a limited effect on the strength and strain at peak stress. Since not all network parameters are evaluated and reported in these studies, the mechanism leading to the increase of the strength with increasing fiber length is unclear. However, it is likely that the crosslink density increases as L_0 increases (Section 4.2.3.2) and this causes the observed increase of S_r.

A note of caution is necessary at this point since, in both experiments and models, once the fiber length becomes comparable to the sample size, strong size effects are expected. Obviously, the strength must increase rapidly when L_0 becomes equal to the sample size and fibers are approximately straight (Kulachenko and Uesaka, 2012). However, this does not apply to networks in which fibers are "infinite" but have large tortuosity, and to networks of staple fibers when the network size is much larger than the fiber length.

An interesting experimental data set obtained with microfibrilated cellulose is presented in Henriksson et al. (2008). These mat-like structures are prepared from pulp via an enzymatic disintegration and beating process and have fibers of different molar mass. The fiber length is evaluated indirectly, by measuring the viscosity of the suspension before network formation. Networks of different L_0 follow the same stress–strain curve which has typical type B behavior (Figure 8.1(b)). The apparent yield stress, the slope of the post-yield regime II branch of the curve, and the Young's modulus are independent of L_0. However, the stress at failure increases proportional to $L_0^{1/2}$. Since the stress–strain curves corresponding to various L_0 overlap, the increase of the strength implies an increase of the total energy absorbed before fracture. Hence, increasing the fiber length increases both the strength and toughness of these materials – a particularly attractive effect.

A similar result is presented in Zhu et al. (2015), who compare cellulose nanopaper with regular paper. The nanopapers considered have fibers of diameters between 10 and 30 nm, while the reference micropaper (regular paper) tested has a mean fiber diameter of about 30 μm. These nanopapers exhibit type B behavior, with strengths two orders of magnitude larger than the micropaper and strains at failure on the order of 6%, which is at least three-times larger than that of the reference paper. This effect may have multiple origins, including the formation of stronger, more cohesive cross-links between nanofibers (compared to the crosslinks in micropaper), and an increase of the number of crosslinks per fiber. A comparison of the mechanical response of mat-like structures made from nanofibers with equivalent structures made from microfibers is also presented in a review article (Benitez and Walther, 2017), where it is reported that nanopapers exhibit larger strength, although differences are not as dramatic as those discussed in Zhu et al. (2015). Since in most of these experimental works network parameters are not controlled or reported, the physical origin of the observed variation of the strength cannot be precisely determined.

8.1.6   Effect of Preferential Fiber Alignment

Fiber alignment leads to elastic anisotropy: The stiffness increases in the direction of alignment and decreases in the directions perpendicular to it (Section 6.1.1.4.7). The expectation based on the results presented in Figure 8.8 is that the strength should follow the trends of the modulus,[2] that is, it should increase in the direction of alignment and should decrease in the directions orthogonal to it. This is because the mean distance between crosslinks projected in the plane orthogonal to the alignment direction is smaller than that probed in any other direction. Hence, the number of crosslinks resisting the applied load per unit area of the plane perpendicular to the load is larger when strength is probed in the direction of alignment. This physical picture linking the projected crosslink density to the strength is implicit in the result of Eq. (8.2), although the respective formula corresponds to isotropic networks.

A collection of data demonstrating these trends is presented in Benitez and Walther (2017). The data is collected from Iwamoto et al. (2011), Sehaqui et al. (2012), and Torres-Rendon et al. (2014), who worked with cellulose nanopapers in which fiber orientation was introduced by various methods. As the degree of alignment increases, both modulus and strength probed in uniaxial tension in the alignment direction increase. The increase of the modulus is more pronounced than that of the strength.

---

[2] Scaling of the strength with network parameters is different from that of the modulus.

The strain at failure decreases at the same time. Similar results are obtained with nonwovens. In Rawal et al. (2013), needle-punched nonwovens with preferentially aligned polymeric microfibers are probed in the direction of fiber alignment and orthogonal to it. The modulus and strength are larger in the direction of alignment, while the strain at failure is lower. Chocron et al. (2008) test commercial polypropylene nonwovens and report a strong effect of the fiber orientation on the strength. The strength measured in the orientation direction is five-times larger than that measured in the transverse direction. The degree of orientation is not reported in this study. Working with crosslinked mats of electrospun gelatin fibers of $\sim$70 nm diameter, Tonsomboon and Oyen (2013) observe similarly that the strength probed in the direction parallel and perpendicular to that of fiber alignment increases and decreases, respectively, with increasing degree of alignment. The strain at failure in these networks turns out to be independent of fiber alignment.

An interesting situation emerges in strongly aligned networks when the fracture toughness anisotropy is large. In this case, cracks growing in the direction orthogonal to the alignment may deflect in the alignment direction. This can be rationalized in terms of the continuum analysis of crack deflection of He and Hutchinson (1989), which indicates that deflection is expected to happen when the ratio of the toughness for cracks growing in the parallel-to-alignment and normal-to-alignment directions is approximately 0.25. This mechanism greatly enhances the toughness evaluated in the normal-to-alignment direction, as is often observed in highly anisotropic structures such as wood. Although not reported to date, the same mechanism may also toughen strongly aligned biological networks, such as tendons. A suggestion was made in the literature on elastomers (Gent et al., 2003) that roughening of the crack plane observed in filled rubbers at high stretches is related to crack deflection. We conjecture that deflection is caused by the mechanism described here, associated with the preferential orientation of the chains.[3]

8.1.7   Effect of Variability of Network Parameters and of Crosslink Properties

The heterogeneity of network materials is not exclusively of structural type. In most biological and other network materials occurring in nature, fibers forming a given network are irregular, have different lengths and different diameters, and may have varying cross-section shapes along their length. In addition, the crosslink strength is not expected to be identical for all crosslinks of a network. This is clearly the case, for example, in paper, in heat sealed nonwovens, and in mats of nanofibers stabilized by cohesion. Therefore, it is of obvious importance to inquire to what extent such variability influences the strength of the respective networks.

This issue was studied primarily by modeling, since in experiments it is difficult to control the distribution of these parameters. Works that address this problem include

---

[3] Note the difference relative to the common crack deflection mechanism in composites in which cracks are deflected by the reinforcement. Here, deflection is caused by toughness anisotropy.

Malakhovsky and Michels (2007), Borodulina et al. (2018), and Deogekar and Picu (2018). In these studies, one structural parameter was rendered stochastic at a time and was selected from a distribution of predefined mean and increasing variance (in separate models). It was sought to determine to what extent such variability influences the mean of the distribution of network strength values.

Numerical studies indicate that fluctuations of fiber diameter do not affect the mean of the strength in any significant way, in agreement with the prediction of Eq. (8.2), which indicates that the strength is independent of fiber properties. In Borodulina et al. (2018) is it found that fluctuations of fiber length lead to variability in crosslink density which, in turn, affects the strength.

According to Eq. (8.2), the strength should be sensitive to the variability of the crosslinks strength. Such an effect is not made specific in Eq. (8.2), which is developed based on models without fluctuations of the fiber and crosslink properties. The implication that f_c appearing in Eq. (8.2) should be the mean of the distribution of crosslink strengths is not necessarily correct. One may rationalize this issue in the following way: it is observed that the peak stress is reached when a fraction $\beta \approx 2\%$ of the crosslinks fail. If the crosslink strength forms a distribution, it is mostly the weakest 2% of the bonds that should fail first. Hence, increasing the variance of the distribution implies that this threshold is reached at lower applied stress, which mandates that the network strength decreases with increasing crosslink strength variability. Numerical results led to apparently contradictory conclusions. Simulations of 2D Delaunay network failure presented in Malakhovsky and Michels (2007) indicate that the network strength decreases as the variability of fiber strength increases. Delaunay networks have large connectivity $\langle z \rangle$ and deform affinely even when the bending fiber deformation mode is softer than the axial mode. Hence, fibers are loaded axially and network rupture may be equally associated with fiber and crosslink failure. Working with cellular 3D networks, Deogekar and Picu (2018) conclude likewise that increasing the variability of the crosslink strength causes a pronounced decrease of the mean network strength. However, the simulations in Borodulina et al. (2018) indicate little effect of the crosslink strength variability on the mean network strength. This apparent discrepancy is possibly caused by the different distribution functions considered in these studies and the different set of system parameters considered.

8.1.8 Effect of Sample Size

Network materials exhibit significant size effects. These are caused by the intrinsic heterogeneity of the network: The more pronounced the heterogeneity, the stronger the size effect.

The relation between heterogeneity and size effects was studied extensively in statistical physics and in mechanics. A summary of the physics perspective is provided in Alava et al. (2009). The mechanics perspective is presented in various books on composite materials, such as Horii and Nemat-Nasser (1993) and Torquato (2002).

These works are not concerned with network materials. However, it so happens that the physics community made substantial progress toward understanding the role of heterogeneity (or "disorder") using discrete lattice models, such as the random fuse model and the random spring model (de Arcangelis et al., 1985). While these models do not reproduce the mechanics of real networks since they are confined to small deformations of a nonevolving network structure, they do provide the conceptual framework based on which the size effect may be analyzed.

As discussed in Chapter 4, networks are mechanically heterogeneous due to the spatial variability of fiber and/or crosslink densities and/or due to the variability of fiber and crosslink strengths. This leads to a size effect that may be interpreted based on the Weibull (1961) and Gumbel (1935) models, for example. This discussion applies to networks without pre-existing cracks or notches and without embedded inclusions. A different perspective pertaining to networks with pre-existing cracks is reviewed in Section 8.2.

Following Section 2.3.7, consider a partition of a network domain of size L in sub-domains of smaller size, L' (the number of sub-domains is $n = L/L'$), each having a cumulative probability of failure at stresses below S_r, given by $P_{L'}(S_r)$. The probability that a sub-domain does not fail at stresses below S_r is $1 - P_{L'}(S_r)$. If the sub-domains are assumed to fail independently, the probability that the entire domain does not fail at stresses below S_r is $1 - P_L(S_r) = (1 - P_{L'}(S_r))^n$. For large n, the probability of failure of the entire body may be approximated:

$$P_L(S_r) = 1 - \exp\left(-nP_{L'}(S_r)\right). \tag{8.4}$$

This establishes the relation between local and global failure probabilities. To make progress, one has to postulate a functional form for the local probability. In the Weibull theory, the function $P_{L'}(S_r)$ is taken to be a power function. Hence, the cumulative probability of failure of the entire body at stress S_r is:

$$P_L(S_r) = 1 - \exp\left(-n(S_r/S_0)^\beta\right), \tag{8.5}$$

where β is the Weibull modulus. The probability density is computed as the derivative of Eq. (8.5) with respect to the stress. The mean strength can be computed based on the probability density and is given by:

$$\overline{S}_r \sim n^{-1/\beta} S_0, \tag{8.6}$$

which is identical to Eq. (2.17) derived for the fiber scale. Equation (8.6) indicates that the strength decreases as a power function of the sample size with exponent equal to the inverse of the Weibull modulus. This celebrated scaling law applies to many quasi-brittle materials.

The underlying concept in Weibull's theory is that the body fails when its weakest component fails. While this is obvious in the case of a fiber (Section 2.3.7), in which all sub-domains are connected in series and subjected to the same stress, it is not necessarily obvious in 2D or 3D, in which sub-domains are not loaded by the same local stress. Nevertheless, the "weakest link" concept seems to apply in many cases provided the sample size is larger than a reference length scale L', which is viewed as a characteristic material length scale (see, e.g., Curtin, 1998).

The Gumbel theory differs from Weibull's by the assumption made about the functional form of the local cumulative failure probability. If $P_{L'}(S_r)$ is taken to be an exponential function, the cumulative probability of failure of the body at stress S_r is:

$$P_L(S_r) = 1 - \exp\left(-n\exp\left(-S_0/S_r\right)\right). \qquad (8.7)$$

This functional form is the modified Gumbel cumulative probability introduced in Duxbury et al. (1994), which is defined for $S_r \in (0, \infty)$. It corrects the deficiency of the original Gumbel distribution which allows for negative strengths. Equation (8.7) leads to a logarithmic dependence of the strength on sample size:

$$\overline{S}_r \sim S_0/\ln n, \qquad (8.8)$$

The Weibull scaling of Eq. (8.6) is more widely used in applications than Eq. (8.8).

Network strength data was interpreted based on the Weibull model by several authors. In Heyden (2000), numerical data for 2D Mikado networks indicates a Weibull modulus $\beta = 1.88$. 3D models developed to represent paper in Kulachenko and Uesaka (2012) also convey the conclusion that strength is represented by the Weibull distribution, but the resulting modulus is larger. The 2D Delaunay networks studied in Malakhovsky and Michels (2007) indicate that $\beta = 1.58$ in the low network heterogeneity case. As described by Eq. (8.6), the strength decreases with increasing network size and the steepness of the decay is controlled by the value of β.

A discussion of the size effect of strength in terms of the structural parameter w is provided in Deogekar and Picu (2018). Figure 8.14 shows the variation of the strength of networks of two different w values with the sample size, L. The data in Figure 8.14(a) corresponds to system N1 shown in the master plot of Figure 8.4(b), while the results in Figure 8.14(b) correspond to system N4. Networks N1 are strongly nonaffine and of

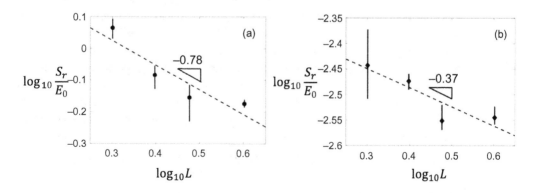

Figure 8.14 Variation of the strength (normalized by the network modulus, E_0) with the sample size for networks N1 and N4 shown in Figure 8.4(b). Reprinted from Deogekar and Picu (2018) with permission from Elsevier

high effective heterogeneity, while networks N4 are almost affine and of lower effective mechanical heterogeneity. Fitting Eq. (8.6) to this data provides the Weibull modulus, which results equal to 1.28 for the nonaffine network N1, and 2.7 for the more affine network N4.

This analysis indicates that more affine networks are less sensitive to the size effect (smaller slope in Figure 8.14(b) and larger β), which correlates with the fact that failure in affine networks is more localized and more brittle. This trend is visible in Figure 8.4(a). A more affine network is approximately mechanically homogeneous (although geometrically it may be strongly heterogeneous) and its behavior is close to that of a nominally homogeneous continuum.

Size effects are often observed in experiments. Popil (2017) reports that size effects are more pronounced in low basis weight paper (low w) compared to the high basis weight case. Networks of large w are less sensitive to the size effect, in agreement with the results shown in Figure 8.14. Similar effects are observed in nonwovens, as discussed in Patel and Kothari (2001), who compare heat sealed and needle punched polymeric nonwovens. The heat-sealed samples are densely crosslinked and are affected less by size effects than the needle punched networks whose strength is controlled by sparse entanglements.

As already pointed out in this section, the scaling of the strength with sample dimensions described by Eqs. (8.6) and (8.8) applies only for L larger than a characteristic length scale L' assumed to be a material property. Consider a situation in which the sample width is varied, while the length is kept unchanged. In experiments performed with paper (Borodulina et al., 2012; Popil, 2017) and with nonwovens (Patel and Kothari, 2001), it is observed that increasing the sample width leads to an increase of the strength. As their width increases, samples transition from effectively being 1D (i.e., characterized by a serial arrangement of sub-domains of varying strengths) to 2D. Fibers have larger kinematic freedom to rearrange during deformation in narrow width samples due to the lack of lateral constraint. This implies that narrow width samples should be more deformable and less strong, which is in agreement with experimental observations. A similar conclusion is reached in Dussi et al. (2020) based on spring network models, which indicate that smaller samples are more ductile than larger sample. These ideas can be related to the occurrence of the self-equilibrated eigenstress induced by network heterogeneity, as discussed in Section 6.1.1.4.12. This eigenstress reduces lateral contraction and filament reorientation in samples which are wide enough. In narrow samples, the boundary conditions preclude the development of the self-equilibrated stress, which allows enhanced lateral contraction and increases ductility. The characteristic length scale L' in paper and nonwovens is expected to be related to the fluctuations of largest length scale present in the material, for example, associated with flocculation. Although not evidenced to date, similar behavior is expected in biological collagen networks.

8.1.9   Empirical Relation between Stiffness and Strength

The relations between network parameters and either strength or stiffness outlined in Sections 8.1.2 (Eq. (8.2)) and 6.1.1.3.1, respectively, suggest that these two material properties may be correlated. The stiffness of an affinely deforming network is proportional to the fiber density, $E_0 \sim \rho$, while its strength is proportional to the crosslink density, $S_r \sim \rho_b$. With Eq. (4.21), it may be implied that, for this type of networks, $S_r \sim E_0$.

These considerations provide the scaling relation, but not the proportionality constant. In fact, inquiring about the magnitude of this constant is an ill-posed problem since the same analysis indicates that the constant should depend on parameters such as the crosslink strength and mean connectivity, which are expected to vary broadly from material to material. In addition, if the deformation is not affine, the relationship between S_r and E_0 is nonlinear. Surprisingly, the analysis of a large number of data sets from the literature indicates that networks may be grouped in three categories based on the S_r/E_0 ratio and these categories align with the type of network behavior defined in Figure 8.1, that is, types A, B, and C. These empirical results are summarized in Figure 8.15.

The S_r/E_0 ratio for materials of type A takes values in the range 0.1–1. The type B materials consistently exhibit values in the range 0.01–0.06. The hyperelastic type C networks have values of S_r/E_0 larger than 2. The material systems considered in this analysis are listed in the caption to Figure 8.15.

Rubbers (type C networks) may exhibit ratios S_r/E_0 larger than 10, but their relation between strength and small strain modulus is not linear (Section 8.4.1.1). Tetra-PEG gels made from end-linked chains of the same length exhibit ratios S_r/E_0 between 3 and 5 (Akagi et al., 2013a). This is to be contrasted with the ratio S_r/E_0 of most metallic and ceramic engineering materials which is much smaller than 0.01.

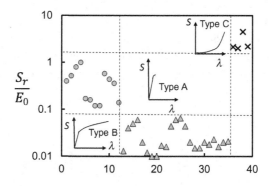

Figure 8.15 Ratio of the uniaxial tension strength to the modulus, S_r/E_0, for a broad range of network materials. The data points are referred to in the order indicated on the horizontal axis and represent: 1–4, paper (Alava and Niskanen, 2006); 5–11, agarose gels of various concentrations (Normand et al., 2000); 12, electrospun polyamide nanofiber mats (Stachewicz et al., 2011); 13–16, polypropylene nonwovens (Chen et al., 2016a, 2016b); 17–19, carbon nanofiber nanopaper, office paper, buckypaper (Mao et al., 2017); 20–22, microfibrilated cellulose nanopapers with different fiber lengths (Henriksson et al., 2008); 23–26, soy protein nanofiber mats of different concentrations (Sinha-Ray et al., 2012); 27–35, carbon nanofiber mats with different porosity and different chemical treatments (Benitez and Walther, 2017); 36–38, needle-punched nonwovens probed in different directions (Chocron et al., 2008); and 39, PDMS (Hajiali and Shojaei, 2017).

8.1.10   Multiaxial Failure

Network materials are generally loaded multiaxially. In fact, many uniaxial experiments with soft materials are actually performed under (partially controlled) multiaxial conditions if samples are held rigidly and not allowed to contract at the clamping sites during deformation. Therefore, it is necessary to extend the results for uniaxial loading discussed in the preceding sections of this chapter to the multiaxial case. Two issues are of particular importance: (i) establishing a failure surface that provides the strength under a generic stress state and (ii) extending the strength–structure relation of Eq. (8.2), which applies to uniaxial tension, to multiaxial loading conditions.

The stress–stretch nonlinear behavior under multiaxial loading is discussed in Section 6.1.3. Here we focus on aspects related to failure. The first interesting observation emerges from Figure 8.16, which shows the strength evaluated in equi-biaxial and equi-triaxial loading, S_r^{EB} and S_r^{ET}, for networks with a broad range of w values. Ratios S_r^{EB}/S_r^U and S_r^{ET}/S_r^U, where S_r^U is the uniaxial strength, are approximately independent of w (Deogekar and Picu, 2021).

A failure surface evaluated for cellular networks of various w values is presented in Deogekar and Picu (2021). Figure 8.17 shows results for plane stress biaxial loading. Data for three w values are provided: $w = -5.2$, which is denoted as network E, and for networks N1 and N4 of Figure 8.4(b). The two axes are normalized by the uniaxial strength for each of these networks. This normalization leads to the collapse of the failure envelopes corresponding to different w. The important consequence of this collapse is that the strength–structure relation of Eq. (8.2) remains valid in the multiaxial case.

The data in the first quadrant of Figure 8.17 indicates that network failure may be described in terms of the maximum stress criterion. This is shown by the dashed line in the first quadrant. However, this criterion does not provide an accurate prediction of

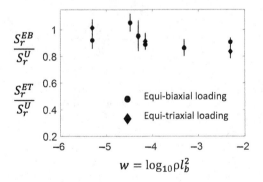

Figure 8.16 Nominal strength of cellular networks in equi-biaxial, S_r^{EB}, and equi-triaxial, S_r^{ET}, loading normalized by the uniaxial strength, S_r^U, for a broad range of w values. Reproduced from Deogekar and Picu (2021) with permission of The Royal Society of Chemistry

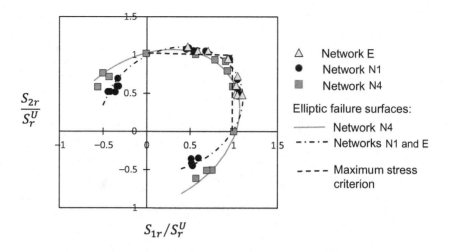

Figure 8.17 Failure surface in the plane of principal nominal stresses (S_1, S_2), with $S_3 = 0$, for networks N1 and N4 of Figure 8.4(b), and network E with $w = -5.2$. The two axes are normalized by the uniaxial strength of the respective networks. The predictions of the maximum stress failure criterion and of the criterion of Eq. (8.9) are shown. Reproduced from Deogekar and Picu (2021) with the permission of The Royal Society of Chemistry

the shear-dominated failure (quadrants 2 and 4). A better approximation is obtained by using an elliptic failure surface of the form:

$$\left(\frac{S_r^{eq}}{\tilde{S}_r^{eq}}\right)^2 + \left(\frac{S_r^m - S_0^m}{\tilde{S}_r^m}\right)^2 = 1, \qquad (8.9)$$

where S_r^{eq} is the von Mises equivalent stress at rupture, S_r^m is the mean stress, while $\tilde{S}_r^{eq}$, $\tilde{S}_r^m$, and S_0^m are fitting constants. This type of failure surface has previously been proposed to describe the failure and yield of porous continua (e.g., Deshpande and Fleck, 2001; Combaz et al., 2011). The fit of Eq. (8.9) is shown along with the data in Figure 8.17. Equation (8.9) also provides an adequate representation of the triaxial part of the failure surface, as discussed in Deogekar and Picu (2021).

Network materials have a large free volume, which renders them pressure sensitive. This is also observed in porous materials. Equation (8.9) captures the pressure sensitivity, which manifests itself primarily in the triaxial and shear loading modes. Pressure sensitivity is less pronounced in affine networks which do not undergo large structural changes accompanied by large volume variation before failure.

While deformation under multiaxial loading was studied in many network materials, a small number of reports provide multiaxial failure data. Figure 8.18 shows a comparison of available experimental and numerical data with the failure surface of Eq. (8.9) (Deogekar and Picu, 2021). The figure shows three data sets referring to plane stress biaxial loading of paper (Gustafsson and Niskanen, 2011), biaxial loading of a IPN (double network) hydrogel (Mai et al., 2018), and biaxial loading of

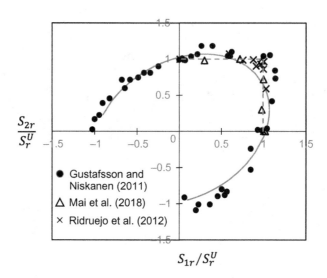

Figure 8.18 Experimental failure surface for paper (Gustafsson and Niskanen, 2011), a polypropylene nonwoven (Ridruejo et al., 2012), and an interpenetrating network gel (Mai et al., 2018) compared with the elliptic failure surface of Eq. (8.9) calibrated with the numerical data presented in Figure 8.17 (continuous line). The prediction of the maximum stress criterion is shown by the dashed line. Reproduced from Deogekar and Picu (2021) with the permission of The Royal Society of Chemistry

polypropylene nonwoven fabrics (Ridruejo et al., 2012). The nonwoven data is obtained numerically using a constitutive model pre-calibrated on experimental data, while the other two data sets are entirely experimental. The nominal strength reported in these references is normalized by the nominal strength in uniaxial tension, S_r^U, which leads to the collapse of the failure surfaces, as also seen in Figure 8.18. The samples used in Gustafsson and Niskanen (2011) are anisotropic. Therefore, in this case the two axes are normalized with S_r^U in the respective directions. The curve labeled "network N4" from Figure 8.17 is added without modification to Figure 8.18 and is seen to provide a good approximation of the experimental data for paper in all loading modes. The other two data sets are restricted to the first quadrant and, within the present accuracy, are described equally well by both Eq. (8.9) and the maximum stress criterion.

Oyen et al. (2004) present a collection of data from the literature referring to the failure of the human fetal membrane. The component providing strength to the fetal membrane is a collagen network. They indicate that the strength in biaxial loading is equal to that measured in uniaxial loading, within the scatter of the experimental data. This is in close agreement with the conclusion emerging from Figures 8.17 and 8.18.

Size effects are present equally in uniaxial and multiaxial loading and their magnitude is likely independent of the type of loading. Under this assumption, the normalization used in Figures 8.17 and 8.18 renders the failure surface size effect-free. In other words, the size effect is accounted for in Eq. (8.9) through the dependence of parameters $\tilde{S}_r^{eq}$, $\tilde{S}_r^m$, and S_0^m on sample size. This remark remains a conjecture at this time since it was not verified experimentally or by modeling.

8.2 Networks with Pre-existing Cracks and Notch Insensitivity

A perspective on the problem of network failure distinct from that outlined in Section 8.1 is to consider network materials with pre-existing crack-like defects. This is the view taken in Fracture mechanics, where the discussion focuses on the growth of pre-existing flaws and not on their nucleation. One may debate the value of such a view in network materials which are not continua. However, there are practical applications in which crack-like defects and holes of size much larger than all characteristic network length scales are present in network materials. In these cases, one may inquire under what conditions material failure is triggered by such defects.

Most experiments are performed with one of the edge cracked configurations shown in Figure 8.19(a) and (b), where a is the length of the pre-defined notch and b is the sample width. In linear elastic Fracture mechanics, the growth of such cracks is rationalized by comparing the stress intensity factor, $K(a,b)$, characterizing the stress field at the crack tip, with the fracture toughness, K_c, which is considered a

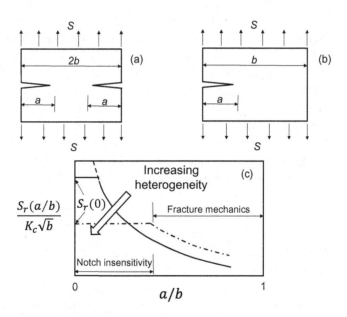

Figure 8.19 Configurations with pre-existing cracks used in experiments are shown in (a) and (b). (c) Schematic representation of the variation of the normalized strength function of ratio a/b for two levels of mechanical heterogeneity. $S_r(0)$ represents the strength of the material without pre-existing cracks.

material property. For both configurations in Figure 8.19 one may write the condition of crack growth:

$$K_c = S_r\sqrt{\pi a}f(a/b) = S_r\sqrt{b}f^*(a/b), \qquad (8.10)$$

where S_r is the stress causing crack growth, and $f^*(a/b) = \sqrt{\pi a/b}f(a/b)$. Function $f(a/b)$ corresponding to K-dominance conditions is tabulated in Fracture mechanics texts (e.g., Tada et al., 1973).

As a/b decreases, the critical stress increases and the value of the strength of the material without pre-existing cracks, $S_r(0)$, is reached in the limit $a/b \to 0$. This is shown in Figure 8.19(c) by the horizontal line which defines the upper limit of the strength S_r. It also defines a crack length, a_c (or a_c/b), below which the material is insensitive to the presence of the crack and fails at the same stress as the virgin network. It is said that for $a < a_c$ the material is notch insensitive.

The strength of the virgin material, $S_r(0)$, is discussed in Section 8.1, where it is shown that $S_r(0)$ decreases as the degree of network heterogeneity increases, while all other network parameters (e.g., the crosslink strength, f_{Xc}, and the crosslink density, ρ_b) are kept constant (Section 8.1.7). At the same time, increasing the degree of material heterogeneity increases the effective material toughness, K_c, which shifts upward the branch of the curve in Figure 8.19(c) predicted by Fracture mechanics (i.e., for $a > a_c$). The combined effect of these two trends leads to an increase of the range of notch insensitivity as the degree of network heterogeneity increases (Figure 8.19(c)).

Therefore, one may identify two regimes: $a/b < a_c/b$, for which the material strength is notch insensitive, and $a/b > a_c/b$, for which Fracture mechanics applies.

Considering a_c a material parameter, one may also take the view that cracks smaller than this length cannot be considered macroscopic cracks and should be viewed as microstructural features. Then, the effective length of a macroscopic crack of physical length a should be taken as $a + a_c$. Alternatively, as suggested by Bazant (2004), a_c represents the size of the fracture process zone present at the tip of any macroscopic crack, and this length may be large in quasi-brittle materials that sustain significant damage accumulation before catastrophic fracture. Then, one rewrites the Fracture mechanics crack growth conditions of Eq. (8.10) as:

$$K_c \sim S_r\sqrt{a + a_c}. \qquad (8.11)$$

This expression may also be obtained from a statistical analysis that takes into account the fact that a_c introduces an internal length scale in the linear elastic fracture problem (which otherwise, has no internal length scale). A brief outline of this derivation is presented in the review of Alava et al. (2009) and in Bazant (2004).

This description based on linear elastic Fracture mechanics holds if the process zone at the crack tip is sufficiently small for K-dominance to apply. Soft network materials exhibit large deformations and large crack opening displacements at the onset of crack propagation. If J-dominance exists, a critical energy release rate criterion is adequate. For example, the results for dense cellulose networks in Isaksson and Dumont (2014) suggest that K-dominance exists in the respective

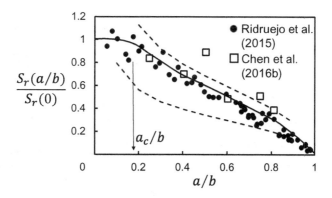

Figure 8.20 Ratio of the strength of materials with and without pre-existing cracks function of the normalized crack length. Data for commercial polymeric nonwovens (Ridruejo et al., 2015) and for fiberglass networks (Chen et al., 2016b) is shown along with the prediction of Eq. (8.10) for two values of K_c (shown with dashed line). The continuous line is added to guide the eye.

material and an interpretation based on Eq. (8.11) is valid. A criterion based on the critical energy release was suggested to apply in the case of a compliant fibrin network (Tutwiler et al., 2020) and is also used to quantify fracture in rubber (see Section 8.4.2). However, it is not entirely understood at this time how network parameters control the size, structure, and evolution of the process zone at the crack tip, and under what conditions one may apply criteria established in classical Fracture mechanics. In the remainder of this subsection we choose to present experimental results in terms of the nominal critical stress at crack growth, which is compared with the nominal strength of the same material without pre-existing cracks.

Based on this background, we next explore experimental results obtained with various network materials. The experimental data presented is adapted to fit the format of Figure 8.19(c). The nominal stress at rupture, S_r, is computed in all cases based on the applied force and the full sample width, $2b$.

Figure 8.20 shows data for networks of microfibers. It combines results for polypropylene nonwovens from Ridruejo et al. (2015) and data for fiberglass nonwovens from Chen et al. (2016b). The vertical axis is normalized by the strength of the material without pre-existing notches. The data for these two (otherwise quite different) nonwovens overlap once the axes are normalized. The resulting curve has the features outlined in the schematic representation of Figure 8.19(c) and a microstructural length scale a_c may be identified in both cases. The material is notch insensitive for crack lengths smaller than a_c. The prediction of the Fracture mechanics-based Eq. (8.10) is shown in Figure 8.20 for two values of K_c (which are arbitrarily selected to provide upper and lower bounds to the data set).

Figure 8.21 shows a similar collection of data for several biological network materials. The data set from Purslow (1985) is obtained by testing cooked beef muscle and covers a broad range of values of a/b. Similar to Figure 8.20, the critical stress of pre-notched samples, $S_r(a/b)$, is normalized by the strength of the material without notches, $S_r(0)$. It is observed that the behavior is notch insensitive for all crack lengths

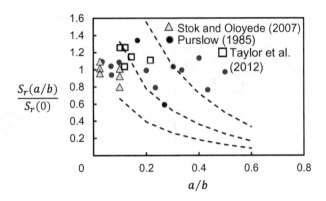

Figure 8.21 Ratio of the strength of materials with and without pre-existing cracks, as a function of the normalized crack length. Data for cartilage (Stok and Oloyede, 2007), raw beef muscle (Taylor et al., 2012), and cooked beef muscle (Purslow, 1985) are shown, along with the prediction of Eq. (8.10) for three values of K_c.

considered. A similar behavior is observed for raw beef muscle, as reported in Taylor et al. (2012), and for cartilage, based on data reported in Stok and Oloyede (2007). The tests on cartilage are performed with sub-millimeter, short edge cracks, which may be considered microstructural cracks and the respective samples are expected to be notch insensitive. A short edge crack in a sample undergoing large deformations opens so much that it effectively disappears during loading and the sample has essentially smooth edges at rupture. The cracks considered in Taylor et al. (2012) are ~3 mm long and are certainly much larger than the intrinsic network length scales. This observation also applies to the results of Purslow (1985), whose tests are performed with cracks as large as 15 mm. All samples considered in these studies are notch insensitive for all crack lengths. The notch insensitivity effect in these biological networks is much stronger than that shown in Figure 8.20 for networks of microfibers.

Figure 8.22 shows results obtained with various mats of nanoscale fibers. These include cellulose nanofiber nanopaper prepared from softwood pulp, bacterial cellulose nanopaper, carbon nanotube-based buckypaper from Mao et al. (2017), and electrospun polyamide nanofiber mats from Stachewicz et al. (2011). The figure also includes results obtained with edge cracked fibrin samples from Tutwiler et al. (2020), where the mean fibrin fiber diameter is 125 nm. The cellulose nanopapers and the buckypaper have area weights of 46, 61, and 63 g/m^2, respectively, while the polyamide fibrous mats have an average density of 1.2 g/cm^3 and fiber diameter of 186 nm. The strength of samples of various crack lengths is normalized with the strength of the corresponding uncracked specimens reported in the respective publications. The polyamide nanofiber networks exhibit notch insensitivity for the entire range of a/b considered. The other networks exhibit weak dependence on a/b. An interesting observation, which is quite different from the behavior shown in Figures 8.20 and 8.21, is

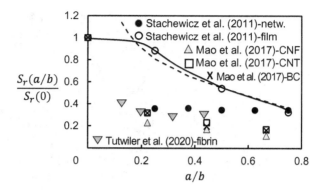

Figure 8.22 Ratio of the strength of materials with and without pre-existing cracks as a function of the normalized crack length. Data for polyamide nanofiber networks and polyamide films (Stachewicz et al., 2011), nanopapers made from cellulose nanofibrils (CNF), bacterial cellulose (BC), and carbon nanotube buckypaper (CNT) (Mao et al., 2017), and for fibrin networks (Tutwiler et al., 2020) are shown along with the prediction of Eq. (8.10) fitted to the large a/b range of the polyamide film data set (dashed line).

that $S_r(a/b)$ is significantly smaller than the un-notched sample strength, $S_r(0)$. This indicates that the presence of the crack reduces the strength, but this effect is weakly dependent on the crack length.

The article by Stachewicz et al. (2011) mentioned in the previous paragraph presents data for polyamide films, in addition to the data for polyamide nanofiber mats. These results are included in Figure 8.22 and exhibit the expected behavior predicted by Fracture mechanics (Figure 8.19(c)). The fit of the larger a/b range of the thin film data with Eq. (8.10) is included in Figure 8.22 for reference.

Observations of insensitivity of the strength to pre-existing cracks of length smaller than a threshold are also made in paper (Krasnoshlyk et al., 2018) and gels (Yang et al., 2019). Papers of high and low density are compared in Krasnoshlyk et al. (2018) and it is observed that, as expected, a_c increases with decreasing the average network density. A dense paper does not exhibit a measurable a_c, while a low-density material (23 g/m²) is insensitive to pre-existing cracks of up to 9 mm in length. In gels, it is observed that increasing structural heterogeneity controlled by increasing the degree of network defectiveness leads to notch insensitivity. a_c of about 1 mm, much larger than any network length scale, is reported for a polyacrylamide gel in Yang et al. (2019).

A reliable prediction of a_c in terms of the network structure and parameters is not available at this time in the literature. However, attempts have been made to relate a_c to the size of the fiber flocs in paper (Krasnoshlyk et al., 2018) and to the ratio of the toughness to the specific work of fracture (area under the stress–strain curve) (Chen et al., 2017); this ratio has units of length. These works place notch insensitivity in relation to material heterogeneity on the network scale. An interesting alternative perspective is due to Balankin (e.g., Balankin et al., 2011). The heterogeneity is considered to have fractal structure (in a certain range of length scales), which is supported by direct measurements of flocculation (Figure 4.6). A theoretical argument

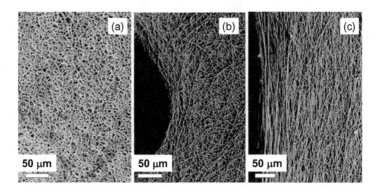

Figure 8.23 Scanning electron micrographs of the notch tip region in a polycaprolactone electrospun mat in the (a) unloaded state and (b, c) at two levels of the applied load. Reprinted from Koh et al. (2013) with permission from Elsevier

is used to show that the crack tip field is not singular if the fractal dimension is larger than a threshold, which is equivalent to requiring the heterogeneity to be sufficiently pronounced in order to ensure notch insensitivity. Since the fractal dimension is related to the exponent of the two-point correlation function of the fiber density, ζ (Eq. (4.23) and Figure 4.6), it is implied that, for $\zeta > 0.5$, the crack field is not singular. However, this theory does not explain the dependence of the fracture behavior on crack length.

Direct observations of the evolution of the fibrous structure in the close vicinity of the crack tip provide empirical evidence of the mechanism leading to notch insensitivity. If fibers have sufficient kinematic freedom to reorient in the direction normal to the crack plane as the crack is loaded in mode I, strong local anisotropy and stiffening of the material in front of the crack along with crack blunting are observed. These effects lead to an increase of the effective material toughness. The development of preferential fiber orientation can be seen in Figure 8.23, which shows a series of scanning electron micrographs of the region immediately in front of a notch in a polycaprolactone electrospun mat at increasing levels of the applied load (Koh et al., 2013). Fiber alignment and crack tip blunting lead to a decrease of the stress concentration, such that the material becomes flaw insensitive.

Koh et al. (2013) compare the behavior of polycaprolactone mats with that of commercial polypropylene nonwovens and with mats of electrospun gelatin nanofibers. Only one value is considered for the a/b ratio and the effect of the microstructure is not studied. Notch insensitivity is observed for the polypropylene nonwoven and the gelatin electrospun scaffolds.

Fiber alignment in the crack tip region and blunting of pre-existing cracks are also observed in paper (Balankin et al., 2011; Isaksson and Dumont, 2014) and in model 2D networks close to the isostatic limit (Zhang et al., 2017). This effect manifests itself if fibers are free to rearrange under load. For this to happen, the network crosslink density and the fiber bending stiffness must be low. Nonaffine networks of small w are more likely to be notch insensitive than networks of large w belonging to the transition or affine regions of the master plot (Figure 6.6). Likewise, decreasing the connectivity, $\langle z \rangle$, is expected to promote notch insensitivity. It becomes apparent that the effect of w on notch insensitivity is related to the ductile-to-brittle transition observed in networks without pre-existing crack as w increases, as discussed in Section 8.1.

A discussion of the effect of crack tip blunting on material toughness is presented in Rawal and Sayeed (2013). These authors compare the strength of nonwoven geotextiles with cracks and with circular holes. It is observed that the strength in the presence of these defects is slightly smaller than the strength of the virgin material, but no difference is observed between samples with cracks and those with circular holes of diameter equal to the length of the cracks. Therefore, the network has the ability to eliminate the stress singularity expected at the tip of a crack in a continuum via large structural rearrangements, as seen in Figure 8.23, leading to notch insensitivity.

8.3   Principles of Network Toughening

8.3.1   General Considerations

A trade-off exists in most materials between strength and toughness. High strength usually implies low toughness and brittle behavior, while high toughness is generally associated with large deformations and lower strength. The behavior described in Section 8.1.2 for network materials follows this trend.

Strength and toughness are related to network parameters in different ways. Equation (8.2) indicates that strength depends on ρ_b and f_{Xc}, but not on fiber properties, such as l_b. On the other hand, w depends on ρ and l_b, while the degree of heterogeneity of the network depends additionally on $\langle z \rangle$. Hence, it is possible to construct a network of low w and of large strength, for example by keeping l_b low and increasing f_{Xc}. Such a network is expected to deform extensively before peak stress, hence exhibiting high toughness. This argument is shown qualitatively in Figure 8.24(a) (networks of low w and high free volume exhibit type C behavior). Consider a reference

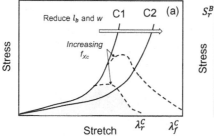
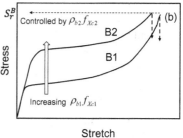

Figure 8.24 Schematic representation of the effect of network parameters on network toughness for (a) type C and (b) type B behaviors. Toughness may be increased in single networks of type C by increasing f_{Xc} (dashed lines in (a)), while ductility increases by decreasing l_b and w at constant f_{Xc} and ρ_b. The toughness of interpenetrating networks with type B behavior depends on $\rho_{b1} f_{Xc1}$ and $\rho_{b2} f_{Xc2}$, as indicated in (b), and on the operation of additional mechanisms discussed in Section 11.3.

network of given w corresponding to the curve labeled C1. Increasing f_{Xc} moves the peak stress up the characteristic curve. The energy absorbed per unit network volume is equal to the area under the stress–strain curve up to failure, as shown by the shaded area in Figure 8.24(a). Therefore, increasing f_{Xc} increases the area under the curve. When w is reduced by reducing l_b, the characteristic curve becomes C2. If f_{Xc} is kept large, the area under the curve evaluated up to peak stress increases in this process. However, toughness depends on the post-peak response. The decay of the stress beyond the peak may be abrupt, if damage becomes localized and a macroscopic crack forms, or gradual, if damage is distributed and failure does not localize. It results that obtaining large toughness without compromising the strength is possible provided the rupture process beyond peak stress remains gradual.

Distributed damage provides kinematic freedom to the network to reorganize during deformation. This process increases the degree of strain stiffening, keeps the tangent stiffness positive, and, hence, ensures deformation stability. High strength does not imply brittle behavior. Ductility depends on the degree of the kinematic constraint experienced by the deforming network.

The post-critical deformation may be stabilized in several ways, of which two are mentioned here: (i) increasing the degree of elastic heterogeneity and (ii) working with interpenetrating double networks. Heterogeneity helps spatially isolate the damage sites and reduces their field-mediated interaction, therefore postponing localization. As discussed in Chapter 6, the degree of elastic heterogeneity of the network may be increased by reducing w. However, controlling w is not the most efficient post-critical stabilization method. Another concept explored for the same purpose is the use of a strong and deformable second network; this issue is discussed in Section 8.3.2.

8.3.2   Tough Gels

Interpenetrating networks (also known as double networks – see also Section 11.3) are composed from two subnetworks which interpenetrate and individually span the entire sample. One of the networks (network 1) is sacrificial, in the sense that it ruptures gradually and dissipates energy. The other network (network 2) is stretchy and strong and is used to ensure that the global deformation is stable (tangent stiffness is positive) while network 1 ruptures. The double network design requires that: (i) network 1 has high crosslink density ρ_{b1} and moderate crosslink (or fiber) strength, f_{Xc1}, (ii) network 2 has lower filament and crosslink densities, ρ_{b2}, and has high crosslink (and fiber) strength, f_{Xc2}; (iii) the filaments of network 2 must have large tortuosity, to sustain stretches much larger than those at which network 1 fails.

The strength of the first subnetwork is defined by $\rho_{b1}f_{Xc1}$, while the strength of the second subnetwork taken independently is proportional to $\rho_{b2}f_{Xc2}$, Eq. (8.2). The double network design requires $\rho_{b1}f_{Xc1} \ll \rho_{b2}f_{Xc2}$. Since $\rho_{b1} > \rho_{b2}$ such to ensure as much internal energy dissipation as possible, it results that one must have $f_{Xc1} \ll f_{Xc2}$. As the material is loaded, the onset of damage in network 1 leads to the occurrence of a shoulder in the stress–stretch curve and, hence, the resulting behavior is of type B. The shoulder marks the transition between regimes I and II

(Figure 8.1(b)). The second subnetwork controls the large strain response and should carry a small fraction of the applied load at stretches at which network 1 starts to rupture. Eventually, the second network (which is essentially swelled by the disintegrating network 1) becomes dominant and undergoes strain stiffening. Final failure is controlled by $\rho_{b2}f_{Xc2}$ and by the synergistic interaction of network 2 with the fragments of the damaged network 1. Figure 8.24(b) shows the effect of these parameters on the appearance of the stress–stretch curve and suggests ways to increase the area under the curve (or total energy dissipated per unit volume before fracture). Further discussion of such gels is provided in Section 11.3.

The method outlined in the previous paragraph is based on the concept that a large number of sacrificial bonds are allowed to rupture and dissipate energy, while the stability of deformation is ensured by the second network. Other internal energy dissipation methods can be envisioned. For example, the crosslinks may be replaced by nanocrystallites or by folded molecular domains. These play the role of compliant crosslinks and unfold gradually during the deformation, dissipating large amounts of energy (Li et al., 2014). Large hysteresis is observed upon unloading. The physical crosslinks are able to reform upon unloading, which provides self-healing properties to the hydrogel (Li et al., 2014).

Another method to enhance toughness is the use of reversible crosslinks which unbind and reform at some other location (dissociative polymers). The resulting deformation is plastic, characterized by large dissipation and self-healing.

These ideas have been used to increase the toughness of gels. Single network gels are soft and brittle and are less than ideally suited for applications in biomechanics. A large amount of work has been dedicated to increasing their toughness and strength. Reviews of these efforts are presented in Myung et al. (2008), Gong (2010), and Zhao (2014). Single network hydrogels have strength of about 100 kPa or smaller and toughness of approximately 1 J/m^2. Tough double network gels achieve strengths above 1 MPa and toughness above 1 kJ/m^2; toughness larger than 10 kJ/m^2 was reported (Li et al., 2014).

Interpenetrating networks based on these principles have been quite successful (Gong, 2010; Sun et al., 2012). The use of nano-crystalline domains that work as crosslinking sites and unravel during deformation was also explored (see e.g., Stauffer and Peppas, 1992). Hydrogels with a combination of chemical and physical crosslinks (Sun et al., 2012, 2013) and with nonpermanent, sliding inter-filament junctions (Ito, 2007) have been developed and shown to possess excellent toughness; see also Figure 6.21(b). An alternative also explored is based on connecting any pair of crosslinks by a bundle of filaments, each filament having different tortuosity (Wang et al., 2010). Upon stretching, the filaments with the smallest tortuosity fail first, but the respective crosslinks remain connected by the remainder of the filaments in the bundle. In this process, energy is dissipated, while the overall structure of the network remains unchanged. This ensures that deformation remains stable while internal energy dissipation takes place.

This area of research is quite active at this time and it is possible that new toughening concepts that go beyond those outlined in this section will be imagined and implemented.

8.3.3   The Lake–Thomas Theory

The toughness of single network gels can be understood in terms of the Lake–Thomas theory (Lake and Thomas, 1967; Sakai et al., 2008). This theory indicates that the fracture energy (energy released per unit crack advance and unit crack front length) is proportional to the product of the number density of molecular strands intersecting the crack plane, the energy required to rupture a strand, and a characteristic length (Eq. (8.15)). In thermal networks, the characteristic length is the contour length of molecular strands between crosslinks, while in athermal networks it may be taken equal to l_c.

This concept is similar to Griffith's theory in linear elastic Fracture mechanics in which crack growth is controlled by the balance between the strain energy stored in the solid and the energy required to create new surfaces (crack faces). In Griffith's theory, the energy sink during fracture is the rupture of bonds across the crack plane and the creation of free surface.

The difference between the two theories may be understood by considering that in network materials, the strands penetrating the crack plane are long, soft springs, while, in brittle nonpolymeric materials (e.g., ceramics or glass), the bonds across the crack plane are stiff, short springs (inter-atomic bonds). The rupture of a molecular strand implies the scission of a single bond. The energy of this bond is of the same order of magnitude as that of any other inter-atomic bond in a brittle material. However, in the soft spring case, large deformations have to be applied before the level of force required to rupture the individual bond is reached. Therefore, the key energy expenditure in the soft spring case is not associated with the bond rupture per se, but rather with the deformation of the entire strand up to the force at which bond scission occurs. The scission of inter-atomic bonds leads to an estimate of the fracture energy of about 1 J/m^2, which is of the same order of magnitude as the energy released by breaking bonds across the crack plane in a brittle solid (the surface energy). The mechanism implied by the Lake–Thomas theory leads to energies one to two orders of magnitude larger. For example, the intrinsic toughness of unfilled elastomers associated exclusively with the rupture of strands penetrating the crack plane is $\sim$50 J/m^2. The energy dissipated by tough gels is yet orders of magnitude larger than this estimate. A further discussion of the Lake–Thomas theory in the context of elastomers is presented in Section 8.4.2.

8.3.4   Networks with Low Free Volume: Epoxy

The networks discussed in the preceding paragraphs are composed from fibers that are free to move relative to each other and reorganize during network deformation. This is an important requirement for toughness. To place it in context, it is useful to discuss the case of a network with low free volume, as for example epoxy.

Epoxy is a highly crosslinked network which is relatively brittle (most epoxies have failure strains of about 4%). The properties of epoxy below the glass transition temperature are not as much controlled by the network as they are controlled by the nonbonded interactions between molecular strands. Crosslinking the resin leads to a significant reduction of volume. The elimination of the free volume allows the molecules to interact in the short range, which leads to a rapid increase of the stiffness. The difference between the bulk modulus of the uncrosslinked resin and that of the crosslinked material at the same temperature may be as large as one order of magnitude (Nawab et al., 2013). Hence, the essential role of crosslinking is to enable the nonbonded interactions which, in turn, define the stiffness and, to some extent, the strength of the material.

Epoxy has limited ability to deform plastically before rupture, and hence has low toughness. This is due to the fact that the molecular strands are not free to slide relative to each other. The relative molecular motion is limited by the small mean segment length, l_c, and by the fact that, at room temperature, the material is below its glass transition temperature.

Several solutions can be adopted to enable local plasticity. A common one is the use of plasticizers. Plasticizers are small molecules added to the resin for the purpose of enhancing polymeric chain mobility. They act either by increasing the free volume, in which case their addition leads to a reduction of T_g and of the stiffness, or by the modification of the chemical interactions of the chains, which may lead to the reduction of T_g, but does not necessarily lower the stiffness. For example, one may say that in hydrogels water works as a plasticizer. It effectively separates the molecular strands, reducing the nonbonded interactions and increasing the free volume (volume not occupied by the polymer). Another method to enable the structural evolution of the network during deformation is the concomitant use of hydrophobic and hydrophilic building blocks. For example, if the network is composed from hydrophilic filaments, hydrophobic chains may be embedded such to swell the network (Nagaoka, 1989); this enables relative filament motion without reducing the stiffness.

Epoxies may be effectively toughened by extrinsic mechanisms, that is, by the addition of inclusions. Toughening using micron size rubber particles is well-known and is used industrially. Sub-micron silica particles with weak interfaces with the epoxy matrix may be used for the same purpose (e.g., Picu et al., 2019). The discussion of toughening mechanisms in composites is beyond the scope of the present overview.

8.4 Failure of Thermal Networks: Fracture of Rubber and Gels

Rubber is one of the first systematically studied network materials. Its resistance to fatigue and to damage accumulation when subjected to large deformations is one of the key reasons for its widespread use in engineering applications. Gels are materials of high current interest due to their applications in bioengineering. The rupture of elastomers and gels is of obvious importance for the respective applications.

As with the athermal networks discussed in Sections 8.1–8.3, in this section we analyze first the rupture of networks without pre-existing macroscopic cracks, followed by a brief review of results pertaining to the growth of cracks in rubber.

8.4.1   Thermal Networks without Pre-existing Cracks

This section addresses the fracture of elastomers without pre-existing detectable damage or macroscopic crack-like features. As with any network material, the rubber network is stochastic and imperfect. In addition, rubber may contain microscopic pores or spatial variations of the crosslink density on the sub-micron scale. These features are considered to be part of the intrinsic structure of the material and are not viewed here as pre-existing defects. The following discussion refers to materials containing such features, but which are free from macroscopic cracks or voids.

In the case of rubber, the key controllable network parameter is the crosslink density, ρ_b. Vulcanized rubber has a stochastic network structure and the strand length between crosslinks is a Poisson distributed stochastic variable. Special rubbers have been developed in which the distribution of strand lengths is controlled. Likewise, gels in which all strands have the same length and specific connectivity, $\langle z \rangle$, have been developed (e.g., the tetra-PEG gels; Akagi et al. (2010)). In this broader sense, the strand length distribution may be regarded as an additional controllable network parameter. The effect of these two parameters on the strength of thermal networks is discussed in the next subsection.

8.4.1.1   Strength–Structure Relation

The affine model of rubber elasticity predicts that the small strain stiffness, G_0, is a linear function of the number density of elastically active molecular strands effectively engaged in the network, $\rho_\#$, as $G_0 \sim \rho_\# k_B T$. In the phantom network model, the proportionality constant depends on connectivity, $\langle z \rangle$ (see the discussion of rubber in Section 6.1.1.3.3). In the absence of dangling ends and loops (in the case of loops, filaments start and end at the same crosslink), $\rho_\# = \rho/l_c$ and the geometric relation (Eq. (4.21)) indicates that the crosslink density, ρ_b, may be expressed as $\rho_b = 2\rho_\#/\langle z \rangle$.

The effect of $\rho_\#$ on the strength of rubber was studied by many authors. Figure 8.25 shows results for vuclanizates of Viton, which is a synthetic fluoropolymer elastomer (Smith, 1967). The crosslinking agent concentration, c_X, was modified in the range 0.38×10^{-4} to 3.2×10^{-4} mol/cm^3. Tests were performed in uniaxial tension at a strain rate of ~ 0.1 s^{-1} and two temperatures of 55°C and 230°C. The data indicates linear scaling of the nominal peak stress, S_r, with c_X at both temperatures. Assuming a linear relation of c_X and ρ_b, the result in Figure 8.25 is in agreement with the prediction of Eq. (8.2) developed for athermal networks.

The strength decreases at larger crosslink densities (Bueche and Dudek, 1963; Choi et al., 1994). The relation between strength and modulus is nonlinear in most elastomers. As the crosslink density increases, the strength and modulus increase approximately in proportion, but beyond a certain degree of crosslinking the strength decreases, while the modulus continues to increase with increasing crosslink density (Choi et al., 1994).

While this result provides a suggestion as to how to construct networks of increased strength, it does not indicate how to control the toughness. Controlling the strand length distribution has offered an unexpected answer to this question. As indicated in the

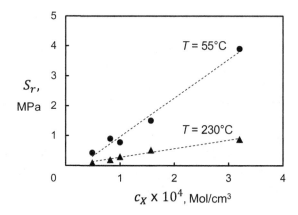

Figure 8.25 Nominal strength of Viton rubber versus crosslinker concentration, c_X, at two temperatures. Data from Smith (1967)

previous paragraphs, rubber obtained by the usual vulcanization process is a stochastic network, with broad distribution of strand lengths. To evaluate the importance of the strand length distribution, networks of single strand length (delta function distribution of lengths) (Akagi et al., 2013b), and bimodal networks composed from a combination of short and long strands (Mark and Erman, 1988; Mark, 1992) have been synthesized and tested. These tests indicate that significant improvements of both strength and toughness can be obtained with bimodal networks composed from short and long strands.

Unimodal networks of short strands are stiff and brittle. They have high strength, but low toughness. Unimodal networks of long strands are much softer and have low strength, although they deform extensively before the peak stress. Increasing both strength and toughness in bimodal networks may be achieved by fulfilling the following requirements: (i) the difference between the length (or molecular weight) of the short and long strands should be large; for example, the short strand length should be two orders of magnitude smaller than the long strand length; (ii) the number density of the short strands must be large ($\sim$95%); (iii) the shorter the short strands are, the stronger the toughening effect is. The difference between the behavior of networks with unimodal and bimodal strand distributions is much larger at low temperatures than at high temperatures (Sun and Mark, 1987). The early work on rubber bimodal networks is reviewed in Mark (1999).

Interestingly, enhancement of ultimate properties was also reported in bimodal networks below T_g. These are systems in which a small weight fraction of long chains is incorporated in an otherwise short chain network as for example a brittle thermoset (Tang et al., 1984) and epoxy (Holmes and Letton, 1994).

8.4.1.2 The Effect of Temperature and Strain Rate

Rubber elasticity is based on the idea that stress in thermal networks is entropic. The concept is described in Section 2.2 and the corresponding formulation is presented in

Section 2.4. The entropic force in thermal filaments is always contractile and tends to bring the two filament ends together. Rubber elasticity assumes that the energy associated with excluded volume interactions does not change during deformation and, hence, the energetic component of stress vanishes.

On the other hand, fracture is an energetic process since it implies bond rupture. The fracture process is controlled by an energy balance, as described in Fracture mechanics. The discrepancy between the mechanism of stress production and that of network rupture needs to be emphasized.

The signature of the entropic nature of stress is the proportionality relation between the relaxed network modulus and temperature, $G_0 \sim \rho_\# k_B T$. As discussed in Section 5.3.2, G_0 enters the hyperelastic constitutive equation of rubber as a scaling parameter, Eq. (5.27). Hence, the theory implies that stress is always proportional to $k_B T$. In view of this discussion, it becomes of interest to explore the variation of rubber strength with the temperature.

Figure 8.26(a) shows the variation of the nominal strength, S_r, with temperature for Viton rubber tested in uniaxial tension at a strain rate of ~ 0.1 s^{-1} (Smith, 1967). Results for three concentrations of the crosslinking agent (crosslink density) are shown. It is observed that the strength decreases fast in a certain range of temperatures, while at higher and lower temperatures it decreases with increasing temperature at a much smaller rate. The experimental data for Viton in Figure 8.26(a) do not show the low temperature range, which is represented schematically by the dashed lines. Similar data is presented for other elastomers such as neoprene, butyl, and natural rubber in a broad range of temperature in Boonstra (1950). The decrease of the strength with increasing temperature shown in Figure 8.26(a) is faster than $1/T$.

Figure 8.26(b) shows the stretch at failure versus temperature corresponding to the data in Figure 8.26(a). λ_r decreases with increasing temperature. This indicates a rapid decrease of the strain energy at failure with increasing temperature.

Figure 8.26(c) shows the variation of the nominal strength and stretch at rupture with the strain rate for SBR rubber. Tests are performed in tension at room temperature (Bekar et al., 2002). Two regimes are observed. At small strain rates, both S_r and λ_r increase. Based on the time–temperature superposition principle (Chapter 9), increasing the strain rate is equivalent to decreasing the temperature and, hence, this regime is similar to that shown in Figure 8.26(a) and (b) for Viton rubber. A second regime appears at high strain rates, where S_r levels off and λ_r decreases. At these rates, the characteristic time of local and structural relaxation mechanisms becomes comparable to, or smaller than, the loading time (inverse of the strain rate), which leads to a strain rate-induced glass transition. The rate corresponding to the onset of this transition may be estimated based on the peak of the loss modulus at high frequencies; in SBR tested at room temperature, the strain rate corresponding to the transition is $\dot{\lambda} \approx 10^3$ s^{-1} (Kluppel, 2008). As the glass transition is approached, the strand mobility decreases. This prevents internal relaxation and the reorganization of the chains, which otherwise stabilizes the deformation. The observed strength reduction is a consequence of the reduced chain mobility. Similar results for SBR are presented in Smith (1958), while a review of strain rate effects in rubbers is presented in Roland (2006).

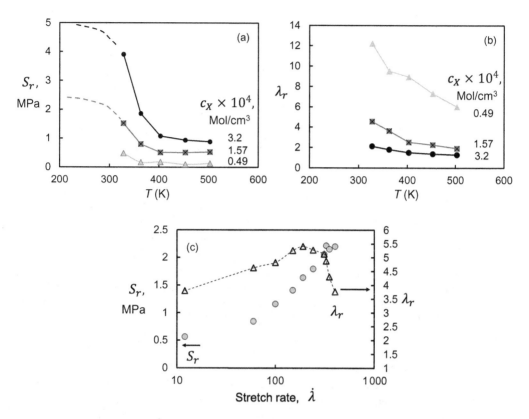

Figure 8.26 Variation with temperature of (a) the nominal strength and (b) stretch at rupture for Viton rubber tested with a strain rate of ~ 0.1 s^{-1} and for three values of c_X (data from Smith, 1967). The dashed lines in (a) show qualitatively the trend at lower temperatures. (c) Nominal strength and stretch at rupture for SBR rubber measured at room temperature, as a function of the strain rate (data from Bekar et al., 2002).

The strong dependence of the strength and fracture energy on strain rate and temperature in elastomers is consistent with viscous dissipation (see also Section 8.4.2). As the temperature increases and/or the strain rate decreases, the contribution of viscous dissipation to S_r and to toughness decreases and both toughness and S_r decrease. The contribution to toughness of the athermal component described by the Lake–Thomas theory remains unchanged. Since dissipation is associated with the nonbonded interactions between strands, swelling the network and removing a fraction of the nonbonded interactions results in a significant decrease of S_r and of the toughness.

The rupture data for rubber at various temperatures and strain rates collapses onto a failure surface when plotted as $\log (S_r T_0 / T)$ versus $\log (\lambda_r - 1)$. T_0 is a reference temperature taken, for example, to be equal to the room temperature, while T is the temperature of the test. As the temperature increases and/or the strain rate decreases, both S_r and λ_r decrease. However, data corresponding to a wide range of temperatures and strain rates fall on a unique curve. The failure surface is approximately linear in

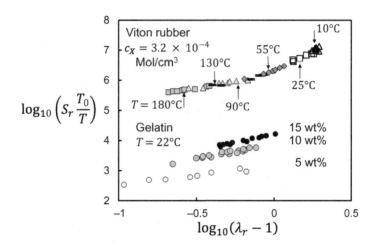

Figure 8.27 Failure surfaces for gelatin gels of 5 wt%, 10 wt%, and 15 wt% concentration and for Viton rubber. The data for gelatin is collected at various strain rates and room temperature. The data for rubber is collected at different strain rates and different temperatures, as indicated in the figure. Data from Smith (1967) and McEvoy et al. (1985)

the high temperature range. The slope of this line depends weakly on the type of network considered (Smith, 1964; Dickie and Smith, 1969) and is independent of the crosslink density (Smith, 1967).

Figure 8.27 shows several examples of failure surfaces for Viton rubber (Smith, 1967) and for gelatin of three concentrations (McEvoy et al., 1985). In the case of gelatin, the data points corresponding to a given concentration are obtained at different strain rates (covering a three orders of magnitude range) and at room temperature ($T = T_0$). The curves for the three concentrations are shown are parallel.

The data set for Viton rubber is obtained at different temperatures and different strain rates. The temperature is indicated in Figure 8.27 and, for each temperature, the data points aligned from left to right correspond to increasing strain rates. The failure surface becomes curved at lower temperatures (right end of the plot). At temperatures approaching the glass transition temperature, the strength levels off and the stretch at failure decreases (Smith, 1963). This indicates the transition from elastomeric to glassy rupture.

The representation in Figure 8.27 provides a unifying perspective on the strength of various types of thermal networks. It indicates that, for a given network and in the range in which the failure surface is linear (in log–log coordinates), one may write:

$$S_r \frac{T_0}{T} = \beta \varepsilon_r^{\alpha}, \tag{8.12}$$

where $\varepsilon_r = \lambda_r - 1$, β is a function of the crosslink density and likely of the strength of crosslinks and filaments, and exponent α, which represents the slope of the linear segment of the failure surface, takes values close to 1. Both strength, S_r, and stretch at rupture (or ε_r) depend on temperature and strain rate. If the crosslink density is modified, S_r, β, and ε_r change in a manner consistent with Eqs. (8.2) and (8.3), as may be inferred from the results presented by Smith (1967). Specifically, as ρ_b increases, S_r increases and ε_r decreases. In the case of gelatin, S_r increases with increasing ρ_b, while ε_r remains approximately constant or increases slightly, depending on the strain rate.

The failure of rubber under multiaxial loading is of obvious practical importance. Hamdi et al. (2006) present plane stress failure data for several rubbers, including natural rubber, styrene-butadiene rubber, and crosslinked polyurethane, obtained using the membrane inflation test. They report failure surfaces in the principal stretch and in the principal Cauchy stress planes normalized by the rupture stretch and stress in uniaxial tension, respectively. The data reported corresponds to the first quadrant of the plane stress space. The failure surface resembles that for athermal networks, shown in Figure 8.17. However, the nominal strength in equi-biaxial tension is about 75% of the uniaxial value, which is a fraction notably smaller than that observed in athermal networks (Section 8.1.10). It results that rubbers are much more sensitive to triaxiality than athermal networks without an embedding matrix and with large free volume. One of the mechanisms leading to this enhanced sensitivity is discussed in Section 8.4.1.3.

8.4.1.3   Cavitation

Cavitation is a form of elastic instability leading to the formation of a void that grows rapidly in a continuum. This may be viewed in two ways: as the unstable growth of a pre-existing cavity or as a bifurcation of the elasticity solution of a continuum without a pre-existing void.

Gent and Lindley (1959) performed one of the best known studies of cavitation in rubber and obtained a linear relationship between the hydrostatic stress at instability, p_c, and the rubber shear modulus, G_0. Using theoretical results for pressurized cavities in neo-Hookean solids from Green and Zerna (1954), they established that the critical pressure is

$$p_c = \frac{5}{2} G_0 = \frac{5}{6} E_0. \tag{8.13}$$

This result indicates the critical hydrostatic (tensile) stress at which an initially spherical cavity in an incompressible hyperelastic solid of new-Hookean type grows unboundedly. The critical pressure is independent of the size of the pre-existing cavity, as long as the outer dimensions of the object in which the void grows are much larger than the initial void size.

Ball (1982) considered the bifurcation problem in nonlinear elastostatics associated with the emergence of a cavity in a body subjected to spherically isotropic tractions. His solution for the set-up considered by Gent and Lindley provides the same critical pressure (Eq. (8.13)) as for the unstable growth of a pre-existing cavity.

The coefficient in Eq. (8.13) relating p_c and G_0 depends on the functional form of the constitutive equation of the material. If the free energy function of the hyperelastic material is given by the Ogden form of Eq. (7.15), the coefficient is a function of the exponents α_i. Specifically, if only one term is retained in the sum defining the free energy function, α_1, the ratio p_c/E_0 takes the values 0.16, 0.53, 0.83, and 1.58 for α_1 equal to 1, 1.5, 2, and 2.5, respectively (Chou-Wang and Horgan, 1989). The case with $\alpha_1 = 2$ corresponds to the neo-Hookean material, for which Eq. (8.13) is recovered.

The abovementioned works make the assumption that loading is symmetric, that is, the stress state is purely hydrostatic. However, this is not the case in most practical situations in which the stress tensor has both hydrostatic and deviatoric components. It becomes of interest to inquire to what extent the cavitation criterion changes under nonsymmetric loading conditions. This problem was investigated by Hou and Abeyaratne (1992) and Ganghoffer and Schultz (1995). For incompressible homogenous neo-Hookean solids subjected to axisymmetric stress states (defined by $\sigma_1 = \sigma_2$ and σ_3), the critical conditions at which a pre-existing cavity grows unboundedly are shown in Figure 8.28. The solution of Eq. (8.13) is recovered in the hydrostatic case ($\sigma_1 = \sigma_2 = \sigma_3$). In the presence of a deviatoric stress component, the critical pressure leading to instability increases. Also, since the far field does not have spherical symmetry, the cavity loses symmetry as it grows.

The effect of material compressibility on cavitation was studied by several authors. The growth of cavities in compressible materials subjected to hydrostatic loading is discussed in James and Spector (1991) and Lopez-Pamies (2009). It is observed that, as the incompressibility restriction is relaxed, cavitation takes place at smaller stresses and larger stretches.

All these studies consider the instability of a symmetric cavity, that is, either spherical, in 3D, or circular, in 2D. Real pre-existing pores are mostly nonsymmetric. The lack of symmetry of the nucleus is expected to lead to some deviations from the instability criteria reviewed here.

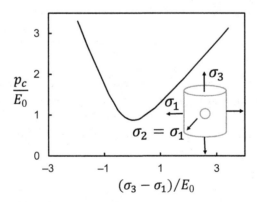

Figure 8.28 The effect of the deviatoric stress component on the critical pressure at cavitation for a spherical cavity in a neo-Hookean solid. Data from Hou and Abeyaratne (1992) and Ganghoffer and Schultz (1995)

Two corrections are needed to render the above results applicable to a wide range of real incompressible materials. First, one must account for the contribution of the surface tension which tends to shrink the void and hence opposes the applied stress. This effect increases linearly with increasing the curvature of the cavity and can be represented by a negative pressure acting on the inner cavity wall, proportional to the surface tension and inversely proportional to the cavity radius. The analysis in Gent and Tompkins (1969) indicates that, for realistic material parameters corresponding to rubber, the surface tension effect is strong only for voids of diameter smaller than ~100 nm.

The second observation of importance is the fact that cavitation requires large deformations of the material in the immediate vicinity of the cavity wall. The local stretches at the wall greatly surpass the maximum extensibility of the network. Therefore, the occurrence of cavitation implies network fracture, and the interplay between these two processes is not captured by the studies discussed in this section, which are focused on the stability of elastic deformation. The fact that the cavitation criteria are verified experimentally indicates that, although some degree of material rupture at the surface of the rapidly growing void is required, the elastic instability is still the limiting process.

Another important limiting aspect is related to the total energy available for void growth. The stress-based criteria of Eq. (8.13) and Figure 8.28 are incomplete in the sense that they predict the conditions under which a void may start growing unstably, but do not ensure that sufficient strain energy is available in the surrounding material to support cavity growth. While the criteria can be used on the local scale to predict the onset of the process, they don't ensure that such cavities will lead to material rupture. To make this inference, a global energy condition must be fulfilled.

Closing the discussion on cavitation, it is worth pointing out the role played by this process in polymer crazing. Crazing is a phenomenon by which fibrils form in a continuum solid polymer under certain stress states. The formation of fibrils is made possible by the nucleation of micropores, that is, by cavitation. The ligaments separating the pores become strongly stretched and form fibrils that can be observed at the microscale. The fibrils dissipate large amounts of energy before rupture. Once they rupture, a crack forms at the location of the craze. For crazing to occur, cavitation must be possible and the fibrils must support large stretches before rupture. Crazing is favored by triaxial stress states. The presence of crystallites separated by amorphous regions in semi-crystalline polymers favors crazing due to the enhanced stress field heterogeneity introduced by the presence of hard crystallites in the softer amorphous matrix.

8.4.2   Propagation of Cracks in Thermal Networks

Crack propagation in elastomers is a subject studied from the early days of Fracture mechanics. In fact, some of the fundamental concepts in Fracture mechanics, such as the notion that the energy release rate is an adequate crack growth criterion, and that under fatigue conditions cracks grow at a rate which is a power function of the amplitude of the energy release rate cycle, have been developed first in relation to fractures in rubber (Thomas, 1955, 1958). These ideas are now used broadly to predict crack growth under static and cyclic conditions in all materials.

Fundamentally, fractures of elastomers and gels are not different from fractures in other materials. However, several differences arise and need to be emphasized. Soft materials sustain large deformations and crack tips do not remain sharp under load (see Figure 8.23). Crack tip blunting eliminates the stress singularity and, hence, the stress intensity factor, which is generally used as a parameter controlling crack growth in linear elastic fracture, is not defined.

Another important detail is that deformation of unfilled elastomers and gels is time dependent. These materials are viscoelastic, present hysteresis when loaded cyclically, but exhibit no plasticity. Time dependence and hysteresis are pronounced in elastomers in which strain-induced crystallization does not take place and increases as the molecular strand length between crosslinks increases (crosslink density decreases). These trends can be intuitively understood based on the observation that the relative motion of molecules is less constrained when the crosslink density is low, and such local rearrangements cause internal energy dissipation.

Although crack growth criteria based on the maximum principal stretch have been investigated, there is ample evidence at this time that an energy release rate criterion is best suited for this purpose. The energy release rate represents the amount of potential energy dissipated per unit of crack front length and per unit crack advance. This energy is extracted from the strain energy stored in the body, is provided by the work performed by boundary tractions, or emerges from both these sources. The energy release rate, G, associated with specific boundary conditions and for given sample geometry, can be computed using methods developed in Fracture mechanics. This represents the driving force for crack growth. Cracks propagate when G is equal to, or larger than a material-specific parameter known as the critical energy release rate, G_c. G_c accounts for all processes leading to material rupture which take place in the vicinity of the crack tip and is the single material parameter describing the material resistance to crack growth. The energy release rate is a global measure not necessarily related to the details of the deformation process in the vicinity of the tip. This situation is also encountered in the fracture of ductile metals, where large plastic deformations take place close to the tip, while the energy release rate remains a well-defined parameter.

G_c accounts for dissipation via two types of processes: (i) bond rupture, leading to the creation of free surfaces and (ii) internal dissipation, taking place primarily in the region of intense deformation in the fracture process zone. The first process is present in all materials. The nature of the dissipation associated with the second process depends strongly on the material. In elastomers, the dissipation is viscoelastic. Therefore, one may write:

$$G_c = G_c^{surf} + G_c^{visc}, \tag{8.14}$$

where the two terms represent the energy associated with the creation of free surface and the energy dissipated by viscoelastic processes, respectively.

G_c^{surf} is usually evaluated with the relation proposed by Lake and Thomas (1967). It represents the energy required to break the strands traversing the fracture surface in a unit area of the respective surface and may be written as:

$$G_c^{surf} \sim \rho_\# n U_{Xc} L_{ee}, \tag{8.15}$$

where $\rho_\#$ is the number density of elastically active strands of the network, n is the number of repeat units (Kuhn segments) along a strand between two successive cross-links, L_{ee} is the mean end-to-end distance of a strand of n repeat units, and U_{Xc} is the energy required to break a molecular strand or its crosslinks. Noting that U_{Xc} is related to the force required to break the filament or the crosslinks, f_{Xc}, as $U_{Xc} \approx f_{Xc} a$, and that $\rho_\# L_{ee} \sim \rho_{\#-2D}$, where $\rho_{\#-2D}$ is the number density of strands penetrating the crack plane, Eq. (8.15) can be rearranged as:

$$G_c^{surf} \sim \rho_{\#-2D} f_{Xc} n a. \tag{8.16}$$

This expression has an intuitive physical meaning: G_c^{surf} is the work performed to stretch up to failure all molecules crossing the unit area of the crack plane, that is, to a length equal to their contour length, na. This is necessary in order to reach axial energetic forces in the strand that may lead to bond scission.

It is useful to return to the comparison made in Section 8.3 with Griffith's theory of linear elastic Fracture mechanics, which applies to brittle materials. In Griffith's theory, the critical energy release rate is equal to the energy required to create free surfaces. The surface energy equals the energy required to break the bonds traversing a unit area of the crack plane, which can be written $\rho_{\#-2D} f_{Xc} a$. Factor n in Eq. (8.16) does not appear in this case since the relative displacement of two atoms to produce bond failure is of the order of their equilibrium distance, a. This energy is much smaller than that of Eq. (8.16). The increased G_c^{surf} in the case of soft materials is due to the large compliance of the bonds traversing the crack plane.

The Lake–Thomas theory predicts that the fracture process is temperature and strain rate independent. Equations (8.15) and (8.16) are supported by experimental observations for rate-independent elastomers (Gent and Tobias, 1982; Tsunoda et al., 2000) and gels (Akagi et al., 2013b).

In situations in which rupture is not entirely brittle and additional dissipation takes place as the crack grows, the second term in Eq. (8.14) becomes important. This is the case in gels and elastomers, primarily in unfilled networks of low crosslink density and in which strain-induced crystallization does not take place. In such cases, the measured G_c is significantly larger than the prediction of the Lake–Thomas theory.

G_c^{visc} is generically a function of temperature and crack velocity (or imposed strain rate). Figure 8.29(a) shows schematically the relationship between crack speed and the applied far-field boundary conditions. Experimental data supporting this representation can be found in the literature (e.g., Knauss, 2015, and references cited therein). Consider the boundary conditions shown in the inset of Figure 8.29(a). The sample is loaded in uniaxial tension with prescribed strain, ε_∞, but the grips are rigid and constrain the Poisson contraction in the horizontal direction. The energy release rate

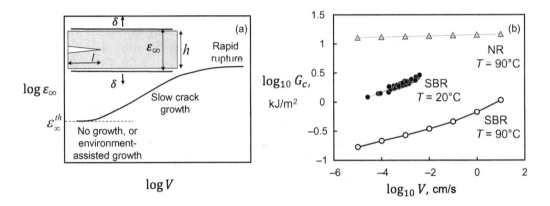

Figure 8.29 (a) Schematic representation of the variation of the crack velocity with the imposed strain in the edge-cracked configuration shown in the inset. The energy release rate for this configuration is $G \sim \varepsilon_\infty^2 h$. (b) Variation of the critical energy release rate with the crack velocity for natural rubber (NR) and styrene-butadiene rubber (SBR) at 90°C and for SBR at 20°C. Data from Thomas (1960, 1994)

for this set-up is independent of the crack length, l, and the driving force for crack extension is constant as the crack grows, $G \sim \left(\sigma_\infty^2/E_0\right)h \sim \varepsilon_\infty^2 h$. If the applied strain ε_∞ is small, the crack remains stationary. In aggressive environments, cracks may grow even at very small loads due to material degradation induced by oxygen and ozone. Beyond a threshold strain ε_∞^{th}, the crack grows at a speed V that increases as a power function of ε_∞: $V \sim \varepsilon_\infty^m$, with exponent $m \gg 1$.

Figure 8.29(b) shows the dependence of the critical energy release rate on crack velocity, $V = \dot{l}$, for natural rubber and styrene-butadiene rubber (SBR) at room temperature and 90°C (Thomas, 1960, 1994). The critical energy release rate for natural rubber is essentially independent of crack velocity. Since natural rubber undergoes strain-induced crystallization, small crystallites form at the crack tip, which renders the material behavior rate-independent. SBR rubber does not crystallize and hence its response is rate sensitive. In this case, G_c is dominated by G_c^{visc}, which increases with increasing V. It is seen that, in agreement with Figure 8.29(a), $G_c^{visc} \sim V^{2/m}$, where $(2/m < 1)$. Since the tests are performed at a set temperature, the range of velocities that can be probed is limited. Time–temperature superposition can be used to collapse curves obtained at various temperatures into a master curve of shape similar to that of the curve in Figure 8.29(a) (Knauss, 2015). The data for SBR tested at room temperature shown in Figure 8.29(b) indicate that G_c increases as the temperature decreases. The various data points in the $T = 20°C$ group correspond to tests performed with specimens of different geometry such as edge cracked, central cracked, and "trousers" (the large deformation version of the standard edge cracked mode III fracture specimen). The fact that the points come together on the same curve indicates that the energy release rate criterion adequately represents fracture under multiple loading conditions and specimen configurations. Experiments performed with acrylamide gels in Tanaka et al. (2000) indicate that the physical picture described here for elastomers applies identically to gels.

8.4.3   Fatigue

The majority of the fatigue studies of network materials have been performed with elastomers. Very little is known about fatigue of nonwovens, biological tissues, and gels. The key results obtained from tests with elastomeric materials are summarized in this section. A number of reviews on this subject are available in the literature (e.g., Lake, 1995, 2003; Mars and Fatemi, 2002).

Fatigue of materials refers to the response to cyclic loading. It is observed that, under such conditions, samples fail at a fraction of their strength measured in a monotonic test. The load applied is characterized based on two parameters: the cycle amplitude and the ratio R of the minimum to the maximum cycle values. The frequency of the cyclic load applied is important if the material behavior is rate sensitive.

When working with elastomers without pre-existing cracks, it is convenient to characterize the load applied in terms of the nominal strain energy density, Ψ, stored in the sample (Poisson et al., 2011). This is in contrast with the more common use of the stress as a measure of the applied load in materials which undergo limited global deformation during cycling. Hence, the load amplitude refers to the amplitude of the strain energy density cycle and, similarly, the R ratio refers to the ratio of the minimum to the maximum values of the strain energy density variation during cycling. When the growth of a pre-existing crack under cyclic loading is monitored, the applied load is characterized in terms of the energy release rate, that is, the thermodynamic driving force for crack growth. Most crack growth tests are performed with loads that oscillate between a maximum value and zero; in this case, the cycle can be described with a single parameter: The maximum energy release rate, $G_{\max}$.

Figure 8.30(a) shows the relationship between the maximum value of the strain energy density of the cycle, $\Psi_{\max}$, and the number of cycles to failure, N_f, for natural rubber and SBR rubber (Lindley, 1972) without pre-existing cracks and for moderate levels of the applied load. This representation is similar to the Wohler curve often used to characterize fatigue in other engineering materials in the absence of pre-existing cracks. Natural rubber has higher toughness (Figure 8.29(b)) and hence the corresponding curve in Figure 8.30(a) is above the SBR curve. However, the two curves converge in the high cycle fatigue limit (large N_f). Note that natural rubber undergoes strain-induced crystallization at large strains and this affects the low N_f (large $\Psi_{\max}$) end of the curve in Figure 8.30(a).

In the presence of a pre-existing crack, one is concerned with the rate of crack growth function of $G_{\max}$. Figure 8.30(b) shows the crack growth per cycle, dl/dN, as a function of $G_{\max}$ for natural rubber and SBR (Lake and Lindley, 1965). The curves exhibit three regimes: in regime 1, $G_{\max}$ is smaller than the threshold value $G_{\max}^{th}$ below which no crack growth is expected. A small growth rate is detectable due to environmental reasons (e.g., oxidation[4]). When tested in vacuum or inert atmosphere,

---

[4] Ozone leads to chain scission, softening and degradation of the network. Oxygen increases the crosslink density and hence increases the network stiffness, rendering it more brittle at the same time (Lake and Lindley, 1964).

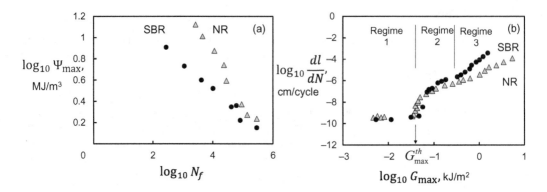

Figure 8.30 (a) Maximum strain energy density per cycle versus the number of cycles to failure, and (b) crack growth rate versus the amplitude of the energy release rate for natural rubber (NR) and styrene-butadiene rubber (SBR). Data from Lindley (1972)

this effect disappears and the crack remains stationary at load levels within regime 1. The threshold G_{max}^{th} has approximately the same value, ~ 50 J/m^2, in most elastomers.[5] Note that this value is at least one order of magnitude smaller than the fracture toughness measured for the same materials. Once $G_{max} > G_{max}^{th}$, the crack grows a finite amount per cycle, with dl/dN increasing fast in the vicinity of G_c (right side of Figure 8.30(b)). The transition regime 2 spans approximately one decade on the G_{max} axis and represents the behavior in the vicinity of the threshold G_{max}^{th}. Regime 3 is characterized by a power law variation of the crack growth rate with G_{max}: $dl/dN \sim G_{max}^{p}$, where $p \approx 2$ for many rubbers. A collection of such curves for several unfilled elastomers is available in Mars and Fatemi (2004).

This behavior is generally observed in most materials, although when working with materials of low compliance, such as metals and alloys, the variable of the horizontal axis is generally the applied stress intensity factor. Exponent p varies in a broader range from material to material, while the threshold separating the nonpropagating regime 1 from regime 2 is a function of the ratio R and of the material microstructure.

A useful effect specific to elastomers which undergo strain-induced crystallization is worth mentioning. In these materials, the threshold G_{max}^{th} increases rapidly with increasing R ($R = G_{min}/G_{max}$) and, hence, the fatigue life increases substantially as the value of G_{min} increases at constant G_{max} (Lindley, 1973). Hence, if a tensile pre-stress is applied to an elastomeric machine component, its fatigue life increases significantly. This effect is used in a number of applications. Its physical origin is related to the dynamics of crystallization in the crack tip region during a cycle. Crystallization takes place once the local stretch is large enough. When the minimum stretch of the cycle in the crack tip region is set below the crystallization limit, the

---

[5] Unfilled PDMS, butadiene rubbers, and natural rubber have thresholds G_{max} in the range 20–70 J/m^2, while chloroprene rubber has a slightly larger threshold in the range 40–100 J/m^2.

region undergoes crystallization and amorphization in each cycle. This causes intense energy dissipation, which favors crack growth. However, if the minimum stretch of the cycle in the vicinity of the tip is larger than the crystallization threshold, the region remains crystalline during cycling and crack growth is slowed down.

The effect of the crosslink density is also of importance. Increasing the crosslink density leads to increased stiffness and reduced hysteresis. The reduction of internal dissipation increases the fatigue life. When the material is too densely crosslinked, it becomes brittle. Therefore, it is observed that an optimal crosslink density exists for which the fatigue life under given loading conditions reaches a maximum (Coran, 1978).

References

Akagi, Y., Katashima, T., Sakurai, H., Chung, U. I. & Sakai, T. (2013a). Ultimate elongation of polymer gels with controlled network structure. *RSC Adv.* **3**, 13251–13258.

Akagi, Y., Matsunaga, T., Shibayama, M., Chung, U. I. & Sakai, T. (2010). Evaluation of topological defects in tera-PEG gels. *Macromolecules* **43**, 488–493.

Akagi, Y., Sakurai, H., Gong, J. P., Chungm, U. I. & Sakai, T. (2013b). Fracture energy of polymer gels with controlled network structure. *J. Chem. Phys.* **139**, 144905.

Alava, M. J. & Niskanen, K. (2006). The physics of paper. *Rep. Prog. Phys.* **69**, 669–723.

Alava, M. J., Nukala, P. K. V. V. & Zapperi, S. (2009). Size effects in statistical fracture. *J. Phys. D: Appl. Phys.* **42**, 214012.

de Arcangelis, L., Redner, S. & Herrmann, H. J. (1985). A random fuse model for breaking processes. *J. Physique (Paris) Lett.* **46**, L585–L590.

Balankin, A. S., Susarrey, O., Mora Santos, C. A., et al. (2011). Stress concentration and size effect in fracture of notched heterogeneous material. *Phys. Rev. E* **83**, 015101(R).

Ball, J. M. (1982). Discontinuous equilibrium solutions and cavitation in nonlinear elasticity. *Phil. Trans. R. Soc. A* **306**, 557–611.

Bazant, Z. P. (2004). Scaling theories for quasi-brittle structural failure. *Proc. Nat. Acad. Sci.* **101**, 13400–13407.

Bekar, I., Fatt, M. S. H. & Padovan, J. (2002). Deformation and fracture of rubber under tensile impact loading. *Tire Sci. Technol.* **30**, 45–58.

Benitez, J. A. & Walther, A. (2017). Cellulose nanofibril nanopapers and bioinspired nano-composites: A review to understand the mechanical property space. *J. Mater. Chem. A* **5**, 16003–16024.

Bergstrom, P., Hossain, S. & Uesaka, T. (2019). Scaling behaviour of strength of 3D-, semi-flexible-, cross-linked fibre network. *Int. J. Sol. Struct.* **166**, 68–74.

Berthier, E., Kollmer, J. E., Henke, S. E., et al. (2019). Rigidity percolation control of the brittle-ductile transition in disordered networks. *Phys. Rev. Mat.* **3**, 075602.

Boonstra, B. S. T. (1950). Tensile properties of natural and synthetic rubbers at elevated and subnormal temperatures. *Rubber Chem. Technol.* **23**, 338–346.

Borodulina, S., Kulachenko, A., Galland, S. & Nygards, M. (2012). Stress–strain curve of paper revisited. *Nordic Pulp Paper Res. J.* **27**, 318–328.

Borodulina, S., Motamedian, H. R. & Kulachanko, A. (2018). Effect of fiber and bond strength variations on the tensile stiffness and strength of fiber networks. *Int. J. Sol. Struct.* **154**, 19–32.

Brandberg, A. & Kulachenko, A. (2020). Compression failure in dense non-woven fiber networks. *Cellulose* **27**, 6065–6082.

Bueche, F. & Dudek, T. J. (1963). Tensile strength of amorphous gum. *Rubber Chem. Technol.* **36**, 1–10.

Burla, F., Dussi, S., Martinez-Torres, C., et al. (2020). Connectivity and plasticity determine collagen network fracture, *Proc. Nat. Acad. Sci.* **117**, 8326–8334.

Chen, C., Wang, Z. & Suo, Z. (2017). Flaw sensitivity of highly stretchable materials. *Exteme Mech. Lett.* **10**, 50–57.

Chen, N., Koker, M. K. A., Uzun, S. & Silberstein, M. N. (2016a). In-situ X-ray study of the deformation mechanisms of non-woven polypropylene. *Int. J. Sol. Struct.* **97–98**, 200–208.

Chen, Y., Ridruejo, A., González, C., Llorca, J. & Siegmund, T. (2016b). Notch effect in failure of fiberglass non-woven materials. *Int. J. Sol. Struct.* **96**, 254–264.

Chocron, S., Pintor, A., Galvez, F., et al. (2008). Lightweight polyethylene nonwoven felts for ballistic impact analysis: material characterization. *Composites B* **39**, 1240–1246.

Choi, I. S., Roland, C. M. & Bissonnette, L. C. (1994). An elastomeric ejection system. *Rubber Chem. Technol.* **67**, 892–903.

Choi, K. J., Spruiell, J. E., Fellers, J. F. & Wadsworth, L. C. (1988). Strength properties of melt blown nonwoven webs. *Polym. Eng. Sci.* **28**, 81–89.

Chou-Wang, M. S. & Horgan, C. O. (1989). Void nucleation and growth for a class of incompressible nonlinearly elastic materials. *Int. J. Solids Struct.* **25**, 1239–1254.

Combaz, E., Bacciarini, C., Charvet, R., Dufour, W. & Mortensen, A. (2011). Multiaxial yield behaviour of Al replicated foam. *J. Mech. Phys. Sol.* **59**, 1777–1793.

Coran, A. Y. (1978). Vulcanization. In *Science and technology of rubber*, F. R. Eirich, ed. Academic Press, New York, p. 291.

Curtin, W. L. (1998). Size scaling of strength in heterogeneous materials, *Phys. Rev. Lett.* **80**, 1445–1449.

Deogekar, S., Islam, M. R. & Picu, R. C. (2019). Parameters controlling the strength of stochastic fibrous materials. *Int. J. Sol. Struct.* **168**, 194–202.

Deogekar, S. & Picu, R. C. (2018). On the strength of random fiber networks. *J. Mech. Phys. Sol.* **116**, 1–16.

Deogekar S. & Picu, R. C. (2021). Strength of stochastic fibrous materials under multiaxial loading. *Soft Matt.* **17**, 704–714.

Deshpande, V. S. & Fleck, N. A. (2001). Multi-axial yield behaviour of polymer foams. *Acta Materialia* **49**, 1859–1866.

Dickie, R. A. & Smith, T. L. (1969). Ultimate tensile properties of elastomers. VI – strength and extensibility of a styrene-butadiene rubber vulcanizate in equal biaxial tension. *J. Poly. Sci. A-2 Poly Phys* **7**, 687–707.

Driscoll, M. M., Chen, B. G., Beuman, T. H., et al. (2016). The role of rigidity in controlling material failure. *Proc. Nat. Acad. Sci.* **113**, 10813–10817.

Dussi, S., Tauber, J. & van der Gucht, J. (2020). Athermal fracture of elastic networks: How rigidity challenges the unavoidable size-induced brittleness. *Phys. Rev. Lett.* **124**, 018002.

Duxbury, P. M., Kim, S. G. & Leath, P. L. (1994). Size effect and statistics of fracture in random materials. *Mater. Sci. Eng. A* **176**, 25–31.

Eriksson, M., Torgnysdotter, A. & Wagberg, L. (2006). Surface modification of wood fibers using polyelectrolyte multilayer technique: Effects on fiber joint and paper strength properties. *Ind. Eng. Chem. Res.* **45**, 5279–5286.

Fang, J. & Li, H. (2012). A facile way to tune mechanical properties of artificial elastomeric proteins-based hydrogels. *Langmuir* **28**, 8260–8265.

Forsstrom, J., Torgnysdotter, A. & Wagberg, L. (2005). Influence of fibre/fibre joint strength and fibre flexibility on the strength of papers from unbleached kraft fibres. *Nordic Pulp Paper Res. J.* **20**, 186–191.

Fukasawa, M., Sakai, T., Chung, U. & Haraguchi, K. (2010). Synthesis and mechanical properties of a nanocomposite gel consisting of a tetra-PEG/Clay network. *Macromolecules* **43**, 4370–4378.

Ganghoffer, J. F. & Schultz, J. (1995). A new theoretical approach to cavitation in rubber. *Rubber Chem. Technol.* **68**, 757–772.

Gent, A. N. & Lindley, P. B. (1959). Internal rupture of bonded rubber cylinders in tension. *Proc. R. Soc. London* **249A**, 195–205.

Gent, A. N., Razzaghi-Kashani, M. & Hamed, G. R. (2003). Why do cracks turn sideways? *Rubber Chem. Technol.* **76**, 122–131.

Gent, A. N. & Tobias, R. H. (1982). Threshold tear strength of elastomers. *J. Polym. Sci. Polym. Phys. Ed.* **20**, 2051–2058.

Gent, A. N. & Tompkins, D. A. (1969). Surface energy effects for small holes or particles in elastomers. *J. Polym. Sci. A-2 Poly Phys* **7**, 1483–1488.

Gong, J. P. (2010). Why are double network hydrogels so tough? *Soft Matt.* **6**, 2583–2590.

Goutianos, S., Mao, R. & Peijs, T. (2018). Effect of inter-fibre bonding on the fracture of fibrous networks with strong interactions. *Int. J. Sol. Struct.* **136–137**, 271–278.

Green, A. E. & Zerna, W. (1954). *Theoretical elasticity*. Clarendon Press, Oxford.

Gumbel, E. J. (1935). Les valeurs extrêmes des distributions statistiques. *Annales de l'Institut Henri Poincaré*, **5**, 115–158.

Gustafsson, P. & Niskanen, K. (2011). Paper as an engineering material. In *Mechanics of paper products*, K. Niskanen, ed. de Gruyter, Berlin, pp. 5–28.

Hajiali, F. & Shojaei, A. (2017). Network structure and mechanical properties of PDMS filled with nanodiamond. *Comp. Sci. Technol.* **142**, 227–234.

Hamdi, A., Abdelaziz, M. N., Hocine, N. A., Heuillet, P. & Benseddiq, N. (2006). A fracture criterion of rubber-like materials under plane stress conditions. *Polym. Testing* **25**, 994–1005.

He, M. Y. & Hutchinson, J. W. (1989). Crack deflection at an interface between dissimilar elastic materials. *Int. J. Sol. Struct.* **25**, 1053–1067.

Henriksson, M., Berglund, L. A., Isaksson, P., Lindstrom, T. & Nishino, T. (2008). Cellulose nanopaper structures of high toughness. *Biomacromolecules* **9**, 1579–1585.

Heyden, S. (2000). Network modelling for the evaluation of mechanical properties of cellulose fiber fluff. PhD Thesis, Lund University, Lund, Sweden.

Holmes, G. A. & Letton, A. (1994). The dynamic mechanical characterization of a 6300 (340/22,500) weight average molecular weight system. *Poly. Eng. Sci.* **34**, 1635–1643.

Horii, M. & Nemat-Nasser, S. (1993). *Micromechanics: overall properties of heterogeneous materials*. North-Holland, Amsterdam.

Hou, H. S. & Abeyaratne, R. (1992). Cavitation in elastic and elastic-plastic solids. *J. Mech. Phys. Solids* **40**, 571–592.

Isaksson, P. & Dumont, P. J. J. (2014). Approximation of mode I crack-tip displacement fields by a gradient enhanced elasticity theory. *Eng. Fract. Mech.* **117**, 1–11.

Ito, K. (2007). Novel cross-linking concept of polymer network: Synthesis, structure, and properties of slide-ring gels with freely movable junctions. *Polym. J.* **39**, 489–499.

Iwamoto, S., Isogai, A. & Iwata, T. (2011). Structure and mechanical properties of wet-spun fibers made from natural cellulose nanofibers. *Biomacromolecules* **12**, 831–836.

James, R. D. & Spector, S. J. (1991). The formation of filamentary voids in solids. *J. Mech. Phys. Solids* **39**, 783–813.

Kluppel, M. (2008). Evaluation of viscoelastic master curves of filled elastomers and applications to fracture mechanics. *J. Phys.: Cond. Matt.* **21**, 035104.

Knauss, W. G. (2015). A review of fracture in viscoelastic materials. *Int. J. Fract.* **196**, 99–146.

Koh, C. T., Strange, D. G. T., Tonsomboon, K. & Oyen, M. L. (2013). Failure mechanisms in fibrous scaffolds. *Acta Biomater.* **9**, 7326–7334.

Krasnoshlyk, V., Rolland du Roscoat, S., Dumont, P. J. J. & Isaksson, P. (2018). Influence of the local mass density variation on the fracture behavior of fiber network materials. *Int. J. Sol. Struct.* **138**, 236–244.

Kulachenko, A. & Uesaka, T. (2012). Direct simulations of fiber network deformation and failure. *Mech. Mater.* **51**, 1–14.

Lake, G. J. (1995). Fatigue and fracture of elastomers. *Rubber Chem. Tech.* **68**, 435–460.

Lake, G. J. (2003). Fracture mechanics and its application to failure in rubber articles. *Rubber Chem. Tech.* **76**, 567–591.

Lake, G. J. & Lindley, P. B. (1964). Ozone cracking, flex cracking and fatigue of rubber. Part 1. Cut growth mechanisms and how they result in fatigue failure. *Rubber J.* **146**, 24–30.

Lake, G. J. & Lindley, P. B. (1965). The mechanical fatigue limit for rubber. *J. Appl. Poly. Sci.* **9**, 1233–1251.

Lake, G. J. & Thomas, A. G. (1967). The strength of highly elastic materials. *Proc. R. Soc. London* **300**, 108–119.

Li, J., Suo, Z. & Vlassak, J. J. (2014). Stiff, strong, and tough hydrogels with good chemical stability. *J. Mater. Chem. B* **2**, 6708–6713.

Lindley, P. B. (1972). Energy for crack growth in model rubber components. *J. Strain Anal.* **7**, 132–140.

Lindley, P. B. (1973). The relation between hysteresis and the dynamic crack growth resistance of natural rubber. *Int. J. Fract.* **9**, 449–462.

Lopez-Pamies, O. (2009). Onset of cavitation in compressible, isotropic, hyperelastic solids. *J. Elast.* **94**, 115–145.

Mai, T. T., Matsuda, T., Nakajima, T., Gong, J. P. & Urayama, K. (2018). Distinctive characteristics of internal fracture in tough double network hydrogels revealed by various modes of stretching, *Macromolecules* **51**, 5245–5257.

Malakhovsky, I., & Michels, M. A. J. (2007). Effect of disorder strength on the fracture pattern in heterogeneous networks. *Phys. Rev. B* **76**, 1–13.

Malho, J. M., Ouellet-Plamondon, C., Ruggeberg, M., et al. (2015). Enhanced plastic deformation of nanofibrillated cellulose film by adsorbed moisture and protein-mediated interactions. *Biomacromolecules* **16**, 311–318.

Mao, R., Goutianos, S., Tu, W., et al. (2017). Comparison of fracture properties of cellulose nanopaper, printing paper and buckypaper. *J. Mater. Sci.* **52**, 9508–9519.

Mark, J. E. (1992). Molecular aspects of rubber-like elasticity. *Angew. Makromol. Chem.* **202**, 1–30.

Mark, J. E. (1999). Improved elastomers through control of network chain distributions. *Rubber Chem. Technol.* **72**, 465.

Mark, J. E. & Erman, B. (1988). *Rubberlike elasticity: A molecular primer*. Wiley & Sons, New York.

Mars, W. V. & Fatemi, A. (2002). A literature survey on fatigue analysis approaches for rubber. *Int. J. Fatigue* **24**, 949–961.

Mars, W. V. & Fatemi, A. (2004). Factors that affect the fatigue life of rubber: A literature survey. *J. Rubber Chem. Tech.* **77**, 391–412.

McEvoy, H., Ross-Murphy, S. B. & Clark, A. H. (1985). Large deformation and ultimate properties of biopolymer gels: 1. Single biopolymer component systems. *Polymer* **26**, 1483–1492.

Myung, D., Waters, D., Wiseman, M., et al. (2008). Progress in the development of interpenetrating polymer network hydrogels. *Polym. Adv. Technol.*, **19**, 647–657.

Nagaoka, S. (1989). Mechanical properties of composite hydrogels. *Poly. J.* **21**, 847–850.

Nawab, Y., Casari, P., Boyard, N. & Jacquemin, F. (2013). Characterization of the cure shrinkage, reaction kinetics, bulk modulus and thermal conductivity of thermoset resin from a single experiment. *J. Mater. Sci.* **48**, 2394–2403.

Normand, V., Lootens, D. L., Amici, E., Plucknett, K. P. & Aymard, P. (2000). New insight into agarose gel mechanical properties. *Biomacromolecules* **1**, 730–738.

Ovaska, M., Betalan, Z., Miskic, A., et al. (2017). Deformation and fracture of echinoderm collagen networks. *J. Mech. Beh. Biomed. Mater.* **65**, 42–52.

Oyen, M. L., Cook, R. F. & Calvin, S. E. (2004). Mechanical failure of human fetal membrane tissues. *J. Mater. Sci.* **15**, 651–658.

Patel, P. C. & Kothari, V. K. (2001). Influence of fibre/fibre joint strength and fibre flexibility on the strength of papers from unbleached kraft fibres. *Indian J. Fiber Text. Res.* **26**, 409–413.

Picu, R. C., Krawczyk, K. K., Wang, Z., et al. (2019). Toughening in nanosilica-reinforced epoxy with tunable filler-matrix interface properties. *Comp. Sci. Technol.* **183**, 107799.

Poisson, J. L., Lacroix, F., Meo, S., Berton, G. & Ranganathan, N. (2011). Biaxial fatigue behavior of a polychloroprene rubber. *Int. J. Fatigue* **33**, 1151–1157.

Popil, R. E. (2017). *The physical testing of paper*. Smithers Group Company, Shrewsbury, UK.

Purslow, P. P. (1985). The physical basis of meat texture: Observations on the fracture behaviour of cooked bovine *M.* Semitendinosus. *Meat Science* **12**, 39–60.

Rawal, A., Patel, S. K., Kumar, V., Saraswat, H. & Sayeed, M. M. A. (2013). Damage analysis and notch sensitivity of hybrid needlepunched nonwoven materials. *Textile Res. J.* **83**, 1103–1112.

Rawal, A. & Sayeed, M. M. A. (2013). Mechanical properties and damage analysis of jute-polypropylene hybrid nonwoven geotextiles. *Geotextiles and Geomembranes* **37**, 54–60.

Ridruejo, A., Gonzalez, C. & Llorca, J. (2012). Failure locus of polypropylene nonwoven fabrics under in-plane biaxial deformation. *Comptes Rendus Mechanique* **340**, 307–319.

Ridruejo, A., Jubera, R., González, C. & LIorca, J. (2015). Inverse notch sensitivity: Cracks can make nonwoven fabrics stronger, *J. Mech. Phys. Sol.* **77**, 61–69.

Roland, C. M. (2006). Mechanical behavior of rubber at high strain rates. *Rubber Chem. Technol.* **79**, 429–459.

Sakai, T., Matsunaga, T., Yamamoto, Y., et al. (2008). Design and fabrication of a high-strength hydrogel with ideally homogeneous network structure from tetrahedron-like macromonomers. *Macromolecules* **41**, 5379–5384.

Sehaqui, H., Ezekiel Mushi, N., Morimune, S., et al. (2012). Cellulose nanofiber orientation in nanopaper and nanocomposites by cold drawing. *ACS Appl. Mater. Interfaces* **4**, 1043–1049.

Shi, L., Long, K., Zhong, Y., et al. (2020). Compressive and shear performance of three-dimensional rigid stochastic fibrous networks: Experiment, finite element simulation, and factor analysis, *J. Eur. Ceram. Soc.* **40**, 115–126.

Sinha-Ray, S., Khansari, S., Yarin, A. L. & Pourdeyhimi, B. (2012). Effect of chemical and physical crosslinking on tensile characteristics of solution-blown soy protein nanofiber mats. *Ind. Eng. Chem. Res.* **51**, 15109–15121.

Smith, T. L. (1958). Dependence of the ultimate properties of a GR-S rubber on strain rate and temperature. *J. Poly. Sci.* **32**, 99–113.

Smith, T. L. (1963). Ultimate tensile properties of elastomers. I. Characterization by a time and temperature independent failure envelope. *J. Polym. Sci.* **1**, 3597–3615.

Smith, T. L. (1964). Ultimate tensile properties of elastomers. II: Comparison of failure envelopes for unfilled vulcanizates. *J. Appl. Phys.* **35**, 27–36.

Smith, T. L. (1967). Ultimate tensile properties of Elastomers. III. Dependence of the Failure Envelope on Crosslink Density. *Rubber Chem. Technol.* **40**, 544–555.

Stachewicz, U., Peker, I., Tu, W. & Barber, A. H. (2011). Stress delocalization in crack tolerant electrospun nanofiber networks, *ACS Appl. Mater. Interfaces* **3**, 1991–1996.

Stauffer, S. R. & Peppas, N. A. (1992). Poly(vinyl alcohol) hydrogels prepared by freeze–thawing cyclic processing. *Polymer* **33**, 3932–3936.

Stok, K. & Oloyede, A. (2007). Conceptual fracture parameters for articular cartilage. *Clinical Biomech.* **22**, 725–735.

Sun, C. C. & Mark, J. E. (1987). The effect of network chain length distribution, specifically bimodality, on strain-induced crystallization. *J. Poly. Sci.* **25**, 2073–2083.

Sun, J. Y., Zhao, X., Illeperuma, W. R. K., et al. (2012). Highly stretchable and tough hydrogels. *Nature,* **489**, 133–136.

Sun, T. L., Kurokawa, T., Kuroda, S., et al. (2013). Physical hydrogels composed of poly-ampholytes demonstrate high toughness and viscoelasticity. *Nature Mater.* **12**, 932–937.

Tada, H., Paris, P. C. & Irwin, G. R. (1973). *The stress analysis of cracks handbook.* Del Research Corporation, St. Louis, MO.

Tanaka, Y., Fukao, K. & Miyamoto, Y. (2000). Fracture energy of gels. *Eur. J. Phys. E* **3**, 395–401.

Tang, M. Y., Letton, A. & Mark, J. E. (1984). Impact resistance of unfilled and filled bimodal thermosets of PDMS. *Colloid Poly. Sci.* **262**, 990–992.

Taylor, D., O'Mara, N., Ryan, E., Takaza, M. & Simms, C. (2012). The fracture toughness of soft tissues. *J. Mech. Beh. Biomed. Mater.* **6**, 139–147.

Thomas. A. G. (1955). Rupture of rubber. II. The strain concentration at an inclusion. *J. Poly. Sci.* **18**, 177–188.

Thomas, A. G. (1958). Rupture of rubber. V. Cut growth in natural rubber vulcanizates. *J. Poly. Sci.* **31**, 467–480.

Thomas, A. G. (1960). Rupture of rubber. VI. Further experiments on the tear criterion. *J. Appl. Polym. Sci.* **3**, 168–174.

Thomas, A. G. (1994). The development of fracture mechanics for elastomers. *Rubber Chem. Tech.* **67**, G50–G60.

Tonsomboom, K. & Oyen, M. L. (2013). Composite electrospun gelatin fiber–alginate gel scaffolds for mechanically robust tissue engineered cornea. *J. Mech. Beh. Biomed. Mater.* **21**, 185–194.

Torquato, S. (2002). *Random heterogeneous materials.* Springer, New York.

Torres-Rendon, J. G., Schacher, F. H., Ifuku, S. & Walther, A. (2014). Mechanical performance of macrofibers of cellulose and chitin nanofibrils aligned by wet-stretching: A critical comparison. *Biomacromolecules* **15**, 2709–2717.

Tsunoda, K., Busfield, J. J. C., Davies, C. K. L. & Thomas, A. G. (2000). Effect of materials variables on the tear behavior of a non-crystalizing elastomer. *J. Mater. Sci.* **35**, 5187–5198.

Tutwiler, V., Singh, J., Litvinov, R. I., et al. (2020). Rupture of blood clots: Mechanics and pathophysiology. *Sci. Adv.* **6**, eabc0496.

Wang, Q., Mynar, J. L., Yoshida, M., et al. (2010). High-water-content mouldable hydrogels by mixing clay and a dendritic molecular binder. *Nature*, **463**, 339–343.

Weibull, W. (1961). *Fatigue testing and analysis of results.* Pergamon Press, Oxford.

Yang, C., Yin, T. & Suo, Z. (2019). Polyacrylamide hydrogels. I. Network imperfection. *J. Mech. Phys. Sol.* **131**, 43–55.

Zhang, L., Rocklin, D. Z., Sander, L. M. & Mao, X. (2017). Fiber networks below the isostatic point: Fracture without stress concentration, *Phys. Rev. Mater.* **1**, 052602(R).

Zhao, X. (2014). Multi-scale multi-mechanism design of tough hydrogels: Building dissipation into stretchy networks, *Soft Matt.* **10**, 672–687.

Zhu, H., Zhu, S., Jia, Z., et al. (2015). Anomalous scaling law of strength and toughness of cellulose nanopaper. *Proc. Nat. Acad. Sci.* **112**, 8971–8976.

9 Time Dependent Behavior

The mechanical behavior of most materials is time dependent. Materials may creep at constant stress, exhibit stress relaxation when held at constant deformation, or, in general, present a behavior dependent on the deformation rate. It is said that the mechanical behavior is strain rate sensitive.

The majority of network materials exhibit strain rate sensitive mechanical behavior. This category includes all thermal molecular networks and some athermal networks, such as collagen-based tissue and artificial collagen constructs. In comparison, network materials made from athermal nanofibers, such as various types of nanopapers, are weakly rate sensitive. Regular cellulose paper is rate sensitive primarily at elevated humidity. Nonwovens made from fibers of diameter larger than one micron are essentially rate insensitive.

A vast literature is dedicated to the time dependent mechanical behavior of polymeric solutions, melts, and solids. Thermoplastics with molecules long enough to be entangled are network materials. In these systems, entanglements establish a transient network with pronounced time dependent behavior. However, as indicated in Chapter 1, these systems fall outside the scope of this book. In this chapter we focus on crosslinked networks with permanent and transient crosslinks. Since networks with transient crosslinks are discussed in dedicated texts (e.g., Tanaka, 2011), only a summary of key aspects of their mechanics is presented in Section 9.5. The reminder of the chapter is organized based on the physical processes that control rate sensitivity.

The time dependent mechanical behavior of network materials emerges from:

(i) The time dependence of the fiber material behavior;
(ii) The time dependence of the matrix, if the network is embedded in a viscoelastic material, or from the transport of solvent in and out of the network, if the embedding medium is a fluid;
(iii) The dissipative interactions of filaments in contact at sites other than the crosslinks. This includes inter-molecular interactions in elastomers and other molecular networks, as well as viscous friction taking place at fiber contacts in athermal networks; and
(iv) The process of crosslink rupture and re-formation in networks with transient crosslinks.

Networks with permanent crosslinks are viscoelastic in the sense that, given enough time, they return to the under-formed configuration upon unloading. In practical terms, the rebound may be infinitely slow, but the network provides the thermodynamic driving force for full relaxation. If the crosslinks break due to the applied stress and thermal fluctuations, the network develops pronounced rate sensitivity. Such materials are viscoplastic. Since the network connectivity is reconfigured during this process, the driving force for the return to the undeformed configuration is lost and the material develops permanent residual strains upon unloading.

This chapter provides details pertaining to each of the mechanisms listed here. Before addressing these issues, it is necessary to review several fundamental concepts in viscoelasticity. For additional details, see texts dedicated to viscoelasticity such as Christensen (1971), Findley et al. (1976), and Haddad (1995).

9.1 Elements of Viscoelasticity

In linear elasticity stress is proportional to the strain. In linear viscoelasticity, stress is evaluated in a strain-controlled experiment as the summation of contributions of strain increments applied throughout the loading history, $d\varepsilon$, weighted by a time-dependent kernel which has units of stiffness. The linearity of the problem makes possible the superposition of these incremental contributions. Specifically, one writes:

$$\sigma(t) = \int_0^t E(t - t')d\varepsilon(t') = \int_0^t E(t - t')\dot{\varepsilon}(t')dt', \tag{9.1}$$

which represents a convolution of the strain rate and the stress relaxation function $E(t)$. In this chapter we work within the framework of the linear, small strain theory and, hence, the Cauchy stress $\boldsymbol{\sigma}$ is identical to the nominal stress, $\mathbf{S}$. $\boldsymbol{\sigma}$ is used throughout the chapter to keep the notation compatible with other descriptions of linear viscoelasticity. Equation (9.1) is written for uniaxial tension. The equivalent form for shear, in which the shear stress is computed based on the shear strain rate and in terms of the shear stress relaxation function, $G(t)$, is more broadly used.

In a stress-controlled experiment, the equivalent equation providing the strain at the current time, t, function of the stress history reads:

$$\varepsilon(t) = \int_0^t C(t - t')\dot{\sigma}(t')dt', \tag{9.2}$$

where $C(t)$ is the time-dependent compliance (or the creep function) measured in a uniaxial test. An equation equivalent to Eq. (9.2) is written for shear loading and the shear compliance function is usually denoted by $J(t)$.

In a uniaxial relaxation experiment, the strain is applied as a step function which increases from 0 to ε_0 at $t = 0$, $\dot{\varepsilon}(t)$ is a Dirac delta function at zero and stress relaxation is described by $E(t)$, that is, $\sigma(t) = E(t)$. In a creep experiment, the stress is a step function at zero, $\dot{\sigma}(t)$ is a delta function, and $\varepsilon(t) = C(t)$.

Equations (9.1) and (9.2) are written as if the response of the material depends only on the difference between the time when the step perturbation is applied and the current time. In other words, kernels are functions of a single variable, $t - t'$, rather than functions of both t and t'. This is possible only under the assumption of time translational invariance of the material response.

Applying the Fourier transform to the convolution of Eq. (9.1) gives $\hat{\sigma} = i\omega\hat{E}\hat{\epsilon}$. The hat indicates the Fourier transform of the respective quantity. All transforms are functions of frequency, ω. Performing the same operation on Eq. (9.2) gives $\hat{\epsilon} = i\omega\hat{C}\hat{\sigma}$. Combining these two equations leads to a relation between the relaxation and creep functions which reads $-\omega^2\hat{E}\hat{C} = 1$. This indicates that $E(t)$ and $C(t)$ are not independent.

The complex modulus, $E(\omega)$, is defined as $E(\omega) = E' + iE'' = i\omega\hat{E}$.[1] The real and imaginary parts of the complex modulus are known as the storage and loss moduli, respectively. They are given by $E'(\omega) = -\omega\mathrm{Im}(\hat{E}(\omega))$ and $E''(\omega) = \omega\mathrm{Re}(\hat{E}(\omega))$.

If relaxation is exponential and $E(t) = E_g \exp(-t/\tau)$, where E_g is the instantaneous (or glassy) modulus and τ is the relaxation time, the storage and loss moduli become:

$$E'(\omega) = E_g \frac{\omega^2}{\omega^2 + (1/\tau)^2}; \quad E''(\omega) = E_g \frac{\omega/\tau}{\omega^2 + (1/\tau)^2}. \tag{9.3}$$

In the limit of high frequencies, $\omega \to \infty$, $E' = E_g$, and $E'' = 0$, while as ω approaches 0, $E' \sim \omega^2\tau^2$ and $E'' \sim \omega\tau$. The curves $E'(\omega)$ and $E''(\omega)$ cross at $\omega = 1/\tau$. The storage and loss moduli result directly from rheometry testing and the relaxation time of the system, τ, may be determined based on the intersection of the storage and loss moduli curves.

Real materials may exhibit multiple relaxation times. In this case, the relaxation function is written as a superposition of N exponential functions, each having a different relaxation time, τ_i, called a Prony series:

$$E(t) = E_g \sum_{i=1}^{N} \Upsilon_i \exp\left(-\frac{t}{\tau_i}\right). \tag{9.4}$$

Since E_g is the value of $E(t)$ at the beginning of relaxation, $t = 0$, $\sum_{i=1}^{N} \Upsilon_i = 1$. Equivalently, the storage and loss moduli may be written as summations of expressions similar to those of Eq. (9.3), weighted by Υ_i.

The continuum version of Eq. (9.4) is based on the idea that relaxation times form a continuum spectrum. Υ_i is replaced by a distribution of relaxation times $\Upsilon(\tau)d\tau$ and $E(t)$ becomes:

$$E(t) = E_g \int_0^{\infty} \Upsilon(\tau) \exp\left(-\frac{t}{\tau}\right) d\tau. \tag{9.5}$$

---

[1] The complex modulus generally refers to the shear modulus, with the equivalent notation $G(\omega) = G' + iG'' = i\omega\hat{G}$. Here we refer to $E(\omega)$ for consistency with the other chapters. The shear representation is more natural when performing rheological testing.

Taking the Fourier transform of this expression and using $E' + iE'' = i\omega\hat{E}$ to compute the storage and loss moduli, one obtains the equivalent of Eq. (9.3):

$$E'(\omega) = E_g \int_0^\infty S_E(\omega_0) \frac{\omega^2}{\omega^2 + \omega_0^2} d\omega_0; \quad E''(\omega) = E_g \int_0^\infty S_E(\omega_0) \frac{\omega\omega_0}{\omega^2 + \omega_0^2} d\omega_0, \quad (9.6)$$

where S_E is the frequency-dependent relaxation spectrum, which is related to the spectrum of relaxation time constants as $S_E(\omega_0) = (1/\omega_0^2)\Upsilon(1/\omega_0)$, with $1/\omega_0 = \tau$. Expressions identical to Eq. (9.6) can be written for $G(t)$.

A key concept in linear viscoelasticity is that of time–temperature superposition. It refers to the observation that the variation of temperature is equivalent to the variation of the strain rate (or frequency) if certain conditions apply. If this is the case, one would be able to probe the time-dependent behavior by testing in a broad range of temperatures and in a limited range of frequencies, and then extend these results to determine the behavior at a reference temperature and in a wider range of frequencies. The same statement can be made if the pressure, rather than the temperature, is the control parameter. The variation of pressure at a set temperature leads to mechanical behavior changes equivalent to those caused by the variation of the probing frequency at set pressure and temperature. Materials in which this equivalence applies are referred to as "thermorheologically simple." Many polymeric network materials are thermorheologically simple if probed at temperatures well above their glass transition temperature.

Several conditions must be fulfilled for a material to be thermorheologically simple: The material should not undergo phase changes or chemical reactions and should remain structurally identical as the temperature or the pressure are varied; internal heat generation should be small enough such that the isothermal test conditions are not modified; and the relaxation modes should not be coupled.

In systems in which relaxation modes taking place on various temporal scales depend on the same internal friction coefficient, time–temperature superposition applies. This is the case in the classical theory of polymer dynamics, in which the monomer-scale friction coefficient controls all relaxation modes of the material.

Figure 9.1 shows an example of time–temperature superposition for an Epon epoxy (Plazek and Rosner, 1998). Figure 9.1(a) shows the creep function obtained at various temperatures above T_g, while Figure 9.1(b) presents the same data mapped to the reference temperature $T_0 = 174°C$ using a temperature-dependent shift factor, a_T. The horizontal axis in Figure 9.1(b) is the reduced time defined by t/a_T. The compliance function $C(t)$ starts from a plateau at short times and ends on another plateau at long times. The difference of compliance between these two plateaus is larger than one order of magnitude. The short time limit compliance $1/E_g$ corresponds to the glassy state and is equivalent to the stiffness obtained in an oscillatory experiment at high frequencies, $E'(\infty)$, Eq. (9.3). The long time limit compliance is equivalent to $1/E_0$, where E_0 corresponds to the fully relaxed state. This behavior is different from that represented by Eq. (9.5), which indicates that the stiffness vanishes at long times. Here, the material relaxes to a plateau corresponding to the intrinsic stiffness of the

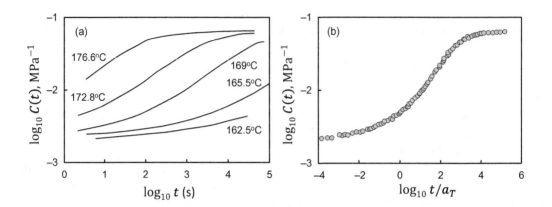

Figure 9.1 (a) Creep function for an Epon epoxy measured after crosslinking and at five temperatures, all above T_g. (b) Data in (a) collapsed to a single $C(t)$ curve using time–temperature superposition, at a reference temperature $T_0 = 174°C$. Data from Plazek and Rosner (1998)

network, E_0. The larger stiffness at short times (lower compliance in the limit $t \rightarrow 0$) is due to the internal friction (nonbonded interactions between molecular strands) which restricts the deformation of the network; see also Section 9.4.

The central issue when attempting to apply time–temperature superposition is the definition of the shift factor, $a_T(T)$.[2] The original ideas leading to the derivation of a_T were based on the concept that relaxation dynamics depends on the available free volume: the larger the free volume, the weaker the nonbonded interactions between relaxing molecules and the faster the relaxation. This applies under fairly general conditions to a broad range of materials, including monatomic glasses, polymers, and polymer-based network materials.

It was proposed that the viscosity is exponentially related to the free volume (Doolittle and Doolittle, 1957) as:

$$\frac{\eta(T)}{\eta(T_0)} = \alpha_1 \exp\left(\frac{\alpha_2 v_0}{v(T) - v_0}\right). \tag{9.7}$$

Here, $v(T) - v_0$ is the temperature-dependent specific free volume, α_1 and α_2 are constants, while $\alpha_1 \eta(T_0) = \eta(T_\infty)$ is the viscosity in the limit of large $v - v_0$. The free volume is also a function of pressure; this dependence is not made explicit here for simplicity. The viscosity is considered representative for the internal kinetics of the material and is related to the way the material relaxes. Its relation to the relaxation time emerges from the observation that, in the zero rate limit, the viscosity may be expressed as the integral of the relaxation modulus, $\eta = \int_0^\infty E(t)dt$. For an exponential

---

[2] See applications related to generic polymeric materials in Bird et al. (1987) and to rubbery materials in Roland (2011).

relaxation $E(t) = E_g \exp(-t/\tau)$, one obtains $\eta = E_g\tau$. Since E_g represents the short time and high frequency limit modulus, it is independent of temperature and hence the temperature dependence of the viscosity emerges from that of the relaxation time. Based on this observation, the shift factor may be expressed as:

$$a_T = \frac{\eta(T)}{\eta(T_0)}.$$ (9.8)

Assuming that the variation of the free volume is a linear function of temperature, Eqs. (9.7) and (9.8) lead to the Vogel–Fulcher–Tammann–Hesse (VFT-H) equation (Ferry, 1980):

$$a_T = \frac{\eta(T)}{\eta(T_0)} = \alpha_1 \exp\left(\frac{\beta_1}{T - T_0}\right),$$ (9.9)

which is often used to define the shift factor. β_1 is treated as a fitting constant.

Equation (9.9) is algebraically equivalent to the Williams–Landel–Ferry (WLF) equation (Williams et al., 1955):

$$a_T = \frac{\eta(T)}{\eta(T_0)} = \exp\left(\frac{-c_1(T - T_0^*)}{c_2 + T - T_0^*}\right),$$ (9.10)

upon the re-definition of the reference temperature, T_0^*, and of the two constants, c_1 and c_2, in terms of α_1, β_1, and T_0 as $c_1 c_2 = \beta_1$, $c_1 = \log(1/\alpha_1)$ and $T_0^* = T_0 - \beta_1/\log\alpha_1$.

Since network materials are generally nonlinear elastic, it is necessary to recall an extension of the concepts presented in the preceding paragraphs to nonlinear viscoelasticity.[3] One of the most widely used formulations of this type is the K-BKZ model (Kaye, 1962; Bernstein et al., 1963) which accounts for the experimentally-observed fact that the relaxation function depends on the magnitude of the applied strain. Clearly, this is not the case in Eqs. (9.1) and (9.2) in which both memory functions E and C have functional forms independent of the applied strain or stress. The K-BKZ version of Eq. (9.1) is:

$$\sigma(t) = \int_0^t E(t - t', \varepsilon(t'))\dot{\varepsilon}(t')dt'.$$ (9.11)

A particular form of this expression is used when elasticity is nonlinear, but the way stress relaxes is independent of the magnitude of strain. In this case, the strain and time-dependent relaxation function is written as a product of two functions, one of time and the other one of strain:

[3] See a full description of nonlinear viscoelasticity in Findley et al. (1976), and including the large deformation formulation (geometric nonlinearity) in Christensen (1971).

$$\sigma(t) = \int\limits_0^t E(t - t')f(\varepsilon)\dot{\varepsilon}(t')dt'. \tag{9.12}$$

The version of K-BKZ of Eq. (9.12) is widely used in the biomechanics literature to describe the viscoelastic deformation of tissue. In this literature it is referred to as the quasi-linear viscoelastic model (QLV) (Fung, 2013). In this model, the elastic stress–strain relation is described with an exponential function, as appropriate for the deformation of tissue (Chapter 6). The time-dependent kernel is represented using a rectangular continuous distribution of relaxation times, that is, the distribution of relaxation times $\Upsilon(\tau)$ in Eq. (9.5) is a constant in the interval bounded by two characteristic times, τ_1 and τ_2, and zero elsewhere. τ_1 and τ_2 are used as fitting parameters and are determined based on experimental data. The QLV model is reviewed in many publications (e.g., Sarver et al., 2003; Nekouzadeh et al., 2007).

9.2   Effect of Fiber Material Viscoelasticity

As a baseline case in the discussion of the viscoelastic behavior of network materials, consider the situation of a network without embedding medium and with permanent, rigid crosslinks. Assume further that the interaction of fibers at sites which are not crosslinks is minimal (the discussion of cases in which such interactions are essential is deferred to Section 9.4). Fibers are assumed to be athermal and made from a viscoelastic material. The rate sensitivity of the network is defined entirely by the behavior of the fiber material and is modulated by the network structure.

The deformation of networks of high density and high crosslink density is affine. Fibers store most of the strain energy in their axial mode (Figure 6.7). Hence, the network-scale stress becomes a superposition of contributions of individual fibers, each being loaded axially, Eqs. (5.14) and (5.32). Fibers oriented in different directions are subjected to different strains. If linear viscoelasticity applies, the relaxation spectrum of the network is identical to the relaxation spectrum of the fiber material and the network structure has no bearing on the overall viscoelasticity. This also applies to nonlinear cases in which the K-BKZ equation may be written in the simpler form of Eq. (9.12). To determine whether this applies in a specific experimental case, it is sufficient to determine whether the network deforms affinely.

An affine network composed from fibers with different relaxation time constants exhibits a spectrum $\Upsilon(\tau)$, Eq. (9.5), which includes all characteristic times of individual fiber materials, with peaks of height weighted by the fraction of the respective fibers in the network (Dhume and Barocas, 2019).

If the network deforms nonaffinely, relaxation is collective and cannot be viewed as a simple superposition of individual fiber contributions. Amjad and Picu (2022) studied the relaxation of athermal networks in which fiber behavior is described by a Maxwell linear viscoelastic model with a single time constant. Relaxation on the network scale exhibits two regimes. At early times stress relaxes exponentially, with a time constant equal to that of individual fibers, in a manner similar to the behavior of an affine network. However, a second regime emerges at later times in which relaxation is described by a stretched exponential function of Kohlrausch type, $E(t) = E(0) \exp\left[-(t/\tau)^n\right]$, n $<$ 1, see also Eq. (9.22). The stretch exponent, n, depends on the structural parameter w which defines the degree of non-affinity of the network (Figure 6.6). The exponent reaches a minimum (slowest relaxation) at the transition between the nonaffine and affine regimes. n decreases further when additional heterogeneity is introduced by allowing the relaxation time of fibers to vary from fiber to fiber. This depicts a physical picture similar to that encountered in polymeric thermal networks in the vicinity of the glass transition. While in the thermal networks case slowing down of relaxation is associated with dynamic heterogeneity, in athermal networks it is introduced by the structural heterogeneity, which is also responsible for non-affine kinematics.

9.3   Effect of Solvent Migration: Poroelasticity

Many network materials are embedded in a fluidic matrix. Since networks undergo a large volume change during nonisochoric deformation, the solvent must flow in and out of the network. The coupled process of deformation and fluid migration is represented by poroelasticity models. The first such model (Biot, 1941) was developed for bedrock infiltrated with fluids (water, oil) and was used extensively in geology (Wang, 2000). The same concept is at the base of the biphasic model used in biomechanics to represent the mechanical behavior of tissue (Mow et al., 1980).

If the solvent is neutral relative to the network, in the sense that the chemical potential within and outside of the material is the same, the contribution of the flow to the mechanics is associated with the drag force which develops as the fluid flows around fibers.

If the solvent is not neutral relative to the network, the free energy of the system changes when a quantity of solvent migrates in or out of the network. This entails an associated thermodynamic force which acts on the network in addition to the drag.

9.3.1   Chemically Neutral Case

Consider first the purely mechanical case in which the solvent is neutral relative to the fibers, that is, the free energy of a solvent molecule within the network is identical to that outside the network. To render ideas more accessible, we analyze first the one-dimensional deformation set-up shown in Figure 9.2. The material is confined in a box of rigid impermeable walls and is compressed in the x direction by a permeable piston. The lateral walls are assumed to be elastically compliant, such that the stress state is uniaxial. The solvent is incompressible. As the piston moves in the negative x direction, the network is compressed and the solvent is expelled into the bath (shown by the gray shade). Since the fluid is incompressible, there is no flow relative to the laboratory frame, rather the fibers move relative to the solvent, with velocity $\dot{u}(x)$.

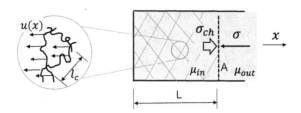

Figure 9.2 Schematic representation of a one-dimensional poroelasticity problem: A network material is compressed by a porous piston moving in the negative x direction, which leads to the partial elimination of the fluid from the network.

A fiber moving relative to a fluid of viscosity η experiences a drag force per unit length of magnitude proportional to $\eta \dot{u}$. The coefficient of proportionality is of order 1 and depends on the shape of the fiber. The drag force per unit volume of the network is $\rho \eta \dot{u}$, where ρ is the network density (total fiber length per unit volume). Therefore, the equilibrium equation for the network in the one-dimensional problem becomes:

$$\frac{\partial \sigma}{\partial x} - \rho \eta \dot{u} = 0, \tag{9.13}$$

where σ is the stress in the network. Equation (9.13) may be written in terms of u, provided a constitutive equation is selected for the network. For simplicity, consider linear elasticity and small deformations, such that $\sigma = E_0 \varepsilon$. This leads to:

$$\frac{E_0}{\rho \eta} \frac{\partial^2 u}{\partial x^2} = \frac{\partial u}{\partial t}. \tag{9.14}$$

Interestingly, the equilibrium equation becomes similar to the diffusion equation, with an effective diffusion coefficient $D_{eff} = E_0/\rho\eta$. Taking into account the relation between the network density and the mesh size (Eq. (4.14)), one may write $D_{eff} \approx E_0 l_c^2/\eta$.

The solution of Eq. (9.14) depends on the boundary conditions. In general, the relaxation is exponential and the time constant involved, τ_P, depends on the size of the sample, L. This reflects the physical condition that the solvent must be transported over a length scale proportional to L in order to leave the sample. One may write:

$$\tau_P \approx \frac{L^2}{D_{eff}} = \frac{\eta}{E_0}\left(\frac{L}{l_c}\right)^2. \tag{9.15}$$

Alternately, one may replace in this equation l_c with $\sqrt{k}$, where k is the intrinsic permeability which appears in Darcy's transport equation discussed in Section 9.3.3.

It is instructive to estimate τ_P for collagen networks, for which one may take, for example, $L = 1$ cm and $l_c \approx 1$ μm. E_0 is in the range of 100 kPa to 1 MPa, while the viscosity of a hyaluronic acid solution at 5 mg/ml and 25°C is about 1 Pa s. This gives τ_P of about 10 s, which is in the range of relaxation times observed for collagen structures (Section 9.4.4).

Clearly, τ_P is a strong function of the sample size. In fact, this dependency is the signature of poroelastic relaxation and may be used to identify the contribution of this mechanism to network behavior.

In many situations it is more convenient to measure material properties by indentation, as opposed to macroscopic uniaxial or shear testing. In this case, solvent redistribution takes place only in the vicinity of the indented site and on length scales comparable with the indenter tip radius. Therefore, in such an experiment L should be taken equal to the indenter tip diameter. For example, if indentation is performed in a collagen structure of the parameters mentioned in the previous paragraph, and with a spherical indenter of 1 mm diameter, τ_P results $\sim$0.1 s. If the indenter tip size is 1 μm, the relaxation time becomes 10^{-6} s. Poroelastic relaxation may be considered instantaneous in such cases.

This discussion is based on the view that poroelasticity is associated with the migration of solvent in and out of the sample. It is interesting to enquire if internal redistribution of solvent, without drainage, may cause a significant poroelastic effect. Since network deformation is generally nonaffine, volumetric strains are spatially variable and hence the fluid may flow within the sample from regions experiencing compression to regions subjected to tension. The average distance over which the fluid flows is of the order of (but larger than) l_c. Assuming that local elasticity is characterized by the network average modulus, E_0, and that $L/l_c \approx 1$, the relaxation time results $\tau_P \approx \eta/E_0$, which is much smaller than 1 s even for very soft, nonaffinely deforming networks. Therefore, internal fluid redistribution may be considered instantaneous on the time scales of macroscopic experiments.

This discussion also answers the question of whether or not a viscous embedding medium may enforce affine deformation on a soft network which deforms nonaffinely in the absence of the matrix. The instantaneous deformation of the material with viscous solvent is affine. In a time comparable to τ_P the network relaxes to a nonaffine deformation pattern and the relaxed stiffness (corresponding to the nonaffinely deforming network) is orders of magnitude smaller than that of the apparent instantaneous (affine) stiffness. If τ_P is very small, affine deformation cannot be enforced. For τ_P to reach 1 s, which would make the mechanism discussed here relevant, the network has to be very soft ($E_0 = 10$–100 Pa) and the viscosity very large ($\eta = 10$–100 Pa s, comparable to honey or sour cream at room temperature), which is possible only for extreme combinations of parameter values. Such combinations may exist in biological applications.

Numerical simulations of the somewhat related problem of permeation of fluid through a fibrous material are presented in Nabovati et al. (2009).

9.3.2 Effect of a Chemical Potential Gradient

While the neutral solvent case discussed in Section 9.3.1 is instructive, in many polymeric and biological network materials some degree of chemical affinity between network fibers and solvent exists. In these cases, the solvent chemical potential inside the network, μ_{in}, is different from that in the bath outside the network, μ_{out}. For example, gels absorb large amounts of water. Similarly, water is absorbed in collagen and collagen/elastin networks, primarily due to the presence of polar molecules (such as glycosaminoglycans) embedded in the host network.

Consider the one-dimensional example of Figure 9.2. The energy required to take one solvent molecule out from the network is $\mu_{out} - \mu_{in}$. If a volume of solvent δV is taken out, the free energy difference is $(\mu_{out} - \mu_{in})\delta V/\Omega$, where Ω is the molecular volume. Since the chemical potential changes only when the molecule crosses the piston, this energy variation may be considered equivalent to a boundary traction which, as the piston moves over a distance $\delta V/A$, performs work equal to the free energy variation. This allows us to evaluate the chemical stress as:

$$\sigma_{ch} = (\mu_{out} - \mu_{in})/\Omega. \tag{9.16}$$

The effective stress experienced by the network (associated exclusively with the variation of the network free energy without the chemical potential contribution), σ_{eff}, is the difference between the applied stress, σ, and the chemical stress:

$$\sigma_{eff} = \sigma - \sigma_{ch}. \tag{9.17}$$

Equation (9.17) indicates an important fact: if $\mu_{out} > \mu_{in}$, the material may sustain stresses larger than those the dry network may carry. The chemical stress shields the structure from damage and this may be an important design criterion for biological materials. Alternately, a network with high affinity for solvent may be sparser (i.e., contain smaller number of fibers) and yet sustain stresses larger than those it would be able to carry if the solvent were neutral. This effect is demonstrated experimentally using biological collagen membranes in Ehret et al. (2017) and in other works.

Further, this discussion indicates the possibility to act on the network mechanically by controlling the difference of the chemical potential $\mu_{out} - \mu_{in}$. The network absorbs solvent and dilates if $\mu_{out} > \mu_{in}$, or it contracts and eliminates solvent if $\mu_{out} < \mu_{in}$. The difference $\mu_{out} - \mu_{in}$ may be controlled by changing the temperature or the pH.

9.3.3   Governing Equations in Three Dimensions

The one dimensional case discussed in Sections 9.3.1 and 9.3.2 is generalized here to three dimensions and to situations in which the chemical potential in the material is position-dependent. The physical processes considered are the same: the network is an elastic structure (not necessarily linear elastic), the chemical potential is a function of position within the network, $\mu(\mathbf{x})$, while the drag mechanism discussed in Section 9.3.1 provides viscous dissipation. Equation (9.17) may be rearranged as:

$$\sigma_{ij} = \frac{\partial \Psi}{\partial \varepsilon_{ij}} + \frac{\mu_{out} - \mu(\mathbf{x})}{\Omega}\delta_{ij}, \tag{9.18}$$

where $\sigma_{eff} = \partial\Psi/\partial\varepsilon$ is the stress carried by the network, and the chemical stress (second term on the right side of Eq. (9.18)) contributes a hydrostatic component. Equation (9.18) becomes the constitutive description of the system. Further, the total stress $\boldsymbol{\sigma}$ must fulfill equilibrium. In the static case (no inertia forces), Eq. (9.13) may be generalized to read $\sigma_{ij,j} - \rho\eta\dot{u}_i = 0$, $i = 1, 2, 3$. The drag force is aligned with the local velocity and enters the equilibrium equation as a body force. Using the constitutive description of Eq. (9.18), the equilibrium equation becomes:

$$\frac{\partial}{\partial x_j} \frac{\partial \Psi}{\partial \varepsilon_{ij}} - \frac{\mu_{,i}}{\Omega} = \rho\eta\dot{u}_i. \tag{9.19}$$

To make Eq. (9.19) useful it is necessary to specify the functional form of Ψ. For a linear elastic network, Eq. (9.19) becomes:

$$G u_{i,jj} + \lambda u_{k,ki} - \frac{1}{\Omega}\mu_{,i} = \rho\eta\dot{u}_i, \tag{9.20}$$

where G is the shear modulus and the Lamé constant λ is related to the Poisson ratio, ν, as $\lambda = 2G\nu/(1 - 2\nu)$.

Equation (9.20) stands for three equations. The unknown fields are the three displacements and the scalar chemical potential field. The fourth equation, which closes the system, represents the mass conservation condition. The continuity equation is $\dot{c} + j_{i,i} = 0$, with c being the number of molecules per unit volume and $\mathbf{j}$ representing the molecular flux. The constitutive law of the transport problem is the Darcy equation, $j_i = -(k/\eta\Omega^2)\mu_{,i}$, which indicates that the flux is proportional to the gradient of chemical potential and the direction of the flux is aligned with that of the gradient. Here, k is the permeability. Combining the two equations, one obtains: $\dot{c}\Omega = (k/\eta\Omega)\mu_{,i}$. The quantity $\dot{c}\Omega = \partial(c\Omega)/\partial t$ is the rate of change of the volume occupied by molecules and is identical to the time derivative of the dilatation strain ε_{kk}. Therefore, it is possible to write:

$$\frac{\partial \varepsilon_{kk}}{\partial t} = \frac{k}{\eta\Omega}\mu_{,ii}. \tag{9.21}$$

Equations (9.20) and (9.21) form a system of four equations for the four unknown fields.

The dissipation mechanism built in this formulation is based on the drag associated with fluid migration. Other dissipative mechanisms may be integrated, for example, those related to intermolecular friction (Section 9.4). Damping leads to viscoelastic behavior. Chemical potential gradients contribute to elasticity, but do not lead directly to time-dependent behavior. The relaxation time constant associated with Eq. (9.20) is identical to τ_P of Eq. (9.15).

A formulation of poroelasticity under finite strains is presented in Buhan et al. (1998).

9.4   Effect of Dissipative Interactions of Filaments

This section is divided into two parts referring to molecular, thermal networks and to athermal networks, respectively. In molecular networks, including entangled polymeric melts, time dependent mechanical behavior is associated with inter-molecular friction caused by mechanisms taking place on the monomer scale.

The time dependence of athermal networks without an embedding matrix is much less pronounced. In cases in which it occurs, it is due to the viscous frictional interactions of fibers in contact.

9.4.1   Time Dependence due to Molecular Friction in Molecular Networks

In the classical mean field view of polymer physics, a representative polymeric chain is subjected to thermal fluctuations and interacts via viscous friction with a homogen-ized background representing the neighboring chains. The relaxation times of the material are identical to those of the representative chain. A vast literature is dedicated to this subject and reviews of key results can be found in most polymer physics books. Here we outline only those aspects that are specific to network materials. In particular, we are interested in the relation between the viscoelastic functions and the network structure and parameters.

One may distinguish between molecular networks without solvent (dry networks) and networks with solvent. As the degree of swelling increases, direct inter-chain interactions become less frequent and hence dissipation via this mechanism decreases. The transport of solvent within and in and out of the network becomes the dominant mechanism in these cases (Section 9.3). To separate the two effects, we focus here on the dry network case. Further, we consider networks above the glass transition temperature, T_g. If the temperature is below the glass transition temperature, only local, monomer-scale relaxations are possible and the dissipation, as well as the strain rate sensitivity of the material behavior, are greatly reduced. A large literature exists on the linear and nonlinear viscoelasticity of elastomers, which interested readers may consult for further details (e.g., Dicke and Ferry, 1966; Payne and Whittaker, 1971; Isono and Ferry, 1984; Gent and Mars, 2013; Ngai et al., 2013).

Figure 9.3(a) shows a schematic representation of the relaxation function $E(t)$ for several types of networks. The continuous line corresponds to a permanently cross-linked network, the dashed line represents an entangled uncrosslinked network, while the dotted line represents an unentangled uncrosslinked melt. The difference between these cases is observed primarily at long times (or low ω). The permanently cross-linked network exhibits a plateau which corresponds to the stiffness of the network without inter-chain frictional interactions. This is denoted here by $E(\infty)$ and is identical to E_0 of Section 9.1 and Chapter 6. In this limit, the stiffness of thermal networks is proportional to the number density of load carrying strands and to the temperature, Eq. (5.25). As the crosslink density ρ_b increases, the stiffness increases and the plateau shifts up, as shown in Figure 9.3(b). The compliance curve in Figure 9.1(b) is equivalent to the stiffness representation in Figure 9.3(b). In melts of entangled polymers, the plateau emerges due to the constraining effect of entangle-ments on chain dynamics. However, at times longer than that required for chains to disentangle (by reptation), the network relaxes fully and $E(t)$ decays to zero.

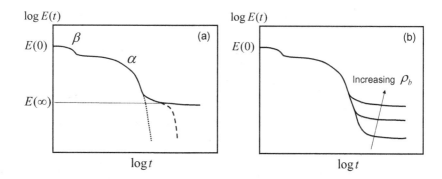

Figure 9.3 Schematic representation of the relaxation modulus. (a) Curves for a crosslinked network (continuous line), an entangled polymeric melt (dashed line) and an unentangled melt (dotted line). (b) Effect of increasing crosslink density.

Polymeric systems which are neither crosslinked nor entangled do not exhibit a long time plateau, as shown schematically in Figure 9.3(a).

The variation of the crosslink density and of the entanglement state have little effect on the relaxation function at times below those at which the $E(\infty)$ plateau emerges. This early relaxation regime is defined by a spectrum of relaxation times associated with segmental relaxation (α modes) and local, monomer-scale relaxation (β modes).[4] Figure 9.3(a) shows two exponential relaxation regimes denoted by α and β.

The ratio $E(0)/E(\infty) = E_g/E_0$ of a permanently crosslinked network may be very large. The difference between the instantaneous and relaxed moduli is due to the contribution of nonbonded interactions between molecules to the instantaneous modulus. $E(0)/E(\infty)$ increases as the density of nonbonded interactions increases relative to that of bonded (crosslinks) interactions. On short time scales, nonbonded interactions produce an effect similar to that of the crosslinks. Hence, the network behaves as if it were densely crosslinked at $t \to 0$ and, based on the discussion in Chapter 6, deforms affinely in this limit. In this sense, the large difference between $E(0)$ and $E(\infty)$ observed here is similar to the difference between the stiffness of nonaffine and affine networks shown in Figure 6.6.

The rate at which the material relaxes from $E(0)$ to $E(\infty)$ depends on temperature, pressure, and the presence of solvent. Increasing the temperature, decreasing the pressure, and the addition of solvent increase the free volume which, in turn, leads to reduced dissipation and faster relaxation. These effects are captured by the Doolittle equation (Eq. (9.7)) and by the VFT-H equation (Eq. (9.9)) for the shift factor, a_T. Therefore, the shift factor accounts for the effect of these parameters on network relaxation. While the VFT-H and WLF functions are most often used to describe a_T

---

[4] Multiple modes with short relaxation times may be observed. These are denoted by $\beta, \gamma, \delta \ldots$ and are generally easier to observe (as a function of the applied frequency) in oscillatory experiments and in dielectric relaxation. A spectrum of relaxation modes is also observed in crystalline solids probed by internal friction experiments and is related to the presence of specific crystal defects (Zener, 1956).

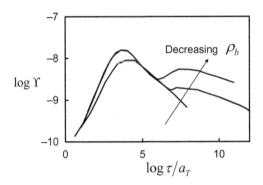

Figure 9.4 Relaxation spectra for Viton rubbers with different crosslink densities. The curves are reduced to slightly different temperatures to render them equidistant to T_g. Adapted from Ngai et al. (2013) with permission from Elsevier

for networks with low crosslink density and for uncrosslinked polymers, increasing ρ_b may render a_T Arrhenius, as observed in a family of Epon epoxies tested above T_g in Ngai et al. (2013). Increasing ρ_b renders the deformation more affine, which makes collective relaxation less relevant. At temperatures significantly above T_g, the shift factor becomes independent of the crosslink density, as reported, for example, for crosslinked polycaprolactones in Izuka et al. (1994).

Figure 9.4 shows the relaxation spectrum, $\Upsilon(\tau)$, Eq. (9.5), for Viton, a fluorinated hydrocarbon elastomer (Ngai et al., 2013) for materials of three crosslink densities. The crosslink densities in the three materials considered are in the ratio 1:1.5:2.8. The curves are reduced to slightly different temperatures, which are equidistant to T_g (5°C above T_g). This is necessary since the glass transition temperature increases with increasing degree of crosslinking. The variation of T_g with the degree of crosslinking is primarily due to shrinkage, that is, the elimination of free volume. The network contracts significantly during crosslinking, which decreases the free volume and brings the chains in closer contact. The difference of T_g between the uncrosslinked resin and the crosslinked state of an epoxy may be as large as 100°C (Plazek and Chay, 1991). For the data in Figure 9.4, T_g varies by 6.5°C between the least to the most crosslinked samples.

The curves in Figure 9.4 show two peaks corresponding to the slower, α, and faster, β, processes. The faster relaxations are independent of ρ_b since the length scale at which β processes take place is much smaller than the network mesh size. However, segmental relaxation causing the slower α processes involves chain segments which may be as large as l_c. Reducing the mesh size leads to constraints imposed on these relaxation modes. Consequently, the height of the large relaxation time peak decreases as ρ_b increases and, for the high ρ_b case, it becomes just a faint shoulder. The proximity to the glass transition temperature makes this effect of ρ_b more pronounced.

The results in Figure 9.4 indicate that decreasing ρ_b leads to more pronounced dissipation and higher effective viscosity. This is observed in other elastomers, including natural rubber (Ferry et al., 1964; Ferry, 1980).

The behavior close to the glass transition temperature is more complex. The time constants of longer wavelength modes become progressively longer and, hence, thermorheological simplicity does not apply. These effects occurring in the vicinity of T_g are more pronounced in molecular networks of large ρ_b.

Since the full relaxation of the network at long times is restricted close to T_g, the plateau modulus $E(\infty)$ develops a nonlinear dependence on temperature. At high temperatures, $T \gg T_g$, $E(\infty)/T$ is independent of temperature, as expected based on the network kinetic theory. At temperatures close to T_g, $E(\infty)/T$ increases as T converges to T_g from above.

The slowing down of relaxation in the vicinity of T_g is accounted for by representing the relaxation modulus as a stretched exponential (known as the Kohlrausch function):

$$E(t) = E_0 \exp\left[-(t/\tau)^n\right], \tag{9.22}$$

where $n \leq 1$ is the stretch exponent. As the coupling of the relaxation modes becomes more pronounced, either due to the reduction of temperature and/or the increase of the crosslink density, exponent n decreases and relaxation becomes slower.

Many efforts have been dedicated to the justification of the phenomenological expression (Eq. (9.22)) based on first principles. The Mode Coupling Theory was developed as a theoretical representation of vitrification, primarily for monatomic glass formers, and it was later extended to polymeric systems (Schweitzer, 1989). Its application to molecular networks (Ngai et al., 1997) is less developed compared with the mode coupling treatment of entangled uncrosslinked polymeric systems.

9.4.2 Time Dependence Caused by Inter-fiber Friction in Athermal Networks

Inter-fiber contacts form dynamically during the deformation of athermal networks. Siding at contacts may take place with Coulombic or viscous friction. Coulomb friction does not introduce time-dependence. However, viscous friction causes dissipation and strain rate sensitivity. Viscous friction is important only in networks of micron and submicron diameter fibers in which the fiber contacts are too small for the mechanism leading to the usual Coulomb friction to operate. Viscous friction is promoted by high humidity and the presence of environmental macromolecular debris covering fiber surfaces.

The mechanics of networks of tortuous fibers in frictional contact with each other presents several interesting aspects (Negi and Picu, 2019). To tackle complexity step by step, we discuss first rate independent frictional contacts. Consider mats of non-crosslinked fibers with large tortuosity and with inter-fiber friction. The frictional force is aligned with the relative displacement of the fibers at contacts and has prescribed magnitude, F_{fric}, which is independent of the relative velocity and of the normal forces at contacts. This is a first step toward dealing with viscous friction. Since fiber tortuosity is large, little strain energy is stored in the system and the macroscopic stress is controlled primarily by frictional dissipation at fiber contacts.

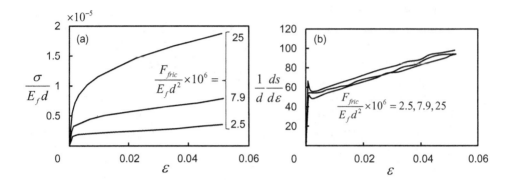

Figure 9.5 (a) Stress–strain curves for mats of uncrosslinked fibers interacting frictionally, for different values of the friction force. σ represents the applied force divided by the mat edge length and d is the fiber diameter. (b) Average incremental sliding at contacts, s, versus strain for the three friction force values in (a). Adapted from Negi and Picu (2019)

Figure 9.5(a) shows stress–strain curves of such mats loaded in uniaxial tension, for three values of the friction force. The stress increases as the friction force increases and all curves exhibit strain hardening. The rate of hardening increases with increasing friction. To understand this behavior, we write the stress in this uniaxial deformation as:

$$\sigma \approx \frac{dW_{fric}}{d\varepsilon} = F_{fric}\frac{d(N_X s)}{d\varepsilon}, \tag{9.23}$$

where N_X is the number density of active contacts and s represents the mean sliding distance at contacts. Equation (9.23) implies that the majority of the work performed by the applied tractions is dissipated, and only a negligible fraction is stored in fibers as strain energy. Simulations of such systems (Negi and Picu, 2019) indicate that the density of contacts is approximately constant during deformation (for strains below 10%) and, hence, $\sigma \sim F_{fric}N_X ds/d\varepsilon$. It results that the key to understanding the stress–strain curves in Figure 9.5(a) is the rate of sliding at contacts, $ds/d\varepsilon$. The strain dependence of this parameter is shown in Figure 9.5(b). Beyond an initial ramp-up, $ds/d\varepsilon$ increases linearly with strain and is independent of the magnitude of the friction force. With the result in Figure 9.5(b), Eq. (9.23) explains the increase of the stress with increasing friction observed in Figure 9.5(a). The result in Figure 9.5(b) also indicates that strain hardening is due to the continuous increase of the rate of sliding with increasing strain.

Adapting Eq. (9.23) to the case of viscous friction, one may write $F_{fric} = \eta_X ds/dt = \eta_X (ds/d\varepsilon)\dot{\varepsilon}$, where η_X is the effective viscosity at contacts and $\dot{\varepsilon}$ is the applied, macroscopic strain rate. Therefore,

$$\sigma \sim \eta_X N_X \left(\frac{ds}{d\varepsilon}\right)^2 \dot{\varepsilon}. \tag{9.24}$$

Considering that $ds/d\varepsilon$ increases linearly with the strain (Figure 9.5(b)), and that N_X is approximately strain-independent, $\eta_X N_X (ds/d\varepsilon)^2$ is a quadratic function of strain. Hence, the macroscopic response is similar to that of a viscous fluid whose viscosity increases quadratically with the strain. Other models for the viscous sliding contact may be used, for example, by assuming that the friction force is proportional to the contact area which, in turn, depends on the angle between fibers (Figure 3.11). However, as long as the result of Figure 9.5(b) remains valid, the functional form of Eq. (9.24) and its physical interpretation should hold.

If the elastic component is important and needs to be accounted for, the effective model of the network becomes a Voigt element (a spring in parallel with a damper) with strain-dependent viscosity defined by Eq. (9.24). At large stretches, fiber tortuosity is gradually pulled out and a larger fraction of the work is stored as strain energy. The fiber elasticity comes gradually into play and becomes dominant at large strains. Since these networks are not crosslinked, they rupture if made from discontinuous fibers.

9.4.3   Superposition of Relaxation Mechanisms

As discussed in the preceding sections of this chapter, multiple mechanisms contribute to defining the time-dependence of the mechanical behavior in complex network materials. These mechanisms include the rate dependence of the fiber material, the transport of fluid through the network, internal dissipation associated with fiber interactions, and the viscosity of the embedding medium. These mechanisms superimpose and may lead to complex relaxation spectra. Based on the current understanding of the synergies of these processes, it is not easy to distinguish their relative contribution to an experimental relaxation spectrum – see also Strange et al. (2013). However, several considerations may provide guidance in this sense.

Consider a situation in which poroelasticity operates together with internal dissipation due to inter-fiber friction. The equilibrium equation (Eq. (9.20)) accounts for dissipation caused by viscous drag associated with fluid transport within the network. The viscosity η in Eqs. (9.15) and (9.20) is that of the embedding fluid. The dissipation due to internal friction between network filaments must be represented by an additional term which, in the case of athermal networks, is given by Eq. (9.24), and, in the case of thermal networks, is of the more complex type discussed in Section 9.4.1. The coupling of this damping with the elasticity of the network provides a relaxation spectrum which is characterized here based on its longest relaxation time, τ_{max}. The key difference between τ_P and τ_{max} is that the first one scales quadratically with the sample size, L, Eq. (9.15), while the second is independent of L. In principle, one may identify experimentally the contribution of the two mechanisms by testing samples of different sizes.

Further, note that the poroelastic relaxation time may be computed either with the relaxed or unrelaxed network elastic constants, and the two estimates are: $\tau_P^{long} = (\eta/E(\infty))(L/l_c)^2$ and $\tau_P^{short} = (\eta/E(0))(L/l_c)^2$. If $\tau_{max} \ll \tau_P^{short}$, internal relaxation is completed by the time solvent transport causes stress relaxation via the poroelastic mechanism, and the relevant relaxation time is τ_P^{long}. If $\tau_P^{long} \ll \tau_{max}$, stress relaxation through the poroelastic mechanism takes place fast, on time scales comparable with τ_P^{short}, and the terminal relaxation time is τ_{max}.

The more interesting and likely more relevant situation is that in which poroelastic relaxation takes place on time scales comparable with those of the internal relaxation. In this case, the effective stiffness of the network changes during poroelastic relaxation. This situation may be represented with a series of Voigt elements (spring, E_i^v, and damper, η_i^v, coupled in parallel, with $i = 1 \ldots n^v$) connected in parallel, all being in series with a spring of stiffness $E(0)$. Each Voigt element represents an individual dissipation mechanism and η_i^v is defined by the respective physics. Since the elastic component is provided in all cases by the network, E_i^v may be taken as identical for all i, $E_i^v = E^v$. This stiffness may be computed by requiring that the overall system stiffness in the fully relaxed state is the dry network stiffness, $E(\infty) = E_0$. It results that $E^v = E(0)E(\infty)/[n^v(E(0) - E(\infty))]$. A somewhat different, but more detailed discussion of such an approach can be found in Hu and Suo (2012). These models may be used to interpret experimental data and to identify the contributions of the various mechanisms to the overall mechanical behavior.

9.4.4   Applications

9.4.4.1   Collagen

Fibrillar collagen encountered in biological materials has a hierarchical structure (Section 2.1.3) and is embedded in an aqueous environment. Collagen-based connective tissue embeds GAG and other macromolecules. Tests of the time-dependent behavior were performed with tissues as well as with reconstituted collagen. Since reconstituted collagen lacks GAG inclusions, the comparison with the relaxation of collagenous tissue reveals the effect of these inclusions. The relaxation modulus is usually fitted with a Prony series (Eq. (9.4)) or is analyzed in terms of the relaxation spectrum, which may be obtained from the measured $E(t)/E(0)$ through a Laplace transform.

The time-dependent behavior of single collagen fibrils is described in Shen et al. (2011) and Svensson et al. (2010). In these studies, collagen fibrils are separated from the dermis of the sea cucumber and from the human patellar tendon, respectively, and have diameters ranging from 10 to 500 nm. Stress relaxation of fibrils may be described by a Prony series with two terms, with time constants of ~8 s and 100 s.

The stress relaxation of reconstituted collagen may be described with a Prony series with three time constants (Pryse et al., 2003; Xu and Li, 2013; Babaei et al., 2015). The values of these constants vary from report to report but are in the range 0.6–8 s, 13–40 s, and 800–1300 s, respectively. In these networks, the time-dependent behavior originates from the response of individual fibers and from solvent transport. Although not definitely established to date, it is likely that the smaller two relaxation times are representative for the fiber behavior, while the longest time is associated with poroelasticity. Also, further investigation is needed to clarify the effect of network heterogeneity on the relaxation spectrum.

Stress relaxation of collagenous tissue may also be described using a Prony series with three terms. In Komatsu (2010), the time constants reported for the rabbit periodontal ligament are 0.4 s, 4 s, and 400 s, in Wagenseil et al. (2003), the time constants for fibroblast-seeded collagen are approximately 10 s, 100 s, and several thousands of seconds, while, in Sarver et al. (2003), only two time constants are reported for sheep tendon relaxation, of 2 s and 1 500 s. No order of magnitude differences between the time constants of connective tissue and those of reconstituted collagen emerge. This does not imply that the increased complexity of the tissue is unimportant, but rather that the contribution of the GAGs cannot be differentiated from that of network fibers and poroelasticity based exclusively on this type of measurement.

Stress relaxation of collagen structures depends on the magnitude of the strain applied at $t = 0$ (Xu and Li, 2013; Nam et al., 2016). For relatively small strain levels, $E(t)/E(0)$ is independent of the initial strain, but relaxation becomes faster (the time constants become smaller) as the initial step strain magnitude increases. In Xu and Li (2013) it is shown that the number of terms in the Prony series needed to represent the relaxation of reconstituted collagen increases from three to four as the magnitude of the initially applied strain increases. Interestingly, the short (order of 1 s) and long (order of 1 000 s) time constants remain unchanged as the initial strain increases, while the relaxation mode characterized by the middle time constant splits up in two modes. This supports the hypothesis that the relaxation mode with longest time constant is poroelastic, while the others emerge from the relaxation of network fibers.

9.4.4.2 Gels

The behavior of gels is time-dependent and two of the mechanisms discussed in this chapter contribute: solvent transport and internal friction. Both mechanisms operate in physical gels (e.g., agarose), while poroelasticity is dominant in chemical gels (e.g., acrylamide). The large free volume of (swelled) chemical gels limits the direct interaction of filaments and the associated dissipation which, in turn, renders poroelasticity more important than internal friction for this type of network material. A large number of publications present data for such materials (e.g., Weiss et al., 1981; Hui and Muralidharan, 2005; Lin et al., 2007; Galli et al., 2009; Hu et al., 2010; Zhao et al., 2010; Strange et al., 2013).

It is of interest to relate the time-dependent behavior of gels to their structure and network parameters. Assuming that poroelasticity is a good descriptor of gel mechanics, the relaxation time is computed with Eq. (9.15), which may be rearranged as $\tau_P = (\eta/E_0)(L^2/k)$. τ_P depends on network parameters via E_0 and the intrinsic permeability, k. The relation between network stiffness, E_0, and network parameters is discussed extensively in Chapter 6. The equivalent relation for k is less well understood, but trends may be inferred from the comprehensive literature data collection presented in Oyen (2014) for agar and acrylamide. It results that $E_0 \sim \rho^x$ (Chapter 6) and $k \sim \rho^{-y}$ (Oyen, 2014), which implies that $\tau_P \sim \rho^{y-x}$. In the case of agar, $x \approx 1.5$ and $y \approx 2.5$, and hence $\tau_P \sim \rho$. For acrylamide, $x \approx 2$ and $y \approx 3$, and again, $\tau_P \sim \rho$. However, a linear relation between the relaxation time and network density cannot be established in general, for all gels. Furthermore, uncertainties of about ± 0.5 apply to both x and y values indicated here, which affects the prediction of the exponent of the τ_P versus ρ relation. Further studies along these lines are needed to obtain accurate scaling relations of this type.

9.4.4.3   Paper

Networks composed from cellulose fibers exhibit time-dependent behavior, including delayed strain recovery, creep, and relaxation. These phenomena are observed at all humidity levels, but increasing humidity greatly increases the creep rate and decreases the relaxation time. It is currently thought that time dependence in paper is associated with the deformation of the amorphous cellulose regions of fibers. Crystalline regions (as in nanocellulose) do not introduce significant time-dependent behavior. Since paper is densely crosslinked, the deformation is approximately affine and, hence, the material response is entirely characterized by the behavior of individual fibers. Paper provides a good example of the mechanism described in Section 9.2.

Creep of paper is reported to be logarithmic under steady state conditions, that is, the creep strain increases in proportion to the logarithm of time (Brezinski, 1956; DeMaio and Patterson, 2007). The initial stage of creep may be described as Andrade creep, with the creep strain increasing proportional to $t^{1/3}$ (Alava and Niskanen, 2006). Mao et al. (2017) show that the rate sensitivity of cellulose nanopaper is associated with amorphous regions of fibers and it is not caused by relative fiber sliding. They report, based on constant strain rate uniaxial tension experiments performed at several temperatures, that the yield stress is proportional to the logarithm of the strain rate. The data in Andersson and Sjoberg (1953) for the strain rate sensitivity of wrapping paper also indicates $\sigma \sim \log \dot{\varepsilon}$. Relaxation experiments accounting for the anisotropy of paper are reported in Uesaka et al. (1980). These, as well as data in Alava and Niskanen (2006), indicate multiple relaxation times, with at least two characteristic time constants, of about 2 s and 200 s. The total time span of these experiments is not large enough to allow determining whether relaxation is best described by a Prony series (Eq. (9.4)) or a Kohlrausch expression (Eq. (9.22)).

9.4.4.4   Nonwovens

Typical nonwovens made from polymeric fibers of diameter 10 μm or larger exhibit weak to moderate strain rate sensitivity. The rate sensitivity is positive, in the sense that increasing the strain rate leads to an increase of the yield and flow stresses (Patel and Kothari, 2001; Jubera et al., 2014). These networks are not embedded in a matrix and, if the crosslinks are rigid and contribute little to the overall deformation, the network-scale rate sensitivity is defined by the properties of the fiber material (Section 9.2). However, it is possible to observe rate independent network-scale behavior even when the rate sensitivity of the fiber material is pronounced. This happens when the deformation is nonaffine and the strain experienced by fibers is significantly smaller than the macroscale applied strain. Since rate sensitivity manifests itself in the plastic deformation range, if the fiber deformation before crosslink failure and network rupture remains predominantly elastic, the network-scale response is rate insensitive, irrespective of the fiber properties. An example of this type is provided in Jubera et al. (2014), who tested polypropylene nonwovens of ~50 μm fiber diameter (a geotextile).

They also tested individual fibers and observed strong rate sensitivity. However, the networks exhibit little to no rate sensitivity. If a more densely crosslinked network with stronger bonds would have been considered, the rate sensitivity of the fibers would have become manifest in the network behavior.

Tests on polycaprolactone (PCL) nonwovens with submicron fibers obtained by electrospinning are reported in Duling et al. (2008). They observe significant rate sensitivity and report logarithmic relaxation: $G(t)/G(0) = 1 - \alpha \log t$ (with α and t_0 being constants and $t > t_0$). Since fibers have sub-micron diameter, it is possible that multiple mechanisms (viscous friction between fibers and time-dependent splitting of fiber bundles) define the time-dependent behavior of these networks.

9.5 Effect of Transient Crosslinks

Pronounced strain rate sensitivity emerges if the crosslinks unbind during deformation. If some of the crosslinks are permanent while others are transient, the network is viscoelastic. If all crosslinks are transient, the network behavior is viscoplastic. Crosslink unbinding is a key process in molecular networks known as associative polymers.

Transient crosslinks between molecular strands may be established by ionic interactions, H-bonds, $\pi-\pi$ stacking, and hydrophobic association. Electrostatic interactions between ions or ionic groups form transient crosslinks in ionomers and polyampholytes.[5] $\pi-\pi$ stacking interactions take place between aromatic rings of neighboring chains. Hydrophobic association leads to the formation of transient crosslinks between hydrophilic chains with hydrophobic segments. In an aqueous environment, the hydrophobic segments group, forming associations that work as crosslinks between polymeric chains.

Crosslink unbinding takes place under the combined effect of thermal fluctuations and applied force. The probability of unbinding, and hence the dissociation rate, depend on the binding energy, E_X, and may be expressed as $\exp\left(-E_X + f_X \delta_X / k_B T\right)$ (Groot et al., 1996). f_X is the force acting on the bond promoting bond scission, while δ_X is an activation length. The binding energy is the key parameter that describes the bond and each type of interaction is characterized by a range of E_X. H-bonds have energies ranging from below 0.2 eV to 7 eV ($\sim$4 to 160 KJ/mol), with the typical range of weak H-bonding being 0.2 eV to 0.8 eV. $\pi-\pi$ stacking has a typical bonding energy of 0.4 eV ($\sim$9 KJ/mol). The energy required to pull a strand out of a hydrophobic association group depends on the length of the hydrophobic domain. Ionic interactions are described by the Coulomb law and the interaction energy is proportional to the charge involved and the distance between charges within the bond. The energy of a single bond is about 0.3 eV (7 KJ/mol). The thermal energy at the ambient temperature is $k_B T = 0.025$ eV, that is, at least 10-times

---

[5] Polyampholytes are polymers containing electrolyte groups, both cationic and anionic. These groups dissociate in water and the polymer becomes charged.

smaller than E_X of any of these bonds. Hence, the probability of unbinding under the action of thermal fluctuations is $\exp(-10) \approx 10^{-5}$ or smaller. The dissociation rate is computed as the product of the probability of unbinding and the frequency of attempt. Since the thermal frequency of attempt is very large, the dissociation rate may result large, although the unbinding probability is small. Furthermore, with δ_X of about 1 Å, a force as small as 100 pN greatly increases the dissociation rate. Therefore, many weak bonds fluctuate, separating and reforming dynamically. This process is irrelevant for network mechanics since it leaves the network configuration unchanged. Relaxation takes place when a crosslink which has broken reforms at a different site, therefore reconfiguring the network. This requires the detached chain segment to diffuse over a distance at least comparable with that between two crosslink sites.

In order to understand the following discussion, it is necessary to recall some of the physical aspects underlying the linear viscoelasticity of polymers. The dynamics of polymeric systems above T_g is discussed in many dedicated texts (e.g., Doi, 1996; Strobl, 1996; Rubinstein and Colby, 2003) and is generally characterized in terms of the complex modulus and its real and imaginary components, the storage and loss moduli, G' and G'' (Section 9.1). We use the shear modulus in this section since many of the experiments probing the linear viscoelasticity of polymers are performed in shear.

Consider a melt (or concentrated solution) of linear polymeric chains of n repeat units. A repeat unit is larger than one monomer and is referred to as a Kuhn segment, that is, a segment which may take any angular position relative to the neighboring segments along the respective chain. We denote the length of the Kuhn segment by a and ρ_a is the number density of such segments. If n is smaller than a threshold known as the entanglement length, n_e, the chains are free to change conformations by performing random walks in three dimensions under the action of thermal fluctuations. Since such systems are generally dense (e.g., melts), chains exchange energy at contacts and these interactions are viewed in the frame of a representative chain as friction with a fixed background (the background represents the other chains). The dynamics of the representative chain under the action of thermal fluctuations and in the presence of friction is described by the Rouse model. This model predicts that relaxation following a perturbation may be represented by the superposition of n modes, each with a characteristic time $\tau_{Ri} = \tau_0(n/i)^2$ representing the relaxation of a segment of length n/i ($i = 1 \ldots n$). τ_0 is the shortest Rouse time ($i = n$) and is proportional to the friction coefficient and inversely proportional to $k_B T$. The relaxation modulus results in the form of a Prony series: $G(t) = (\rho_a k_B T/n)\sum_{i=1}^{n} \exp(-t/\tau_{Ri})$. The storage and loss moduli predicted by the Rouse model result from $G(t)$ by using the procedure described in Section 9.1, as a superposition of expressions of the form (Eq. (9.3)), each corresponding to one of the τ_{Ri}. For the range of frequencies $1/\tau_0 n^2 < \omega < 1/\tau_0$, the storage and loss moduli scale as $G'(\omega) \sim G''(\omega) \sim \omega^{1/2}$. At frequencies smaller than $1/\tau_0 n^2$, the system relaxes as a simple fluid with $G'(\omega) \sim \omega^2$ and $G''(\omega) \sim \omega$.

If the chains are longer than the entanglement length, $n > n_e$, their motion is perturbed by topological constraints imposed by the neighboring chains. Segments shorter than n_e relax unhindered, while those longer than n_e are significantly slowed down. A chain cannot relax fully unless it diffuses along its contour over a distance equal to its contour length, that is, na. This motion is known as reptation. For frequencies larger than $1/\tau_e = 1/\tau_0 n_e^2$, the Rouse relaxation of segments shorter than the entanglement length prevails and $G'(\omega) \sim G''(\omega) \sim \omega^{1/2}$. A rubbery plateau appears in G' at frequencies smaller than $1/\tau_e$. The storage modulus corresponding to the plateau is $\rho_a k_B T/n_e$, that is, it is equal to the stiffness of a crosslinked thermal network composed from segments of length n_e; note that $\rho_a/n_e = \rho_\#$. The plateau extends to frequencies smaller than the longest Rouse time of the chain, $\tau_0 N^2$. The reptation model (Doi and Edwards, 1986) associates the longest relaxation mode with the diffusion of the representative chain out of the confining tube formed by its neighbors and predicts a terminal relaxation time of $\tau_{rept} \sim \tau_0 n^3/n_e = \tau_e (n/n_e)^3$. The rubbery plateau extends in the frequency range $1/\tau_{rept} < \omega < 1/\tau_e$; this range increases as the chain length increases. The terminal relaxation characterized by $G'(\omega) \sim \omega^2$ and $G''(\omega) \sim \omega$ is observed at frequencies smaller than $1/\tau_{rept}$.

We turn now to the analysis of polymers with transient crosslinks (or "stickers") and consider first unentangled polymeric systems, that is, cases with $n < n_e$. The stickers are spaced along the chain by n_X and their lifetime is τ_X. For simplicity, we consider that all crosslinks have the same lifetime, which is also independent of the applied force.

The linear viscoelastic response is characterized by the G' function shown schematically in Figure 9.6(a). On short time scales, the material behaves similar to a crosslinked network of segment length $n_X a$. The network strands perform Rouse motion on times smaller than the corresponding Rouse time $\tau_{RX} = \tau_0 n_X^2$ (frequencies larger than $1/\tau_0 n_X^2$). If the lifetime of the crosslinks is longer than this relaxation time, $\tau_X > \tau_{RX}$, a plateau appears in G' at frequencies smaller than $1/\tau_{RX}$. The modulus corresponding to the plateau reflects the physical nature of the network: it is composed

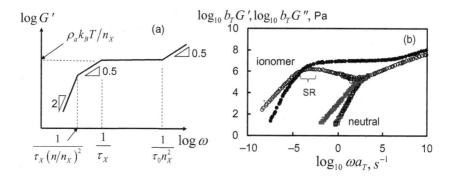

Figure 9.6 (a) Schematic representation of $G'(\omega)$ for an unentangled polymer with transient crosslinks. (b) Corresponding experimental data obtained with an ionomer (circles) and its neutral counterpart (squares). SR marks the sticky Rouse regime. Filled and open symbols indicate G' and G'', respectively. Adapted from Chen et al. (2013)

from strands of length n_X and of number density ρ_a/n_X. Therefore, the plateau modulus is $(\rho_a/n_X)k_BT$. The low frequency limit of the plateau is $1/\tau_X$ (Figure 9.6(a)).

On time scales longer than τ_X, chains perform "sticky Rouse" motion. This indicates Rouse relaxation with a baseline time step of τ_X. This should be compared with the normal Rouse motion on scales larger than n_X in the absence of stickers, for which the reference time step is $\tau_{RX} = \tau_0 n_X^2$. Baxandall (1989) has shown that stress relaxation modulated by the release mechanism at transient crosslinks is Rouse-like. Therefore, for $\omega < 1/\tau_X$, a second Rouse regime characterized by $G'(\omega) \sim G''(\omega) \sim \omega^{1/2}$ appears. It extends down to a frequency corresponding to the longest sticky Rouse time, $\tau_X(n/n_X)^2$. The terminal relaxation characterized by $G'(\omega) \sim \omega^2$ appears at even lower frequencies.

If the crosslink lifetime is comparable with τ_{RX}, $\tau_X \approx \tau_{RX}$, the plateau disappears and sticky Rouse motion takes place at frequencies lower than $1/\tau_X \approx 1/\tau_{RX} = 1/\tau_0 n_X^2$. This regime extends down to the frequency $1/\left[\tau_X(n/n_X)^2\right] \approx 1/\left[\tau_{RX}(n/n_X)^2\right] = 1/(\tau_0 n^2) = 1/\tau_R$. Hence, in this case, the sticky Rouse regime ends at a frequency identical to that of the longest Rouse mode of the equivalent system without stickers. This makes observing sticky Rouse motion in experiments with materials for which $\tau_X \approx \tau_{RX}$ difficult. The same arguments apply to cases in which the transient crosslinks are even shorter lived, $\tau_X < \tau_{RX}$.

The presence of the plateau is conditioned by the existence of a sufficient density of stickers (at least two stickers per chain) of a long enough lifetime. In real networks, the polydispersity of segment lengths causes broad transitions from one regime of the complex modulus to another. This makes difficult the observation of regimes defined over a narrow frequency range.

Figure 9.6(b) shows experimental storage and loss moduli for the ionomer poly (tetramethylene oxide) (PTMO-100%Na) and for its neutral counterpart (PTMO-0% Na) (Chen et al., 2013). The ionomer exhibits a broad plateau, while the neutral material exhibits Rouse relaxation followed by the terminal regime. The sticky Rouse regime is somewhat visible at the low ω end of the plateau (labeled by SR in Figure 9.6(b)). Further examples of sticky Rouse behavior are provided in Annable et al. (1993), Weiss and Zhao (2009), and Zhang et al. (2018).

Networks of semiflexible filaments with transient crosslinks exhibit similar phenomenology. A Rouse-like regime with $G'(\omega) \sim G''(\omega) \sim \omega^{1/2}$ appears at frequencies below $1/\tau_X$, and G'' has a maximum at $\omega = 1/\tau_X$, indicating maximum dissipation. G' exhibits a plateau for $\omega > 1/\tau_X$. At even higher frequencies, the Rouse regime seen in Figure 9.6(a) is replaced by a regime in which $G' \sim G'' \sim \omega^{3/4}$, which is a signature of the behavior of individual filaments described by the worm-like chain model (Section 2.4.4). Such behavior is observed in actin networks crosslinked with the transient crosslinker α-actinin-4 (Broedersz et al., 2010); see also Lieleg et al. (2008).

A similar analysis can be made for entangled polymeric systems with transient crosslinks (see also, Zhang et al., 2018). Consider that the number of Kuhn segments along a chain between stickers is n_X and the entanglement length is n_e, such that $n_X > n_e$. Chain segments of length equal to, or shorter than, n_e perform Rouse motion

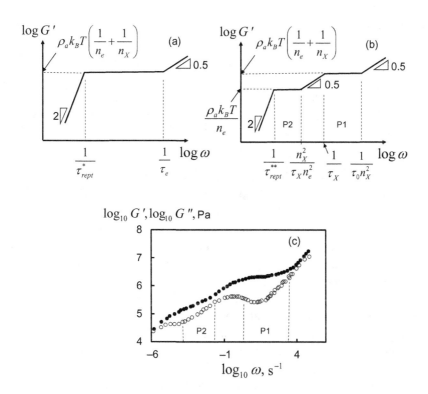

Figure 9.7 Schematic representation of $G'(\omega)$ for entangled polymers with transient crosslinks with (a) sparse transient crosslinks with $n_X > n_e$, and (b) dense transient crosslinks with $n_X < n_e$. P1 and P2 denote the two rubbery plateaus described in the text. (c) Storage and loss moduli for an entangled ionomer. Filled and open symbols indicated G' and G'', respectively. Data from Kim et al. (1994)

unhindered by the presence of stickers and entanglements. Hence, for $\omega > 1/\tau_e$, chains perform Rouse dynamics with $G'(\omega) \sim G''(\omega) \sim \omega^{1/2}$, just like in the equivalent system without stickers. However, reptation is hindered by the transient crosslinks and chains may diffuse along their contour only on the τ_X time scale. The effective reptation time results $\tau^*_{rept} = \tau_X (n/n_X)^2 (n/n_e)$. A plateau in G' emerges for frequencies smaller than $1/\tau_e$ and larger than $1/\tau^*_{rept}$. The value of the modulus corresponding to this plateau is defined by the number of load carrying strands. These may be counted accounting for both stickers and entanglements. The number of pinning points (stickers and entanglements) per chain is $n/n_e + n/n_X$, and the length of a strand between two pinning points is $n/[n/n_e + n/n_X]$. Hence, the plateau modulus results $\rho_a k_B T(1/n_e + 1/n_X)$. Figure 9.7(a) shows schematically the functional form of $G'(\omega)$.

Consider now the situation in which the density of transient crosslinks is larger than that of entanglements and $n_X < n_e$ (Figure 9.7(b)). A representative chain performs unhindered Rouse motion on time scales shorter than τ_{RX}. Therefore, for frequencies larger than $1/\tau_{RX} = 1/\tau_0 n_X^2$, chain motion is Rouse-like and $G'(\omega) \sim G''(\omega) \sim \omega^{1/2}$.

A plateau appears in $G'(\omega)$ for $\omega < 1/\tau_{RX}$, corresponding to the network stabilized by both stickers and entanglements. The stiffness corresponding to this plateau is $\rho_a k_B T (1/n_e + 1/n_X)$. The lower frequency bound of the plateau is controlled by the lifetime of stickers, $1/\tau_X$. As the constraint imposed by stickers is eliminated on time scales larger than τ_X, the chain performs sticky Rouse motion up to the point when entanglements are engaged. This occurs on the time scale of the sticky Rouse mode corresponding to n_e, $\tau_X(n_e/n_X)^2$. A second plateau emerges at longer time scales, bounded above by $1/\left[\tau_X(n_e/n_X)^2\right]$ and below by $1/\tau_{rept}^{**}$, where $\tau_{rept}^{**} = \tau_X(n_e/n_X)^2(n/n_e)^3 = \tau_{rept}^*$. Since stickers are fully relaxed on these time scales, the stiffness corresponding to this plateau is controlled exclusively by entanglements and reads $\rho_a k_B T/n_e$. The terminal relaxation is observed at frequencies smaller than $1/\tau_{rept}^{**} = 1/\tau_{rept}^*$.

Figure 9.7(c) shows experimental storage and loss moduli for a poly(styrene-co-sodium methacrylate) ionomer of large molecular weight (Kim et al., 1994). The two rubbery plateaus are visible and, in general, the features described in relation to the schematic representation in Figure 9.7(b) are easily identifiable.

The presence of transient crosslinks introduces thermorheological complexity. Results in the literature indicate that time–temperature superposition applies in some, but not in all cases. Time–temperature superposition applies in the two extreme regimes, at very high and very low frequencies, while lack of collapse on a master curve is observed in the middle range of frequencies. The failure of thermo-rheological simplicity is due in part to the fact that the various dissipation mechanisms producing rate sensitivity on the material scale depend on temperature in different ways. Sticker unbinding is Arrhenius-dependent on temperature, while the internal friction-controlled relaxation of the polymeric system is described by the WLF equation (Eq. (9.10)). This is supported by experimental observations which indicate that the shift factor a_T is described by the WLF equation in some temperature range and by the Arrhenius equation in some other range (Annable et al., 1993; Chen et al., 2013).

The glass transition temperature and the viscosity are larger in the material with stickers than in the equivalent polymer without transient crosslinks. The viscosity of the network with stickers may be larger by several orders of magnitude. It is observed that T_g increases in ionomers by a factor proportional to the ratio of the charge of the ionomer ion to the separation of charges in the ionic pair (taken to be proportional to the ionic radius) (Eisenberg, 1971). The viscosity increases exponentially with the same ratio and with ionic concentration (Weiss and Zhao, 2009).

The nonlinear rheology of polymers with transient crosslinks exhibits features not observed in systems without stickers. This topic is outside the scope of the present review. Details can be found in the works dedicated to this subject (e.g., Serero et al., 2000; Berret et al., 2001; Pellens et al., 2004; Tripathi et al., 2006).

A large number of molecularly-informed and continuum models have been developed to represent the time-dependent mechanical behavior of these networks (Tanaka and Edwards, 1992a, 1992b; Vaccaro and Marrucci, 2000; Abhilash et al., 2012; Ahmadi et al., 2015; Vernerey et al., 2017) to mention just a few.

Before closing this section, it is necessary to mention the new class of transient networks known as vitrimers. Vitrimers are polymers with transient crosslinks in which crosslink unbinding takes place through a bond exchange reaction, that is, a crosslink may unbind only provided it exchanges strands with a neighboring crosslink. The first vitrimer system based on the transesterification reaction was created in 2011 (Montarnal et al., 2011) using an epoxy base and carboxylic acids as a hardener. The bond-exchange reaction ensures that the total density of crosslinks remains constant as the reaction proceeds. Likewise, the connectivity $<z>$ of the network remains unchanged. Hence, these are network materials in which the network topology changes in time, but the properties of the network (connectivity, density of crosslinks, strand length distribution) remain unchanged. An intense activity is currently dedicated to the development of vitrimer systems for specific applications, as recyclable alternatives to traditional thermosets.

References

Abhilash, A. S., Purohit, P. K. & Joshi, S. P. (2012). Stochastic rate-dependent elasticity and failure of soft fibrous networks. *Soft Matt.* **8**, 3004–3016.

Ahmadi, M., Hawke, L. G. D., Goldansaz, H. & van Ruymbeke, E. (2015). Dynamics of entangled linear supramolecular chains with sticky side groups: Influence of hindered fluctuations. *Macromolecules* **48**, 7300–7310.

Alava, M. & Niskanen, K. (2006). The physics of paper. *Rep. Prog. Phys.* **69**, 669–723.

Amjad, S. N. & Picu, R. C. (2022). Stress relaxation in network materials: the contribution of the network. *Soft. Matter* **18**, 446–454.

Andersson, O. & Sjoberg, L. (1953). Tensile studies of paper at different rates of elongation. *Svensk Pappertid.* **56**, 615–624.

Annable, T., Buscall, R., Ettelaire, R., & Whittlestone, D. (1993). The rheology of solutions of associating polymers: Comparison of experimental behavior with transient network theory. *J. Rheol.* **37**, 695–726.

Babaei, B., Davarian, A., Pryse, K. M., Elson, E. L. & Genin, G. M. (2015). Efficient and optimized identification of generalized Maxwell viscoelastic relaxation spectra. *J. Mech. Beh. Biomed. Mater.* **55**, 32–41.

Baxandall, L. G. (1989). Dynamics of reversibly cross-linked chains. *Macromolecules* **22**, 1982–1988.

Bernstein, B., Kearsley, E. A. & Zapas, L. J. (1963). A study of stress relaxations with finite strain. *Trans. Soc. Rheol.* **7**, 391–410.

Berret, J. F., Sereo, Y., Winkelman, B., et al. (2001). Nonlinear rheology of telechelic polymer networks. *J. Rheol.* **45**, 477–492.

Biot, M. A. (1941). General theory of three-dimensional consolidation. *J. Appl. Phys.* **12**, 155–164.

Bird, R. B., Armstrong, R. C. & Hassager, O. (1987). *Dynamics of polymeric liquids. Vol. 1: Fluid mechanics.* Wiley, New York.

Brezinski, J. P. (1956). The creep properties of paper. *Tappi J.* **39**, 116–128.

Broedersz, C. P., Depken, M., Yao, N. Y., et al. (2010). Cross-link-governed dynamics of biopolymer networks. *Phys. Rev. Lett.* **105**, 238101.

Buhan, P. D., Chateau, X. & Dormieux, L. (1998). The constitutive equations of finite strain poroelasticity in the light of micro-macro approach. *Euro. J. Mech. A* **17**, 909–921.

Chen, Q., Tudryn, G. J. & Colby, R. H. (2013). Ionomer dynamics and the sticky Rouse model. *J. Rheol.* **57**, 1441–1462.

Christensen, R. M. (1971). *Theory of viscoelasticity*. Academic Press, New York.

DeMaio, A. & Patterson, T. (2007). Rheological modeling of the tensile creep behavior of paper. *J. Appl. Poly. Sci.* **106**, 3543–3554.

Dhume, R. Y. & Barocas, V. H. (2019). Emergent structure-dependent relaxation spectra in viscoelastic fiber networks in extension. *Acta Biomater.* **87**, 245–255.

Dicke, R. A. & Ferry, J. D. (1966). Dynamic mechanical properties of cross-linked rubbers. III. Dicumyl peroxide vulcanizates of natural rubber. *J. Phys. Chem.* **70**, 2594–2600.

Doi, M. (1996). *Introduction to polymer physics*. Clarendon Press, Oxford.

Doi, M. & Edwards, S. (1986). *The theory of polymer dynamics*. Clarendon Press, Oxford.

Doolittle, A. K. & Doolittle, D. B. (1957). Studies in Newtonian flow. V. Further verification of the free-space viscosity equation. *J. Appl. Phys.* **28**, 901–905.

Duling, R. R., Dupaix, R. B., Katsube, N. & Lannutti, J. (2008). Mechanical characterization of electrospun polycaprolactone (PCL): A potential scaffold for tissue engineering. *J. Biomech. Eng.* **130**, 011006.

Ehret, A. E., Bircher, K., Stracuzzi, A., et al. (2017). Inverse poroelasticity as a fundamental mechanism in biomechanics and mechanobiology. *Nature Comm.* **8**, 1002.

Eisenberg, A. (1971). Glass transition in ionic polymers. *Macromolecules* **4**, 125–128.

Ferry, J. D. (1980). *Viscoelastic properties of polymers*, 3rd ed. Wiley, New York.

Ferry, J. D., Manke, R. G., Maekawa, E., Oyanagi, Y. & Dicke, R. A. (1964). Dynamic mechanical properties of crosslinked rubbers. I. Effects of crosslink spacing in natural rubber. *J. Phys. Chem.* **68**, 3414–3418.

Findley, W. N., Lai, J. S. & Onaran, K. (1976). *Creep and relaxation of nonlinear viscoelatic materials*. North-Holland, Amsterdam.

Fung, Y. C. (2013). *Biomechanics: Mechanical properties of living tissues*. Springer, New York.

Galli, M., Comley, K. S. C., Shean, T. A. V. & Oyen, M. L. (2009). Viscoelastic and poroelastic mechanical characterization of hydrated gels. *J. Mater. Res.* **24**, 973–979.

Gent, A. N. & Mars, W. V. (2013). Strength of rubber. In *Science and Technology of Rubber*, 4th ed., J. E. Mark, B. Erman & M. Roland, eds. Academic Press, Amsterdam, pp. 473–516.

Groot, R. D., Bot, A. & Agterof, W. G. M. (1996). Molecular theory of the yield behavior of a polymer gel: Application to gelatin. *J. Chem. Phys.* **104**, 9220–9233.

Haddad, Y. M. (1995). *Viscoelasticity of engineering materials*. Chapman & Hall, London.

Hu, Y. & Suo, Z. (2012). Viscoelasticity and poroelasticity in elastomeric gels. *Acta Mech. Sinica* **25**, 441–458.

Hu, Y., Zhao, X., Vlassak, J. J. & Suo, Z. (2010). Using indentation to characterize the poroelasticity of gels. *Appl. Phys. Lett.* **96**, 121904.

Hui, C. Y. & Muralidharan, V. (2005). Gel mechanics: A comparison of the theories of Biot and Tanaka, Hocker, and Benedek. *J. Chem. Phys.* **123**, 154905.

Isono, Y. & Ferry, J. D. (1984). Stress relaxation and differential dynamic modulus of carbon black-filled styrene-butadiene rubber in large shearing deformations. *Rubber Chem. Technol.* **57**, 925–943.

Izuka, A., Winter, H. H. & Hashimoto, T. (1994). Temperature dependence of viscoelasticity of polycaprolactone critical gels. *Macromolecules* **27**, 6883–6888.

Jubera, R., Ridruejo, A., Gonzalez, C. & Llorca, J. (2014). Mechanical behavior and deformation micromechanisms of PP nonwoven fabrics as a function of temperature and strain rate. *Mech. Mater.* **74**, 14–25.

Kaye, A. (1962). Non-Newtonian flow in incompressible fluids. College of Aeronautics, Cranfield, CoA Note No. 134.

Kim, J. S., Yoshikawa, K. & Eisenberg, A. (1994). Molecular weight dependence of the viscoelastic properties of polystyrene-based ionomers. *Macromolecules* **27**, 6347–6357.

Komatsu, K. (2010). Mechanical strength and viscoelastic response of the periodontal ligament in relation to structure. *J. Dental Biomech.* **1**, 1–18.

Lieleg, O., Claessens, M. M. A. E., Luan, Y. & Bausch, A. R. (2008). Transient binding and dissipation in cross-linked actin networks. *Phys. Rev. Lett.* **101**, 108101.

Lin, W. C., Shull, K. R., Hui, C. Y. & Lin, Y. Y. (2007). Contact measurement of internal fluid flow within poly(n-isopropylacrylamide) gels. *J. Chem. Phys.* **127**, 094906.

Mao, R., Meng, N., Tu, W. & Peijs, T. (2017). Toughening mechanisms in cellulose nanopaper. *Cellulose* **24**, 4627–4639.

Montarnal, D., Capelot, M., Tournilhac, F. & Leibler, L. (2011). Silica-like malleable materials from permanent organic networks. *Science* **334**, 965–968.

Mow, V. C., Kuei, S. C., Lai, W. M. & Armstrong, C. G. (1980). Biphasic creep and stress relaxation of articular cartilage in compression: Theory and experiments. *J. Biomech. Eng.* **102**, 73–84.

Nabovati, A., Llewellin, E. W. & Sousa, A. C. M. (2009). A general model for the permeability of fibrous porous media based on fluid flow simulations using the lattice Boltzmann method. *Composites A* **40**, 860–869.

Nam, S., Hu, K. H., Butte, M. J. & Chaudhuri, O. (2016). Strain-enhanced stress relaxation impacts nonlinear elasticity in collagen gels. *Proc. Nat. Acad. Sci.* **113**, 5492–5497.

Negi, V. & Picu, R. C. (2019). Mechanical behavior of nonwoven non-crosslinked fibrous mats with adhesion and friction. *Soft Matt.* **15**, 5951–5964.

Nekouzadeh, A., Pryse, K. M., Elson, E. L. & Genin, G. M. (2007). A simplified approach to quasi-linear viscoelastic modeling. *J. Biomech.* **40**, 3070–3078.

Ngai, K. L., Capaccioli, S. & Plazek, D. J. (2013). The viscoelastic behavior of rubber and dynamics of blends. In *Science and Technology of Rubber*, 4th ed., J. E. Mark, B. Erman & M. Roland, eds. Academic Press, Amsterdam, pp. 193–284.

Ngai, K. L., Plazek, D. J. & Rendell, R. W. (1997). Some examples of possible descriptions of dynamic properties of polymers by means of the coupling model. *Rheol. Acta* **36**, 307–319.

Oyen, M. L. (2014). Mechanical characterization of hydrogel materials. *Inter. Mater. Rev.* **59**, 44–59.

Patel, P. C. & Kothari, V. K. (2001). Effect of specimen size and strain rate on the tensile properties of heat-sealed and needle-punched nonwoven fabrics. *Indian J. Fiber Text. Res.* **26**, 409–413.

Payne, A. R. & Whittaker, R. E. (1971). Low strain dynamic properties of filled rubber. *Rubber Chem. Technol.* **44**, 440–478.

Pellens, L., Corrales, R. G. & Mewis, J. (2004). General nonlinear rheological behavior of associative polymers. *J. Rheol.* **48**, 379–393.

Plazek, D. J. & Chay, I. C. (1991). The evolution of the viscoelastic retardation spectrum during the development of an epoxy resin network. *J. Polym. Sci. B* **29**, 17–29.

Plazek, D. J. & Rosner, M. (1998). The calculation of the tear energy of elastomers from their viscoelastic behavior. *Rubber Chem. Technol.* **71**, 679–689.

Pryse, K. M., Nekouzadeh, A., Genin, G. M., Elson, E. L. & Zahalak, G. I. (2003). Incremental mechanics of collagen gels: New experiments and a new viscoelastic model. *Ann. Biomed. Eng.* **31**, 1287–1296.

Roland, C. M. (2011). *Viscoelastic behavior of rubbery materials.* Oxford University Press, Oxford.

Rubinstein, M. & Colby, R. H. (2003). *Polymer physics.* Oxford University Press, Oxford.

Sarver, J. J., Robinson, P. S. & Elliott, D. M. (2003). Methods for quasi-linear viscoelastic modeling of soft tissue: application to incremental stress–relaxation experiments. *J. Biomech. Eng.* **125**, 754–758.

Schweitzer, K. S. (1989). Microscopic theory of the dynamics of polymeric liquids: General formulation of a mode-mode coupling approach. *J. Chem. Phys.* **91**, 5802–5821.

Serero, Y., Jacobsen, V., Berret, J. F. & May, R. (2000). Evidence of nonlinear chain stretching in the rheology of transient networks. *Macromolecules* **33**, 1841–1847.

Shen, Z. L., Kahn, H., Ballarini, R. & Eppell, S. (2011). Viscoelastic properties of isolated collagen fibrils. *Biophys. J.* **100**, 3008–3015.

Strange, D. G., Fletcher, T. L., Tonsomboon, K., et al. (2013). Separating poroviscoelastic deformation mechanisms in hydrogels. *Appl. Phys. Lett.* **102**, 031913.

Strobl, G. R. (1996). *The physics of polymers: Concepts for understanding their structure and behavior.* Springer, Berlin.

Svensson, R. B., Hassenkam, T., Hansen, P. & Magnusson, S. P. (2010). Viscoelastic behavior of discrete human collagen fibrils. *J. Mech. Beh. Biomed. Mater.* **3**, 112–115.

Tanaka, F. (2011). *Polymer physics: Applications to molecular association and thermoreversible gelation.* Cambridge University Press, Cambridge.

Tanaka, F. & Edwards, S. F. (1992a). Viscoelastic properties of physically cross-linked networks. Transient network theory. *Macromolecules* **25**, 1516–1523.

Tanaka, F. & Edwards, S. F. (1992b). Viscoelastic properties of physically cross-linked networks. Dynamic mechanical moduli. *J. Non-Newt. Fl. Mech.* **43**, 273–288.

Tripathi, A., Tam, K. C. & McKinley, G. H. (2006). Rheology and dynamics of associative polymers in shear and extension: Theory and experiments. *Macromolecules* **39**, 1981–1999.

Uesaka, T., Murakami, K. & Imamura, R. (1980). Two-dimensional linear viscoelasticity of paper. *Wood Sci. Technol.* **14**, 131–142.

Vaccarro, A. & Marrucci, G. (2000). A model for the nonlinear rheology of associating polymers. *J. Non-Newt. Fl. Mech.* **92**, 261–273.

Vernerey, F. J., Long, R. & Brighenti, R. (2017). A statistically-based continuum theory for polymers with transient networks. *J. Mech. Phys. Sol.* **107**, 1–20.

Wagenseil, J. E., Wakatsuki, T., Okamoto, R. J., Zahalak, G. I. & Elson, E. L (2003). One-dimensional viscoelastic behavior of fibroblast populated collagen matrices. *J. Biomech. Eng.* **125**, 719–725.

Wang, H. W. (2000). *Theory of linear poroelasticity with applications to geomechanics and hydrogeology.* Princeton University Press, Princeton, NJ.

Weiss, N., van Vliet, T. & Silverberg, A. (1981). Influence of polymerization initiation rate on permeability of aqueous polyacrylamide gels. *J. Polym. Sci.: Polym. Phys.* **19**, 1505–1512.

Weiss, R. A. & Zhao, H. (2009). Rheological behavior of oligomeric ionomers. *J. Rheol.* **53**, 191–213.

Williams, M. L., Landel, R. F. & Ferry, J. D. (1955). The temperature dependence of relaxation mechanisms in amorphous polymers and other glass-forming liquids. *J. Amer. Chem. Soc.* **77**, 3701–3707.

Xu, B. & Li, H. (2013). An experimental and modeling study of the viscoelastic behavior of collagen gel. *J. Biomech. Eng.* **135**, 054501.

Zener, C. (1956). *Elasticity and anelasticity of metals.* University of Chicago Press, Chicago, IL.

Zhang, Z., Chen, Q. & Colby, R. H. (2018). Dynamics of associative polymers. *Soft. Matt.* **14**, 2961–2977.

Zhao, X., Huebsch, N., Mooney, D. J. & Suo, Z. (2010). Stress–relaxation behavior in gels with ionic and covalent crosslinks. *J. Appl. Phys.* **107**, 063509.

10 Networks with Fiber Surface Interactions
Networks of Fiber Bundles

In the preceding chapters it is considered that fibers interact exclusively at contacts and at crosslinks. The interaction at contacts is of excluded volume type: it is repulsive and imposes the restriction that volumes occupied by fibers do not overlap. However, in many practical situations, other types of interactions may take place between fibers and these lead to drastic modifications of the network structure and mechanical behavior.

Surface interactions between fibers lead to the development of a variety of structures (Section 10.1). Many of these are transient and the resulting network behavior is fluid-like. Examples of this type are fluidic suspensions of fibers and of molecules of large persistence length. A broad literature on suspensions and colloids exists, and a review of this subject is beyond the scope of the present overview.

Networks of fiber bundles develop at higher densities and in the presence of strong interactions; their structure is significantly different from the configurations discussed in the preceding chapters. The mechanical behavior of networks of bundles stabilized by inter-fiber surface interactions is of primary concern here, as this topic aligns with the solid mechanics focus of the book. Filament bundling in thermal fibrous systems renders the resulting network athermal. This observation motivates the purely athermal treatment of networks of bundles presented in this chapter.

10.1 Classification of Fiber Surface Interactions

Surface interactions between filaments taking place at sites other than those of crosslinks have diverse physical origins. We classify them here in three categories: excluded volume, attractive, and frictional interactions. The first two control the relative motion of fibers in the direction orthogonal to the contact surface, while the third modifies fiber relative motion in the tangential direction.

Attractive interactions lead to the self-organization of the network into complex structures with rich mechanical behavior. Excluded volume interactions are important primarily in compression and their effect is discussed in Sections 6.1.2 and 6.2. Friction is important in dense networks subjected to deviatoric strain states. When compressing uniaxially a dense network, contacts form at a rate that increases with increasing degree of compression, but the effect of friction is not significant since relative fiber sliding is minimal. However, friction is critical in non-crosslinked networks that self-organize under the action of attractive surface interactions. Therefore, friction is more important for the systems discussed in this chapter than for those analyzed in Chapter 6.

Various types of forces lead to attractive interactions between filaments. These include cohesive forces, surface tension, and depletion forces.

Cohesion[1] refers to the interaction of two surfaces due to dispersive or electrostatic forces. Dispersive forces act between surfaces which are not charged, but either carry permanent charge dipoles, or support the formation of transient dipoles induced by the fluctuation of other dipoles in the counter-surface. Forces develop between dipoles, and between a dipole and its induced dipole, leading to the attraction of the respective surfaces. van der Waals interactions are of dispersive type and decay fast as the distance between dipoles increases; specifically, the interaction energy decays as the sixth power of the distance and the attractive force decays as the seventh power of the distance. Therefore, cohesion is short ranged.

For the present purpose, cohesion is quantified based on the work of separation of the two surfaces in contact, $\delta\gamma$, that is, the work required to break the contact and create two free surfaces. This is discussed in Section 3.2.2 in the context of contacts between fibers within fiber bundles. Two cylindrical fibers in longitudinal contact form a contact surface of width $2a_X$ (Figure 3.4). The relation between the force applied, the work of separation, and a_X is provided by Eq. (3.9) if the JKR theory applies, and by Eq. (3.11) outside the range of validity of the JKR model. The quantity of interest when evaluating the mechanics of fiber separation/bundling is $\delta\bar{\gamma} = \delta\gamma 2a_X$, which is the cohesive energy per unit length of contact. $\delta\bar{\gamma}$ enters the non-dimensional parameter Ψ_{coh} defined by Eq. (3.13), which turns out to be the essential parameter that quantifies the effect of surface interactions on the mechanics of fiber networks.

In molecular networks it is common to encounter transient crosslinks between filaments in longitudinal contact, located at interaction sites discretely distributed along the filament length. Let u_b be the binding energy of one of these "stickers" and λ_b their spacing along the filament. The energy penalty for unbinding a pair of such filaments may be expressed in terms of $\delta\bar{\gamma}$, which can be written as $\delta\bar{\gamma} = u_b/\lambda_b$. Hence, the mechanics of a network of filament bundles with discrete stickers is similar to the mechanics of a network of filaments with cohesion.

Surface tension has strong effects on soft materials in general, and fibrous assemblies in particular. In these materials, the strain energy variation may be smaller than the variation of the surface energy during deformation. Therefore, surface tension may reshape soft materials and create special effects not commonly encountered in stiffer solids. For example, when bringing a soft material in contact with a liquid, the surface

---

[1] The term "adhesion" is typically used to denote attractive interactions between surfaces of different types, while "cohesion" refers to the attraction of surfaces of the same type. Here we use "cohesion" since filaments of a network are generally made from the same material but acknowledge that interacting filaments may also be made from different materials.

tension of the liquid may lead to large local deformations of the solid (Style et al., 2017). Immersing a fibrous material in water followed by partially draining the water or allowing the material to dry leads to fiber bundling under the action of surface tension forces. A familiar demonstration of this effect is lifting long hair out of water; as water drains out, the hair becomes bundled under the action of surface tension. Cohesion may maintain the integrity of bundles upon drying. This effect was used to develop nanostructures with interesting geometries by simply drying arrays of parallel nanopillars fixed at one end to a substrate, after being exposed to a liquid (Roman and Bico, 2010; Style et al., 2017). The surface tension of the drying front deforms the pillars and brings them in contact with each other. If the elastic forces are too weak to break the contact, the nanopillars remain in cohesion-stabilized configurations. Forces induced by capillarity are orders of magnitude longer ranged than those due to cohesion and, hence, are effective in assembling nanofilaments and even microfibers.

The effect of surface tension on elastic bodies may be quantified in terms of the elasto-capillarity length $L_{EC} = \sqrt{E_f I_f/\delta\bar{\gamma}}$, which is discussed in Chapter 3 in relation to Eq. (3.13). Small values of L_{EC} indicate strong cohesion and hence strong effects of the surface forces on the elasticity of structures. This characteristic length must be compared with a geometric characteristic length of the structure of interest in order to determine the importance of surface force effects. Non-dimensional parameters such as Ψ_{coh} become useful when evaluating interactions of this type. In the case of isolated fiber clusters that do not form a self-organized network, the characteristic geometric length of importance is the fiber length, L_0, and hence $\Psi_{coh} = (L_0/L_{EC})^2$. In the case of networks of fiber bundles, the characteristic geometric length is the mean segment length of the network, l_c, and the useful non-dimensional parameter is $\Psi_{coh} = (l_c/L_{EC})^2$.

Depletion forces arise in suspensions in which athermal fibers are dispersed in a polymeric solution or melt; fiber dimensions are much larger than the molecules. Since polymeric chains cannot enter the volume occupied by the fibers, they retract from the fiber surface, leaving a region of low polymer concentration in the vicinity of the surface. The polymer density is graded, being largest away from fibers and smallest close to the fiber surface. The retraction of polymer chains from impenetrable walls is an entropic effect. A polymer residing in the close vicinity of a hard wall cannot assume all configurations it takes in the absence of the wall and, hence, its entropy is smaller than that of the same chain located away from the wall. An entropic force results, which pushes the chain away from the wall and leads to the formation of the depleted layer. Considering a colloidal suspension in which all fibers are surrounded by such low polymer concentration regions, when two fibers approach each other, the depleted layers merge and the osmotic pressure of the polymer forces the respective fibers in contact. Depletion forces lead to the formation of filament clusters in suspensions. The range of depletion forces is controlled by the thickness of the depleted layer which, in turn, is proportional to the gyration radius of the polymeric chains. Therefore, the range of depletion forces may be adjusted within a range by modifying the polymer molecular weight.

The effect of these interactions is to organize the system of filaments into structures. Fiber suspensions are encountered in practice, both in engineering and biology.

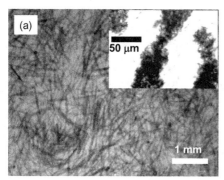
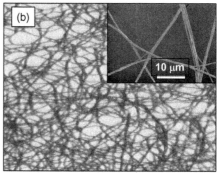

Figure 10.1 Examples of fiber organization due to surface interactions. (a) Suspension of polyamide fibers at rest (reprinted with permission from Bounoua et al. (2016a). Copyright (2016), The Society of Rheology). The inset shows phase separated multiwall carbon nanotubes in suspension under weak shear (reprinted from Hobbie (2010) with the permission of Springer, Copyright (2010)). (b) Mat of poly-acrylonitrile nanofiber bundles, with the inset showing a detail of the main figure (courtesy of I. Chasiotis).

Suspensions of athermal fibers are common in the pulp and paper industry and in the manufacturing of fiber-reinforced composites. Solutions of stiff molecules of high persistence length subjected to Brownian motion, and suspensions of self-assembled micelle are common in biology and the chemical industry. Figure 10.1(a) shows a suspension of polyamide fibers of 15 μm diameter and 500 μm length which is mostly isotropic in the quiescent state, although some degree of fiber bundling is observed as the density increases (Bounoua et al., 2016a). The inset shows a suspension of carbon nanotubes that aggregate and phase separate from the solvent under weak shear flows (Hobbie, 2010). Surface forces may also lead to the self-organization of filaments deposited on substrates without a matrix. Examples include mats produced by electro-spinning polymeric fibers of nanoscale diameter and various types of nanopapers, such as bacterial cellulose and carbon nanotube buckypaper. Figure 10.1(b) shows the emergence of a network of fiber bundles in an electrospun mat of poly-acrylonitrile fibers of sub-micron diameter. The network is not embedded in a matrix. The inset shows a detail of the main figure demonstrating fiber bundling due to cohesive fiber interactions.

Thermal filaments are brought in contact by thermal fluctuations, which favors the formation of bundles and networks of bundles. Athermal filaments are not subjected to thermal fluctuations. Since the range of surface interactions is small, which is the case with cohesion, it is necessary to bring the fibers in longitudinal contact by other means in order for structures of the type shown in Figure 10.1(b) to result. This can be done mechanically or by the exposure to a liquid followed by drying, that is, assisted by capillary forces.

The volume fraction, f, and the fiber aspect ratio, L_0/d, are key parameters controlling the self-organization of non-crosslinked fibers. Considering suspensions of fibers of length L_0 and number density $\rho_\#$, a geometric argument similar to that of Section 4.2.3.2 can be used to estimate that fiber interactions become important when

$\rho_\# L_0^3 \geq 1$, that is, when the volume occupied by a fiber is shared with other fibers. This threshold marks the transition from dilute to semi-dilute solutions and can be written as $\rho_\# L_0^3 = (4f/\pi)(L_0/d)^2 \geq 1$. Fibers of small L_0/d and with strong interactions tend to self-organize, forming isolated bundles. Long, slender fibers (large L_0/d) self-organize forming networks of fiber bundles. Semi-dilute solutions take various configurations, ranging from isotropic distributions of filaments to the formation of a network of filaments stabilized by surface interactions, as described in Section 10.3.1.1. A similar discussion is presented in Section 10.3.2.1 regarding the self-organization of dry networks without fluidic matrix.

The aspect of central concern here is the effect of fiber surface interactions on the mechanical behavior. Section 10.2 discusses the effect of cohesive interactions on the mechanics of crosslinked networks, with primary focus on those aspects of the mechanical behavior introduced by cohesion. Section 10.3 presents the broader class of non-crosslinked networks, including fiber suspensions (Section 10.3.1) and free-standing dry networks of filaments (Section 10.3.2). A brief review of key results is presented, with focus on the solid-like response of networks of fiber bundles that emerge at large fiber volume fractions in both suspension and the dry states. Such networks are qualitatively different from those discussed in Chapter 6 since they exhibit solid-like behavior similar to that of crosslinked networks at small strains, and fluid-like behavior at large strains. Networks of fiber bundles are a distinct type of network material and hence are given special attention in this chapter.

10.2   Crosslinked Networks with Fiber Surface Interactions

The mechanical behavior of crosslinked networks is discussed in Chapter 6. A special case of such structures, in which fibers interact by cohesion (or other surface interactions), is considered in this section. Cohesive interactions lead to the formation of filament bundles. The degree of bundling is limited by the constraints imposed on fiber kinematics by the presence of the crosslinks. Therefore, the size of bundles is smaller in crosslinked networks compared to non-crosslinked networks of same inter-fiber cohesion (Section 10.3).

Elastomers belong to this category. Rubber networks are crosslinked at specific sites, as dictated by the chemistry of the molecular backbone, while the molecular strands interact cohesively along their entire length.

The discussion here does not address specifically any material system; rather, it is sought to demonstrate the interplay between strain and cohesive energy contributions to global stress in a simplified context. To this end, the numerical results presented in Negi and Picu (2019a) are summarized. This study considers a two-dimensional athermal network in which fibers are allowed to form bundles, while the connectivity of the network defined by the crosslinks is preserved at all times. The network is crosslinked before cohesion is activated, such that the initial network connectivity is not controlled by cohesion. Once cohesive interactions are enabled, fiber bundles form and the network shrinks (Figure 10.2), much like the shrinkage of a thermoset during

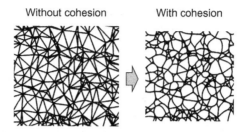

Without cohesion   With cohesion

Figure 10.2 Isostatic crosslinked network before (left) and after (right) activating the cohesive interactions between fibers. Cohesive interactions lead to network shrinkage. Reprinted from Negi and Picu (2019a) with permission of Elsevier

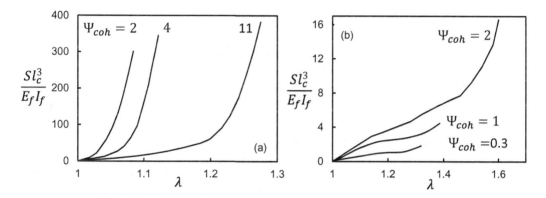

Figure 10.3 Nominal stress–stretch curves for (a) isostatic and (b) sub-isostatic planar networks with various cohesion strengths characterized by Ψ_{coh}. Networks are considered isostatic or sub-isostatic based on their state in the absence of cohesion. All networks have finite stiffness at infinitesimal strains once cohesion is enabled. The nominal (2D) stress is normalized by $E_f I_f / l_c^3$. The effect of cohesive interactions is to soften the isostatic networks (a) and to stiffen the sub-isostatic ones (b). Data from Negi and Picu (2019a)

curing. The process stops when the resistance associated with the elastic deformation of the network and excluded volume interactions balances the forces produced by fiber surface interactions. The key constitutive parameter defining the structure is $\Psi_{coh} = (l_c/L_{EC})^2 = \delta \bar{\gamma} l_c^2 / E_f I_f$.

Two fundamentally different states of the reference crosslinked network before the activation of surface interactions are considered. An isostatic network with large connectivity, $\langle z \rangle$, is studied first (Figure 10.2). Further, a sub-isostatic network which has vanishing moduli in the absence of cohesion is considered.

Figure 10.3(a) shows nominal stress–stretch curves corresponding to the uniaxial tension of networks of isostatic type with increasing strength of cohesive interactions. Increasing Ψ_{coh} has two effects. The effective network modulus in the linear elastic regime I decreases monotonically with increasing the strength of cohesion. At the same time, the range of the linear elastic regime I increases as Ψ_{coh} increases. Hence, cohesion renders networks softer and postpones the onset of strain stiffening.

As may be expected based on these results, the behavior at stretches corresponding to regime I is controlled by cohesion. The strain energy variation is minimal and the network response in this regime is mostly controlled by fiber bundling and un-bundling. In the unloaded configuration fibers are strongly deformed and the structure carries strain energy; see the right panel of Figure 10.2. The strain energy does not vary significantly when the network is subjected to small strains. This is in sharp contrast with the behavior of the same network in the absence of cohesion, which is entirely controlled by the variation of the strain energy.

The strain stiffening regime II emerges at larger stretches. This regime is primarily controlled by the strain energy. The nonlinear response is exponential and has the same functional form as that of the network without surface interactions.

Figure 10.3(b) shows similar curves for networks which are sub-isostatic in the absence of cohesive interactions. In the limit $\Psi_{coh} \to 0$, the small strain stiffness vanishes. Cohesion leads to the strong collapse of the network, which acquires non-zero stiffness in this process. As Ψ_{coh} increases, the small strain stiffness of the network with cohesive interactions increases – a trend opposite to that observed in the case of the isostatic networks shown in Figure 10.3(a). Nevertheless, as in the isostatic case, the response at relatively small stretches is primarily controlled by cohesive interactions and not by the variation of the strain energy.

The interesting result in Figure 10.3(b) is the emergence of softening under far field uniaxial tension. This effect results for all values of Ψ_{coh}. At larger stretches, the softening regime ends and strain stiffening is observed. The large stretch limit of the curve is controlled by the network structure and is only marginally affected by Ψ_{coh}. This behavior is qualitatively similar to that of elastomers and results here from the interplay of cohesion and strain energy in purely athermal networks. In the absence of cohesive interactions and for the same sub-isostatic network, the initial regime of vanishing stiffness is followed directly by strain stiffening.

While the examples discussed here are somewhat abstract, they demonstrate the interplay between cohesive and strain energy during network deformation. Cohesion introduces special effects not observed in the crosslinked networks without fiber surface interactions discussed in Chapter 6.

10.3   Non-crosslinked Networks with Fiber Surface Interactions

Surface interactions play an important role in non-crosslinked network materials. Due to the larger kinematic freedom of filaments in such structures (compared with the crosslinked networks case discussed in Section 10.2), surface interactions may lead to more drastic network reorganization and to the emergence of self-organized structures of fiber bundles. These range from isolated fiber clusters dispersed in a liquid, to a system-spanning network of fiber bundles. Once such a percolated network is stabilized, the material develops solid-like behavior at small strains, while at large strains it either ruptures or exhibits liquid-like behavior. The reminder of this chapter presents an overview of the structure and mechanical behavior of semi-dilute fiber suspensions and of self-organized networks of fiber bundles without a matrix. The emphasis is on the solid-like behavior emerging from the underlying network structure.

10.3.1   Fiber Suspensions

Given their practical importance in diverse fields, ranging from composite processing to the food and cosmetics industries, the rheology of suspensions of fibers in liquids has been studied extensively. Here we focus on the emergent behavior associated with the self-organization of fibers into super-structures, and necessarily leave out many other important results. The general behavior of dilute and semi-dilute suspensions is reviewed in several works (Zirnsak et al., 1994; Larson, 1999; Petrie, 1999) which the interested reader may consult.

10.3.1.1   Structure

The problem considered in this section is that of an ensemble of fibers of length L_0 dispersed in a liquid of viscosity η_s at volume fraction f. The fiber number density is denoted by $\rho_\#$ and the density is $\rho = \rho_\# L_0$. Surface interactions are characterized by the cohesive energy per unit length of longitudinal contact, $\delta\bar{\gamma}$. If the fiber bending stiffness is finite, Ψ_{coh} is a more convenient measure of the strength of surface interactions. We discuss the fiber structures that develop in the suspension and the emerging mechanical behavior.

The key interactions controlling the mechanics of the suspension are hydro-dynamic, surface, and excluded volume interactions. Surface interactions are ener-getic, while excluded volume interactions are entropic.

Fibers are organized by these forces and structures such as those shown in Figure 10.4 emerge. Figure 10.4(a) shows an isotropic assembly with random fiber orientations, while Figure 10.4(b) and (c) show structures known as nematic and smectic, respectively. Such alignment is also encountered in fiber bundles. Bundles with smectic order have length equal to the fiber length, while in the nematic case, the bundle structure does not limit the bundle length.

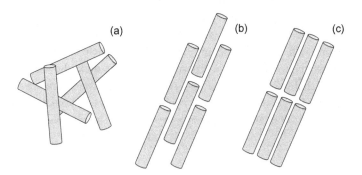

Figure 10.4 (a) Isotropic, (b) nematic, and (c) smectic fiber structures.

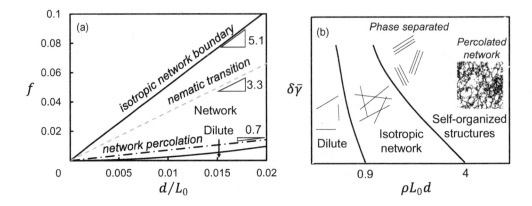

Figure 10.5 (a) Bounds for the existence of various phases in suspensions of rigid fibers with $\delta\bar{\gamma} = 0$ obtained from geometric considerations. (b) Schematic representation of the phase diagram for suspensions of fibers with $\delta\bar{\gamma} > 0$ reported in Groot and Agterof (1994) and Zilman and Safran (2003).

Geometric arguments provide guidelines for the type of fiber organization expected for specific values of the two geometric parameters, f and L_0/d. In Section 10.1 it is indicated that fibers interact when their associated volumes overlap, that is, when $\rho_\# L_0^3 \geq \alpha$. Based on geometry alone, $\alpha \approx 1$, but experiments indicate larger α values, for example, $\alpha \approx 30$ in Mori et al. (1982). Therefore, the transition from the dilute regime, in which fibers evolve independently, and the semi-dilute regime takes place at $f_c = (\pi\alpha/4)(d/L_0)^2 \approx 24(d/L_0)^2$. On the other hand, one may use for the same purpose Eq. (4.41), which provides the volume fraction at percolation for a three-dimensional random distribution of cylindrical rods with overlap (see also Figure 4.10). This expression, $f_c = 0.705d/L_0$, provides the lower bound for the existence of a random non-crosslinked network within the liquid suspension. The upper bound for the existence of a random network is given by the close packing limit of Eq. (4.42): $f_c = 5.1d/L_0$. As long as the network is isotropic (un-bundled) and fibers are rigid, the volume fraction cannot be increased beyond this limit. These bounds are shown in Figure 10.5(a) in the plane of the geometric parameters, f and L_0/d. The figure includes one more curve which corresponds to the onset of nematic order according to Onsager's theory (Onsager, 1949). This entropic transition from isotropic to nematic fiber arrangements is predicted to take place when $f = 3.3d/L_0.$[2] The transition to nematic ordering renders the upper bound of the isotropic structures unreachable in practical terms. A complete phase diagram for rigid sphero-cylinders (rods with hemispherical caps) for a broad range of volume fractions and aspect ratios is presented in Bolhuis and Frenkel (1997). The phase diagram predicts nematic and

---

[2] The result applies to rigid rods of same length. See also the effect of polydispersity of the fiber length in Speranza and Sollich (2003).

several smectic phases at higher volume fractions and reproduces the Onsager result in the limit $d/L_0 \to 0$.

A larger number of phases occur in the presence of energetic surface interactions. Zilman and Safran (2003) developed a phase diagram for stiff rods with attractive surface interactions, while Groot and Agterof (1994) studied flexible associative filaments. They define conditions in which a network spanning the problem domain forms. In an extension of their work on rigid rods with excluded volume, Bolhuis et al. (1997) evaluated phase diagrams for rigid sphero-cylinders with attraction. They identified isotropic, nematic, and smectic phases, along with domains of phase coexistence. The nature of the resulting structure depends on the fiber aspect ratio and on the range of attraction.

A schematic representation of the results reported in Groot and Agterof (1994) and Zilman and Safran (2003) is shown in Figure 10.5(b) in the $\delta\bar{\gamma}$ versus $\rho L_0 d \sim f(L_0/d)$ plane. Two boundaries define three domains in this diagram. The left boundary corresponds to the dilute to semi-dilute transition, while the right boundary corresponds to the emergence of self-organized structures. The intersection of these boundaries with the horizontal axis corresponds to the case without attractive surface interactions. The two values indicated on the horizontal axis in Figure 10.5(b) represent the corresponding boundaries shown in Figure 10.5(a). As attractive interactions are enabled ($\delta\bar{\gamma} > 0$), the two transitions take place at smaller $f(L_0/d)$.

In the domain to the right of the transition from isotropic to self-organized structures, the density, fiber length, and the strength of the attractive interactions are large enough to enable the formation of bundles and networks of bundles. This regime is of interest here since it leads to solid-like behavior of the suspension.

In suspensions, fiber structures may be destroyed by shear flow with large shear rates, and the suspension remains in the disordered state upon the abrupt cessation of the flow. However, a small strain rate perturbation applied to these metastable configurations leads to the disorganization followed by rapid re-organization of the fiber structures at the end of the perturbation (Rahatekar et al., 2006). It is also observed that the initial formation of self-organized carbon nanotube structures in suspension is favored by weak perturbations of small strain rate (Hobbie and Fry, 2006). These observations point to the importance of activation in the emergence of self-organized structures. In thermal systems, activation is provided by thermal fluctuations. However, in athermal networks activation must be produced mechanically.

10.3.1.2 Mechanical Behavior

The viscosity of suspensions increases with increasing filler volume fraction and aspect ratio. In the dilute limit, the relative viscosity, $\eta_r = (\eta - \eta_s)/\eta_s$ (where η_s is the viscosity of the solvent) is approximately linear with f. When accounting for hydrodynamic interactions in suspensions of athermal rods, η_r becomes a nonlinear function of f proportional to $(L_0/d)^2$ (Shaqfeh and Fredrickson, 1990; Larson, 1999). At larger volume fractions, the relative viscosity is approximately quadratic in f (Sundararajakumar and Koch, 1997; Lindstrom and Uesaka, 2009). Larson (1999) notes that the effect of fillers is more pronounced in suspensions of thermal filaments than in suspensions of athermal fibers subjected to steady shear flow. This is due to the fact that thermal fluctuations have a randomizing effect on the orientation of thermal filaments, which is absent in the case of athermal fibers. Furthermore, the effect of athermal fibers on extensional flow is more pronounced than that on the shear flow.

Beyond the effect of hydrodynamics interactions, surface interactions introduce significant changes in the rheology of suspensions. Likewise, the fact that fibers are generally deformable and not rigid rods adds another layer of complexity. In general terms, surface interactions have two effects: (i) they lead to shear thinning and (ii) they may produce solid-like behavior which becomes manifest at small deformations.

Shear thinning refers to the nonlinear variation of the shear stress with the shear strain rate, and, specifically, to the decrease of the viscosity with increasing strain rate. Shear thinning increases in magnitude with increasing volume fraction, f, and becomes more pronounced as the strength of surface interactions increases. For example, shear thinning is moderate in suspensions of rigid rods with cohesive interactions (Chaouche and Koch, 2001), but it is more pronounced when fibers are flexible (e.g., in the case of agar fibers (Wolf et al., 2007) and cellulose pulp (Cui and Grace, 2007)) and even more pronounced when the strength of cohesion increases (e.g., in carbon nanotube suspensions (Fan and Advani, 2007)).

Fiber aggregation promoted by attractive surface interactions and fiber compliance leads to the formation of transient structures which have solid-like behavior at small strains. These suspensions exhibit elastic response and a yield stress (Wierenga et al., 1998; Servais et al., 1999), and may also exhibit a plateau in the storage modulus, G', in some range of probing frequencies. This behavior is of interest here since it is indicative of the organization of fibers into transient networks within the suspension.

Figure 10.6 shows the dependence of the shear stress on the shear strain rate for a suspension of polyamide fibers of aspect ratio $d/L_0 = 1/33$ (Bounoua et al., 2016a)

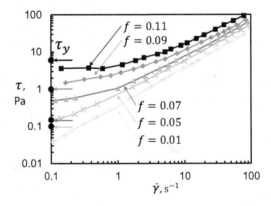

Figure 10.6 Shear stress–shear strain rate curves for suspensions of polyamide fibers of aspect ratio $d/L_0 = 0.03$ at several values of the volume fraction. The yield stress measured under stress-control in separate experiments is shown by the horizontal arrows and filled circles on the vertical axis, which correspond to $f = 0.05, 0.07, 0.09$, and 0.11 from bottom to top. Data from Bounoua et al. (2016a)

with several fiber volume fractions. For this aspect ratio, the geometric analysis of Section 10.3.1.1 (Figure 10.5(a)) indicates that the transition between the dilute and semi-dilute states takes place at $f \approx 0.02$. The curves in Figure 10.6 are obtained under strain rate control. For suspensions of $f > 0.01$, the stress is approximately independent of the strain rate at low rates, while at larger rates it becomes proportional to the strain rate. Therefore, the viscosity increases with increasing rate at small strain rates and asymptotes to a constant value at larger rates. The dilute suspension with $f = 0.01$ exhibits Newtonian behavior with constant viscosity at all rates reported.

These suspensions also exhibit yielding, which may be observed in a stress-controlled experiment. The yield stress for the respective materials is indicated in Figure 10.6 by the horizontal arrows pointing to the vertical axis. It is seen that the plateau of the stress at small strain rates in the rate-controlled experiment is close to the yield stress values resulting from the stress-controlled experiment. Both effects are caused by self-organized fiber structures stabilized by surface interactions, which induce solid-like behavior at small perturbations.

Similar results are obtained with other suspensions, including carbon nanotube suspensions with stronger surface interactions (Hobbie, 2010). The manifestation of the self-organized fiber structures is the occurrence of the yield point and the development of a plateau in $G'(\omega)$ as the filler volume fraction increases. The yield stress, τ_y, and the plateau modulus, G_0, characterize the solid-like behavior of such suspensions.

A model accounting for friction between fibers predicts a quadratic dependence of the yield stress on volume fraction, $\tau_y \sim f^2$ (Bounoua et al., 2016b). Accounting for aggregation adds a correction which becomes important as the volume fraction increases: $\tau_y \sim f^2 / \left(1 - (f/f_0)^2\right)$ (Bounoua et al., 2016a). Likewise, the plateau modulus is also expressed as a power function of f.

Experimental data obtained with a broad range of suspensions lead to exponents of the yield stress–volume fraction relation in the range 2–4. For example, carbon nanotube suspensions with $d/L_0 < 10^{-3}$ tested in Rahatekar et al. (2009) lead to $\tau_y \sim f^{2.1}$ (Figure 10.7). Suspensions of nanofibrillated cellulose studied by Martoia et al. (2016) are characterized by $\tau_y \sim f^{3.1}$. In Bennington et al. (1990) it is reported that bleached cellulose fiber suspensions and mechanically processed pulps are characterized by $\tau_y \sim f^{2.72}$ and $\tau_y \sim f^{3.56}$, respectively. The chemically processed pulp has longer, more flexible and less rough fibers than the mechanically processed one. The same article reports yield stress data for suspensions of nylon fibers of various aspect ratios, which lead to $\tau_y \sim f^{2.73}$.

Measuring the yield stress in monotonic stress-controlled shear tests is generally difficult due to the very small values of τ_y. However, detecting the emergence of a plateau of $G'(\omega)$ (measured in oscillatory shear) as f increases is much easier, even at relatively low f. Such plateau is indicative of the development of an elastic network.

Results from multiple sources indicate that $G_0 \sim f^\beta$. Large exponents, close to $\beta = 7$, are reported in Hobbie and Fry (2006) and Yearsley et al. (2012) for carbon nanotube suspensions of relatively low volume fraction. The data of Rahatekar et al. (2009) and Hobbie (2010) exhibits two exponents, with $\beta \approx 7$ at smaller filler

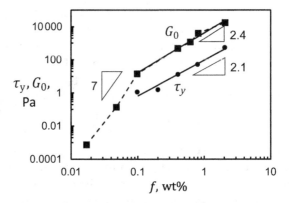

Figure 10.7 Yield stress and plateau modulus for suspensions of carbon nanotubes function of the filler concentration. Data from Rahatekar et al. (2009) and Hobbie (2010)

concentrations, and with $\beta = 2.4$ at larger concentrations. This data set is shown in Figure 10.7, along with the corresponding yield stress results. It is interesting to observe from the results in Figure 10.7 that, in the high concentration range, both the plateau modulus and the yield stress scale with the filler concentration in a similar manner. Therefore, these two parameters are proportional, which implies that the strain at yield is independent of concentration. This is similar to the observation made in athermal networks that the strain at the transition from the linear elastic regime I to the nonlinear, strain stiffening regime II is independent of the fiber density and fiber properties.

An interesting data set is provided in Lieleg et al. (2007) who work with actin solutions to which the actin binding protein fascin is added at various concentrations. Above a certain concentration of fascin, actin organizes in a network of bundles. At given fascin concentration, the plateau modulus of the solution is a power function of the concentration with exponent 2.4, in agreement with the result shown in Figure 10.7 for carbon nanotube suspensions. The same data set indicates that the modulus scales with the fourth power of the bundle diameter (which was determined directly, by fluorescence microscopy), which suggests that the modulus is proportional to the bending rigidity of the bundle, $E_f I_b$, as expected for both networks of semi-flexible filaments and for athermal networks.[3]

The large exponent of 7 observed at lower filler concentrations is considered in the literature on suspensions as being indicative of fractal filler organization.[4] For such structures, the exponent β (in $G_0 \sim f^\beta$) is related to the fractal dimension of the structure, D_p, as $\beta = (3 + D_b)/(3 - D_p)$, where D_b is the dimension of the backbone forming the fractal (Shih et al., 1990). Hence, the exponent $\beta = 7$ obtained experimentally (Figure 10.7) suggests the existence of a percolated fractal structure of

---

[3] A network of semi-flexible fiber bundles becomes gradually more athermal as the bundle size increases. See also Section 10.3.2.2.
[4] A 3D network in the vicinity of the percolation threshold has fractal structure.

dimension $D_\rho = 2.42$. The fractal structures discussed here are viewed as preliminary stages of the development of a self-organized network of bundles. The change of the exponent from $\beta = 7$ to $\beta \approx 2$ occurs as the structure moves away from the vicinity of the percolation threshold into the range of well-developed networks of larger fiber volume fraction. An exponent close to 2 can be rationalized based on results presented in Section 6.1.1.3.1 for crosslinked networks. This is discussed further in Section 10.3.2.2 in relation to self-organized networks of bundles without a matrix.

In summary, results obtained with a diverse set of suspensions of fibers with surface interactions support the concept that, at high enough filler concentrations, the fibers self-organize creating a network of bundles. The presence of the network conveys solid-like behavior to the suspension, consistent with expectations from network theory (see Chapter 6).

10.3.2 Networks of Fiber Bundles without a Matrix

This section is dedicated to the self-organization of fibers into networks of fiber bundles in the dry state, that is, without matrix. Nanoscale filaments with strong surface interactions tend to organize in bundles which, in turn, may form networks of bundles. The necessary conditions for the occurrence of such networks is that the constituent fibers have sub-micron diameter (nanofibers) and hence small bending stiffness, and their surface interactions are sufficiently strong. These two conditions taken together are equivalent to requiring that Ψ_{coh} is sufficiently large.

A typical example of such a network is buckypaper, which is obtained by depositing carbon nanotubes on a substrate. The nanotubes have strong cohesive interactions and form bundles during the aerosol phase, before deposition, and after being deposited in the mat. Buckypaper is strongly stabilized by cohesive interactions. It typically exhibits a brittle mechanical response, since the network lacks the kinematic freedom to reorganize during stretching. Collagen is another example of a multiscale structure composed from bundles of fibrils organized hierarchically (Figure 2.3). In general terms, nanofibers forming a dry mat may either remain unbundled as in the as-deposited state – for example in the case of bacterial cellulose – or may undergo self-organization driven by surface interactions leading to the formation of bundles – as in the case of carbon nanotubes. This section focuses on the structures emerging from these processes and compares the mechanical behavior of networks of bundles with that of networks of fibers discussed in Chapter 6.

10.3.2.1 Structure

The problem considered in this section is that of an ensemble of fibers of length L_0 forming a mat. The density measured in two dimensions, in the projection onto the plane of the mat, is ρ_{2D}, where the subscript 2D is used to distinguish the projected density from the real density ρ evaluated in three dimensions. Surface interactions are enabled and are characterized by $\delta\bar{\gamma}$ and Ψ_{coh}.

The essential fiber interactions controlling the mechanics of the fiber assembly are friction, surface, and excluded volume interactions. We are interested in outlining the

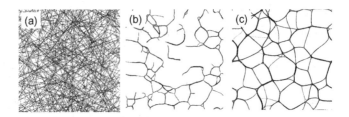

Figure 10.8 Fiber structures of quasi-two-dimensional mats: (a) network of fibers without bundling or self-organization, (b) bundles with internal smectic order which separate and do not form a percolated network, and (c) network of fiber bundles. Reproduced from Picu and Sengab (2018) with permission of The Royal Society of Chemistry

conditions in which these interactions lead to the formation of a network of bundles. The problem stated here is similar to that of Section 10.3.1.1, with the exception that hydrodynamic interactions are not present, while friction becomes essential.

Three types of structures emerge and are shown in Figure 10.8. Figure 10.8(a) shows a regular network of fibers without self-organization, while Figure 10.8(b) and (c) show structures that result upon fiber bundling. If smectic order emerges, bundles separate into segments of length L_0, and the resulting structure is not a percolated network (Figure 10.8(b)). However, if the internal structure of bundles remains nematic, conditions exist for the formation of a network of bundles, as shown in Figure 10.8(c).

The process by which an as-deposited network (Figure 10.8(a)) becomes a network of bundles (Figure 10.8(c)) has three stages: activation, self-organization, and arrest. The activation step can be understood in terms of the discussion in Section 3.3.2 related to Figure 3.6(a). Two filaments in contact whose axes form an angle θ need to rotate at the contact point in order to allow the surface forces to become engaged. Once activated, surface forces cause the evolution from the configuration in the left panel of Figure 3.6(a) to that in the right panel. If surface interactions are short ranged, an external agent is needed to bring the fibers into initial contact (i.e., render the fibers tangent at point B in Figure 3.6(a)). This is generally not difficult since as-deposited networks are subjected to the environment and flow of air or of a liquid may lead to such activation.

Self-organization leads to fiber bundling. If two initially nonparallel fibers are allowed to come in contact along their length, they may zip, as shown in Figure 3.6. Since such bundles form at multiple sites in the network of Figure 10.8(a), topological conditions require that nodes of bundles should form. Nodes of connectivity $z = 3$ and $z = 4$ are shown in Figure 10.9(a) and (b), respectively. As discussed in Picu and Sengab (2018), in quasi-2D, mat-like structures, nodes with $z = 4$ are metastable and tend to split into two nodes with $z = 3$ (Figure 10.9(b)). It results that the $z = 3$ node is the stable configuration in a quasi-two-dimensional geometry, while an equivalent node with tetrahedral symmetry can be envisioned in a three-dimensional setting. These nodes perform the role of crosslinks in networks of bundles. Their structural stability is discussed in Picu and Sengab (2018) and Picu and Negi (2019). Since the

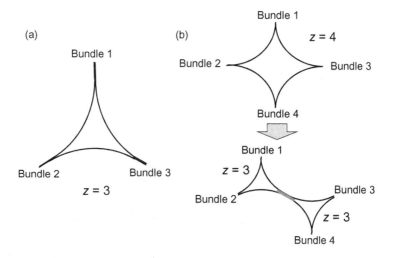

Figure 10.9 (a) Triangular node connecting three coplanar fiber bundles. (b) The transformation of a fiber bundle node with $z = 4$ into a more stable structure with two triangular nodes with $z = 3$.

crosslink connectivity in quasi-2D structures is by necessity $z = 3$, networks of bundles are of cellular type.

Two mechanisms make the structural evolution of networks of bundles possible: (i) The motion of triangular nodes and (ii) the axial motion of fibers within bundles. Certain topological rules must be fulfilled in order to allow node motion (Picu and Sengab, 2018). The exact same rules arrest structural evolution once a sufficiently large number of crosslinks become of triangular type, $z = 3$. It is of interest to observe that foams are also stabilized by triangular fluidic structures that form at the intersection of cell walls. These are known as Plateau triangles. However, the mechanics of the surface tension-controlled Plateau triangles is different from that of the triangular nodes formed by bundles, whose structure is controlled by the interplay of cohesion and fiber bending energies.

The motion of fibers in the direction of the bundle axis may be driven by gradients of cohesion and strain energy (Picu and Sengab, 2018). This evolution leads to smectic order and to the emergence of structures such as that in Figure 10.8(b). Friction may effectively inhibit this motion, particularly if fibers are long.

Denoting by ρ_{bdl} and l_{c_bdl} the density (total length of bundles per unit volume) and the mean bundle segment length of the network of bundles, mass conservation requires $\rho_{bdl}\bar{n} = \rho$, where ρ is the fiber density and $\bar{n}$ is the mean number of fibers per bundle. This expression applies equally in two and three dimensions. Furthermore, in the 2D projection, the network evolves from the Mikado-like structure in Figure 10.8(a) to the cellular structure in Figure 10.8(c). Using Eq. (4.6), which applies to Mikado networks, and Eq. (4.11), which applies to 2D cellular networks, one infers:

$$\rho_{bdl}^{2D}l_{c\_bdl}^{2D} \approx (2/\pi)\rho_{2D}l_c^{2D}. \qquad (10.1)$$

The average pore size of the mat measured in the 2D projection is proportional to the mean segment length both before, $d_p \sim l_c^{2D}$, and after bundling, $d_{p\_bdl} \sim l_{c\_bdl}^{2D}$. Using Eq. (10.1) it results that:

$$d_{p\_bdl} \sim d_p\bar{n}, \qquad (10.2)$$

that is, the pore size increases linearly with the mean bundle size.

Based exclusively on energetic arguments, one may infer that networks of bundles evolve until the entire population of fibers forms a single large bundle of length L_0. However, this is not what is observed in experiments. Therefore, mechanisms that restrict network evolution must operate. This issue is discussed in Groot (2013), Picu and Sengab (2018), and Sengab and Picu (2018), where three such mechanisms are proposed. Friction taking place between filaments within bundles and at inter-bundle contacts limits both the onset of bundling and the evolution of the network.[5] The effect of friction increases with increasing filament length. Another mechanism is based on the development of a far field stress as the network evolves. Since the network volume tends to decrease during network self-organization (e.g., Figure 10.2), any boundary conditions that restrict shrinkage lead to the development of tensile stresses. Once large enough, this far field hydrostatic stress inhibits the further evolution of the network. A third mechanism leading to the arrest of network self-organization is of topological nature. In Picu and Sengab (2018) it is shown that structural evolution must stop once triangular nodes with $z = 3$ (Figure 10.9(a)) form at a majority of bundle intersections.

Figure 10.10 shows the phase diagram of self-organized networks with surface interactions corresponding to dry quasi-2D mat-like athermal structures (Picu and Sengab, 2018; Sengab and Picu, 2018). It shows stable configurations resulting from the structural evolution of the ensemble of fibers in Figure 10.8(a) after activation. The diagram is drawn in the plane of the two non-dimensional parameters, Ψ_{coh} and $\rho_{2D}L_0$. At densities below the percolation threshold $\rho_{2D}L_0 = 5.637$ (Eq. (4.38)), a percolated network does not form and the problem posed here is not defined. At larger densities, the initial structure either remains in the unbundled state (Figure 10.8(a)), if Ψ_{coh} is low, or evolves by filament bundling (Figure 10.8(b) and (c)), if Ψ_{coh} is larger than a threshold. The boundary separating the nonevolving from the evolving structures is defined by the equation:

---

[5] The nature of friction between nanofibers requires discussion. However, a detailed analysis of this topic is beyond the scope of the present review. It suffices to indicate that friction between nanoscale objects is not of Coulomb type and a viscous model is more appropriate for such situations. In this representation, the friction force increases in proportion to the contact area. If filaments have parallel axes, this area is large (see Figure 3.4) and scales linearly with the length of the longitudinal contact. Friction forces in these cases are large and are expected to prevent the axial motion of fibers within bundles. If filaments cross at an angle, the contact area is a function of the respective angle but remains finite and is independent of L_0. These issues are discussed in Negi and Picu (2019b).

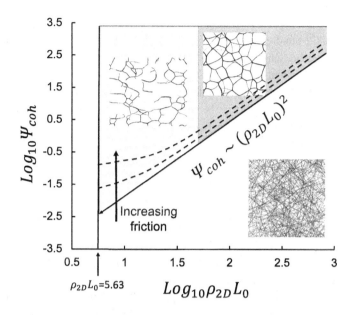

Figure 10.10 Phase diagram for self-organized networks of fiber bundles with surface interactions. Adapted from Picu and Sengab (2018) and Sengab and Picu (2018)

$$\Psi_{coh} = A(\rho_{2D}L_0)^2 + B(\rho_{2D}L_0)\Psi_{fric}, \tag{10.3}$$

where A and B are constants and $\Psi_{fric} = F_{fric}L_0^2/E_fI_f$ is the normalized friction force at crossed contacts between filaments. The boundary corresponding to zero friction is shown in Figure 10.10 by a continuous, inclined line. Two positions of the boundary corresponding to increasing friction are shown by a dashed line. The boundary shifts up as friction increases.

If conditions for filament bundling exist, the two types of structures in Figure 10.8(b) and (c) develop at lower and larger $\rho_{2D}L_0$, respectively.

Increasing the filament length, while keeping all other parameters unchanged, corresponds to a trajectory in the diagram parallel to the boundary separating the nonevolving and evolving structures (Eq. (10.3)). This promotes the emergence of cellular networks of bundles, provided $\delta\bar{y}/E_fI_f = 1/L_{EC}^2$ is sufficiently large (above the continuous, inclined line). If the density increases, while all other parameters are constant, the corresponding trajectory is horizontal, going from left to right. Hence, increasing ρ_{2D} inhibits the formation of networks of bundles.

The evolution of the pore size during bundling is of importance in applications ranging from filtration to the fabrication of scaffolds for tissue engineering. A definite conclusion on this issue is not available at this time. In Li and Kroger (2012) it is suggested that the pore size of stable structures is proportional to the elasto-capillary length, $d_{p\_bdl} \sim L_{EC}$, while a weaker scaling, $d_{p\_bdl} \sim L_{EC}^{1/2}$, is suggested by DeBenedictis et al. (2020).

Before closing this section, it must be noted that networks of bundles tend to be athermal. This result may be inferred by evaluating the variation of the ratio between the persistence length of bundles and the mean segment length, as the network evolves and the bundle size, $\bar{n}$, increases. Consider on one hand a network of thermal filaments of persistence length $L_p = E_f I_f / k_B T$ having the un-bundled structure of Figure 10.8(a) with mean segment length l_c^{2D}. On the other hand, consider the equivalent bundled structure with $L_{p\_bdl} = E_f I_b / k_B T$ and mean bundle segment length $l_{c\_bdl}^{2D}$. Since the moment of inertia of the bundle may be written $I_b = \bar{n}^2 I_f$ (Section 2.5.1), it result that $L_{p\_dbl} = L_p \bar{n}^2$. Equation (10.1) indicates that $l_{c\_bdl}^{2D} \sim \bar{n} l_c^{2D}$. Hence, $L_{p\_bdl} / l_{c\_bdl}^{2D} \sim \bar{n} L_p / l_c^{2D}$. The equivalent derivation for 3D networks leads to $L_{p\_bdl} / l_{c\_bdl} \sim \bar{n}^{3/2} L_p / l_c$. Therefore, the thermal nature of the bundles is quickly lost as the bundle size, $\bar{n}$, increases.

10.3.2.2 Mechanical Behavior

Networks of bundles exhibit elastic response at small strains but may flow at large strains due to the fact that fibers may slide relative to each other within bundles. Reaching larger stretches before fracture is conditioned by strong cohesion and the presence of fibers with a large aspect ratio. Effective load transfer between fibers is necessary in order to stabilize plastic flow. Frictional inter-fiber interactions ensure good load transfer. While the overall physical picture is relatively clear, only a few details are available at this time.

Networks of fiber bundles have cellular structure and their elasticity is similar to that of crosslinked cellular networks. Both the elastic moduli of cellular networks and the upper limit of the linear elastic deformation range (denoted here by σ_y) scale as $E_0, \sigma_y \sim E_f I_f / l_c^4$ (Gibson and Ashby, 1988). For a network of bundles, this relation becomes $E_0, \sigma_y \sim E_f I_b / l_{c\_bdl}^4$. Considering that the relation $\rho_{bdl} l_{c\_bdl}^2 = const$ holds in 3D, imposing mass conservation, $\rho_{bdl}\bar{n} = \rho$, and assuming that $l_{c\_bdl} \sim d_{p\_bdl} \sim L_{EC}$ (Li and Kroger, 2012), it is possible to infer:

$$\frac{E_0}{E_f}, \frac{\sigma_y}{E_f} \sim \rho^2 I_f \sim \left(\rho L_0^2\right)^2 \left(\frac{d}{L_0}\right)^4 \quad \text{if } I_b \sim \bar{n}^2, \tag{10.4a}$$

$$\frac{E_0}{E_f}, \frac{\sigma_y}{E_f} \sim \rho I_f / L_{EC}^2 \sim \left(\rho L_0^2\right)\left(\frac{d}{L_0}\right)^4 \Psi_{coh} \quad \text{if } I_b \sim \bar{n}. \tag{10.4b}$$

The case with $I_b \sim \bar{n}^2$ represents bundles in which fibers do not slide axially during bundle bending, while the case with $I_b \sim \bar{n}$ represents the opposite situation. In the first case, for given network geometry, the modulus is independent of Ψ_{coh} and L_0. Note that the network geometry depends on these parameters, as shown in Figure 10.10. Also, since E_0 and σ_y are proportional to each other, the strain at which the linear elastic regime ends is independent of the network parameters.

Equation (10.4a) is supported by results obtained with carbon nanotube networks (Rahatekar et al., 2009; Hobbie, 2010) (Figure 10.7) and by the data in Lieleg et al. (2007) on actin bundled with fascin, which indicate that the plateau modulus is a power function of the concentration with exponent close to 2. The results in DeBenedictis et al. (2020) indicate that the modulus scales linearly with $\delta\bar{\gamma}$ (i.e., with Ψ_{coh}), which is possible if Eq. (10.4b) applies. In the computational study of Li and

Kroger (2012) it is reported that the modulus decreases with increasing L_0. These apparent differences can be understood based on the observation that bundles obtained under various conditions have different degrees of internal packing, which influences whether filaments are able to slide in the bundle axial direction or not.

The experimental study of Piechocka et al. (2016) is particularly relevant in this sense. They work with networks of fibrin bundles which are treated with a bundling inhibitor agent added at various concentrations. It is observed that, as bundles become looser and likely of smaller $\bar{n}$, the linear elastic modulus decreases. In the terms of Eq. (10.4), the treatment with the bundling inhibitor is equivalent to switching from the situation of tightly packed bundles of Eq. (10.4a) to that of loosely packed bundles of Eq. (10.4b). Equation (10.4b) predicts a softer response than Eq. (10.4a).

Several differences relative to the mechanics of crosslinked networks with surface interactions (Section 10.2) are worth emphasizing. The numerical study of the response of a cellular network of bundles subjected to relatively small strains reported in Negi and Picu (2020) indicates that nodal triangles are exceptionally stable under imposed external loads. Their function is similar to that of crosslinks in regular, un-bundled, and crosslinked networks. While bundles undergo zipping and unzipping, the deformation is mainly controlled by the structure of the cellular network and global stress is primarily caused by the variation of the strain energy. These observations are supported by the experimental data in Piechocka et al. (2016) where it is seen that modifying the degree of bundling changes E_0, but does not affect the nonlinear mechanics of the network and the strain at the onset of the nonlinear regime. This is in contrast with the behavior of crosslinked networks of bundles described in Section 10.2. In the respective case, the small strains behavior is controlled by cohesion, while the network defines the response at large strains.

The behavior beyond the linear elastic regime is controlled by the nonlinear deformation of the cellular network and by bundle stretching sustained by the relative sliding of filaments within bundles (Xie et al., 2011; Stallard et al., 2018). Friction and cohesion play the central role in this process. Cohesion stabilizes the bundles, while friction allows load transfer between fibers, which increases the maximum stress sustained by the structure. Plastic deformation takes place by axial bundle stretching. Bundles rupture during stretching by fiber pull-out. These competing processes lead to stress–strain curves of the type shown schematically in Figure 10.11. If the cellular network is sparse (low ρ_{bdl}) and nonaffine, the nonlinear elastic regime II is entered before the axial deformation mode is engaged. Hence, sliding of fibers within bundles is not activated at the beginning of the deformation process, when the primary network deformation mode is bending. Once bundles enter the axial mode and begin to deform plastically, two possibilities exist: if L_0 is small, rupture takes place at small plastic strains; if L_0 is large and cohesion is strong, some degree of plastic flow may be sustained at approximately constant stress (see also Sections 2.5.3 and 3.3.3).

Most networks of bundles, such as buckypaper, are dense enough for the deformation to be approximately affine at small strains. In these cases, the axial deformation mode is engaged even at infinitesimal strains. If L_0 is small or fibers are weak, rupture takes place at small strains. If L_0 is large, global plastic deformation is sustained as

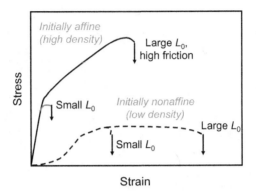

Figure 10.11 Schematic representation of stress–strain curves of cellular networks of fiber bundles.

bundles are stretched. This process must be stabilized by strong cohesive interactions. In the presence of friction, an apparent strain hardening may be observed (Figure 10.11), which is associated with the collapse of pores and network densification during stretching (network densification leads to enhanced frictional dissipation). This phenomenology is observed in models of carbon nanotube structures loaded up to failure (Wang et al., 2018).

Networks of fiber bundles are much less studied at this time compared to the common networks in which surface interactions are negligible. It is likely that new behaviors, beyond those presented in this overview, will be uncovered in the future. This is an active area of research at this time.

References

Bennington, C. P. J., Kerekes, R. J. & Grace, J. R. (1990). The yield stress of fiber suspensions. *Canad. J. Chem. Eng.* **68**, 748–757.

Bolhuis, P. & Frenkel, D. (1997). Tracing the phase boundaries of hard spherocylinders. *J. Chem. Phys.* **106**, 666–687.

Bolhuis, P. G., Stroobants, A., Frenkel, D. & Lekkerkerker, H. N. W. (1997). Numerical study of the phase behavior of rodlike colloids with attractive interactions. *J. Chem. Phys.* **107**, 1551–1564.

Bounoua, S., Lemaire, E., Ferec, J., Ausias, G. & Kuzhir, P. (2016a). Shear thinning in concentrated rigid fiber suspensions: Aggregation induced by adhesive interactions. *J. Rheol.* **60**, 1279–1300.

Bounoua, S., Lemaire, E., Ferec, J., et al. (2016b). Apparent yield stress in rigid fibre suspensions: The role of attractive colloidal interactions. *J. Fluid Mech.* **802**, 611–633.

Chaouche, M. & Koch, D. L. (2001). Rheology of non-Brownian rigid fiber suspensions with adhesive contacts. *J. Rheol.* **45**, 369–382.

Cui, H. & Grace, J. R. (2007). Flow of pulp fibre suspension and slurries: A review. *Int. J. Multiphase Flow* **33**, 921–934.

DeBenedictis, E. P., Zhang, Y. & Keten, S. (2020). Structure and mechanics of bundled semiflexible polymer networks. *Macromolecules* **53**, 6123–6134.

Fan, Z. & Advani, S. G. (2007). Rheology of multiwall carbon nanotube suspensions. *J. Rheol.* **51**, 585–604.

Gibson, L. J. & Ashby, M. F. (1988). *Cellular solids: Structure and properties*. Cambridge University Press, Cambridge.

Groot, R. D. (2013). Mesoscale simulation of semiflexible chains. II. Evolution dynamics and stability of fiber bundle networks. *J. Chem. Phys.* **138**, 224904.

Groot, R. D. & Agterof, W. G. A. (1994). Monte Carol study of associative polymer networks. I. Equation of state. *J. Chem. Phys.* **100**, 1649–1656.

Hobbie, E. K. (2010). Shear rheology of carbon nanotube suspensions. *Rheol. Acta* **49**, 323–334.

Hobbie, E. K. & Fry, D. J. (2006). Nonequilibrium phase diagram of sticky nanotube suspensions. *Phys. Rev. Lett.* **97**, 036101.

Larson R. G. (1999). *The structure and rheology of complex fluids*. Oxford University Press, Oxford.

Li, Y. & Kroger, M. (2012). Computational study of entanglement length and pore size of carbon nanotube buckypaper. *J. Appl. Phys.* **100**, 021907.

Lieleg, O., Claessens, M. M. A. E., Heussinger, C., Frey, E. & Bausch, A. R. (2007). Mechanics of bundled semiflexible polymer networks. *Phys. Rev. Lett.* **99**, 088102.

Lindstrom, S. B. & Uesaka, T. A. (2009). Numerical investigation of the rheology of sheared fiber suspensions. *Phys. Fluids* **21**, 083301.

Martoia, F., Dumont, P. J. J., Orgeas, L., Belgacem, M. N. & Putaux, J. L. (2016). Micromechanics of electrostatically stabilized suspensions of cellulose nanofibrils under steady state shear flow. *Soft Matter* **12**, 1721–1735.

Mori, Y., Ookubo, N., Hayakawa, R. & Wada, Y. (1982). Low-frequency and high-frequency relaxations in dynamic electric birefringence of poly(g-benzyl-L-glutamate) in m-cresol. *J. Poly. Sci. Poly. Phys. Ed.* **20**, 2111–2124.

Negi, V. & Picu, R. C. (2019a). Mechanical behavior of cross-linked random fiber networks with inter-fiber adhesion, *J. Mech. Phys. Sol.* **122**, 418–434.

Negi, V. & Picu, R. C. (2019b). Mechanical behavior of nonwoven non-crosslinked fibrous mats with adhesion and friction. *Soft Matt.* **15**, 5951–5964.

Negi, V. & Picu, R. C. (2020). Mechanical behavior of cellular networks of fiber bundles stabilized by adhesion. *Int. J. Sol. Struct.* **190**, 119–128.

Onsager, L. (1949). The effects of shape on the interaction of colloidal particles. *Ann. N. Y. Acad. Sci.* **51**, 627–659.

Petrie, C. J. (1999). The rheology of fibre suspensions. *J. Non-Newt. Fluid Mech.* **87**, 369–402.

Picu, R. C. & Negi, V. (2019). Mechanics of random networks of nanofibers with inter-fiber adhesion. In *Mechanics and physics of solids at micro and nano-scales*, I. R., Ionescu. S., Queyreau, R. C., Picu, & O. U. Salman, eds. Wiley, Hoboken, NJ, pp. 157–184.

Picu, R. C. & Sengab, A. (2018). Structural evolution and stability of non-crosslinked fiber networks with inter-fiber adhesion. *Soft. Matter* **14**, 2254–2266.

Piechocka, I. K., Jansen, K. A., Broedersz, C. P., et al. (2016). Multiscale strain stiffening of semiflexible bundle networks. *Soft Matter*, **12**, 2145–2156.

Rahatekar, S. S., Koziol, K. K. K., Butler, S. A., et al. (2006). Optical microstructure and viscosity enhancement for an epoxy resin matrix containing multiwall carbon nanotubes. *J. Rheol.* **50**, 599–610.

Rahatekar, S. S., Koziol, K. K., Kline, S. R., et al. (2009). Length-dependent mechanics of carbon nanotube networks. *Adv. Mater.* **21**, 874–878.

Roman, B. & Bico, J. (2010). Elasto-capillarity: Deforming an elastic structure with a liquid droplet. *J. Phys.: Condens. Matter* **22**, 493101.

Sengab, A. & Picu, R. C. (2018). Filamentary structures that self-organize due to adhesion. *Phys. Rev. E* **97**, 032506.

Servais, C., Manson, J. A. E. & Toll, S. (1999). Fiber–fiber interaction in concentrated suspensions: Disperse fibers. *J. Rheol.* **43**, 991–1004.

Shaqfeh, E. R. G. & Fredrickson, G. H. (1990). The hydrodynamic stress in a suspension of rods. *Phys. Fluid A* **2**, 7–24.

Shih, W. H., Shih, W. Y., Kim, S. I., Liu, J. & Aksay, I. A. (1990). Scaling behavior of the elastic properties of colloidal gels. *Phys. Rev. A* **42**, 4772–4779.

Speranza, A. & Sollich, P. (2003). Isotropic-nematic phase equilibria in the Onsager theory of hard rods with length polydispersity. *Phys. Rev. E* **67**, 061702.

Stallard, J. C., Tan, W., Smail, F. R., et al. (2018). The mechanical and electrical properties of direct-spun carbon nanotube mats. *Extreme Mech. Lett.* **21**, 65–75.

Style, R. W., Jagota, A., Hui, C. Y. & Dufresne, E. R. (2017). Elastocapillarity: Surface tension and the mechanics of soft solids. *Annu. Rev. Condens. Matter Phys.* **8**, 99–118.

Sundararajakumar, R. R. & Koch, D. L. (1997). Structure and properties of sheared fiber suspensions with mechanical contacts. *J. Non-Newt. Fluid Mech.* **73**, 205–239.

Wang, Y., Xu, H., Drozdov, G. & Dumitrica, T. (2018). Mesoscopic friction and network morphology control the mechanics and processing of carbon nanotube yarns. *Carbon* **139**, 94–104.

Wierenga, A., Philipse, A. P., Lekkerkerker, H. N. & Boger, D. V. (1998). Aqueous dispersions of colloidal boehmite: Structure, dynamics, and yield stress of rod gels. *Langmuir* **14**, 55–65.

Wolf, B., White, D., Melrose, J. R. & Frith, W. J. (2007). On the behaviour of gelled fibre suspensions in steady shear. *Rheol. Acta* **46**, 531–537.

Xie, B., Liu, Y., Ding, Y., Zheng, Q. & Xu, Z. (2011). Mechanics of carbon nanotube networks: Microstructural evolution and optimal design. *Soft Matter* **7**, 10039–10047.

Yearsley, K. M., Mackley, M. R., Chinesta, F. & Leygue, A. (2012). The rheology of multiwall carbon nanotube and carbon black suspensions. *J. Rheol.* **56**, 1465–1490.

Zilman, A. G. & Safran, S. A. (2003). Role of crosslinks in bundle formation, phase separation and gelation of long filaments. *Europhys. Lett.* **63**, 139–145.

Zirnsak, M. A., Hur, D. U. & Boger, D. V. (1994). Normal stresses in fiber suspensions. *J. Non-Newt. Fluid Mech.* **54**, 153–193.

11 Composite Networks

The previous chapters address the mechanical behavior of network materials composed from fibers of identical type (having identical mechanical properties); these networks do not embed other entities, such as reinforcing fibers and particles, and are not embedded in a matrix. They are mechanically heterogeneous because they have stochastic connectivity and fiber segment lengths but are not heterogeneous from the point of view of their composition. Several cases in which material properties of fibers are allowed to vary across the network are discussed in Sections 6.1.1.3.9 and 8.1.7.

In many practical situations, network materials are composites. This term is used here to indicate any configuration different from those described in the preceding paragraph. Composite networks embed particles or fiber inclusions, may be embedded in a solid matrix, and/or may contain fibers of different types.

The cytoskeleton embeds cellular organelles and is a complex network composed from filaments of several types, for example, F-actin and microtubules. The extracellular matrix embeds cells, while the collagen network of articular cartilage embeds large polysaccharide molecules known as glycosaminoglycans. According to the present definition, all these network materials are composites.

Some engineering network materials, such as epoxy, chemical gels, and PDMS, are not composite networks according to the definition used here. Others are composites by design, such as thermoplastic elastomers (e.g., based on styrene-butadiene-styrene, SBS, copolymer), in which the components segregate into an elastomeric phase and a more rigid phase that plays the role of network crosslinks (the styrene component). Natural rubber presents a somewhat similar situation as it develops nanocrystals when it is subjected to large deformations. Yet, in other cases, nominally pure networks are modified by the addition of particles. Reinforcement of elastomers with carbon black is an established technology which conveys enhanced mechanical properties to the rubber.

The field of polymer composites, including those materials in which the matrix is a molecular network, is broad. Many thermosets and an even larger number of thermoplastics are reinforced by the addition of particles or fibers. This general field of research is, for the most part, outside the scope of the present discussion.

The objective of this chapter is to outline the mechanisms of reinforcement that are based on those features that distinguish networks from continua. To this end, the following types of materials are discussed (Figure 11.1):

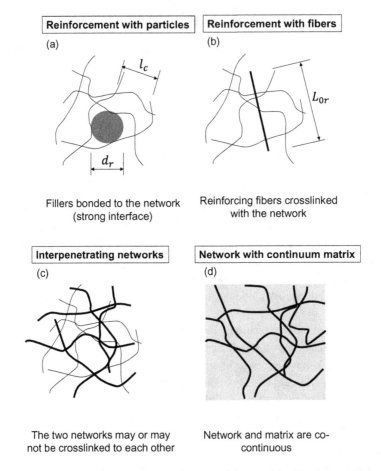

Figure 11.1 Four types of network reinforcement discussed in this chapter: (a) networks reinforced with particles, (b) networks reinforced with stiff staple fibers, (c) interpenetrating networks, and (d) networks embedded in a matrix.

(i)   Networks reinforced with particles of size comparable with the mean segment length, l_c, or the mean pore size of the network,

(ii)   Networks reinforced with staple fibers with material properties significantly different from those of the base network fibers,

(iii) Interpenetrating networks (IPN), and

(iv)   Networks embedded in a continuum.

When reinforcing networks with particles it is relevant to compare the diameter of the reinforcement, d_r, with the network mesh size. While networks have multiple characteristic lengths (see Chapter 4), the reference characteristic length is the mean distance between crosslinks, the mesh or the pore size. Here we take the mean segment length, l_c, as a proxy for the mesh size. If $d_r \gg l_c$, the network appears as a continuum matrix on the scale of the filler. The field perturbation introduced by fillers is defined on a length scale comparable with d_r and, if $d_r \gg l_c$, the discreteness of the network becomes irrelevant on this length scale. Such problems can be addressed using homogenization theory, since the network may be treated as a continuum on length scales equal and larger than the filler size. On the other hand, if $d_r \ll l_c$, fillers are too small to effectively reinforce the network. It results that, in order to exploit effects associated with the discreteness of the network, these two characteristic lengths should be comparable, that is, $d_r \approx l_c$ (Figure 11.1(a)). Such situations are discussed in Section 11.1.

Networks may be reinforced by incorporating stiffer staple fibers. The field of fiber-reinforced plastics is well-developed. As with particle reinforcement, situations in which fillers are much larger – both in length and diameter – than the length scale of the network are outside the scope of the present discussion. Fiber composites with epoxy matrix are used extensively. However, the fiber diameter is typically orders of magnitude larger than any characteristic length scale of the epoxy network, an observation which places these systems in the realm of classical composites. The discussion in Section 11.2 focuses on situations in which the reinforcing fiber dimensions are comparable to those of the base network fibers in Figure 11.1(b).

A network may be reinforced by the interaction with another network that spans the same problem domain. Interpenetrating networks exhibit interesting properties whose physical origins are not entirely understood at this time. Two networks that span the same problem domain (Figure 11.1(c)) may be crosslinked to each other or may just be entangled as such to interact topologically during deformation. Results pertaining to these systems are discussed in Section 11.3.

A network embedded in a continuum matrix (Figure 11.1(d)) behaves differently from the equivalent dry network (without matrix) and the differences are outlined in Section 11.4. This configuration may be viewed as a limit case of the interpenetrating networks problem in which the two networks have significantly different characteristic length scales, that is, $l_{c1} \gg l_{c2}$, a case in which the network with the smaller mean segment length behaves as a continuum on the scale of the network with larger l_c.

The prediction of the effective properties of classical composites is a mature field and the reinforcement mechanisms are well understood. This cannot be said about network materials in which filler dimensions are comparable with the network length scales; such composites remain a subject of investigation. The discussion in the subsequent sections of this chapter summarizes the present understanding in this area and sets the stage for further developments. This addresses a number of current needs. While the reinforcement of elastomers and epoxy with nanoscale fillers has been studied extensively, effective reinforcement of gels and biomaterials is still to be achieved. Most single network gels, collagen constructs (e.g., reconstituted collagen), and electrospun nonwovens used as scaffolds for tissue growth have mechanical properties inferior to those of the tissues they are intended to replace. In a more general sense, being able to effectively reinforce networks is desirable since it aids in controlling the composite network properties in a broad range.

11.1   Networks Reinforced with Particles

The reinforcement of network materials by the addition of particles was studied from the early days of the rubber industry. One of the most successful such developments has been the reinforcement of elastomers with carbon black and with silica, which led to industrially viable products. An important lesson learned from these works is that interesting effects result at sufficiently large filler volume fractions, f_r, and when filler dimensions are sub-micrometric.

More recently, efforts have been made to reinforce gels with nanoparticles. The results of these works are not particularly encouraging when the filling fraction is kept low and moderate.

Most nanocomposites of this type are based on thermal networks. Very few attempts have been made to reinforce athermal networks with particles. However, interest in this area is emerging, particularly driven by biomedical applications of athermal networks such as reconstituted collagen.

A qualitative understanding of the effect of fillers on network mechanics can be obtained by considering the length scales involved. The two length scales d_r and l_c characterize individual fillers and the network, respectively. The filler volume fraction is related to another important length scale – the mean wall-to-wall distance between fillers, l_{ww} – which may be evaluated using:

$$\frac{l_{ww}}{d_r} = \left(\frac{\pi}{6f_r}\right)^{1/3} - 1. \tag{11.1}$$

The filling fraction at which the mean wall-to-wall distance becomes equal to the filler diameter is 6.5%. Keeping the volume fraction below 1%, as is usually done in nanocomposites in order to ensure miscibility, implies that $l_{ww} > 2.7d_r$. In addition to d_r, l_c, and l_{ww}, one must consider length scales that emerge from the interaction of fillers and matrix, as discussed in this section.

As expected, based on continuum homogenization theory, the properties of the filled network should depend decisively on the difference between the mechanical properties of fillers and matrix. Soft network materials are usually reinforced with much stiffer particles. The optimal contrast between the properties of the filler and the properties of the matrix required for reinforcement depends on how the network behaves. Strong reinforcement may be obtained in a network with large free volume, which undergoes large volume changes during deformation, with fillers which are incompressible, but not rigid, for example, with liquid-filled vesicles. However, in a volume preserving networks, such as rubber, fillers which are merely incompressible should have limited effect.

11.1.1   Conditions Required for Reinforcement

Three conditions considered essential for an effective reinforcement of network materials are:

(a) If the network is thermal, it must be above the glass transition temperature, T_g. Below T_g, the strong and dense nonbonded interactions between molecular strands effectively screen the interaction of fillers. The energetic interactions of molecules at sites which are not crosslinks become long-lived and perform mechanically in a manner similar to the network crosslinks. As the temperature decreases from above to below T_g, the effective mean segment length, l_c, decreases, becoming comparable with the Kuhn segment length, and, consequently, the ratio of the filler diameter to the effective segment length, d_r/l_c, increases. In addition, the network largely loses its thermal character and its deformation is controlled by the elasticity of the nonbonded interactions and the axial deformation mode of network filaments. Since below T_g the effective crosslink density is large, the network deforms affinely. Hence, below T_g, the material becomes similar to a regular composite with a continuum matrix and the degree of reinforcement is limited to that predicted by homogenization theory. Maintaining the temperature above T_g is required in order to retain the mechanical signature of the underlying thermal network.

(b) The network crosslink density should be kept relatively low. This requirement is mechanistically similar to that discussed in (a). It ensures that l_c is large and the network has sufficient kinematic freedom to preserve its signature mechanical behavior. However, decreasing the crosslink density implies a reduction of the modulus and strength, which goes against the objective of reinforcement. Reducing the crosslink density is beneficial only as far as the increased mobility of network filaments enables other reinforcement mechanisms, as for example those discussed in Section 11.3 in the context of interpenetrating networks.

(c) Fillers should be well bonded to the network. A strong filler-matrix interface is required for proper load transfer. This was recognized in the early literature on reinforced elastomers where fillers were divided into "reinforcing" and "non-reinforcing" functions of the state of their interface with the rubber. A strong filler-matrix interface also promotes the formation of an "interphase," that is, a region of the matrix which is subjected to the constraining mechanical effect of the stiffer filler and has different mechanical behavior than the bulk matrix.

11.1.2 Mechanisms of Reinforcement

11.1.2.1 The Reference Case

The baseline reinforcement effect is the prediction of homogenization theory. This description considers continuum behaviors for both filler and matrix and does not account for the interplay of length scales associated with the presence of fillers in a discrete matrix. The theory is adequate for situations in which the filler size is much larger than any internal length scale of the matrix, which is the case with most of the current network materials-based composites. It also does not apply to situations in which the network undergoes large volume changes during deformation (Figures 6.29 and 6.31).

Homogenization results obtained within linear elasticity (Hashin and Shtrikman, 1961; Horii and Nemat-Nasser, 1993) apply to networks of type A and B (Figure 6.1) when assuming linear elastic behavior for the network and reinforcement, and a network Poisson ratio smaller than or equal to 0.5. Bounds for the behavior of composite materials with hyperelastic matrix, relevant for filled type C networks, are presented in Ponte Castaneda (1989). Theoretical and numerical results for filled

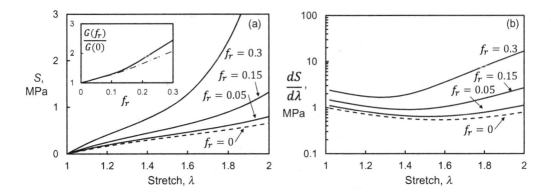

Figure 11.2 (a) Nominal stress–stretch curves predicted by the homogenization theory in Lopez-Pamies et al. (2013) for composites of a hyperelastic matrix and rigid spherical particles. Curves correspond to the unfilled matrix ($f_r = 0$) and to three non-zero values of the filler volume fraction, f_r. The inset shows the small strain shear modulus of the composite (continuous line) normalized by the shear modulus of the unfilled matrix, versus f_r. The dash-dot line represents the linear elastic Hashin–Shtrikman lower bound. (b) Data in (a) replotted as tangent stiffness versus stretch.

incompressible elastomeric continua are presented in Lopez-Pamies et al. (2013) and Avazmohammadi and Ponte Castaneda (2014).

To demonstrate this baseline reinforcement effect, Figure 11.2 shows the response to uniaxial tension of an incompressible elastomer filled with rigid spherical particles at several volume fractions, as predicted by the homogenization theory developed in Lopez-Pamies et al. (2013). The matrix is incompressible and hyperelastic with constitutive behavior fitted to that of a silicone elastomer. Figure 11.2(a) shows nominal stress–stretch curves for composites with 5%, 15%, and 30% filler volume fraction and for the unfilled matrix. The inset to Figure 11.2(a) shows the dependence on f_r of the composite small strain shear modulus, normalized by the shear modulus of the matrix. The inset includes the lower bound of the composite modulus predicted with the linear elastic Hashin–Shtrikman theory. The shear modulus (continuous line) follows the lower bound for $f_r < 0.15$ and increases slightly above this limit for larger volume fractions. Overall, the increase of stiffness due to the presence of rigid fillers is moderate.

The curves in Figure 11.2(a) are replotted in Figure 11.2(b) as tangent stiffness versus stretch. It becomes apparent that the filled networks enter the strain stiffening regime at smaller stretches than the unfilled matrix. This effect is confirmed by experiments with filled elastomers.

Several other mechanisms have been proposed in the literature to explain effects that go beyond those captured by homogenization theory. These were introduced in the literature on filled elastomers in the 1960–1980 period and are summarized in review articles, such as Payne (1974), Wagner (1976), and Medalia (1978). These concepts have been revisited and applied to polymer-based nanocomposites in the early 2000s. A brief review is presented next.

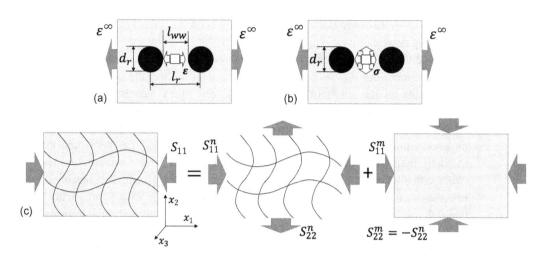

Figure 11.3 Schematics showing (a) strain amplification due to the presence of rigid inclusions, (b) development of a multiaxial strain state in the network volume confined by rigid inclusions, and (c) development of a tensile biaxial stress state in the $x_2 - x_3$ plane in an unfilled network embedded in a volume preserving matrix subjected to compression in the direction x_1.

11.1.2.2 Strain Amplification

The concept of strain amplification is shown schematically in Figure 11.3(a) for uniaxial loading. It implies that the local strain experienced by the matrix, ε, is larger than that applied on the scale of the composite, ε^∞, simply because fillers are rigid. This effect is captured by continuum homogenization models and was described in the early rubber literature (see e.g., Mullins and Tobin, 1965).

An upper bound prediction is obtained by assuming that the two phases are connected in series. In this approximation, the applied relative displacement corresponding to ε^∞ and length l_r, that is, $\varepsilon^\infty l_r$, results exclusively from matrix deformation over the length l_{ww}. The local strain results $\varepsilon = \varepsilon^\infty / \left(1 - (6f_r/\pi)^{1/3}\right)$, which is a rapidly increasing function of f_r. An alternative mean field description of the same effect is based on the idea that the average strain over the composite volume must be ε^∞, while the strain within fillers vanishes. This leads to the rule of mixtures equation for the strain, which implies that $\varepsilon = \varepsilon^\infty / (1 - f_r)$. The difference between these two predictions is large. However, both indicate that the local strain must be larger than the applied strain.

Strain amplification does not contribute to stiffening if the constitutive behavior of the matrix is linear. In such cases, the matrix stiffness is independent of the applied strain and no stiffening results from the fact that the local strain is larger than the applied, global strain. However, if the matrix has hyperelastic behavior, the stiffness increases with the strain and, hence, an amplification of the local strain implies an amplification of the composite stiffness. The faster the matrix strain stiffens, the more pronounced is the reinforcement amplification. This effect is accounted for by the homogenization theory leading to the results in Figure 11.2, provided the continuum constitutive model used for the matrix reproduces the multiaxial behavior of the underlying network.

An additional effect appears when the wall-to-wall distance becomes comparable to the filler dimensions, $l_{ww} \approx d_r$. If fillers are much stiffer than the network, the degree of triaxiality of the strain in the region between fillers increases. This is shown schematically in Figure 11.3(b). The network confined between fillers is constrained to deform triaxially, even for uniaxial far fields. As discussed in Section 6.1.3, networks enter the strain stiffening regime at much smaller stretches when loaded triaxially than when subjected to uniaxial loading. This effect increases the rate of strain stiffening of the composite and contributes to reinforcement. This mechanism is expected to make a strong contribution to the reinforcement in networks with large free volume, which are able to undergo large volumetric changes during deformation. A limited impact is expected in networks constrained to deform isochorically, for example, due to the presence of a volume-preserving matrix (as in hydrogels) or to the absence of free volume (elastomers).

Networks with large free volume not embedded in a volume preserving matrix generally soften in compression (Figure 6.43). Also, their Poisson effect is weak in compression (Figure 6.46). Therefore, both effects described here are expected to be weak in compression in such composites.

However, if the network is constrained to preserve its initial volume due to the presence of an isochorically-deforming matrix, a biaxial stretch state develops during uniaxial compression in the plane normal to the compressive load. With λ being the stretch in the compression direction, the transverse stretches emerging from the incompressibility condition are $1/\sqrt{\lambda}$. The network without a matrix has low Poisson ratio in compression and does not expand much in the plane orthogonal to the compression direction; hence the matrix effectively loads the network in tension in the respective directions (Figure 11.3(c)). The network strain stiffens rapidly under biaxial tension (Figure 6.53). This effect occurs with or without inclusions. In the presence of rigid fillers, it is amplified, particularly at large filler volume fractions close to the filler jamming density. An example of this type is discussed in van Oosten et al. (2019). Note that if fillers are merely incompressible, but have low shear modulus, as for example in the case of vesicles, no amplification is expected due to the presence of inclusions in the already volume preserving network.

11.1.2.3 Filler Structures

Strong enhancement of mechanical properties results in elastomers upon loading at high filler volume fractions. This is generically attributed to the formation of filler structures. Several types of filler configurations may be envisioned:

(i)   Fillers form a percolated network. In this case, the volume fraction f_r is above the percolation threshold, but below the jamming limit. Fillers are in direct contact and form a (possibly fractal) structure that spans the problem domain (Figure 11.4(a)). Such "chains" of fillers are able to carry load and hence produce significant reinforcement.

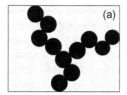 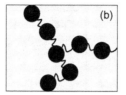 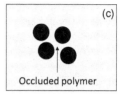

Figure 11.4 Schematic representation of filler structures: (a) Branched (possibly fractal) filler structure; (b) percolated filler structure in which fillers are not necessarily in contact, but are sufficiently close to be connected by molecular strands; and (c) matrix confined by fillers (occluded polymer).

(ii) Fillers separated by small wall-to-wall distances form a percolated network. This configuration is similar to that shown in Figure 11.4(a), except that the average wall-to-wall distance between fillers forming a percolated path does not vanish and neighboring particles are not necessarily in direct contact. However, neighboring fillers are connected by network segments of length close to the mean segment lengths (Figure 11.4(b)). Such short connectors transmit loads effectively.

(iii) Clusters of fillers embed a substantial fraction of the matrix. Since particles within clusters are not close packed, each cluster incorporates a certain elastomer volume. This matrix, also referred to in the elastomers literature as "occluded volume," is screened from the far field since it is surrounded by stiff particles. Consequently, each such cluster behaves as an effective inclusion of stiffness smaller than that of fillers, but larger than the stiffness of the matrix, and of volume much larger than that of individual particles (Figure 11.4(c)).

Ample evidence of the existence of filler structures in various network materials was accumulated over the years. Direct microscopic evidence of carbon black structures in rubbers is presented in Oono (1978). Structures form at filler volume fractions between 15% and 25% of carbon black in SBR rubber. The effect of such structures is observed in rheological experiments as an increase of the plateau modulus, G', with increasing f_r, at small strain amplitudes. In Aranguren et al. (1994), who work with silica-filled PDMS, the effect of filler structures is visible at $f_r = 0.15$, but it is not observed at $f_r = 0.08$.

Filler structures are destroyed as the applied strain amplitude increases, which leads to softening at high amplitudes. This effect was summarized by Payne and co-workers (Payne and Whittaker, 1971; Payne, 1974).

The structures of type (ii) require further discussion. These situations may be described in a broader sense in terms of the interaction of the characteristic length scales of fillers and network. Strong reinforcement is expected once the characteristic length scale of the filler distribution, l_{ww}, becomes comparable with the characteristic length of the network, l_c, or, in the case of molecular networks, the gyration radius of the elastomer molecular chains, R_g. If filaments are strongly bonded to fillers, filler structures of type (ii) in which individual molecules connect neighboring fillers develop.

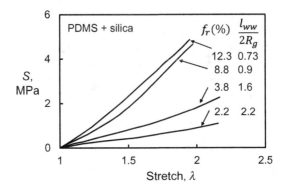

Figure 11.5 Nominal stress–stretch curves for PDMS reinforced with precipitated nano-silica at various volume fractions. The mean wall-to-wall distance normalized by the chain radius of gyration is shown for each f_r. Data from Yuan and Mark (1999)

This effect is demonstrated in Figure 11.5, which shows nominal stress–stretch curves obtained in uniaxial tensile tests performed with PDMS filled with silica nanoparticles (Yuan and Mark, 1999). In this material, silica is not blended in the polymer before crosslinking, as most nanocomposite processing methods require, but rather it is allowed to precipitate in-situ. This creates small silica fillers with a much more uniform spatial distribution than that obtained by blending. The curves indicate strong increases of the small strain modulus, E_0, and of the strain stiffening rate. More remarkable is the rapid increase of the reinforcement effect observed once filler structures develop. The values of the volume fraction and of the ratio of the mean wall-to-wall distance, l_{ww}, to the polymer gyration radius corresponding to each curve are shown in Figure 11.5. Once $l_{ww}/2R_g$ decreases to approximately 1, and conditions for the formation of a percolated filler structure are fulfilled, a step increase of the stiffness is observed. It should be noted that such an obvious jump is only exceptionally observed and it is likely due in this case to the good dispersion obtained by the precipitation method used to produce the composite. The reinforcement effect obtained in these materials goes significantly beyond the reference prediction of homogenization theory (Figure 11.2).

The increase of strain stiffening in filled networks belonging to the type (ii) composites defined in this section is generally attributed to the non-Gaussian behavior of the chains. Within the framework generally used for the mechanics of molecular networks, it is assumed that chains are loaded axially, in tension, and the network undergoes an approximately affine deformation. Then, the chain-scale behavior reflects directly in the behavior of the network (see Section 5.3.2). With linear elastic chains, network nonlinearity is entirely geometric and emerges exclusively from the reorientation of filaments during loading. However, if the chains behave nonlinearly (e.g., Langevin chains, Section 2.4.1.2), the network-scale nonlinearity has a constitutive component related to the nonlinearity of the chains, and a geometric component associated with chain reorientation. Therefore, this model implies that using shorter

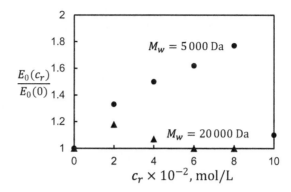

Figure 11.6 Normalized small strain stiffness versus clay concentration in two tetra-PEG-clay composites with different molecular weights of chain segments between crosslinks. The vertical axis is normalized with the small strain stiffness of the corresponding unfilled network. Data from Fukasawa et al. (2010)

chains, which have a more pronounced nonlinear response, leads to faster stiffening both in unfilled and filled networks.

Whether this physical picture holds or not is less of a concern from the standpoint of applications. The fact is that networks with shorter molecular strands are reinforced more effectively than networks of long molecules. This is reported in the early literature on carbon black-reinforced elastomers (Voet, 1980) and further reported in more recent works. This effect is also observed in tetra-PEG networks reinforced with exfoliated clay (Fukasawa et al., 2010). Tetra-PEG form regular networks with monodisperse molecular strands whose molecular weight is closely controlled. These networks are different from those obtained by vulcanization of rubber, in which molecular segments between crosslinks have polydisperse lengths. Figure 11.6 shows the increase of the small strain stiffness obtained upon the addition of clay relative to the stiffness of the unfilled network, function of filler concentration, c_r. The variable on the horizontal axis is proportional to f_r. Two molecular weights are considered: 5 kDa and 20 kDa. It is observed that the reinforcement obtained at the same filler concentration is more pronounced for the network of smaller molecular weight.

Other effects emerged recently from studies of networks with controlled architecture. An example is shown in Figure 11.7 and refers to the same tetra-PEG system discussed in relation to Figure 11.6 (Fukasawa et al., 2010). This data set indicates the effect of network density on reinforcement and on network strength. While in most other networks with random crosslinking increasing the density implies a reduction of l_c, in this type of network the strand length between crosslinks is pre-defined and is independent of the polymer concentration of the gel. Increasing the polymer concentration leads to the increase of the crosslink number density which, according to Eq. (8.2), increases the network strength. This is observed in Figure 11.7 for the unfilled tetra-PEG network; see also Figure 8.10(c). The addition of exfoliated clay does not increase the small strain stiffness significantly, but almost doubles the strength and stretch at failure of all networks considered. The clay particles stabilize

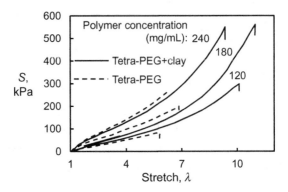

Figure 11.7 Nominal stress–stretch curves for unfilled tetra-PEG gels of three different concentrations, and for the same gels filled with exfoliated clay, demonstrating significant increases of strength and ductility upon the addition of fillers. Data from Fukasawa et al. (2010)

the deformation and favor strain stiffening, but the details of the mechanism behind this exceptional effect are not entirely understood at this time.

It should be pointed out that simultaneous increases of strength and ductility upon filling were also observed in carbon black-filled elastomers (Medalia and Kraus, 1994). The increase of the strength can be explained based on theories reviewed in this chapter (including the formation of a network of fillers and the presence of interphases), but these concepts do not explain the increased ductility.

11.1.2.4 Interphases

It was recognized early in the development of polymer-based composites that the presence of hard inclusions modifies the physical properties of the polymer located in the close vicinity of fillers. Two situations may be envisioned. If the polymer–filler interaction is repulsive, the interface is weak. Since the filler volume is inaccessible to the polymers, an entropic force develops which retracts the chains from the interface, causing the formation of a low-density region in the vicinity of fillers. This entropic effect occurs because the filler restricts the number of configurations a polymer may take in its vicinity, and hence polymers located close to the interface have lower entropy than those located away from the interface. In the opposite case in which the polymer–filler interaction is strongly attractive, polymers are packed against the interface, and this leads to ordering and the formation of alternate dense and less dense layers of polymers close to the filler surface. This denser surface layer also effectively tethers the chains to the filler surface, causing the formation of a polymeric brush around each filler. Therefore, in both cases described, although the polymer in the vicinity of fillers is chemically identical to that far from fillers, it has different mobility and, hence, different mechanical properties. This physically modified polymer layer is called "interphase."

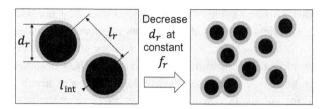

Figure 11.8 Schematic representation of reinforced networks with filler-matrix interphases. As the filler diameter decreases, while the filling volume fraction is kept constant, the volume fraction occupied by interphases increases.

The presence of interphases was postulated in the early literature on carbon black-filled elastomers. The concept was used in the newer literature on nanocomposites with a thermoplastic (non-crosslinked) matrix, where the thickness and mechanical properties of interphases were studied numerically and experimentally.

The importance of interphases in nanocomposites emerges from the observation that, as filler dimensions decrease, the volume fraction of the interphase increases. Referring to the schematic in Figure 11.8 and considering l_{int} to be the thickness of the interphase, the volume fraction of the interphase material is $f_{int} = f_r(6l_{int}/d_r)$, where $f_r = (\pi/6)(d_r/l_r)^3$ is the filler volume fraction. If fillers are large, l_{int}/d_r is small and f_{int} is negligible. However, when filler dimensions are small and d_r is comparable with l_{int}, f_{int} may become larger than f_r, and the composite properties are defined by the interphase properties.

An additional effect resulting from the presence of interphases is related to percolation (Figure 11.4(b)). Percolation of the reinforcement phase in the composite takes place at a certain volume fraction of the respective phase. In the presence of interphases, it is possible to compute the filling fraction corresponding to interphase percolation. The result depends on l_{int}. The joint volume fraction of the interphases and fillers can be computed (for the spherical filler geometry) as $f_{r+int} = f_r(1 + 6l_{int}/d_r)$. The amplification factor $(1 + 6l_{int}/d_r)$ increases with l_{int}/d_r and, hence, percolation of interphases takes place at relatively small f_r.

The importance of this discussion hinges on the interphase thickness l_{int}. In both crosslinked and un-crosslinked polymeric matrices, l_{int} is related to a length scale of the matrix. In thermoplastics, l_{int} scales with the chain gyration radius, while, in thermosets, it scales with the characteristic length scale of the network, for example, average molecular weight of network segments. The average wall-to-wall distance between fillers and l_{int} are controlled independently via f_r and the characteristic length scale of the network, respectively. Rendering these two lengths closer to each other increases the contribution of the interphase to the composite behavior. Further, creating reinforcing interphases requires attractive filler–polymer interactions. The stronger these interactions are, and the longer the strands tethered to the filler surface, the thicker is the interphase.

Interphases have been evidenced so far primarily in thermal networks. An example of an interphase occurring in athermal networks is presented in Islam and Picu (2019).

In this work, an athermal, nonaffine, and bending dominated network is reinforced with spherical fillers much stiffer than the matrix and having a diameter comparable with the mean segment length of the network, $d_r \approx l_c$. Fibers are bonded rigidly to the filler surface, resulting in a strong interface. It is observed that the network in the vicinity of fillers becomes axially-dominated, and hence it shifts to a stiffer local deformation mode (Figure 6.6). The effective properties of these interphases are modified by the mechanical constraint imposed by the rigid filler.

The emergence of axially dominated interphases produces significant network-scale stiffening, even at small filler volume fractions. Interestingly, the stiffening effect depends on network properties. Obviously, the effect described here is absent in a dense network which is axially dominated and deforms affinely. The mechanism applies to athermal networks which deform nonaffinely and store most of the strain energy in the bending deformation mode of fibers. The presence of interphases does not modify the functional form of strain stiffening and, hence, leaves unchanged the nonlinear aspects of network deformation.

Figure 11.9(b) shows the reinforcement effect for networks of the same density (hence, of the same l_c) filled with rigid spherical fillers (Figure 11.9(a)) having $d_r = l_c$. The vertical axis shows the ratio of the small strain stiffness of the filled and unfilled networks, while the horizontal axis represents the fiber aspect ratio (note that $l_b/L_0 \sim d/L_0$). It is observed that the affine networks, which deform predominantly in their axial mode (Figure 6.6), are weakly sensitive to the presence of fillers. In fact, the level of reinforcement at large l_b/L_0 is compatible with that shown in Figure 11.2(a) and predicted by homogenization theory. The more interesting case is that of the

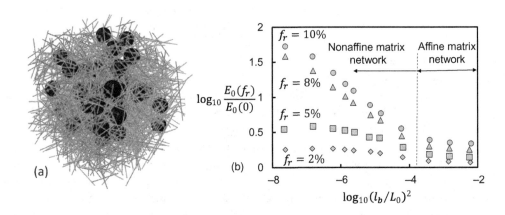

Figure 11.9 Effect of reinforcing particles on athermal networks: (a) Configuration showing a fibrous network reinforced with spherical fillers of $d_r = l_c$, and (b) reinforcement effect function of the fiber aspect ratio, $l_b/L_0 \sim d/L_0$. The vertical axis shows the small strain modulus of filled networks normalized by the small strain modulus of the equivalent unfilled networks. Results are shown for several filling fractions, f_r. Soft, nonaffine networks can be reinforced effectively due to the formation of interphases within which the deformation mode shifts from bending to axial. Reproduced with permission from Islam and Picu (2019), Copyright (2019) by the American Physical Society

nonaffine networks made from thin and bendy fibers (small l_b/L_0). Stiffening of more than one order of magnitude is predicted at small f_r in nonaffine networks due to the formation of interphases.

11.1.2.5 Other Mechanisms

Several other mechanisms influence the mechanics of reinforced networks. When filler dimensions are comparable with the mean segment length of the network, $d_r \approx l_c$, and the particle–matrix interface is strong, it may be envisioned that fillers act as effective crosslinks for the network. Then, the filled network may be mapped to an equivalent unfilled network in which a subset of the crosslinks has high connectivity. This increases the mean connectivity of the network, which may lead to a shift of the network behavior from nonaffine to affine, which implies stiffening (see Section 6.1.1.3.4). This mechanism was explored in Islam and Picu (2019) and it was concluded that the increase of the mean network connectivity associated with the presence of reinforcing particles is modest, since the number density of fillers is much smaller than the number density of regular, low connectivity network crosslinks. It results that the effectiveness of this reinforcement mechanism is limited. However, the possibility of controlling the mechanical behavior of unfilled networks by using multiple types of crosslinks, with different functionality and hence different z, remains valid. The viability of this mechanism of network reinforcement awaits adequate assessment.

Jamming of reinforcing particles may take place as the filling fraction increases. The strict jamming threshold for monodisperse spheres is reached at $f_r = 0.64$ (Donev et al., 2004). In the presence of interphases, the jamming limit should be evaluated based on $f_{r+\mathrm{int}}$. Given that $f_{r+\mathrm{int}} = f_r(1 + 6l_{\mathrm{int}}/d_r)$, as discussed earlier in this section, and considering that the amplification factor $(1 + 6l_{\mathrm{int}}/d_r)$ may take large values, the jamming threshold may be reached in these networks at filling fractions significantly below 64%. In the jamming limit, the response of the material is expected to be much stiffer as it is controlled by direct filler–filler or interphase–interphase interactions.

Manufacturing composites by filler–matrix blending at filling fractions anywhere close to the filler jamming threshold is, obviously, impossible. This mechanism becomes important if particles are created by precipitation after or during network crosslinking. The stiffening mechanism based on filler jamming may be activated when the material is loaded in shear with a superimposed compressive stress. In this case, a rapid increase of the tangent stiffness measured in shear is expected. This trend was observed in several recent works (van Oosten et al., 2019; Shivers et al., 2020).

Reinforcement associated with the presence of particles is observed if the filler–matrix interface is strong. Weak filler–matrix interfaces may be used to advantage in some cases, particularly to increase the composite toughness. Toughening results in this case from cavitation. Weak interfaces allow easy debonding and, once this takes place and voids nucleate at fillers, the effective local behavior becomes similar to that of a porous matrix. If the matrix accommodates large deformations, voids grow and may reach the cavitation instability. As discussed in Section 8.4.1.3, local network rupture or plastic deformation are required to sustain cavitation.

This mechanism is demonstrated in an epoxy network reinforced with silica nanoparticles in Picu et al. (2019). In this work, the state of the filler–matrix interface is controlled and adjusted from weak to strong by exposure to ultraviolet radiation. The composite toughness is substantially larger in the weak interface case compared with the strong interface case. The key conditions for achieving this effect are: (i) Filler diameter must be in the range 100–400 nm. Fillers of smaller diameter (below 100 nm) produce voids whose growth is prevented by the surface tension effect. (ii) The embedding network must sustain plastic deformation. This is challenging in densely crosslinked networks below the glass transition temperature. Plasticity may be enhanced by the addition of plasticizers, by increasing the mean segment length of the network, and by proximity to T_g. It is important to note that only small scale, local plasticity is needed for this toughening mechanism to operate. (iii) The mechanism is enhanced if l_{ww} is reduced.

11.2   Networks Reinforced with Staple Fibers

Fibers of finite length are widely used to reinforce thermoplastic and thermoset polymers. Reinforcing fibers are usually large relative to the length scales of the network matrix. Such situations belong to the broad class of classical composites.

As in Section 11.1, we focus here on situations in which the dimensions of the reinforcing fibers are not very different from those of the base network fibers (Figure 11.1(b)). Further, in general, one is interested in modifying the properties of the base network by adding a small number of reinforcing fibers. This requires that the mechanical properties of the reinforcement should be significantly different from those of the base network fibers. To render the narrative more fluid, the base network fibers are called in this section "B fibers," while the reinforcement fibers are referred to as "R fibers." As in Section 11.1, an "r" subscript refers to the R fibers.

11.2.1   Conditions Required for Reinforcement

Several conditions need to be fulfilled for effective reinforcement:

(i)   The R and B fiber cross-sectional dimensions should be comparable. This implies that the diameter of the R fibers, d_r, is comparable with d, or l_c. A contrasting example is that of continuous fiber composites with an epoxy matrix in which the fiber diameter is several orders of magnitude larger than the mean segment length of the epoxy network.

(ii) The properties of R fibers should be significantly different from those of B fibers. In athermal networks, if the B network is bending dominated, the R fibers should have large bending stiffness. For the same reason, in thermal networks R filaments should have a much larger persistence length than B filaments.

(iii) The R fibers should be crosslinked with the base network. In addition, the R fibers may or may not be crosslinked with each other. If R fibers establish crosslinks with R fibers, a percolated R network may or may not result – as a function of the density of R fibers. Any percolated network of R fibers is much stiffer than the B network and, if such a structure develops, it becomes dominant and controls the mechanical behavior of the composite. The more interesting situation is that in which the density of the R fibers is smaller than the percolation threshold for the R network. Exceptional properties may result in this case, whether the R fibers establish crosslinks with R fibers or not, as discussed in the remainder of this section. These considerations also apply to cases in which R and B are entangled and not crosslinked. If the density of R filaments is small, R will not become entangled with R, although R may be entangled with B.

(iv) The reinforcement fibers should be evenly distributed in the B network, should not phase separate, and should not form bundles.

Since the percolation of reinforcement fibers is important in this problem, it is necessary to recall results presented in Chapter 4 related to the conditions in which fibers of finite length percolate. In 2D, the critical area fraction at percolation of ribbon-like fibers, f_{cr}, is given by Eq. (4.39): $f_{cr}L_{0r}/d_r = 5.637$, where L_{0r}/d_r is the aspect ratio of the R fibers. With ρ_r and ρ being the densities (total length of fiber per unit area) of R and B fibers, respectively, and using Eq. (4.6), the condition that reinforcement fibers do not form a percolated structure becomes $\rho_r < \rho_{rc} = 3.58\rho/(L_{0r}/l_c)$. This introduces the ratio of the R fiber length to the mean segment length of the base network, L_{0r}/l_c, as an important parameter of these systems. Since this ratio may be large, ρ_r is expected to be significantly lower than ρ, which is also desirable given that, generally, reinforcement is to be achieved with the smallest number of reinforcing fibers possible.

The 3D percolation condition is given by Eq. (4.41), $f_{cr}L_{0r}/d_r = 0.705$, which applies to straight cylindrical R fibers. With $\rho l_c^2 = \beta$ (β is of order 1, Eq. (4.14)), the condition of no percolation of the reinforcement fibers becomes $\rho_r < \rho_{rc} = (0.89/\beta)\rho(l_c/d_r)(l_c/L_{0r})$.

11.2.2 Mechanism of Reinforcement

Two mechanisms provide reinforcement in networks that fulfill conditions (i) to (iv) listed in Section 11.2.1: The formation of a percolated network of R fibers, and the occurrence of interphases. The second mechanism is less obvious than the first and requires further discussion. Both mechanisms are demonstrated next based on data from Bai et al. (2011) and Shahsavari and Picu (2015).

Consider a 2D crosslinked Mikado B network of parameters ρ and l_b, reinforced by R fibers of bending stiffness at least two orders of magnitude larger than the B fibers, and of density ρ_r. For simplicity, the length of the reinforcing fibers is taken equal to that of the B fibers. The fiber length and fiber modulus are considered to be the units of length and stress in this problem. The reinforcement fibers are randomly distributed and oriented and are crosslinked to the base network. Two cases are distinguished, in which R fibers (1) are, and (2) are not crosslinked to R fibers, which are denoted by RX and RNX, respectively.

Figure 11.10 shows the small strain modulus of reinforced networks of given ρ and various l_b, as a function of ρ_r normalized by the percolation density of the R fibers, ρ_{rc} (Shahsavari and Picu, 2015). In all cases reported here, $\rho = 5\rho_{rc}$. The stiffness of the base network without reinforcement scales with $I_f \sim d^4 \sim l_b^4$, and, hence, E_0 varies in a broad range when $\rho_r = 0$.

When the reinforcing fibers are crosslinked to both R and B fibers (RX), the variation of E_0 with ρ_r/ρ_{rc} exhibits two stages of rapid increase. The first of these occurs below percolation, at $\rho_r/\rho_{rc} \approx 0.3$, while the second corresponds to percolation of the R network and occurs in the vicinity of $\rho_r/\rho_{rc} = 1$. For values of ρ_r larger than percolation, all curves converge to that representing the stiffness of the R network, which becomes dominant over the B network.

When reinforcing fibers are not crosslinked to each other (RNX), but are crosslinked to the base network fibers, only the first stage of rapid increase of E_0 is visible. The reinforcement effect decreases as l_b increases and the base network becomes less nonaffine. This indicates that soft nonaffine networks can be reinforced more effectively by the addition of stiff staple fibers than the stiffer, close to affine networks.

The particularly interesting effect shown in Figure 11.10 is the occurrence of the first stage of a rapid increase of E_0 at R fiber densities well below the percolation of

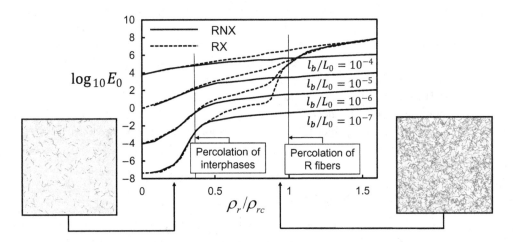

Figure 11.10 Variation of the small strain modulus of a 2D Mikado network, E_0, reinforced with stiff staple fibers function of the density of the reinforcing fibers, ρ_r. The horizontal axis is normalized by the percolation threshold of the reinforcing fibers, ρ_{rc}. Curves corresponding to base networks of the same density and increasing l_b/L_0 are shown. The RX case, in which the R fibers establish crosslinks with both B and R fibers, is shown with a dashed line, while the opposite case in which the R fibers are crosslinked only with B fibers is shown with a continuous line. The two vertical lines indicate percolation of the R network and of interphases, respectively. Adapted with permission from Shahsavari and Picu (2015), Copyright (2019) by the American Physical Society

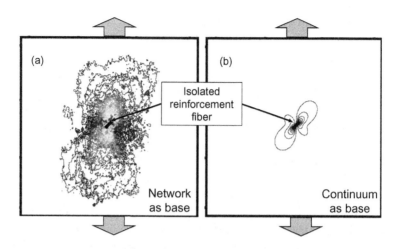

Figure 11.11 Difference between the displacement in the vertical direction (far field loading direction) in the presence of a single reinforcing stiff fiber, and without the reinforcing fiber for: (a) The case in which the reinforcing fiber is crosslinked to a soft base network of $l_b/L_0 = 10^{-6}$, and (b) the case in which the base network is replaced with a linear elastic continuum having elastic constants identical to those of the base network in (a). The analysis is performed in the linear range. The contours in both figures indicate the same values of the perturbation field. Reprinted with permission from Shahsavari and Picu (2015), Copyright (2015) by the American Physical Society

reinforcement, ρ_{rc}. The physical origin of this effect is the formation of interphases in the vicinity of the reinforcing fibers. A B fiber crosslinked to an R fiber is subjected to the kinematic restrictions emerging from the fact the R fibers are almost rigid and the crosslinks are of weld type (transmit both forces and moments). The kinematics of such B fibers is modified by the presence of the R fiber. The softer the B network, the more pronounced this modification, and the longer ranged is the respective effect (thicker interphase). The effect of the interphase is demonstrated in Figure 11.11, which shows the perturbation displacement field introduced in the base network by the presence of a single reinforcing fiber placed in the center of a network of B fibers. The perturbation field is computed as the difference between the displacement field (in the direction of the far-field load) with and without the reinforcing fiber. Two cases are presented: Figure 11.11(a) shows the perturbation displacement field for a soft base network with $l_b/L_0 = 10^{-6}$, while Figure 11.11(b) shows the perturbation field introduced by the same reinforcing fiber glued to a linear elastic continuum having the same elastic properties as the base network. The number of contours in Figure 11.11(a) and (b) is the same and the contours represent the same value of the field. The difference between the two images is striking. The perturbation in the network is much longer ranged than that in the continuum. While this is somewhat expected since the soft base network is reinforced on scales at which it does not behave like a continuum, the observation implies that reinforcing fibers added to soft networks interact at large distances. Therefore, the rapid increase of the stiffness observed at low values of ρ_r/ρ_{rc} in Figure 11.10 is due to the percolation of interphases associated with R fibers.

The difference between the perturbation fields shown in Figure 11.11 decreases as the base network becomes more affine. This is in agreement with the reduction of the reinforcement effect in networks of increasing l_b/L_0 seen in Figure 11.10. This conclusion is also reached in Raisanen et al. (1997), who studied the reinforcement of densely crosslinked (and close to affine) paper with cellulose fibers of different properties. They observe that stiff reinforcing fibers increase the stress in neighboring fibers.

The nonlinear behavior of reinforced 3D networks of thermal semiflexible filaments is studied numerically in Huisman et al. (2010). They report that the presence of reinforcing fibers does not change the functional form of strain stiffening (of type C base networks), but leads to a reduction of the stretch marking the onset of strain stiffening, λ_{c1} (Figure 6.1). λ_{c1} decreases continuously as ρ_r increases. Likewise, Gavrilov et al. (2013) report that the strain at the onset of strain stiffening decreases with increasing the carbon nanotube loading fraction in an elastomeric matrix. They report a rapid increase of the stiffness as ρ_r and/or L_{0r} become sufficiently large, which in this case is due to the percolation of the reinforcing phase.

This effect is explored experimentally in Lin et al. (2011), who worked with F-actin gels reinforced with microtubules. The persistence length of F-actin is about 17 μm, while that of the microtubules is about 5 mm, and hence the stiffness difference between the two types of filaments is large. The microtubules are not crosslinked to the F-actin network or to each other but have a diameter comparable with the characteristic length of the base network. The effect of the microtubules on the small strain stiffness of the composite is weak, which is likely due to the lack of crosslinking and limited load transfer from the base network to the reinforcing filaments. However, the presence of reinforcing filaments leads to increased strain stiffening relative to the unreinforced matrix.

Experimental works in which the four conditions of reinforcement outlined in this section are fulfilled are rarely reported in the literature. Much more often reported is the failure to reinforce networks at levels beyond those predicted for classical composites with continuum matrices. Three works that achieve reinforcement compatible with the theoretical results presented here are Zhang et al. (2010), Hassanzadeh et al. (2016), and Wang et al. (2017).

Wang et al. (2017) study chitosan gels reinforced with chitin nanowhiskers at several volume fractions, all below the percolation threshold of fillers. The whisker diameter is 14 nm, and the aspect ratio is approximately 15. The nanowires are hydrogen bonded to the base network, are athermal, and are much stiffer in both axial and bending modes relative to the molecular strands forming the gel. The reinforced gels exhibit the stress–strain curves shown in Figure 11.12 for filler volume fractions $f_r = 0.7\%$, 2.8%, and 3.5%. The strength increases by a factor of about 3 for this range of filler volume fractions. Even more interesting is the observation that the strain at failure doubles at the same time and hence gains in both strength and ductility result due to reinforcement.

A supramolecular copolymer gel (PEO-PPO-PEO) reinforced with cellulose nanowhiskers is studied in Zhang et al. (2010). The reinforcing fibers have a diameter of 15 nm and aspect ratio of approximately 15 and are well-bonded to the base network.

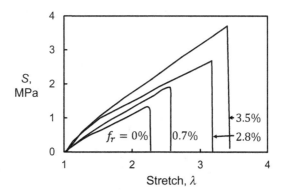

Figure 11.12 Stress–strain curves for chitosan-based gels reinforced with chitin nano-whiskers of $f_r = 0.7\%, 2.8\%$ and 3.5%. Data from Wang et al. (2017)

Their volume fraction is below the percolation threshold and hence no second network of nano-whiskers forms. A rapid increase of the modulus is reported, with an increase relative to the unfilled matrix of 3- and 50-times for filler volume fractions of 0.06% and 0.3%, respectively. Clustering at higher filling fractions leads to the reduction of the composite stiffness.

In Hassanzadeh et al. (2016), chitin nanofibers are used to reinforce gelatin. The gelatin is treated with methacrylic anhydride (GelMA) and chemically crosslinked by exposure to UV radiation. Chitin forms long fibers of diameter 10 nm which organize in a percolated and hydrogen bonded network. The gel concentration is 1 wt%, while the concentration of chitin fibers is varied in the range 1–3 wt%. The presence of the percolated chitin network has a strong effect on the mechanical properties of the composite: Both stiffness and strength of the gel reinforced with 1 wt% chitin increase by more than one order of magnitude relative to the unfilled gel, while the strain at failure remains unchanged. Further increasing the chitin concentration to 3 wt% causes a reduction of the strength and stiffness and the increase by a factor of 2 of the failure strain. This effect is likely due to bundling of chitin fibers, but direct observations of bundling are not reported.

The magnitude of these effects is impressive, but these examples are rather isolated. Further control of the network structure and of the nature of interactions between the base network and reinforcing fibers is needed before such results may be obtained systematically.

11.3   Interpenetrating Networks (IPN)

The alternative of reinforcing networks by the interaction with other networks defined on a similar length scale was studied intensely over the last decade due to a series of results indicating spectacular increases of strength and toughness in hydrogels. These network materials are referred to in the literature as "double networks." However, the

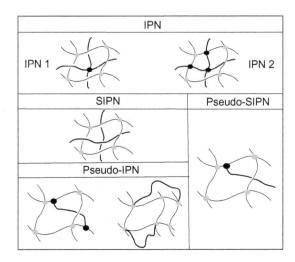

Figure 11.13 Schematic representation of various types of IPNs.

term "interpenetrating network" (IPN) is more encompassing and therefore is adopted here. It refers to situations in which multiple networks share and span the same spatial domain. Each individual network may be internally crosslinked and/or entangled. The interpenetrating networks may be inter-crosslinked or just entangled. The solvent used is generally a good solvent for the polymer species forming the two networks.

Figure 11.13 shows a classification of IPNs. While, in general, an arbitrary number of networks may co-exist, only two networks are shown in each schematic in order to preserve clarity. In the strict sense, IPNs are those in which each network is internally crosslinked and the networks forming the IPN interpenetrate and entangle (case IPN-1 in Figure 11.13). No inter-crosslinks are present in such structures. IPN-2 networks are those in which individual networks are internally crosslinked and are also inter-crosslinked. The case in which one network is internally crosslinked, while the other is only entangled, and inter-crosslinks do not form is referred to as semi-IPN (or SIPN). The term pseudo-IPN refers to situations in which filaments of a second network are grafted onto the primary network, which is itself internally crosslinked. These filaments may be long and entangled and may graft on the first network in either of the two configurations shown in Figure 11.13. An intermediate type denoted as pseudo-SIPN is discussed in the literature, in which the chains of the second network are grafted on the first network at one of their ends only, while the other end is free.

11.3.1 General Considerations

To set-up a conceptual framework to assist in the understanding of the behavior of IPNs, we discuss several idealized cases to which the results presented in Chapter 6 (pertaining to single networks) apply. Several situations may be identified for this purpose:

- Case 1: IPNs without inter-crosslinks in which the constituent networks taken separately deform affinely. In this case, the two networks are independent and are loaded in parallel. The IPN stress is the sum of the stresses produced by each component network. Therefore, the small strain stiffness results as the sum of the stiffnesses of the two constituent networks:

$$\mathbf{S}^{IPN} = \mathbf{S}^{(1)} + \mathbf{S}^{(2)}$$
$$E_0^{IPN} = E_0^{(1)}\left(\rho^{(1)}\right) + E_0^{(2)}\left(\rho^{(2)}\right) \cdot \tag{11.2}$$

  The superposition of Eq. (11.2) implies that the constitutive behavior of the IPN, including the nonlinear response, is dictated by the constitutive behaviors of the component networks. Since most affinely deforming, dense, and/or densely crosslinked single networks are brittle, the resulting IPN is expected to be brittle.
- Case 2: IPNs without inter-crosslinks in which at least one of the constituent networks tested separately is nonaffine. Nonaffinity favors the formation of load-transmitting contacts between the two networks. Such contacts play the role of inter-crosslinks. The probability of their formation depends on the free volume present in the network. With large free volume (e.g., in swollen gels), the probability of load transfer between networks is low and, even though each component network is nonaffine, they deform independently and the summation of their responses defines the IPN behavior (Eq. 11.2).

  If the free volume is low, load transfer takes place and the reciprocal constraints may shift the deformation mode from nonaffine to affine. This situation is in-between that of two independent nonaffine networks acting in parallel (Eq. (11.2) applies) and that of a single crosslinked network of density $\rho_{eff} = \rho^{(1)} + \rho^{(2)}$. In the small strain limit, the stiffness of nonaffinely deforming networks scales as $E_0 \sim \rho^{x+1}$ (see Section 6.1.1.3.1 and Figure 6.6). If the two nonaffine networks act in parallel, Eq. (11.2) implies that $E_0^{IPN} = \alpha_1 \rho^{(1)x+1} + \alpha_2 \rho^{(2)x+1}$, where α_1 and α_2 are functions of the respective fiber parameters (Section 6.1.1.3.1). In this case, the IPN strain stiffening is of the same functional form as that of the constituent networks. In the other limit, in which the two networks are fully coupled, two possibilities emerge: If $\rho_{eff} = \rho^{(1)} + \rho^{(2)}$ is sufficiently large and topological interactions impose global affinity, $E_0^{IPN} \sim \rho_{eff} = \rho^{(1)} + \rho^{(2)}$ and strain stiffening is sub-exponential. However, if ρ_{eff} is not as large as to impose global affinity, $E_0^{IPN} \sim \rho_{eff}^{x+1} = \left(\rho^{(1)} + \rho^{(2)}\right)^{x+1}$ and strain stiffening has a functional form similar to that of the constituent networks.
- Case 3: IPNs with inter-crosslinks. This case is similar to case 2. The considerations presented for case 2 apply.

It is observed that microfiber nonwovens composed from two interpenetrating networks which are created by spinning athermal fibers concurrently on the same substrate follow the rule of mixtures (Amoroso et al., 2012). Their mechanical response can be predicted by Eq. (11.2), in which the contributions of the two networks are weighted by the respective volume fractions.

Consider now the situation in which the crosslink density of each network, ρ_b, is low and the strands between crosslinks are long and entangled. Recall that, in networks, entanglements are trapped, that is, strands cannot disentangle by diffusion (reptation) due to the presence of crosslinks at both their ends. It may be considered that the mechanical contribution of entanglements to the mechanics of the network is similar to that of the crosslinks. Therefore, the effective crosslink density of a single network with trapped entanglements may be computed as the sum of the number density of real crosslinks, ρ_b, and the number density of entanglements, ρ_e. It is useful to write ρ_e in terms of the density of the network, ρ. To this end, recall from the discussion of thermal networks in Section 4.2.3.4.2 that the entanglement length, L_e, scales with the density of the polymer in solution as $L_e \sim \rho^{-\varsigma/(3\varsigma-1)}$, where ς is the exponent of the power law relation between the end-to-end length of a strand and its contour length.[1] $1/\varsigma$ is the fractal dimension of the strand. The entanglement number density is computed as $\rho_e = \rho/L_e$. For athermal solvents with purely repulsive polymer–solvent interactions, $\varsigma = 3/5$ and $\rho_e \sim \rho^{1.75}$. For ideal chains with random walk statistics (e.g., in a θ-solvent or in a melt), the chain size is independent of the density and $\varsigma = 1/2$. The mesh size and L_e scale with the density as $L_e \sim \rho^{-2/3}$,[2] and the entanglement number density scales with the density (ignoring dangling ends and chains not connected to the network) as $\rho_e \sim \rho^{5/3}$. Further, using Eq. (4.21) and likewise assuming no dangling ends or free chains, it is possible to express ρ_b in terms of ρ. For a two-network IPN in which both networks are sparsely crosslinked and have long segments between crosslink sites, it results that trapped entanglements play an important role. The number density of crosslinks is $\rho_b^{(1)}+\rho_b^{(2)}$, while the number density of entanglements is proportional to $\left(\rho^{(1)}+\rho^{(2)}\right)^{1.75}$ in the athermal solvent limit and $\left(\rho^{(1)}+\rho^{(2)}\right)^{5/3}$ in θ-conditions. As the overall network density increases, the contribution of entanglements to the total effective crosslink density becomes dominant. As discussed in Chapter 8, the crosslink density controls the network strength. However, rupture of IPNs is more complex than in single networks, as pointed out next in relation to hydrogels.

11.3.2 Hydrogel IPNs

The principles behind the design of tough hydrogels based on the IPN idea are outlined in Section 8.3 (see also Section 6.1.1.4.3).[3] To summarize, the requirements for achieving enhanced strength and toughness in IPNs are:

- Network 1 must be stiff and brittle and must have a high density of sacrificial bonds of relatively low strength. Network 1 may be chemically crosslinked, may be

[1] In Section 11.4, the discussion refers to the mesh size, ξ, and the contour length between entanglements viewed as the mean segment length of the network denoted by l_c. The entanglement length is the length of the end-to-end vector of the average entangled segment and is equal in this context to the mesh size defined by entanglements, $L_e = \xi$.

[2] Note the difference relative to the prediction of $L_e \sim \rho^{-\varsigma/(3\varsigma-1)}$. This is due to the contribution of ternary contacts, as discussed in polymer physics texts (e.g., Rubinstein and Colby, 2003).

[3] While hydrogel IPNs are of high interest today, IPNs which are not hydrogels have been studied for a long time (see, Sperling, 1981).

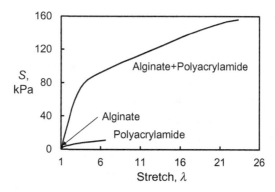

Figure 11.14 Nominal stress–stretch curves for an IPN of alginate–polyacrylamide, and for the component gels. Reprinted from Sun et al. (2012) with permission from Springer Nature, Copyright (2012)

crosslinked by the presence of molecular folded domains or nanocrystallites (physically crosslinked) as in Li et al. (2014), or may be ionically stabilized (Sun et al., 2012; Takahashi et al., 2019).

- Network 2 must be sparsely crosslinked and highly deformable, and its crosslinks must be strong (chemical). Alternately, network 2 may be entangled, as is the case with SIPN and pseudo-SIPN (Figure 11.13).
- Network 1 should restrict the excessive swelling of network 2. In single networks, the toughness scales inversely with the square of the swelling ratio.[4] Hence, a network which is too swelled and has strongly stretched chains becomes brittle.

These principles are demonstrated by the three examples provided next. In Sun et al. (2012), network 2 is a soft and highly deformable, covalently crosslinked polyacrylamide network, while network 1 is ionically crosslinked alginate. Alginate is rather brittle. The acrylamide of the concentration used in this IPN has a strength of about 10 kPa and breaks at a stretch of ∼7. Figure 11.14 shows the nominal stress versus stretch curves for the IPN and the individual gels tested separately. The IPN ruptures at a stretch of about 23 and has a strength of 160 kPa, which are much larger than the stretch and strength of network 2. A yield point develops at a stress of 80 kPa, which is much larger than the failure stress or network 1. The energy absorbed by the IPN before fracture is $8\,000$ J/m^2. The synergy between the two networks is spectacular.

While here we refer to the softening seen in Figure 11.14 at a stretch of about 2 as a yield point, it should be noted that these materials exhibit very small residual strains upon unloading from states beyond yield, and hence this softening is not akin to the common yield point in metals. The first network ruptures at the yield point, and this

---

[4] You may have experienced that vegetables commonly used in the kitchen, such as cucumbers and carrots, become less brittle upon dehydration.

causes softening. Unloading before the yield point leads to negligible hysteresis. Pronounced hysteresis is observed when unloading from beyond yield, but the residual strains are insignificant since the elastic rebound of the intact network 2 forces the damaged network 1 back to the initial dimensions (Na et al., 2006; Ahmed et al., 2014).

In IPNs in which network 1 has crosslinks that may reform upon rupture, as is the case with ionic bonds, crystallites, or other transient physical crosslinking, the material self-heals after a deformation to strains beyond the yield point. Self-healing implies the re-formation of the crosslinks of network 1 and, since network 2 did not sustain damage during the pre-deformation, the material recovers its initial properties (Sun et al., 2012; Li et al., 2014, 2018; Takahashi et al., 2019).

Figure 11.15(a) demonstrates the effect of adjusting the parameters of the sacrificial network 1. The data comes from Takahashi et al. (2019) and refers to an IPN in which network 1 is formed by ionically bonded polyelectrolytes, while network 2 is a stretchable, chemically crosslinked polyacrylamide. These stress–stretch curves correspond to systems in which network 2 is identical, while the concentration of network 1 is increased. As network 1 develops upon increasing the concentration of its fibrils, a yield point appears. The yield stress scales approximately linearly with the concentration of network 1, which suggests that it is proportional to (but larger than) the strength of network 1. Further, strain hardening beyond yield is independent of the concentration of network 1. This parameter is controlled by the crosslink density of network 2, which is kept constant in these experiments. Furthermore, as the network 1 concentration increases, the swelling ratio of the hydrogel decreases, indicating the constraining effect that a stiffer network 1 has on network 2. It is remarkable that the ductility increases monotonically as the concentration of network 1 increases. This

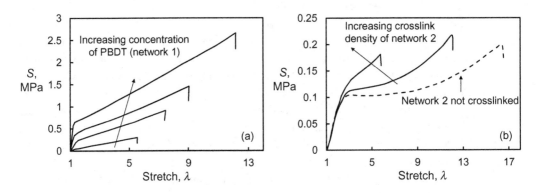

Figure 11.15 Nominal stress–stretch curves for (a) a polyelectrolyte–polyacrylamide IPN in which the polyacrylamide network is kept unchanged and the concentration of network 1 increases (reproduced from Takahashi et al. (2019) with permission of The Royal Society of Chemistry), and (b) double network hydrogels in which network 1 is kept unchanged and the degree of crosslinking of network 2 is modified (reproduced from Es-haghi and Weiss (2016a) with permission from Elsevier).

stabilization effect, which may be due to restricted swelling, leads to strengths close to 3 MPa and stretches at failure above 10.

Similar stabilization of deformation leading to enhanced strength and ductility (relative to the pure matrix) are observed in some soft composites reinforced with networks of nanofibers, as shown in Figure 11.18.

Weiss and collaborators (Es-haghi et al., 2014; Es-haghi and Weiss, 2016a) emphasized the importance of grafting the chains of network 2 on network 1, indicating that if such grafting does not occur, no reinforcement is obtained in pseudo-SIPN. In Es-haghi and Weiss (2016a) the authors discuss the role of cross-linking of the second network in a pseudo-SIPN network. Figure 11.15(b) shows stress–stretch curves obtained with hydrogels of given network 1 and in which network 2 either is or is not self-crosslinked. The pseudo-SIPN in which network 2 is not crosslinked exhibits yielding and necking, followed by deformation at constant stress before stiffening occurs at large stretches. The post-yield response was observed to be rate sensitive in these hydrogels. When network 2 is crosslinked internally, necking is eliminated and strain hardening is observed right after the yield point (compare with Figure 11.15(a)). The post-yield response is not rate sensitive in this case. The observed strain hardening is associated with the deformation of network 2; however, it takes place at stress and stretch levels much larger than those sustained by the unreinforced network 2.

A small number of studies considered hydrogels composed from multiple IPNs (see, e.g., Es-haghi and Weiss, 2016b). The current technology used to produce such networks and the understanding of their structure and mechanical behavior are not yet mature.

To place the behavior of IPN hydrogels in perspective, we recall that single network gels, such as alginate and gelatin, have toughness of about 30–50 J/m^2 and strength below 100 kPa. Biological network materials have much larger toughness and strength. For example, articular cartilage has toughness of about 1 000 J/m^2 and strength of about 10 MPa, while skin has toughness of 10 000 J/m^2 and strength of the order of 10 MPa. Other collagen-based membranes within the human body exhibit similar properties. For example, the amnion has toughness of $\sim$8 000 J/m^2 and strength of $\sim$1 MPa. IPN hydrogels with toughness as high as 14 000 J/m^2 and strength of several MPa have been demonstrated (Li et al., 2014), which brings these artificial constructs closer to the performance of biological network materials.

This discussion leads one to inquire about the mechanism controlling the synergy between the deformation of the two networks. In particular, the most spectacular aspect of the described behavior is the stabilization of the deformation in the regime beyond the yield point, as seen in Figures 11.14 and 11.15 (see also Figures 11.7 and 11.18, which show similar behavior in other contexts). In the context of IPNs, it was postulated that the effect is due to the swelling of network 2 produced by the debris of network 1 after it fragments, that is, after the yield point. Experiments have been performed with hydrogels that contain long free polymers which entangle but are not crosslinked to either network. The results of these studies appear to be contradictory (Tsukeshiba et al., 2005; Nakajima et al., 2009; Es-haghi and Weiss, 2017), with examples both confirming and infirming that such additions stabilize the deformation and increase strength and ductility. It is adequate to state here that the mechanism responsible for the synergistic effect that makes IPN hydrogels so interesting is still unknown.

11.4   Networks Embedded in a Continuum Matrix

Many biological and engineering networks are embedded in an elastic or viscoelastic continuum matrix. The matrix may be considered to behave as a continuum when its characteristic length scales are much smaller than the smallest length scale of the network, such as the mesh size or l_c. It is of interest to analyze the interplay of matrix and network mechanics as they define the behavior of the composite.

This problem is somewhat similar to that of continuous fiber composites in which a woven fiber structure is embedded in a polymeric matrix. A typical example of this type is provided by the carbon fiber composites in which bundles of carbon fibers are woven and embedded in epoxy. Carbon fibers have a diameter of about 10 μm, which is much larger than the mesh size of epoxy. Therefore, the matrix may be considered a continuum on the scale of individual carbon fibers. The difference between this example and the problem discussed here is the fact that, in the present case, the reinforcing network is not woven and is stochastic.

The effect of the matrix on network mechanics manifests itself both on the scale of individual fibers and on the composite scale:

- On the local scale, the matrix restricts the bending deformation mode of fibers. As the matrix stiffness increases, the network transitions from the bending to the axial deformation mode. This renders it less nonaffine and stiffer.
- On the global scale, the matrix restricts the volume variation of the network. While the unembedded network may exhibit a large Poisson effect and may undergo large volume changes as it is subjected to large tensile deformations, isotropic continua have Poisson ratios bounded above by 0.5. The matrix imposes tractions on the network to constrain its volume variation, as shown schematically in Figure 11.3(c).

The analysis presented in Zhang et al. (2013) sheds light on the expected relation between the composite stiffness, the matrix stiffness, and various network parameters. Consider that the matrix effectively constrains the bending mode of fibers and restricts fiber deformation to the axial mode. Figure 11.16 shows a representative fiber of the network embedded in a continuum matrix. The elastic strain field in the matrix results as the superposition of the far field and the perturbation introduced by the presence of the fiber. Since the fiber deforms axially, its effect on the matrix may be represented by a force dipole. The perturbation of the field in the matrix due to the fiber is evaluated using Green's functions. The equilibrium configuration and the value of the force carried by the fiber results from the minimization of the total potential energy. The effective stiffness of the composite, E_c, may be obtained by equating

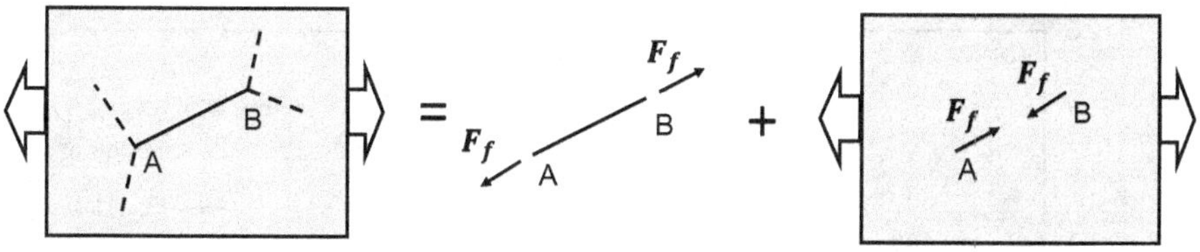

Figure 11.16 Schematic used to derive the composite stiffness, Eq. (11.3).

the energy of a homogenized continuum representing the reinforced matrix to the actual energy of the system and results:

$$E_c = E_m + E_0^{aff} \frac{\eta}{1+\eta}, \tag{11.3}$$

where $\eta = (E_m/E_f)(l_c/2\alpha A_f)$. α is a parameter which depends on the Poisson ratio of the matrix and on an internal (constant) length representing the cut-off used to eliminate the singularity in the vicinity of the point force, and has units of inverse length. E_0^{aff} is the network modulus evaluated with the affine model and, specifically, $E_0^{aff} \sim \rho E_f A_f$ (Eq. 5.38) for an athermal network. Equation (11.3) is derived for elastic matrices of nonvanishing stiffness and becomes more accurate as E_m increases, that is, when the matrix effectively restricts the fiber bending modes. In this limit, ratio $\eta/(1+\eta)$ is close to 1, both network and matrix deform affinely, and the composite behaves as if the network and matrix are loaded in parallel.

Multiple experimental results support the view that the composite moduli may be approximated by considering the matrix and network coupled in parallel. Ansari et al. (2015) report results obtained with networks of cellulose nanofibers embedded in an unsaturated polyester resin. The network density is varied. The variation of the relative composite stiffness, $(E_c - E_m)/E_m$, is shown in Figure 11.17(a) versus the weight fraction of the nanocellulose network (proportional to ρ). A linear function of ρ results for weight fractions below 20%, which is in agreement with the prediction of Eq. (11.3). This implies that the network deforms affinely (the network stiffness is proportional to ρ) and supports the idea that network and matrix are coupled in parallel. Figure 11.17(a) includes similar data from Zimmermann et al. (2004), who studied cellulose nanofiber networks embedded in a hydroxypropyl cellulose matrix. This data set leads to conclusions similar to those based on the results from Ansari et al. (2015). The departure observed at high ρ may be due to insufficient infiltration of the matrix in the denser networks. Numerical studies of linear elasticity of athermal networks embedded in a matrix reported in Lin et al. (2019) also indicate that $(E_c - E_m)/E_m$ varies linearly with the fiber volume fraction. The effect of the fiber preferential orientation on all elastic constants of the respective co-continuous composite is also discussed in Lin et al. (2019).

Figure 11.17(b) shows results from Burla et al. (2019) and from Rombouts et al. (2014). In Burla et al. (2019) the authors work with collagen networks embedded in hyaluronan gels. The network is of set density, while the matrix stiffness is varied by

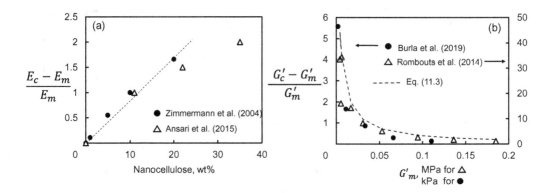

Figure 11.17 (a) Variation of the relative composite stiffness with the network weight fraction for networks of cellulose nanofibers embedded in polyester and in hydroxypropyl cellulose matrices (data from Zimmermann et al. (2004) and Ansari et al. (2015)). (b) Variation of the relative composite storage modulus with the storage modulus of the matrix at fixed network density, for collagen networks embedded in hyaluronan gels of various concentrations (data from Burla et al., 2019) and for interpenetrating protein networks of significantly different mesh size (data from Rombouts et al., 2014). The prediction of Eq. (11.3) is shown by the dashed line.

adjusting the concentration of the gel from 3 mg/ml to 7 mg/ml. This range of gel concentrations corresponds to a variation of the matrix storage modulus, G'_m, from close to 0 to about 120 Pa. Figure 11.17(b) shows $\left(G'_c - G'_m\right)/G'_m$ versus G'_m at constant network density, ρ. Equation (11.3) predicts a monotonically decaying function of the matrix stiffness of the form $\left(G'_c - G'_m\right)/G'_m \sim 1/\left(\zeta_1 + \zeta_2 G'_m\right)$, where ζ_1 and ζ_2 are constants that depend on network parameters. This prediction is shown by the dashed line in Figure 11.17(b). Note the difference between this prediction and that of the parallel coupling model: When the matrix stiffness is close to zero, $\left(G'_c - G'_m\right)/G'_m = G_0/G'_m \to \infty$ if the network and matrix are simply coupled in parallel, while $\left(G'_c - G'_m\right)/G'_m \to 1/\zeta_1$ is predicted by Eq. (11.3), which is in better agreement with the data in Figure 11.17(b).

The figure also includes results from Rombouts et al. (2014), who tested interpenetrating gels composed from two protein networks. One of these contains coarse filaments with large l_c, while the other has much smaller l_c and thinner filaments. The second network behaves as a continuum matrix for the coarser network. The results of these tests agree qualitatively with those in Burla et al. (2019) and both provide support for the physical interpretation of the effect of the matrix on network mechanics outlined in this section. The numerical results in Lin et al. (2019) are compatible with the variation of $(E_c - E_m)/E_m$ versus E_m/E_f shown in Figure 11.17(b) and with Eq. (11.3).

The representation of the composite behavior by using the rule of mixtures is supported by results on co-gels of agar and gelatin (Strange and Oyen, 2012) and collagen and fibrin (Lai et al., 2012). Co-gels are created by varying the concentration of each component and the stiffness is measured in uniaxial tests for a broad range of composite parameters, that is, individual gel concentrations and volume fractions of the two gels. The measured stiffness is compared with predictions based on the assumption that the two networks act independently and are coupled either in parallel or in series (upper and lower bounds of the stiffness, respectively). At times beyond a transient caused by the redistribution of water within the gel, the composite stiffness follows the upper bound in all cases, supporting the parallel coupling interpretation.

The parallel coupling model implies that the stiffer component controls the response of the composite. This is expected to apply not only in the small strain limit, to which the results in Figure 11.17 refer, but also to the nonlinear response measured at larger strains. Specifically, the composite is expected to reproduce the nonlinear behavior of the stiffer component. This is confirmed in Lai et al. (2012), who performed tests with collagen–fibrin co-gels. In these systems, the collagen component is much stiffer than fibrin, has coarser filaments, and a larger mesh size. Once the fraction of collagen becomes large enough, the nonlinear response of the composite follows the upper bound evaluated by coupling the collagen and fibrin networks in parallel. Burla et al. (2019) reached a similar conclusion for their collagen networks embedded in hyaluronan matrix. Both pure collagen and pure hyaluronan strain stiffen exponentially, but the matrix has a larger regime I to II transition strain. The presence of the matrix delays the transition to regime II of the collagen network and, as the matrix stiffness increases, the composite nonlinear response shifts gradually from that of the network to that of the matrix.

The matrix has an important influence on the composite strength, ductility, and toughness due to a synergistic effect. The matrix provides effective load transfer between fibers and renders network deformation more affine relative to the unembedded network. Hence, the network is more homogeneously loaded, which increases its strength. On the other hand, fibers bridge cracks that may form in the matrix which, in turn, greatly delays fracture.

To demonstrate this trend in the context of embedded stochastic networks, Figure 11.18(a) shows a stress–stretch curve for alginate reinforced with electrospun

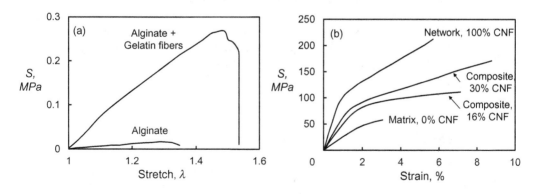

Figure 11.18 (a) Stress–stretch curves for alginate reinforced with uncrosslinked electrospun gelatin fibers and for the unreinforced alginate matrix (data from Tonsomboon and Oyen, (2013). (b) Stress–strain curves for unsaturated polyester reinforced with nanocellulose fibers at several weight fractions of cellulose nanofibers (CNF) (adapted from Ansari et al., 2015).

gelatin fibers of approximately 100 nm diameter, along with the equivalent curve for unreinforced alginate (Tonsomboon and Oyen, 2013). In this example, the gelatin nanofibers are not crosslinked and are simply embedded in the alginate matrix. The strength of the composite is one order of magnitude larger than that of the matrix, while ductility increases upon reinforcement by a factor of ~2.

Figure 11.18(b) shows stress–strain curves of an unsaturated polyester matrix reinforced with a cellulose nanofiber network, at various filling fractions (Ansari et al., 2015). The behavior of this composite is of type B. The yield point is associated with the onset of plastic deformation of the nanocellulose network. As the filling fraction increases, the yield stress increases, while the yield strain remains approximately constant, which is expected based on the affine deformation assumption (Section 6.1.4). More interestingly, the composite exhibits strong strain hardening beyond the yield point and the rate of strain hardening increases with increasing loading fraction. The physical mechanisms responsible for this behavior are still to be clarified. The fibers stabilize the plastic deformation and the strength and strain at failure are several times larger than the corresponding parameters of the unfilled matrix.

The improvement of the ultimate properties upon reinforcement is more pronounced if the matrix is tough. This point is made in Illeperuma et al. (2014) and Murai et al. (2019). The first group uses steel wool to reinforce two types of gels: Alginate and a hybrid alginate–polyacrylamide hydrogels. Alginate is quite brittle, while the hybrid gel is tough. The reinforcing fibers have tens of microns cross-sectional dimensions and are not crosslinked. Therefore, the strength of the network without matrix is zero. The composite exhibits strength much larger than that of the unreinforced matrix. However, the strength increase is substantially larger when the tough matrix is used.

Murai et al. (2019) work with an IPN reinforced with a cellulose nanofiber network. The reinforcing network in this case has characteristic lengths much closer to the matrix network length scales than in the material used in Illeperuma et al. (2014). The strengths of both the unembedded cellulose network and unreinforced IPN are below 1 MPa, the cellulose network is brittle, while the unreinforced IPN breaks at a stretch of 8. The composite has strength of 6 MPa and ductility similar to that of the unreinforced IPN matrix. The resulting toughness is orders of magnitude larger than that of the two components taken separately.

References

Ahmed, S., Nakajima, T., Kurokawa, T., Haque, M. A. & Gong, J. P. (2014). Brittle–ductile transition of double network hydrogels: Mechanical balance of two networks as the key factor. *Polymer* **55**, 914–923.

Amoroso, N. J., D'Amore, A., Hong, Y., et al. (2012). Microstructural manipulation of electrospun scaffolds for specific bending stiffness for heart valve tissue engineering. *Acta Biomat.* **8**, 4268–4277.

Ansari, F., Skrifvars, M. & Berglund, L. (2015). Nanostructured biocomposites based on unsaturated polyester resin and a cellulose nanofiber network. *Comp. Sci. Technol.* **117**, 298–306.

Aranguren, M. I., Mora, E., Macosko, C. W. & Saam, J. (1994). Rheological and mechanical properties of filled rubber: Silica-silicone. *Rubber Chem. Technol.* **67**, 820–833.

Avazmohammadi, R. & Ponte Castaneda, P. (2014). On the macroscopic response, microstructure evolution and macroscopic stability of short-fiber-reinforced elastomers at finite strains: I: Analytical results. *Phil. Mag.* **94**, 1031–1067.

Bai, M., Missel, A. R., Klug, W. S. & Levine, A. J. (2011). The mechanics and affine–nonaffine transition in polydisperse semiflexible networks. *Soft Matt.* **7**, 907–914.

Burla, F., Tauber, J., Dussi, S., van der Gucht, J. & Koenderink, G. H. (2019). Stress management in composite biopolymer networks. *Nature Phys.* **15**, 549–553.

Donev, A., Torquato, S., Stillinger, F. H. & Connelly, R. (2004). Jamming in hard sphere and disk packings. *J. Appl. Phys.* **95**, 989–999.

Es-haghi, S. S., Leonov, A. I. & Weiss, R. A. (2014). Deconstructing the double network hydrogels: The importance of grafted chains for achieving toughness. *Macromolecules* **47**, 4769–4777.

Es-haghi, S. S. & Weiss, R. A. (2016a). Finite strain damage–elastoplasticity in double network hydrogels. *Polymer* **103**, 277–287.

Es-haghi, S. S. & Weiss, R. A. (2016b). Fabrication of tough hydrogels from chemically cross-linked multiple neutral networks. *Macromolecules* **49**, 8980–8987.

Es-haghi, S. S. & Weiss, R. A. (2017). Do physically trapped polymer chains contribute to the mechanical response of a host double network hydrogel under finite tensile deformation? *Macromolecules* **50**, 8267–8273.

Fukasawa, M., Sakai, T., Chung, U. I. & Haraguchi, K. (2010). Synthesis and mechanical properties of a nanocomposite gel consisting of tetra-PEG/Clay network. *Macromolecules* **43**, 4370–4378.

Gavrilov, A. A., Chertovich, A. V., Khalatur, P. G. & Khokhlov, A. R. (2013). Effect of nanotube size on the mechanical properties of elastomeric composites. *Soft Matt.* **9**, 4067–4072.

Hashin, Z. & Shtrikman, S. (1961). Note on a variational approach to the theory of composite elastic materials. *J. Franklin Inst.* **271**, 336–341.

Hassanzadeh, P., Kazemzadeh-Narbat, M., Roesnzweig, R., et al. (2016). Ultrastrong and flexible hybrid hydrogels based on solution self-assembly of chitin nanofibers in gelatin methacryloyl (GelMA). *J. Mater. Chem. B* **4**, 2539–2543.

Horii, M. & Nemat-Nasser, S. (1993). *Micromechanics: Overall properties of heterogeneous materials*. North-Holland, Amsterdam.

Huisman, E. M., Heussinger, C., Storm, C. & Barkema, G. T. (2010). Semiflexible filamentous composites. *Phys. Rev. Lett.* **105**, 118101.

Illeperuma, W. R. K., Sun, J. Y., Suo, Z. & Vlassak, J. J. (2014). Fiber-reinforced tough hydrogels. *Extreme Mech. Lett.* **1**, 90–96.

Islam, M. R. & Picu, R. C. (2019). Random fiber networks with inclusions: The mechanism of reinforcement. *Phys. Rev. E* **99**, 063001.

Lai, V. K., Lake, S. P., Frey, C. R., Tranquillo, R. T. & Barocas, V. H. (2012). Mechanical behavior of collagen–fibrin co-gels reflects transition from series to parallel interactions with increasing collagen content. *J. Biomech. Eng.* **134**, 011004.

Li, H., Wang, H., Zhang, D., Xu, Z. & Liu, W. (2018). A highly tough and stiff supramolecular polymer double network hydrogel. *Polymer* **153**, 193–200.

Li, J., Suo. Z. & Vlassak, J. J. (2014). Stiff, strong and tough hydrogels with good chemical stability. *J. Mater. Chem. B* **2**, 6708–6713.

Lin, X., Zhu, H., Yuan, X., Wang, Z. & Bordas, S. (2019). The elastic properties of composites reinforced by a transversely isotropic random fiber network. *Comp. Struct.* **208**, 33–44.

Lin, Y. C., Koenderink, G. H., MacKintosh, F. C. & Weitz, D. A. (2011). Control of non-linear elasticity in F-actin network with microtubules. *Soft Matt.* **7**, 902–906.

Lopez-Pamies, O., Goudarzi, T. & Danas, K. (2013). The nonlinear elastic response of suspensions of rigid inclusions in rubber: II: A simple explicit approximation for finite-concentration suspensions. *J. Mech. Phys. Sol.* **61**, 19–37.

Medalia, A. I. (1978). Effect of carbon black on dynamic properties of rubber vulcanizates. *Rubber Chem. Technol.* **51**, 437–523.

Medalia, A. I. & Kraus, G. (1994). Reinforcement of elastomers by particulate fillers. In *Science and technology of rubber*, J. E. Mark, B. Erman & F. R. Eirich, eds. Academic Press, San Diego, pp. 387–418.

Mullins, L. & Tobin, N. R. (1965). Stress softening in rubber vulcanizates. Part I. Use of a strain amplification factor to describe the elastic behavior of filler-reinforced vulcanized rubber. *J. Appl. Polym. Sci.* **9**, 2993–3009.

Murai, J., Nakajima, T., Matsuda, T., et al. (2019). Tough double network elastomers reinforced by the amorphous cellulose network. *Polymer* **178**, 121686.

Na, Y. H., Tanaka, Y., Kawauchi, Y., et al. (2006). Necking phenomenon of double network gels. *Macromolecules* **39**, 4641–4645.

Nakajima, T., Furukawa, H., Tanaka, Y., et al. (2009). True chemical structure of double network hydrogels. *Macromolecules* **42**, 2184–2189.

Oono, R. (1978). Distribution of carbon black in SBR. *Rubber Chem. Technol.* **51**, 278–284.

van Oosten, A. S. G., Chen, X., Chin, L. K., et al. (2019). Emergence of tissue-like mechanics from fibrous networks confined by close-packed cells. *Nature* **573**, 96–101.

Payne, A. R. (1974). Hysteresis in rubber vulcanizates. *J. Polym. Sci.: Poly. Symp.* **48**, 169–196.

Payne, A. R. & Whittaker, R. E. (1971). Low strain dynamic properties of filled rubbers. *Rubber Chem. Technol.* **44**, 440–478.

Picu, R. C., Krawczyk, K. K., Wang, Z., et al. (2019). Toughening in nanosilica-reinforced epoxy with tunable filler-matrix interface properties. *Comp. Sci. Technol.* **183**, 107799.

Ponte Castaneda, P. (1989). The overall constitutive behavior of nonlinearly elastic composites. *Proc. R. Soc. London A* **422**, 147–171.

Raisanen, V. I., Heyden, S., Gustafsson, P. J., et al. (1997). Simulation of the effect of a reinforcement fiber on network mechanics. *Nordic Pulp Paper Res. J.* **12**, 162–166.

Rombouts, W. H., Giesbers, M., van Lent, J., de Wolf, F. A. & van der Gucht, J. (2014). Synergistic stiffening in double-fiber networks. *Biomacromolecules* **15**, 1233–1239.

Rubinstein. M. & Colby, R. H. (2003). *Polymer physics*. Oxford University Press, Oxford.

Shahsavari, A. S. & Picu, R. C. (2015). Exceptional stiffening in composite fiber networks. *Phys. Rev. E* **92**, 012401.

Shivers, J. L., Feng, J., van Oosten, A. S. G., et al. (2020). Compression stiffening of fibrous networks with stiff inclusions. *Proc. Nat. Acad. Sci.* **117**, 21037–21044.

Sperling, L. H. (1981). *Interpenetrating polymer networks and related materials*. Springer, Boston, MA.

Strange, D. G. T. & Oyen, M. L. (2012). Composite hydrogels for nucleus pulposus tissue engineering. *J. Mech. Beh. Biomed. Mater.* **11**, 16–26.

Sun, J. Y., Zhao, X., Illeperuma, W. R. K., et al. (2012). Highly stretchable and tough hydrogels. *Nature* **489**, 133–136.

Takahashi, R., Ikai, T., Kurokawa, T., King, D. R. & Gong, J. P. (2019). Double network hydrogels based on semi-rigid polyelectrolyte physical networks. *J. Mater. Chem. B* **7**, 6347–6354.

Tonsomboon, K. & Oyen, M. L. (2013). Composite electrospun gelatin fiber-alginate gel scaffolds for mechanically robust tissue engineered cornea. *J. Mech. Beh. Biomed. Mater.* **21**, 185–194.

Tsukeshiba, H., Huang, M., Na, Y. H., et al. (2005). Effect of polymer entanglement on the toughening of double network hydrogels. *J. Phys. Chem. B* **109**, 16304–16309.

Voet, A. (1980). Reinforcement of elastomers by fillers: review of period 1967–1976. *J. Poly. Sci.: Macromolec. Rev.* **15**, 327–373.

Wagner, M. P. (1976). Reinforcing silicas and silicates. *Rubber Chem. Technol.* **49**, 703–774.

Wang, Q., Chen, S. & Chen, D. (2017). Preparation and characterization of chitosan based injectable hydrogels enhanced by chitin nanowhiskers. *J. Mech. Beh. Biomed. Mater.* **65**, 466–477.

Yuan, Q. W. & Mark, J. E. (1999). Reinforcement of PDMS networks by blended and in-situ generated silica fillers having various sizes, size distributions and modified surfaces. *Macromol. Chem. Phys.* **200**, 206–220.

Zhang, L., Lake, S. P., Barocas, V. H., Shephard, M. S. & Picu, R. C. (2013). Crosslinked fiber network embedded in an elastic matrix. *Soft Matt.* **9**, 6398–6405.

Zhang, X., Huang, J., Chang, P. R., et al. (2010). Structure and properties of polysaccharide nanocrystal-doped supramolecular hydrogels based on Cyclodextrin inclusion. *Polymer* **51**, 4398–4407.

Zimmermann, T., Pohler, E. & Gaiger, T. (2004). Cellulose fibrils for polymer reinforcement. *Adv. Eng. Mater.* **6**, 754–761.

Index

Printed in the United States
by Baker & Taylor Publisher Services